Microelectronics Processing

ADVANCES IN CHEMISTRY SERIES **221**

Microelectronics Processing

Chemical Engineering Aspects

Dennis W. Hess, EDITOR
University of California–Berkeley

Klavs F. Jensen, EDITOR
University of Minnesota

American Chemical Society, Washington, DC 1989

Library of Congress Cataloging-in-Publication Data

Microelectronics Processing: Chemical Engineering Aspects
Dennis W. Hess, editor, Klavs F. Jensen, editor
p. cm.—(Advances in chemistry series, ISSN 0065-2393; 221)
Bibliography: p.

Includes index.

ISBN 0-8412-1475-1

1. Microelectronics-Materials. 2. Integrated Circuits—Design and Construction. 3. Surface Chemistry.

I. Hess, Dennis W. II. Jensen, Klavs F., 1952– III. Series

TK7874.M4835 1989
621.381-dc19

89-6862
CIP

PRINTED IN THE UNITED STATES OF AMERICA

Advances in Chemistry Series

M. Joan Comstock, *Series Editor*

FOREWORD

The ADVANCES IN CHEMISTRY SERIES was founded in 1949 by the American Chemical Society as an outlet for symposia and collections of data in special areas of topical interest that could not be accommodated in the Society's journals. It provides a medium for symposia that would otherwise be fragmented because their papers would be distributed among several journals or not published at all. Papers are reviewed critically according to ACS editorial standards and receive the careful attention and processing characteristic of ACS publications. Volumes in the ADVANCES IN CHEMISTRY SERIES maintain the integrity of the symposia on which they are based; however, verbatim reproductions of previously published papers are not accepted. Papers may include reports of research as well as reviews, because symposia may embrace both types of presentation.

Lawrence A. Casper, SPECIALTY EDITOR

Honeywell, Inc.

This book was acquired for publication through the efforts of Lawrence A. Casper acting in behalf of the American Chemical Society as a Specialty Editor. Dr. Casper developed the concept for this book and assisted in the content development and manuscript review.

ABOUT THE EDITORS

DENNIS W. HESS is a professor of chemical engineering at the University of California–Berkeley. He received his B.S. in chemistry from Albright College and the M.S. and Ph.D. in physical chemistry from Lehigh University. From 1973 to 1977, he performed basic and applied research in microelectronic-device fabrication at Fairchild Semiconductor. Since 1977, he has been on the faculty at the University of California, Berkeley, where he served as assistant dean of the College of Chemistry from 1982 to 1987 and as acting vice chairman of the Department of Chemical Engineering from August to December 1988. He was a visiting professor at the University of Minnesota during the 1983–1984 academic year. He was chairman of the 1988 Gordon Conference on the Chemistry of Electronic Materials. His research interests are in the areas of thin-film science and technology, plasma chemistry, and plasma physics. He is the author and coauthor of over 90 technical publications. He is a divisional editor for the *Journal of the Electrochemical Society* and an associate editor for *Chemistry of Materials*.

KLAVS F. JENSEN is a professor of chemical engineering and materials science and a fellow of the Supercomputer Institute at the University of Minnesota. He received his undergraduate education at the Technical University of Denmark and his Ph.D. from the University of Wisconsin–Madison. He has been a visiting professor at the IBM T. J. Watson Research Center, Massachusetts Institute of Technology, and the Technical University of Aachen. His research interests revolve around the chemistry of and transport phenomena related to electronic materials processing, including (1) organometallic chemical vapor deposition of thin films; (2) plasma and laser processing of thin films, including metal–polymer interfaces; (3) large-scale simulations of chemically reacting flows; (4) dynamics of chemically reacting systems; and (5) reaction and transport in porous media. He is the coauthor of more than 80 technical publications and has been a plenary lecturer at international conferences on the processing of electronic materials. He is the recipient of the 1982 Electrochemical Society Young Authors' Award in solid-state science and technology and the 1984 National Science Foundation Presidential Young Investigator Award. In addition, he received a Camille and Henry Dreyfus Foundation Teacher–Scholar Grant (1985), a Guggenheim Fellowship (1987), and the Allan P. Colburn Award of the American Institute of Chemical Engineers (1987).

CONTENTS

INDEXES

PREFACE

ELECTRONIC MATERIALS PROCESSING is an intensely interdisciplinary field that reaches beyond the traditional boundaries of electrical engineering, materials science, physics, chemistry, and chemical engineering. The goal of electronic materials processing is to design and control processes and processing sequences in order to manufacture materials with specific characteristics. Because chemicals are invariably the starting point for these efforts, chemical engineers can play a pivotal role in achieving this goal. Although processing requirements are quite different from those encountered in the chemical industry, chemical engineers can take advantage of their background in chemical kinetics, reactor design, heat transfer, fluid flow, and mass transfer to attack the problems that limit the production of materials and devices. However, to effectively accomplish this task, chemical engineers must broaden their knowledge base to include the electronic and optical properties of materials.

Electronic materials encompass a wide variety of solids and their applications. Nevertheless, the area that has become synonymous with electronic materials is microelectronics. This situation has arisen because of the rapid and pervasive development and growth of microelectronic devices or integrated circuits (ICs). Although there are literally hundreds of individual steps that compose the manufacture of an IC, essentially each one is a chemical process. Thus, this book emphasizes the fundamental chemical engineering principles involved in the fabrication of ICs. This volume is intended to be a tutorial tool rather than a comprehensive review. Additional details on specific topics can be obtained from the extensive list of references at the end of each chapter.

Acknowledgements

We are most appreciative of the efforts of the authors and anonymous reviewers who spent countless hours writing, reading, and rewriting chapters. In addition, we wish to express our appreciation to Robin Giroux and A. Maureen Rouhi at ACS Books, who, through their persistence, encouragement, and frequent prodding, managed to bring this book to fruition.

Finally, we would like to thank Larry Casper for serving as the catalyst behind this project and for initially suggesting that this undertaking was worthwhile.

DENNIS W. HESS
Department of Chemical Engineering
University of California
Berkeley, CA 94720–9989

KLAVS F. JENSEN
Department of Chemical Engineering
and Materials Science
University of Minnesota
Minneapolis, MN 55455

March 28, 1989

1

Microelectronics Processing

Dennis W. Hess[1] and Klavs F. Jensen[2]

[1]Department of Chemical Engineering, University of California, Berkeley, CA 94720–9989
[2]Department of Chemical Engineering and Materials Science, University of Minnesota, Minneapolis, MN 55455

This chapter gives an overview of issues related to electronic materials processing, with particular emphasis on the fabrication of silicon-based integrated circuits. Selected aspects of solid-state physics are reviewed as a background for processing issues. Unit operations in integrated-circuit manufacturing are summarized, and an example of a process sequence for the fabrication of a metal–oxide–semiconductor is given to provide a perspective of the specific processes described in subsequent chapters.

SINCE 1980, FIRMS SPECIALIZING IN THE PROCESSING of electronic materials have employed 15–30% of chemical engineering graduates at all degree levels. Like other specialized areas for chemical engineers (e.g., biotechnology, waste treatment, food processing, and pharmaceuticals), the electronic materials field makes use of the broad multidisciplinary scientific and engineering background of chemical engineers. In particular, the processing of materials and the fabrication of electronic, magnetic, and optical devices require a knowledge of chemistry, solid-state physics, materials science, chemical kinetics, reactor design, heat transfer, fluid flow, and mass transport. Numerous examples of these topics will be evident in succeeding chapters. The overall goal of this book is to demonstrate that the manufacture of solid-state materials and devices is simply chemical processing.

The microelectronic (or semiconductor) industry is not the only solid-state-processing industry that extensively uses chemists and chemical engineers, although this particular industry often gets the most attention because of the extraordinary and rapid success it has achieved. Industries

0065–2393/89/0221–0001$09.25/0

involved in optical coatings; solar-energy conversion; wear-resistant layers; solid-state batteries; electrochemistry; and magnetic, optical, and photonic devices are intensely interested in chemists and and chemical engineers for process research and development and product manufacture. The scientific background and process technology needed for microelectronic-device fabrication is used in essentially all of the industries just mentioned.

This book will concentrate on the chemistry and fundamental chemical engineering principles needed for integrated-circuit (IC) manufacture. Integrated circuits are currently used in consumer items, such as hand-held calculators, digital watches, microwave ovens, and automobiles, and in microprocessors for communication, defense, education, medicine, and space exploration. Naturally, new application areas are continually being developed.

To appreciate the rapid development of process technology, the progression of the IC industry must be considered first. (For summaries of the historical development of this field, *see* references 1 and 2.) A central theme in the IC industry is the simultaneous fabrication of hundreds of monolithic ICs (or chips) on a wafer (or slice) of silicon (or other material such as gallium arsenide), which is typically 100–150 mm in diameter and 0.75 mm thick. In silicon technology, chip areas generally range from a few square millimeters to over 100 mm^2. A large number (often more than 100) of individual process steps, which are precisely controlled and carefully sequenced, are required for the fabrication.

As an example of Si technology, Figure 1 illustrates a packaged 1-megabit dynamic-random-access-memory (DRAM) chip on a 150-mm-diameter Si substrate containing fabricated chips. Each of the chips will be cut from the wafer, tested, and packaged like the chip shown on top of the wafer. The chip is based on a 1-μm minimum feature size and contains 2,178,784 active devices. It can store 1,048,516 bits of information, which corresponds to approximately 100 typewritten pages.

ICs are built-up from layers of thin (usually between 1 nm and 2 μm) conducting, insulating, and semiconducting films. Each film has a pattern etched into it, so that an exactly registered array of these layers forms individual components such as transistors, resistors, diodes, and capacitors. These components are interconnected by patterned conducting films ("wires") to yield circuits. Figure 2 illustrates a typical process sequence starting from the Si raw material, through Si wafer production, device manufacture, wafer section, and, finally, to packaging of the final chip (3).

The application areas for ICs have expanded continuously, because greater circuit complexity can be realized as the number of components on a single chip increases (Figure 3) (*4*). From the inception of the semiconductor industry in 1959 until the early 1970s (Figure 3, segment A), the number of components on a silicon chip doubled every year. The pace slowed a bit from 1972 to 1984, when the number of components on a chip doubled

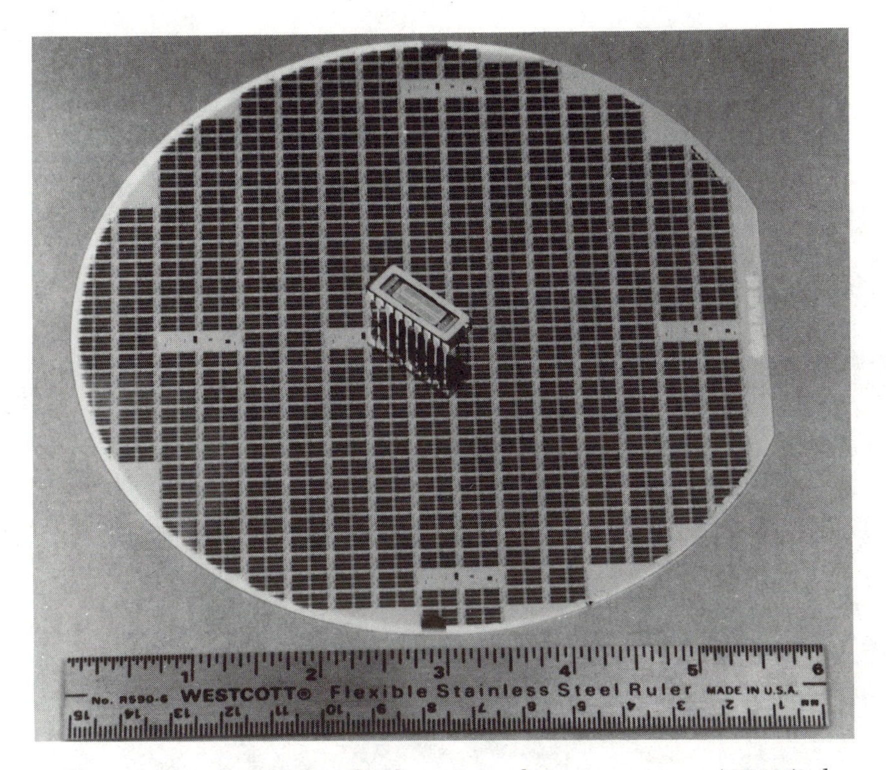

Figure 1. A packaged 1-megabit dynamic-random-access-memory (DRAM) silicon chip on a processed 150-mm-diameter Si wafer. (Used by courtesy of G. B. Larrabee, Texas Instruments.)

every 1.5 years. Projections to the year 2015 (Figure 3, segments CEF and CDG) depend upon the minimum feature (pattern) size assumed for the circuits. That is, if the patterned-film dimensions decrease, then component sizes also decrease, and more components can be fabricated on a silicon chip of fixed area. The minimum feature size of mass-produced (production) circuits has steadily dropped over the past 25 years (Figure 4). Markers placed along or near the line in Figure 4 serve as indicators of relative pattern sizes.

The different length scales and small feature sizes involved in microelectronic devices are demonstrated graphically in Figure 5 by comparison of the size of device features to that of a ladybug. The magnification increases for each picture in clockwise direction starting at the upper left picture, which shows a ladybug on a 75-mm-diameter wafer, and ending at the lower left picture, which shows 2-μm metal lines of an individual electrical device.

To appreciate the fabrication sequences and, thus, the future directions and needs of IC process technology, the operation of simple solid-state devices must be understood. The following section gives a short introduction

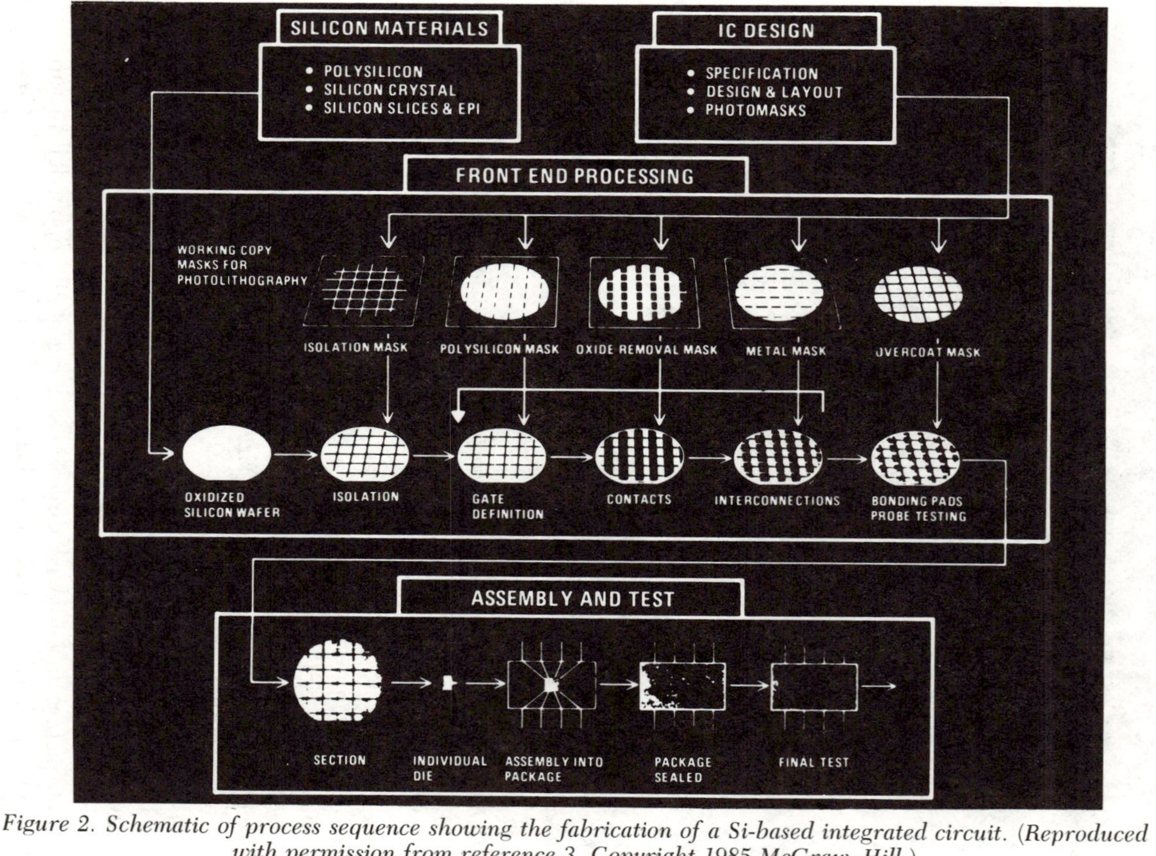

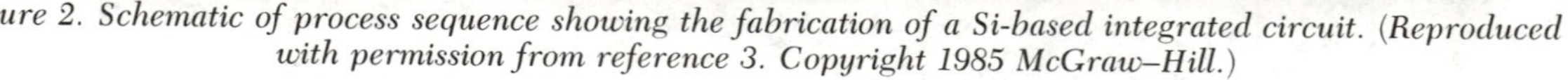

Figure 2. Schematic of process sequence showing the fabrication of a Si-based integrated circuit. (Reproduced with permission from reference 3. Copyright 1985 McGraw–Hill.)

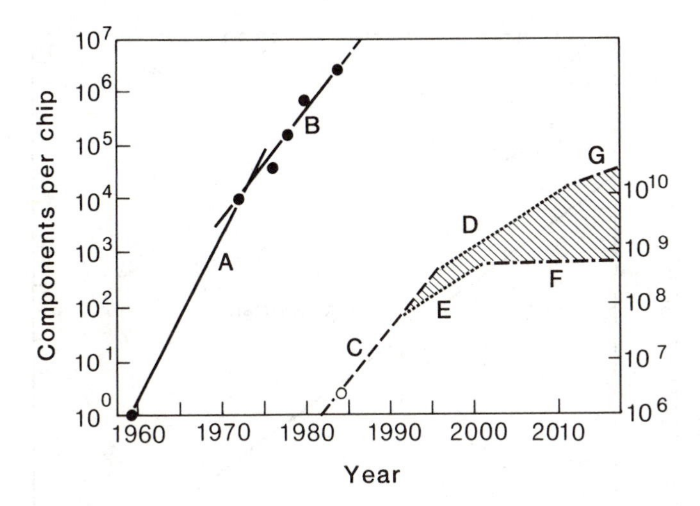

Figure 3. Components per silicon chip versus calendar year. Segments A and B are history; segments CDG and CEF are projections based upon 0.5- and 0.25-μm minimum feature sizes, respectively. (Reproduced with permission from reference 4. Copyright 1984 IEEE.)

to solid-state physics. Additional detail may be found in solid-state and device physics textbooks by, for instance, Warner and Grung (2), Ashcroft and Mermin (5), Kittel (6), Sze (7, 8), and Grove (9).

Overview of Solid-State Physics

Crystal Structure. Solids can be classified into three groups according their conductivity. *Insulators* such as SiO_2 typically have conductivities less than 10^{-8} $(\Omega\text{-cm})^{-1}$, whereas *conductors* (e.g., metals) have conductivities greater than 10^3 $(\Omega\text{-cm})^{-1}$. Solids with intermediate conductivities are called *semiconductors*. These include elemental semiconductors (e.g., Si) or compound semiconductors made up of atoms from groups II to VI of the periodic table, such as ZnSe (group II–VI compound semiconductor), GaAs (group III–V compound semiconductor), SiC (group IV–IV compound semiconductor), and PbTe (group IV–VI compound semiconductor).

A solid can have an amorphous, polycrystalline, or single-crystalline physical structure. In single crystals, the atoms are arranged in three-dimensional periodic arrays. The crystal structure of Si is the diamond lattice, in which each Si atom is tetrahedrally coordinated with four other Si atoms. This structure can be viewed as two interpenetrating face-centered-cubic (fcc) lattices displaced from each other by one-quarter of the distance along the cube diagonal (Figure 6a). Compound semiconductors of the type AB (e.g., GaAs) typically have a zinc blende lattice (Figure 6b), which is simply the diamond structure with one fcc lattice composed of A atoms (e.g., Ga)

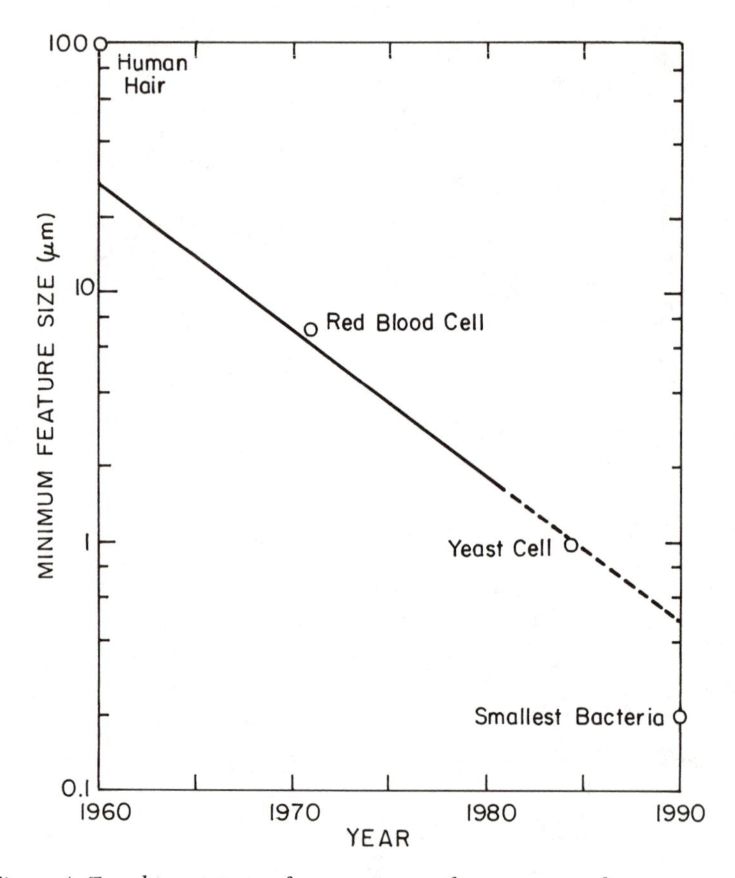

Figure 4. Trend in minimum feature size on silicon integrated circuits versus year of circuit introduction. Markers (○) give an indication of relative feature size.

and the other fcc lattice composed of B atoms (e.g., As). The long-range order persists in polycrystalline solids but is restricted to small crystallites separated by imperfections or defects called grain boundaries. Amorphous materials have no long-range lattice structure, but the nearest neighbor (i.e., short-range) bonding is regular. For instance, insulators such as SiO_2 and Si_3N_4 are deposited as amorphous films.

The specific application of a material generally determines the particular structure desired. For example, hydrogenated amorphous silicon is used for solar cells and some specialized electronic devices (*10*). Because of their higher carrier mobility (*see* Carrier Transport, Generation, and Recombination), single-crystalline elemental or compound semiconductors are used in the majority of electronic devices. Polycrystalline metal films and highly doped polycrystalline films of silicon are used for conductors and resistors in device applications.

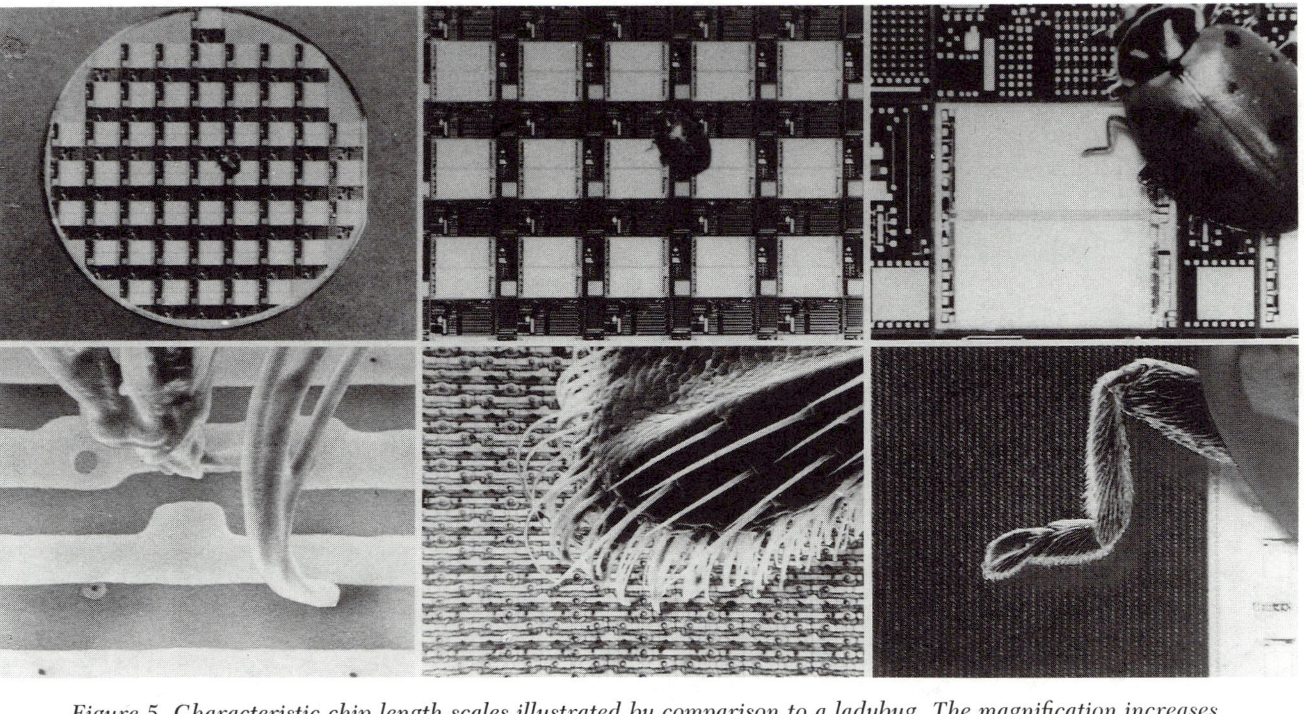

Figure 5. Characteristic chip length scales illustrated by comparison to a ladybug. The magnification increases for each picture in the clockwise direction starting with the upper left corner, which shows a ladybug on 62-kilobit random-access-memory chips on a 75-mm-diameter silicon wafer. The final magnification in the lower left corner shows the metal lines (2 μm) at the device level. (Reproduced with permission from reference 3. Copyright 1985 McGraw–Hill.)

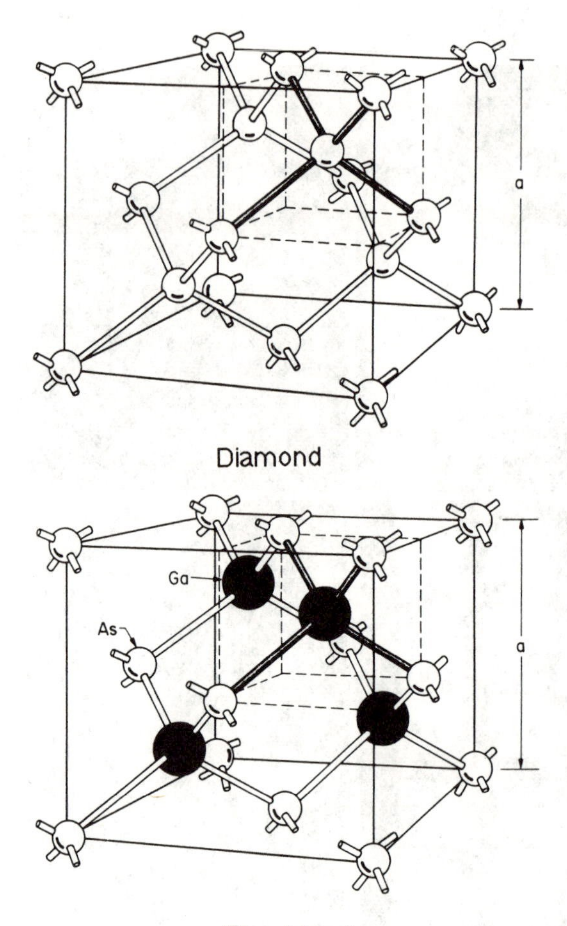

Figure 6. Top, diamond lattice (Si); bottom, zinc blende lattice (GaAs). (Reproduced with permission from reference 8. Copyright 1985 Wiley.)

Energy Bands. Electrons make up the chemical bonds between atoms in a solid. In silicon, this bonding is primarily covalent, whereas in compound semiconductors (group II–VI compounds in particular), the bonds also have substantial ionic character. The electrons participating in these bonds are termed *valence electrons*. Free electrons created by breaking bonds or *doping* (*see* Chapter 6) are available for current flow and are known as *conduction electrons*.

When a bond is broken and a free electron is removed, a *hole* or electron vacancy is created. When this vacancy is filled by an electron from a neighboring bond, the result is a hole moving in a direction opposite to that of the electron. Thus the hole may be considered as a (fictitious) particle with a positive charge. These free electrons and holes are responsible for the operation of electronic devices.

Electrons associated with gas molecules have discrete energy levels, because they are essentially unperturbed by other molecules. When atoms are brought together to form a solid, their electron wave functions interact, and the resulting discrete energy states form nearly continuous bands of allowed electron energies. The energy levels within each band are filled by pairing electrons according to the Pauli exclusion principle. The highest energy filled or partially filled band is called the *valence band*, and the next highest (often empty) band is called the *conduction band*. Between these bands of allowed energy levels are gaps of forbidden energy levels. The energy difference between the valence band maximum and the conduction band minimum is the *band gap*, E_g.

At 0 K the valence band is completely filled and the conduction band is empty for insulators and semiconductors. In insulators such as SiO_2, a large amount of energy is required to break a bond and promote an electron to the conduction band, that is, the band gap is large ($E_g > 5$ eV). Thus, thermal vibrations with energy of the order kT (in which k is the Boltzmann constant and T is temperature; $kT = 0.026$ eV at 300 K) cannot generate conduction electrons (or holes), and the material cannot conduct current. Semiconductors have smaller band gaps ($E_g < 2$–3 eV), so there will be a finite probability of promoting an electron from the valence band to the conduction band and creating a hole in the valence band; thus the material can transport electrons and conduct current.

Energy band gaps for selected semiconductors are summarized in Table I. On the basis of the nature of the transition from the valence band to the conduction band, semiconductors are classified as *direct* or *indirect*. In a direct semiconductor, the transition does not require a change in electron momentum, whereas in an indirect semiconductor, a change in momentum is required for the transition to occur. This difference is important for optical devices such as lasers, which require direct-band-gap materials for efficient radiation emission (7, 8). As indicated in Figure 7, Si is an indirect semiconductor, whereas GaAs is a direct semiconductor.

Conductors (e.g., metals) have partially filled conduction bands or overlapping conduction and valence bands. Because of the easily accessible energy levels (i.e., no band gap), electrons can readily move to higher energy levels and conduct current.

Carrier Concentrations. ***Intrinsic Carriers.*** The number of available carriers depends on thermally generated conduction band electrons and valence band holes, as well as carriers produced from the incorporation (intentional or unintentional) of impurities. In an intrinsic semiconductor, the thermally generated carriers dominate. The number of electrons (n) in the conduction band can then be calculated from the integral over the density

Table I. Energy Band Gaps and Transport Properties for Semiconductors

	Band Gap (eV)		Mobility at 300 K ($cm^2/V{-}s$)		Band[a]	Effective Mass m_o[b]				
						Electrons[c]			Holes[d]	
Semiconductor	300 K	0 K	Electrons	Holes	Type	$m_n \parallel$	m_n	$m_n \perp$	m_{hl}	m_{hh}
Group IV										
C	5.47	5.48	2000	2100	I	1.4		0.36	0.3	1.1
Si	1.12	1.17	1450	300	I	0.92		0.19	0.15	0.54
Ge	0.66	0.74	3900	1800	I	1.6		0.08	0.04	0.28
Group IV–IV										
α-SiC	2.50	3.03	900	25	I	0.7		0.25	—[e]	—[e]
Group III–V										
AlN	6.2	6.28		10	D					
AlP	2.5		50	450	I	3.67		0.21	0.21	0.51[f]
AlAs	2.16	2.22	300	200	I	1.1		0.19	0.15	0.41[f]
AlSb	1.62	1.69	200	400	I	1.8		0.26	0.12	0.34[f]
GaN	3.36	3.50	380		D					
GaP	2.27	2.35	160	135	I	1.12		0.22	0.12	0.45
GaAs	1.42	1.52	9200	400	D		0.067		0.076	0.50
GaSb	0.75	0.82	3750	700	D		0.042		0.05	0.28
InP	1.34	1.42	5900	150	D		0.077		0.12	0.60
InAs	0.36	0.42	33,000	450	D		0.024		0.026	0.35
InSb	0.17	0.23	77,000	850	D		0.014		0.016	0.45

Group II–VI								
ZnO	3.35	3.42	200	180	D	0.28		(0.59) [g]
ZnS	3.68	3.84	165	10	D	0.38	0.2	1.7
ZnSe	2.7	2.82	540	30	D	0.15	0.15	1.47
ZnTe	2.26	2.39	340	100	D	0.12	0.15	0.63
CdS	2.42	2.56	400	50	D	0.21		(0.65) [g]
CdSe	1.75	1.83	800		D	0.13		(0.45) [g]
CdTe	1.49	1.61	1000		D	0.095	0.13	0.72
Group IV–VI								
PbS	0.42	0.29	600	700	I	(0.25) [h]		(0.25)
PbTe	0.31	0.19	6000	4000	I	(0.17) [h]		(0.20)
Group I–VII								
γ-CuCl	3.17	3.39			D			
γ-CuBr	0.91	3.08			D			
AgCl	2.99	3.25			I			
AgBr	2.44	2.68			I			

NOTE: The table is based on data from references 7 and 23. Blank entries mean no or insufficient data.

[a] The abbreviations I and D stand for indirect and direct, respectively.

[b] Effective mass is given in terms of the free electron mass, m_o.

[c] The abbreviations $m_n \parallel$ and $m_n \perp$ denote the longitudinal and transverse effective electron masses, respectively. m_n is the effective mass for isotropic conduction band minimum.

[d] The abbreviations m_{hl} and m_{hh} denote the light and heavy effective hole masses, respectively.

[e] Values are not reported because of the nonparabolic nature of the valence band.

[f] ∥ <100>; *see* reference 23 for other directions.

[g] Value in parentheses is the overall value for the effective hole mass.

[h] Value in parentheses is the overall value for the effective electron mass.

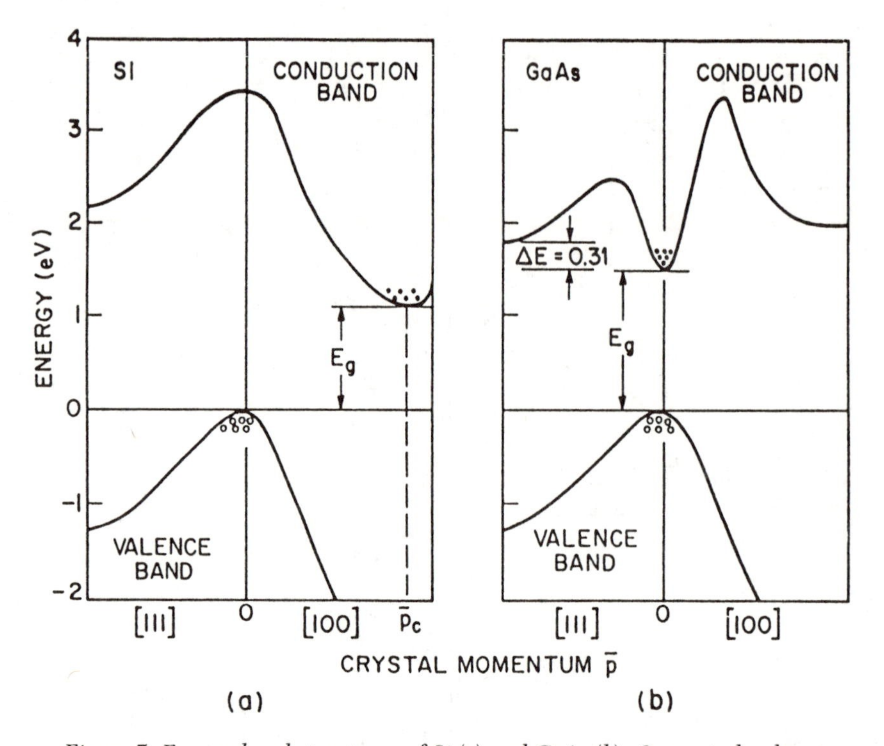

Figure 7. Energy band structures of Si (a) and GaAs (b). Open circles denote holes in the valence bands, and closed circles denote electrons in the conduction bands. Crystal direction is indicated by the bracketed Miller indices. (Reproduced with permission from reference 8. Copyright 1985 Wiley.)

of states, $N_e(E)$, times the probability that an electronic state is occupied by an electron, $F(E)$, that is,

$$n = \int_{E_{cb}}^{E_{ct}} N_e(E)F(E)dE \tag{1}$$

in which E_{cb} and E_{ct} are the energies at the bottom and top of the conduction band, respectively.

The probability distribution, $F(E)$, is the Fermi–Dirac distribution function

$$F(E) = \frac{1}{1 + \exp[(E - E_F)/kT]} \tag{2}$$

in which T is temperature, k is the Boltzmann constant, and E is electron energy. E_F is the Fermi level, which is the energy at which the probability of electron occupancy is exactly 0.5. In an intrinsic semiconductor at reasonable temperatures, the few available electrons are confined near the conduction band minimum, and equivalently, the holes are found primarily in levels near the valence band maximum. Therefore, a simple quadratic approximation may be used to express the density of states for the electrons (5, 7):

$$N_e(E) = \frac{m_n^{3/2}}{\pi^2 \bar{h}^3} \sqrt{2|E - E_g|} \quad (3)$$

In equation 3, m_n is the effective mass of the electron, $\bar{h}$ is the Planck constant divided by 2π, and E_g is the band gap. Unlike the free electron mass, the effective mass takes into account the interaction of electrons with the periodic potential of the crystal lattice; thus, the effective mass reflects the curvature of the conduction band (5). This curvature of the conduction band with momentum is apparent in Figure 7. Values of effective masses for selected semiconductors are listed in Table I. The different values for the longitudinal and transverse effective masses for the electrons reflect the variation in the curvature of the conduction band minimum with crystal direction. Similarly, the light- and heavy-hole mobilities are due to the different curvatures of the valence band maximum (5, 7).

Insertion of equation 3 into equation 1, approximation of the Fermi distribution by a classical Maxwell–Boltzmann distribution, and integration of equation 1 yield the expression for the total number of electrons in the conduction band:

$$n = 2\left(\frac{m_n kT}{2\pi \bar{h}^2}\right)^{3/2} e^{(E_F - E_c)/kT} \quad (4)$$

in which E_c is the energy of the conduction band minimum.

Similarly, the concentration of holes (p) in the valence band can be expressed as

$$p = 2\left(\frac{m_p kT}{2\pi \bar{h}^2}\right)^{3/2} e^{(E_v - E_F)/kT} \quad (5)$$

in which m_p is the effective mass of the hole and E_v is the energy of the valence band maximum.

For an intrinsic (undoped) semiconductor, the number of holes in the valence band equals the number of electrons in the conduction band, because the only free carriers available arise from the promotion of electrons from the valence band (a process that creates holes) to the conduction band. Therefore, the following relationship exists

$$n_i = n = p \tag{6}$$

in which n_i is the intrinsic carrier concentration. Combination of equation 6 with equations 4 and 5 gives the intrinsic Fermi level E_{F_i} as a function of temperature:

$$E_{F_i} = \frac{1}{2} E_g + \frac{3}{4} kT \ln \frac{m_p}{m_n} \tag{7}$$

Because kT is small at room temperature (0.026 eV) and because the effective masses usually do not differ substantially, the Fermi level is positioned near the middle of the band gap. From the law of mass action,

$$np = n_i^2 \tag{8}$$

the intrinsic carrier concentration may be derived as a function of material properties:

$$n_i = 2 \left(\frac{kT}{2\pi \hbar^2} \right)^{3/2} (m_n m_p)^{3/4} e^{-E_g/2kT} \tag{9}$$

At room temperature the value of n_i for Si is approximately 1.45×10^{10} electrons per cm^3.

Extrinsic Carriers. When sufficient dopant is added to a semiconductor so that the resulting carrier concentration exceeds the intrinsic level, the semiconductor is called extrinsic. Dopants that donate free electrons to the conduction band are called *donors*, and dopants that accept electrons and release holes to the valence band are called *acceptors*. Desired dopants introduce energy levels within the forbidden gap that are slightly below the conduction band edge or slightly above the valence band edge; these are known as shallow dopants. P and As are shallow donors in Si, whereas B is a shallow acceptor in Si. Dopants that introduce energy states in the middle of the band gap, the so-called deep levels, are typically not useful for device purposes, because they do not ionize significantly at room temperature. Examples of impurities that create deep levels in Si are C, Au, and Pt.

To calculate the number of carriers resulting from the doping of a semiconductor, the degree of ionization of the donors must be known. Most shallow donors in Si are completely ionized at room temperature, and the number of electrons equals the number of donors (n_D):

$$n = n_D \tag{10}$$

At equilibrium the number of holes may then be computed from the law of mass action (equation 8), because both electrons and holes are always present in a semiconductor. The Fermi level can be determined from equation 4.

Carrier Transport, Generation, and Recombination. When an electric field is applied to a semiconductor, electrons are accelerated in the direction opposite to the field with a drift velocity, v_n, proportional to the field, E:

$$v_n = -\mu_n E \tag{11}$$

The proportionality factor is the *electron mobility*, μ_n, which has units of square centimeters per volt per second. The mobility is determined by electron-scattering mechanisms in the crystal. The two predominant mechanisms are lattice scattering and impurity scattering. Because the amplitude of lattice vibrations increases with temperature, lattice scattering becomes the dominant mechanism at high temperatures, and therefore, the mobility decreases with increasing temperature.

Theory predicts that the mobility decreases as $T^{-3/2}$ because of lattice scattering (8). But because electrons have higher velocities at high temperatures, they are less effectively scattered by impurities at high temperatures. Consequently, impurity scattering becomes less important with increasing temperature. Theoretical models predict that the mobility increases as $T^{3/2}/n_I$, in which n_I is the total impurity concentration (8). The mobility is related to the electron diffusivity, D_n, through the Einstein relation

$$D_n = \frac{kT}{q} \mu_n \tag{12}$$

in which q is the electronic charge.

By analogy to equation 11, the drift velocity for holes (v_p) in the valence band may be expressed as

$$v_p = \mu_p E \tag{13}$$

in which μ_p is the hole mobility and the minus sign is missing because holes drift in the same direction as the electric field. The hole mobility has the same characteristics as the electron mobility and is related to the hole diffusivity by the appropriate Einstein relation. Because of the larger effective mass of holes, the hole mobility is less than the electron mobility.

In a semiconductor, the total current flowing due to an applied field is the sum of the contributions of electrons (I_n) and holes (I_p), as given by equation 14

$$I = I_n + I_p = (qn\mu_n + qp\mu_p)E \tag{14}$$

in which the enclosed quantity is the conductivity, σ, which is the reciprocal of the semiconductor resistivity. Mobilities are weak functions of temperature compared with the exponential dependence of the carrier concentrations, n and p (equations 4 and 5). Therefore, an estimate of the band gap energy, E_g, may be obtained by plotting ln σ against $1/T$ for an intrinsic semiconductor. The slope is then $-E_g/2k$.

Like the performance of chemical reactors, in which the transport and reactions of chemical species govern the outcome, the performance of electronic devices is determined by the transport, generation, and recombination of carriers. The main difference is that electronic devices involve charged species and electric fields, which are present only in specialized chemical reactors such as plasma reactors and electrochemical systems. Furthermore, electronic devices involve only two species, electrons and holes, whereas 10–100 species are encountered commonly in chemical reactors. In the same manner that species continuity balances are used to predict the performance of chemical reactors, continuity balances for electrons and holes may be used to simulate electronic devices. The basic continuity equation for electrons has the form

$$\frac{\partial n}{\partial t} = \frac{1}{q}\nabla \cdot I_n + G_n - R_n \tag{15}$$

in which t is time, G_n is the rate of generation of electrons, R_n is the rate of recombination of electrons, and I_n is the electron current density (i.e., flux). The electron current density is defined as

$$I_n = q\mu_n nE + qD_n\nabla n \tag{16}$$

which is the sum of drift components due to an electric field and diffusion components due to concentration gradients. Equation 15 is very similar to the continuity balance for neutral species for a binary system (*10*)

$$\frac{\partial c_A}{\partial t} = -\nabla \cdot N_A + R_A \tag{15a}$$

in which c_A is the concentration of species A, R_A represents the reaction terms, and N_A is the molar flux of species A. The flux is defined by equation 16a

$$N_A = x_A(N_A + N_B) - cD_{AB}\frac{\partial x_A}{\partial z} \tag{16a}$$

in which x_A is the mole fraction of species A, N_B is the molar flux of species B, c is the total molar concentration, D_{AB} is the binary molar diffusivity, and z is the spatial coordinate. In equation 16a the first term is the flux due to molar flow (drift), and the second term represents diffusion.

Generation typically occurs by carrier injection across a junction or by optical excitation. Photoexcitation is accomplished by shining a light source on the semiconductor. If the energy $h\nu$ is greater than the band gap energy of the semiconductor, the photon can be absorbed and an electron–hole pair is generated. Direct recombination of electrons and holes dominate in direct-band-gap semiconductors such as GaAs, but this process rarely occurs in indirect-band-gap semiconductors because conservation of momentum cannot be satisfied. In indirect-band-gap semiconductors, indirect recombination involving localized states in the gap predominate (*7, 8*). Recombination is often described in terms of carrier lifetimes, τ, which correspond to the time constant in an exponential decay of carrier concentration (*8*).

A continuity equation similar to equation 15 can be derived for holes:

$$\frac{\partial p}{\partial t} = -\frac{1}{q}\nabla \cdot I_p + G_p - R_p \tag{17}$$

In equation 17, I_p is the hole current density, G_p is the rate of generation of holes, and R_p is the rate of recombination of holes. The hole current density is defined by equation 18

$$I_p = q\mu_p pE - qD_p\nabla p \tag{18}$$

in which D_p is the hole diffusivity.

In addition to the electron and hole continuity equations, Poisson's equation must be satisfied.

$$\epsilon\nabla E = q(p - n + \sum \text{concentrations of ionized impurities}) \tag{19}$$

In equation 19, ϵ is the dielectric permittivity of the semiconductor.

With the appropriate boundary conditions, equations 17–19 form a complete description of carrier generation, recombination, and transport that can be used to simulate device performance. Many examples are given in the books by Warner and Grung (*2*), Sze (*7, 8*), and Shur (*12*), with particular emphasis on device applications. A complete treatment of electronic and optical devices is outside the scope of this chapter. However, to gain some insight into the important issues in device fabrication, a brief review of the basic features of p–n junctions and transistors is useful.

p–n Junctions. The p–n junction is the basic element of most electronic devices, including metal–oxide–semiconductor field effect transistors (MOSFETs), bipolar transistors, and solar cells. As illustrated in Figure 8, the junction is the region between p- and n-type-doped semiconductor materials. This junction can be formed by ion implantation followed by annealing, dopant diffusion, or epitaxial growth of a layer of one doping type (e.g., p type) on the substrate (e.g., n type). These individual processing steps are discussed in subsequent chapters. Electrically acceptable junction properties cannot be realized by physically pressing n- and p-type semiconductors together because of the numerous defects at the interface. Nevertheless, from the standpoint of describing the operation of a p–n junction, the joining of uniformly doped p- and n-type semiconductors (Figures 8a and 8b) can be considered.

When p- and n-type semiconductors are brought together, holes from the p side diffuse into the n side, and electrons from the n side diffuse to the p side. This process continues until a steady state is reached in which electric-field-generated drift and carrier diffusion are balanced and no net current flow occurs. At this electrochemical equilibrium condition, the Fermi level is constant (8). As electrons leave the n-type material, they leave behind uncompensated positive donor atoms that create a positive space charge. Similarly, uncompensated negative acceptor atoms that remain in the p-type material create a region of negative space charge. The result is an electric field directed from the positively charged donors toward the negatively charged acceptors.

The built-in potential V_b is given by equation 20

$$V_b = \frac{kT}{q} \ln \left[\frac{n_A n_D}{n_i^2} \right] \tag{20}$$

in which n_A, n_D, and n_i are the concentrations of ionized acceptors, ionized donors, and intrinsic carriers, respectively. A voltage applied to the junction changes the balance between diffusion and drift currents. If a positive ($V > 0$) bias is applied to the p side, the energy barrier for electron movement across the junction is reduced and electrons are injected from the n side into the p side. The net current across the junction is given by:

$$I = I_R(e^{qV/kT} - 1) \tag{21}$$

In equation 21, I_R is a constant current independent of the applied bias. Thus, the current increases exponentially with the applied voltage. This

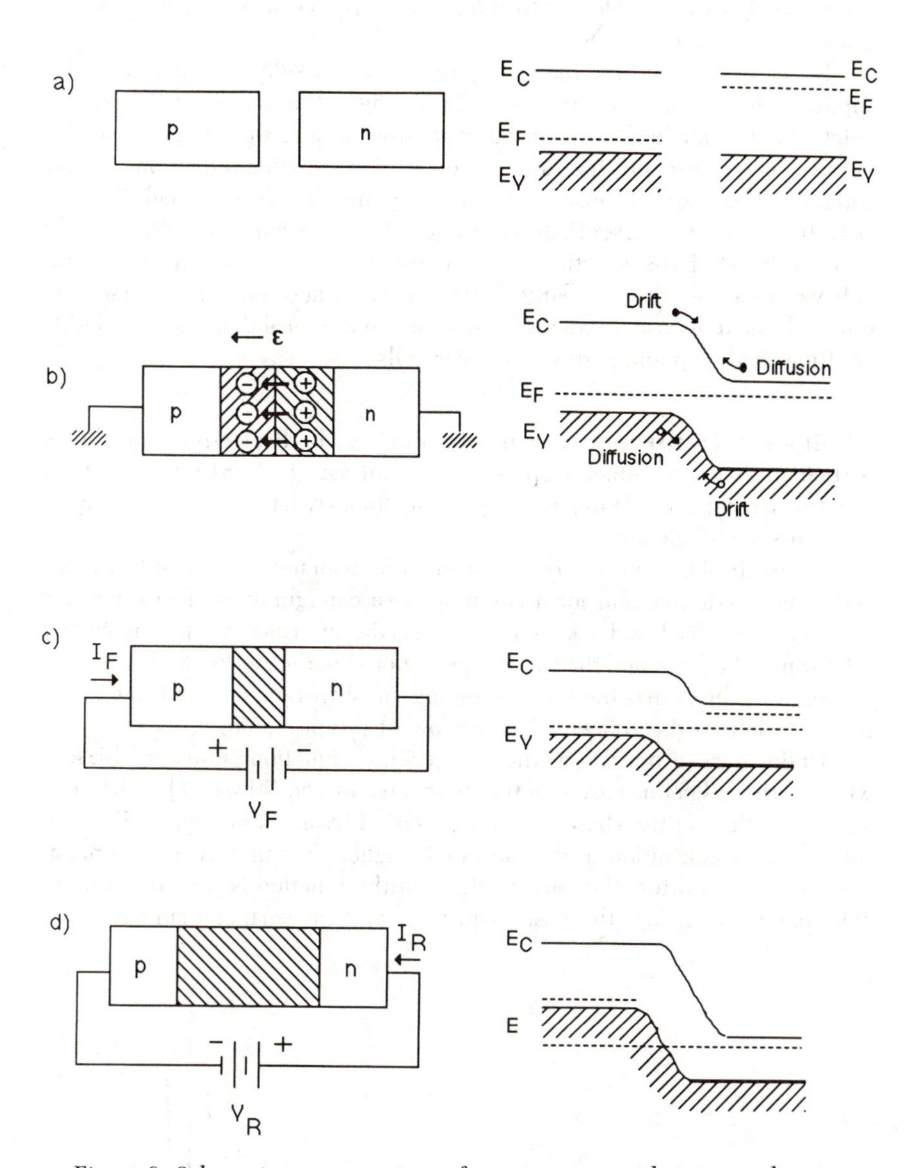

Figure 8. Schematic representations of p–n junctions and corresponding energy band diagrams under various conditions: (a) uniformly doped p-type and n-type semiconductors before junction is formed, (b) thermal equilibrium, (c) forward bias, and (d) reverse bias. Abbreviations are defined as follows: E_c, *electron energy at conduction band minimum;* E_v, *electron energy at valence band minimum;* I_F, *forward current;* V_F, *forward voltage;* V_R, *reverse voltage; and* ε, *electric field.*

situation (Figure 8c) is referred to as *forward bias*, because this bias direction is for enhanced current flow.

In the case of *reverse bias*, (Figure 8d), the n side is positive and the applied voltage is negative ($V < 0$). This configuration increases the barrier height for the forward electron current (from n side to p side) while the reverse current remains the same. The junction equation (equation 21) also applies to this case. Because $V < 0$, the exponential term is small; if $qV > 4kT$, the current is essentially constant ($-I_R$). The current–voltage (I–V) characteristics of the junction are illustrated in Figure 9. The rectifying behavior of a p–n junction (large forward current and minimal reverse current) is basic to its use in diodes, transistors, and optical devices (e.g., light-emitting diodes, photodiodes, and solar cells).

Bipolar Transistors. In bipolar devices, both electrons and holes participate in the conduction process, in contrast to MOSFETs, in which only one carrier type dominates. Bipolar technology has been used typically for high-speed-logic applications.

Figure 10 shows a three-dimensional view of an n–p–n bipolar transistor and a schematic diagram for a common base configuration. The transistor consists of two back-to-back p–n junctions; the intermediate p-type region is known as the *base*, and the two n-type regions are the *emitter* and *collector*. The emitter region has the highest doping (heavy n-type doping is indicated by n^+), whereas the collector has the lowest n-type doping.

Under normal operation, the emitter–base junction is forward biased, whereas the collector–base junction is reverse biased (Figure 11). The voltage across the emitter–base junction is varied by an input signal. Because the donor concentration in the emitter is higher than the acceptor concentration in the emitter, the current through the junction is primarily due to electrons injected into the base. The base width is smaller than the mean

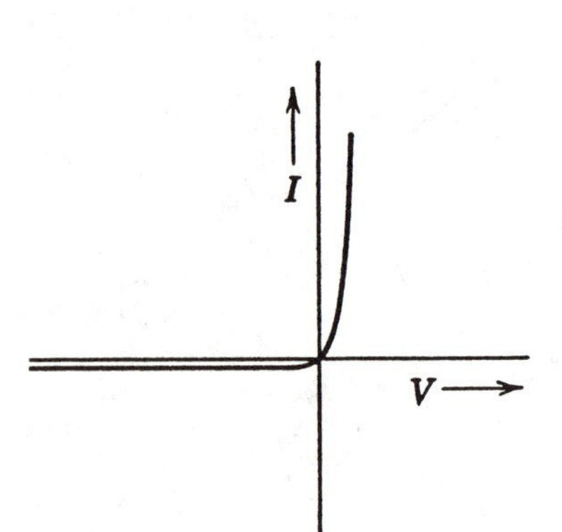

Figure 9. Current–voltage (I–V) characteristics of a Si p–n junction.

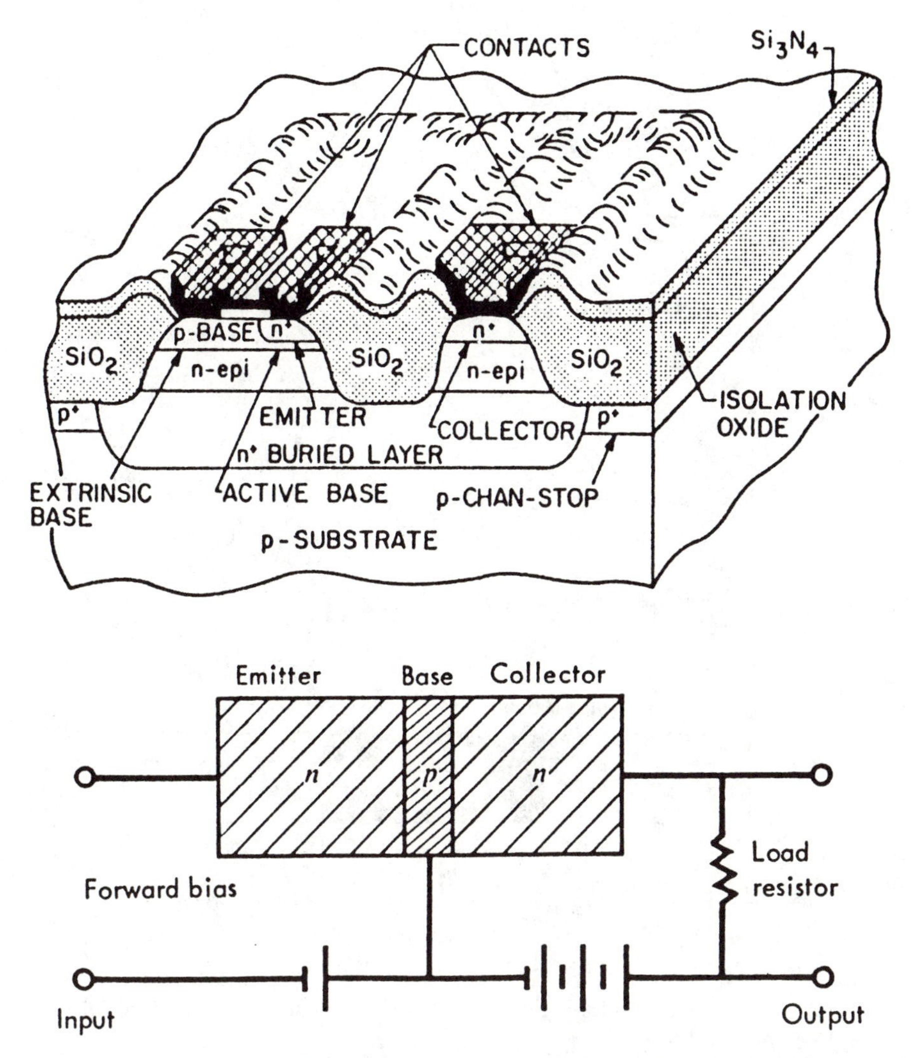

Figure 10. Top, three-dimensional view of an oxide-isolated bipolar transistor. (Reproduced with permission from reference 13. Copyright 1988 McGraw–Hill.) Bottom, schematic of a common base n–p–n transistor circuit. Abbreviations are defined as follows: n-epi, n-type-doped epitaxial-grown silicon; and p-CHAN-STOP, p-type channel stop.

free path of the electrons, and thus, essentially all of the injected electrons flow across the collector junction. Therefore, the collector current, I_c, is only slightly less than the emitter current, I_e, and any change in the emitter current is mirrored in the collector current. However, because of the load resistor and power supply (Figure 10, bottom), the variation in voltage across the collector–base junction can be much larger than the corresponding vari-

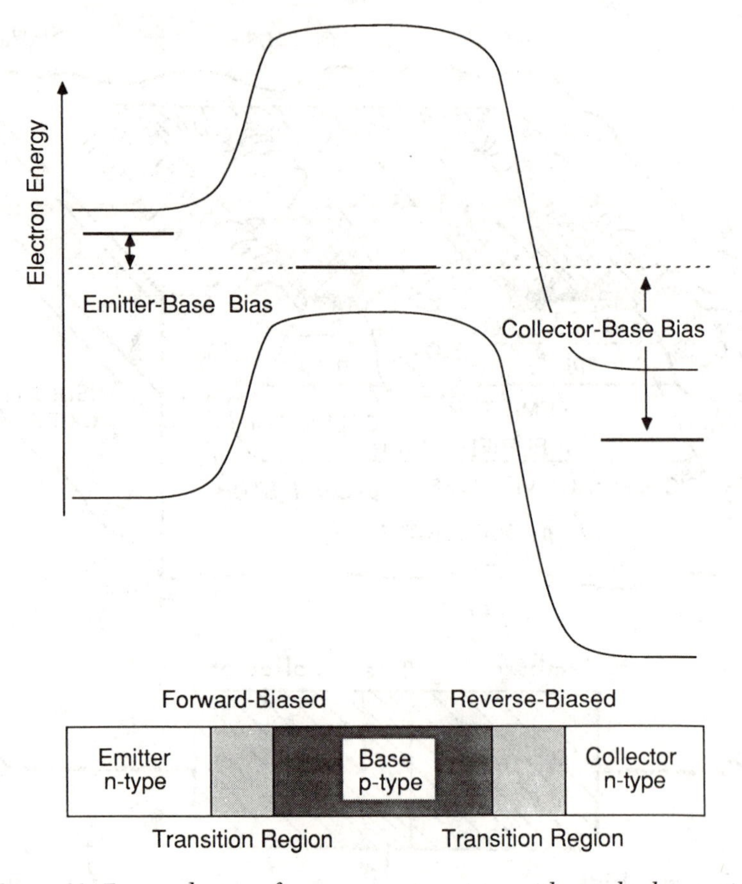

Figure 11. Energy diagram for an n–p–n transistor with standard operating bias voltages.

ation over the emitter–base junction. Thus, the transistor acts as a voltage and power amplifier.

The voltage gain is due to the injection of electrons from the emitter region, through the base, and into the collector region, where the voltage drop is large compared to a small emitter–base voltage (Figure 11). Thus, the base width must be smaller than the electron mean free path, or recombination will occur and the p–n junction will act as two separate junctions. Furthermore, few or no impurities should exist in the base; otherwise impurity scattering and carrier recombination will destroy device performance. Indeed, both feature size and purity constraints are issues that will be brought up repeatedly in subsequent chapters.

The function of the complementary p–n–p transistor may be explained by analogy to the n–p–n transistor by reversing polarities and considering hole rather than electron transport (7, 8). In addition, a large variety of operating regimes and circuit configurations are possible. Quantitative analyses of these systems are given in textbooks on device physics (*2, 7, 8, 12*).

MOSFETs. The metal–oxide–semiconductor field effect transistor (MOSFET or MOS transistor) (*8*) is the most important device for very-large-scale integrated circuits, and it is used extensively in memories and microprocessors. MOSFETs consume little power and can be scaled down readily. The process technology for MOSFETs is typically less complex than that for bipolar devices. Figure 12 shows a three-dimensional view of an n-channel MOS (NMOS) transistor and a schematic cross section. The device can be viewed as two p–n junctions separated by a MOS capacitor that consists of a p-type semiconductor with an oxide film and a metal film on top of the oxide.

Understanding the behavior of a MOS capacitor is useful in understanding the operation of a MOS transistor. When a negative voltage is applied to the conductor or metal, the energy bands in the p-type semiconductor

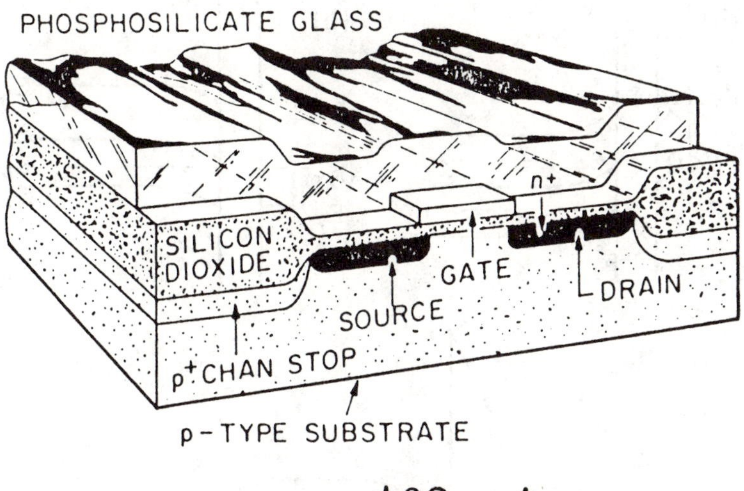

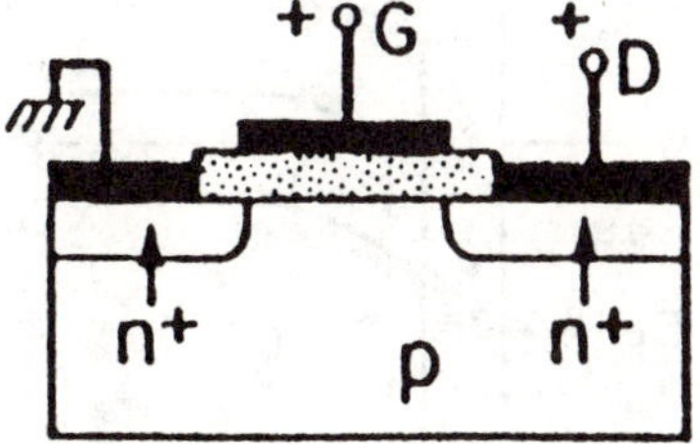

Figure 12. Top, three-dimensional view of an NMOS transistor. (Reproduced with permission from reference 13. Copyright 1988 McGraw–Hill.) Bottom, schematic of an NMOS transistor. (Reproduced with permission from reference 8. Copyright 1985 Wiley.) p^+ CHAN denotes heavily p-type-doped channel.

bend upward at the semiconductor–oxide interface, and holes accumulate near the interface. The situation is called *accumulation* (for accumulation of majority carrier; Figure 13a). If a small positive voltage is applied, the holes are repelled or depleted from the semiconductor surface and the bands bend down. This situation is called *depletion* (Figure 13b). With further increases in applied positive voltage, the intrinsic energy level crosses the Fermi level. Under this condition, more electrons than holes are present near the semiconductor surface, that is, electrons rather than holes have become the majority carrier in the p-type semiconductor surface. This situation is called *carrier inversion* (Figure 13c).

The operation of the NMOS transistor shown schematically in Figure 12 can be considered in the light of the previous discussion of a MOS capacitor. When no voltage is applied to the gate, the source and drain electrodes correspond to p–n junctions connected through the p region; therefore only a small reverse current can flow from source to drain. On the

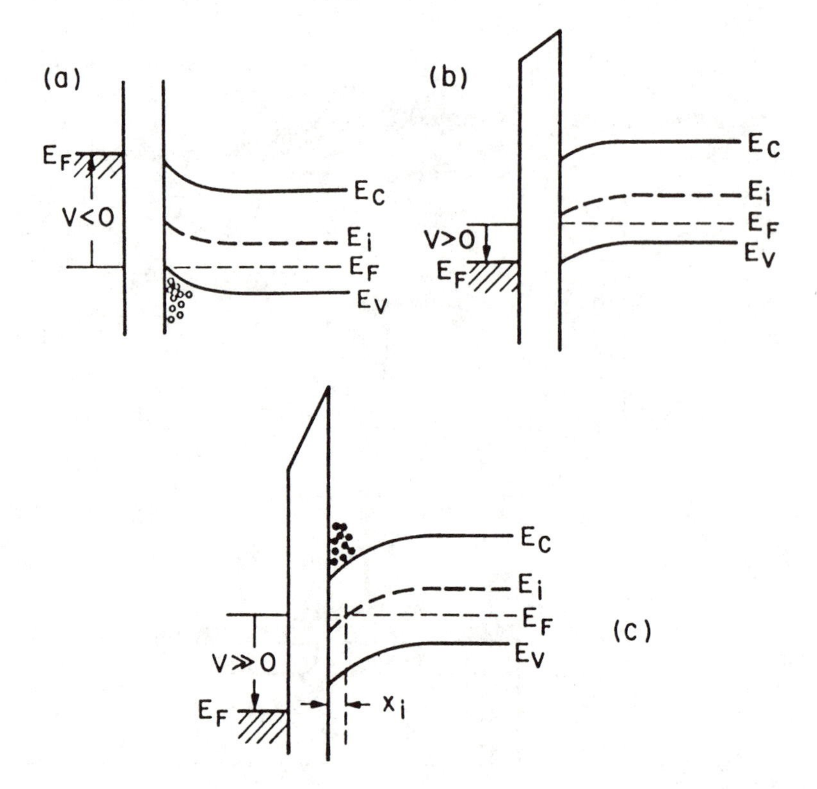

Figure 13. Energy band diagrams and charge distributions of an ideal MOS capacitor using p-type Si: (a) accumulation, (b) depletion, and (c) inversion. E_i *denotes the intrinsic Fermi level. (Reproduced with permission from reference 8. Copyright 1985 Wiley.)*

other hand, if a large positive voltage is applied to the gate, an inversion layer is formed adjacent to the semiconductor–oxide interface, analogous to the inversion case for the MOS capacitor. The semiconductor inversion layer connects the heavily doped (n^+) source and drain regions with an n channel so that a current can easily flow. The conductance of this channel may be modulated by changing the gate voltage. If a small drain voltage is applied, electrons will flow from the source to the drain, with the channel acting as a resistor. Thus the drain current is proportional to the applied voltage. Further increases in the drain voltage eventually lead to a point at which the width of the inversion layer becomes zero. At this "pinch-off" point, the drain current becomes saturated and becomes independent of the applied drain voltage.

Three other major MOS transistor configurations are available, in addition to the n-channel enhancement mode (normally off) just described (Figure 14). In the n-channel depletion mode, the device is fabricated (by appropriate doping) so that with zero bias on the gate an n channel exists between the source and drain. In this case a negative voltage must be applied to the gate to modulate the conductance. If this voltage is sufficiently large, the channel is depleted of electrons and the device turns off. Devices with p channels are similar to n-channel devices, except that p-channel devices are typically slower because of the lower mobility of holes compared with electrons.

NMOS and PMOS (p-channel MOS) transistors are used side by side in complementary metal–oxide–semiconductor (CMOS) technology to form logic elements. These structures have the advantage of extremely low power consumption and are important in ultralarge-scale integration (ULSI) and very-large-scale integration (VLSI) (*13*).

Quantitative descriptions of MOS devices are available (*2*, *8*, *9*, *12*). This short summary of solid-state physics was intended to illustrate the importance of carrier concentrations, transport, generation, and recombination in device performance. These properties, in turn, depend critically on material parameters resulting from a large number of chemical process sequences.

Unit Operations

Although more than 100 individual process steps are used in the manufacture of even simple integrated circuits, the fabrication sequence invokes many of the same operations numerous times. A list of "unit operations" that compose the technological arsenal for the fabrication of solid-state materials and devices can be made. Clearly, these unit operations are distinctly different from those associated with traditional chemical manufacture. Nevertheless, the purpose of defining such a list is the same: to establish the necessary chemical and physical operations so that a complicated process may be designed and carried out from individual, more easily controlled

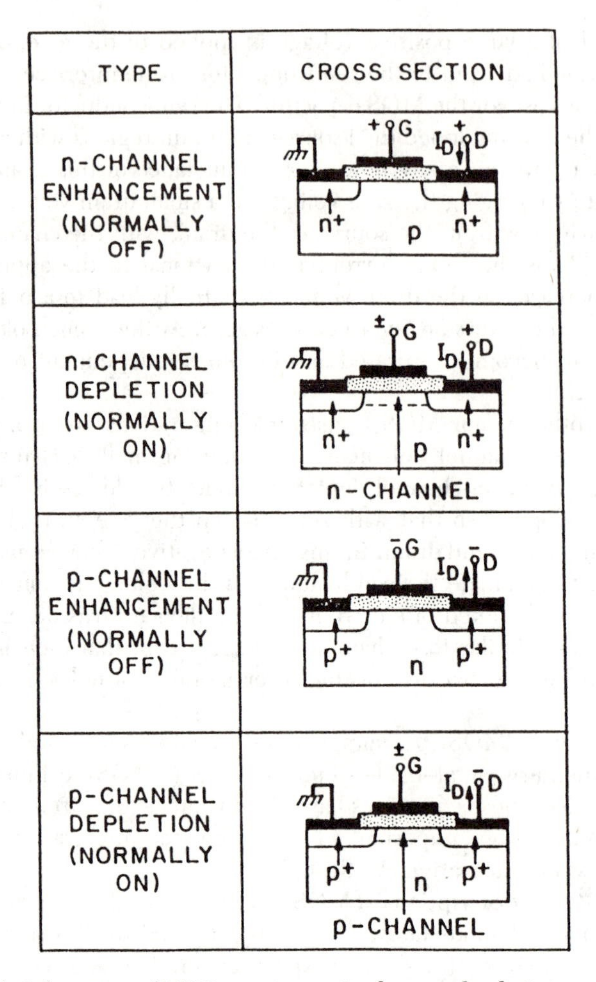

Figure 14. Schematics of MOS transistors. I_D *denotes the drain current.* (*Reproduced with permission from reference 8. Copyright 1985 Wiley.*)

and better understood steps common to all manufacturing sequences. For instance, the fabrication of solid-state devices can be described by various combinations of the operations given in List I.

Overview of Unit Operations. To maximize the electron or hole (carrier) mobility and thus device speed, ICs are built in single-crystal substrates. Methods of bulk crystal growth are therefore needed. The most common of these methods are the Czochralski and float-zone techniques. The Czochralski technique is a crystal-pulling or melt-growth method, whereas the float-zone technique involves localized melting of a sintered bar of the material, followed by cooling and, thus, crystallization.

List I. Unit Operations in Microelectronic Device Fabrication

- Bulk crystal growth
  - Czochralski
  - Float zone
- Chemical reactions with surfaces
  - Cleaning
  - Oxidation
  - Etching
- Thin-film formation
  - Evaporation
  - Sputtering
  - Epitaxy
  - Chemical vapor deposition
  - Spin casting
- Lithography
- Semiconductor doping
  - Solid-state diffusion
  - Ion implantation

After growth, the single-crystal boule is sliced into wafers or disks by diamond-coated wires or saws (*14*). The individual slices are subsequently chemomechanically polished (*14–16*), and sorted according to electrical resistivity. Before the wafers are subjected to high-temperature processes, they are exhaustively cleaned in various acids, solvents, or both to remove organic, inorganic, and particulate contaminants (*17*). Other chemical reactions performed on solid surfaces include the thermal oxidation of silicon to form amorphous SiO_2 and liquid- or plasma-etching techniques to define patterns in film or substrate materials.

Because ICs are built-up of thin-film layers, a variety of film deposition methods (*18–20*) are needed to satisfy specific requirements at particular points in the overall process sequence (e.g., film composition, low deposition temperature, uniform step coverage, cleanliness, deposition rate, film morphology, and crystal structure). In general, the methods can be classified as either physical (evaporation, sputtering, and spin coating) or chemical (chemical vapor deposition), although combinations such as reactive evaporation or reactive sputtering are sometimes invoked. The combinations involve the addition of a reactive gas such as O_2 or N_2 to an evaporative or sputtering (plasma) atmosphere so that the film and target are oxidized or nitrided during deposition (e.g., TiN can be formed from sputtering Ti in an Ar–N_2 atmosphere).

Lithography is the process of generating a stencil or pattern in a radiation-sensitive material. The stencil subsequently serves as a mask to permit etching only in selected areas of the film or substrate. During the fabrication of devices or ICs, the doping type (p or n) and level (resistivity) must be altered. (For instance, transistors and diodes require p–n junctions, and polycrystalline silicon resistors demand precisely controlled dopant concentrations.) Dopants can be introduced into silicon or other semiconductors by bringing a vapor containing the dopant into contact with the semiconductor surface at an elevated temperature. Solid-state diffusion then drives the dopant into the solid to the required depth. Alternatively, gaseous dopant ions can be generated, accelerated, and, finally, implanted into a solid surface.

Example: NMOS Fabrication. The individual steps listed in List I can be sequenced to give a simple process for the fabrication of an NMOS transistor (Figures 12 and 15) Although the example is a MOS transistor, the techniques also apply to the fabrication of bipolar transistors, diodes, capacitors, resistors, and ICs.

First, a p-type silicon wafer is immersed in a sulfuric acid–hydrogen peroxide mixture to remove major organic contaminants. Acid residues are removed by rinsing the wafer in deionized (DI) water (>18 MΩ-cm, a bacterial count of <50/L, and $<10^3$ particles per liter larger than 0.3 μm. Residual organic residues and metals are removed by immersion in an ammonium hydroxide–hydrogen peroxide bath at 75–80 °C, followed by a DI-water rinse. A short dip in a hydrofluoric acid solution followed by a DI-water rinse is often used to remove the "native" oxide layer grown when silicon is exposed to air. The remaining atomic and ionic contaminants are then removed from the wafer surface by immersion in a hydrochloric acid–hydrogen peroxide solution at 75-80 °C. After another DI-water rinse, the wafer is dried with hot nitrogen gas. The cleaned substrate is thermally oxidized at temperatures above 700 °C in a resistance-heated quartz tube into which oxygen, water vapor, or both are flowing (Figure 15a).

Because the diffusivities of normal silicon dopants (boron, phosphorus, and arsenic) are orders of magnitude larger in Si than in SiO_2, a SiO_2 layer can be used as a mask against dopant incorporation. Therefore, selected areas of the p-type Si substrate (source and drain regions) can be doped by generating openings or patterns in the SiO_2 film. The precise patterning is accomplished by lithography.

The lithographic procedure involves coating the wafer with a radiation-sensitive material (usually a polymer) called a *resist*. Selected areas are then exposed to radiation (UV photons, electrons, or X-rays). When UV radiation is used, this selective exposure is achieved by a chromium-on-glass photomask (Figure 15b). For positive resists, the radiation renders the exposed regions more soluble in a developer solution than the unexposed areas (Figure 15c). Thus, differential solubility permits the formation of a resist stencil or mask so that etching of the film beneath occurs only in regions where

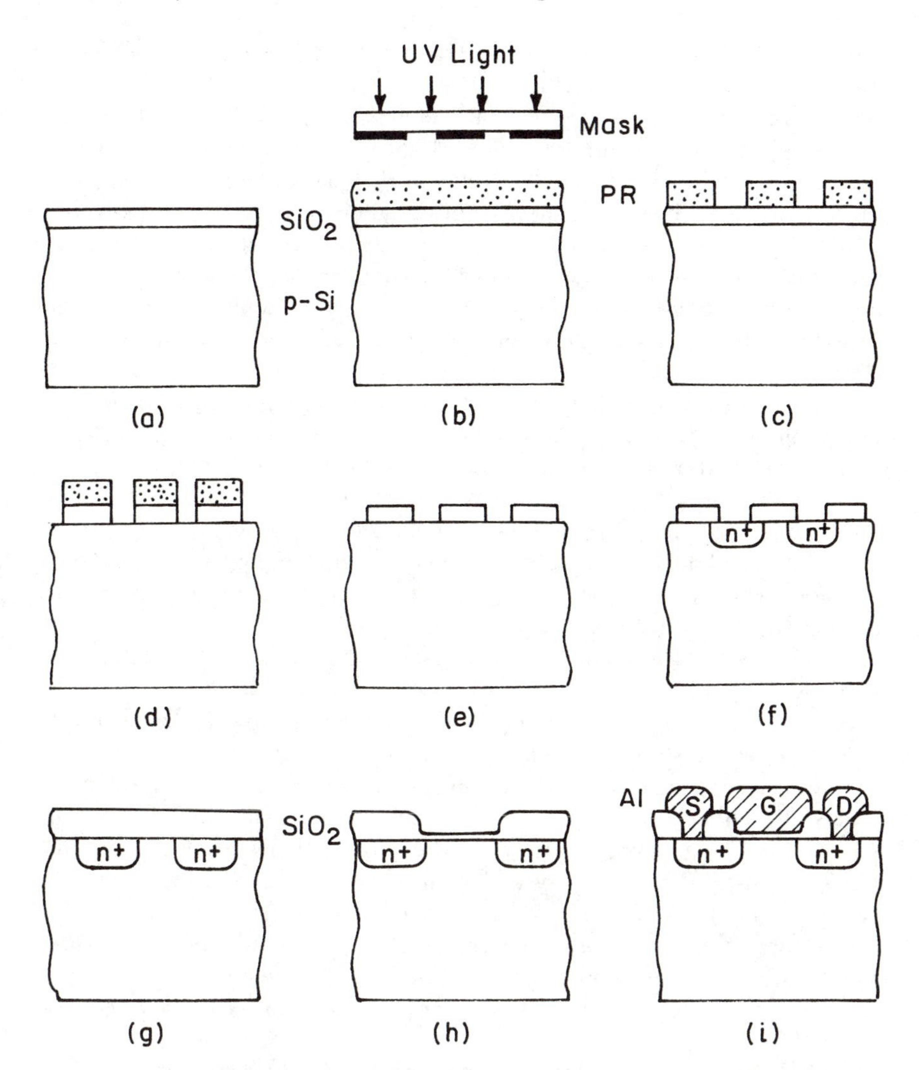

Figure 15. Simplified process sequence for the fabrication of an NMOS transistor: (a) substrate preparation, (b) selective exposure of substrate, (c) mask formation by differential solubility, (d) etching, (e) stripping of resist, (f) doping, (g) reoxidation of silicon surface, (h) formation of gate oxide, and (i) metal deposition and patterning. Abbreviations are defined as follows: p-Si, p-type silicon; PR, photoresist; S, source; G, gate; and D, drain.

the resist has been removed. The wafers then go through an etching step to transfer the resist pattern into the underlying SiO_2 film (Figure 15d). At present, etching is performed in a radio frequency (rf) glow discharge (plasma) containing fluorine species, although hydrofluoric acid solutions have been used in the past. The remaining resist is then stripped in either organic solvents or an oxygen plasma. The result of this lithographic–etching process is a patterned layer (Figure 15e).

The Si regions exposed by the SiO_2 etch are doped with phosphorus or arsenic to form source and drain areas (Figure 15f). Until the early 1970s, this process was accomplished by predeposition and drive-in diffusion. For the past 15 years, however, selective introduction of dopant atoms into the Si lattice has been performed by ion implantation. As in the case of predeposition and diffusion, a patterned SiO_2 layer can be used to attain selective doping. Alternatively, because ion implantation is a low-temperature process relative to predeposition, polymer resist layers can serve as implantation masks.

The next step in the fabrication of the n-channel transistor involves the formation of the thin (10–50-nm) gate oxide between the source and drain regions. First, the SiO_2, resist, or both remaining from the previous step are removed, and the Si surface is reoxidized to a thickness of ~300–500 nm (Figure 15g). By a lithographic–etching process, an opening is formed in the SiO_2, and the gate oxide is grown in this region (Figure 15h). Windows are then etched in the oxide over the source and drain regions in preparation for metal deposition (Figure 15i).

The electrical contact of the source, gate, and drain regions is established by the deposition of a conductor film. These conductors must satisfy a multitude of requirements, including low electrical resistivity, low contact resistance, good step coverage, good adhesion, and high electromigration resistance (*21*, *22*). Because no single material meets all these requirements, a number of conductors are used in a process, on the basis of the specific needs (e.g., contact barriers and interconnections). For instance, aluminum, heavily doped polycrystalline silicon, tungsten, and tungsten silicide are sometimes used in an individual IC process. These films are subsequently patterned to isolate the connections (Figure 15j) and to provide the interconnections among the various circuit elements (transistors, resistors, diodes, etc.) on the chip. In this manner, an electrical circuit is formed.

After the final metallization layer is patterned, the transistor and, on a larger scale, the IC are electrically complete. Before the wafer is diced or sectioned into individual chips and the circuits are subsequently packaged, a protective or passivation coating such as phosphorus-doped SiO_2, plasma-deposited silicon nitride, or polyimide is deposited over the finished IC. This layer protects the surface of the device from impurities such as alkali ions and water vapor, which can degrade the electrical properties or promote metal corrosion, and from mechanical damage due to handling. Finally, another lithographic step opens windows in the passivation coating so that leads from a package can be connected to the metallization pattern. These package leads connect the IC functions to the outside world.

Organization of This Volume

This volume addresses many of the unit operations listed in List I and depicted in Figure 15. The chemistry and chemical engineering principles behind these operations are described, and future directions and needs are suggested. The final chapter indicates the problems associated with the extreme purity and cleanliness demanded by IC processing.

Abbreviations and Symbols

c	total molar concentration
c_A	concentration of species A
D_n	electron diffusivity
D_p	hole diffusivity
D_{AB}	binary molar diffusivity
E	electron energy or electric field
E_c	electron energy at minimum of conduction band
E_{cb}	electron energy at bottom of conduction band
E_{ct}	electron energy at top of conduction band
E_F	Fermi level
E_{Fi} or E_i	intrinsic Fermi level
E_g	band gap
E_v	electron energy at maximum of valence band
$F(E)$	probability distribution, Fermi–Dirac distribution
G_n	rate of generation of electrons
G_p	rate of generation of holes
h	Planck constant
$\hbar$	Planck constant divided by 2π
I	total current
I_c	collector current
I_D	drain current
I_e	emitter current
I_F	forward current
I_n	electron current
I_p	hole current
I_R	reverse current
k	Boltzmann constant
m_n	effective mass of electron
m_p	effective mass of hole
n	concentration of electrons
n_A	concentration of ionized acceptors
n_D	concentration of ionized donors
n_i	intrinsic carrier concentration

n_I concentration of impurity
n_p concentration of holes
N_A molar flux of species A
N_B molar flux of species B
N_e density of electronic states
N_I total concentration of impurity
q electronic charge
R_A rate of reaction
R_n recombination rate of electrons
R_p recombination rate of holes
t time
T temperature
v_n drift velocity of electrons
v_p drift velocity of holes
V potential
V_b built-in potential
V_F forward voltage
V_R reverse voltage
x_A mole fraction of species A
x_i thickness of inversion layer
z spatial coordinate
ε electric field
ϵ_o dielectric permittivity of vacuum
ϵ_s dielectric permittivity of solid
μ_n electron mobility
μ_p hole mobility
ν frequency
σ conductivity
τ carrier lifetime

Acknowledgment

We thank G. B. Larrabee of Texas Instruments for the use of Figures 1, 2, and 5.

References

1. Deal, B. E.; Early, J. M. *J. Electrochem. Soc.* **1979,** *126*, 20c.
2. Warner, R. M., Jr.; Grung, B. L. *Transistors—Fundamentals for the Integrated Circuit Engineer;* Wiley: New York, 1983, Chapter 1.
3. Larrabee, G. B. *Chem. Eng.* **1985,** 92(*12*), 51–59.
4. Meindl, J. D. *IEEE Trans. Electron Devices* **1984,** *ED–31*, 1555.
5. Ashcroft, N. W.; Mermin, D. N. *Solid State Physics;* Saunders College: Philadelphia, 1976.
6. Kittel, C. *Introduction to Solid State Physics;* Wiley: New York, 1976.

7. Sze, S. M. *Physics of Semiconductors and Devices;* Wiley: New York, 1981; 2nd. ed.
8. Sze, S. M. *Semiconductor Devices—Physics and Technology;* Wiley: New York, 1985.
9. Grove, A. S. *Physics and Technology of Semiconductor Devices*; Wiley: New York, 1967.
10. Pankove, J. I., Ed. *Semiconductors and Semimetals, Vol. 21,* Academic: New York, 1984.
11. Bird, R. B.; Stewart, W. E.; Lightfoot, N. E. *Transport Phenomena;* Wiley: New York, 1960.
12. Shur, M. *Physics of Semiconductor Devices;* Prentice Hall: Englewood Cliffs, NJ, 1988.
13. Hillenenius, S. J.; *VLSI Process Integration in VLSI Technology;* Sze, S. M., Ed.; McGraw–Hill: New York, 1988; 2nd ed.
14. Bonora, A. C. In *Semiconductor Silicon 1977;* Huff, H. R.; Sirtl, E., Eds.; Electrochemical Society: Pennington, NJ, 1977; p 154.
15. Feng-Wei, L.; Guo-Chen, C.; Guang-Yui In *Semiconductor Silicon 1986,* Huff, H. H.; Abe, T.; Kolbersen, B., Eds.; Electrochemical Society: Pennington, NJ, 1986; p 183.
16. Schnegg, A.; Grundner, M.; Jacob, H. In *Semiconductor Silicon 1986,* Huff, H. H.; Abe, T.; Kolbersen, B., Eds.; Electrochemical Society: Pennington, NJ, 1986; p 183.
17. Kern, W. *Semiconductor International;* Cahners: Des Plaines, IL, 1984, p 94.
18. *Handbook of Thin Film Technology;* Maissel, L.; Glang, R., Eds.; McGraw–Hill: New York, 1970.
19. *Thin Film Processes;* Vossen, J. L.; Kern, W., Eds.; Academic: New York, 1978.
20. *Deposition Technologies for Films and Coatings;* Bunshah, R. F., Ed.; Noyes: New Jersey, 1982.
21. Learn, A. J. *J. Electrochem. Soc.* **1976,** *123*, 894.
22. Pauleau, Y. *Solid State Technol.* **1987,** *30*(2), 61.
23. *Landolt-Börnstein Semiconductors, Vol. 22*; Madelung, O., Ed.; Springer–Verlag: Heidelberg, 1987.

RECEIVED for review November 16, 1988. ACCEPTED revised manuscript February 2, 1989.

2

Theory of Transport Processes in Semiconductor Crystal Growth from the Melt

Robert A. Brown

Department of Chemical Engineering and Materials Processing Center, Massachusetts Institute of Technology, Cambridge, MA 02139

The quality of large semiconductor crystals grown from the melt for use in electronic and optoelectronic devices is strongly influenced by the intricate coupling of heat and mass transfer and melt flow in growth systems. This chapter reviews the present state of understanding of these processes, starting from the simplest descriptions of solidification processes to the detailed numerical calculations needed for quantitative modeling of processing with solidification. Descriptions of models for the vertical Bridgman–Stockbarger and Czochralski crystal growth techniques are included as examples of the level of understanding of industrially important techniques.

MELT CRYSTAL GROWTH OF SEMICONDUCTOR MATERIALS is a mainstay of the microelectronic industry. Boules of nearly perfect crystals are grown by a variety of techniques for controlled solidification and are used as substrates in almost all device fabrication technologies. The variety of crystalline materials produced in this manner ranges from the ubiquitous silicon to more-exotic materials like GaAs, InP, and CdTe. Although the processing conditions for each of these materials differ in some details, the solidification systems are similar in that the system dynamics and the quality of the crystal produced are governed by the same set of concepts describing the transport processes, thermodynamics, and materials science. These underpinnings make melt crystal growth one of the "unit operations" of electronic materials processing.

Revised and reprinted with permission from *AIChE J.* **1988**, *34*, 881–911

0065–2393/89/0221–0035$16.35/0
Published 1989 American Chemical Society

Many review papers and several books (*1–3*) have focused on the science and technology of crystal growth. The purpose of this chapter is not to duplicate these works but to focus on the fundamental transport processes that occur in melt crystal growth systems, especially advances in understanding that have occurred during the last decade of vigorous research. The chapter also accentuates the features and research issues that are common to many of the techniques used today in laboratories and industrial production.

A thorough understanding of heat and mass transport in melt growth processes is a prerequisite to optimization of these systems for control of crystal quality, as measured by the degree of crystallographic perfection of the lattice and the spatial uniformity of electrically active solutes in it. This discussion concentrates on the role of transport processes in controlling the stability of melt growth techniques and in setting solute segregation in the crystal. Several review papers have also discussed these aspects (*4*, *5*).

The connection between processing conditions and crystalline perfection is incomplete, because the link is missing between microscopic variations in the structure of the crystal and macroscopic processing variables. For example, studies that attempt to link the temperature field with dislocation generation in the crystal assume that defects are created when the stresses due to linear thermoelastic expansion exceed the critically resolved shear stress for a perfect crystal. The status of these analyses and the unanswered questions that must be resolved for the precise coupling of processing and crystal properties are described in a later subsection on the connection between transport processes and defect formation in the crystal.

The simple models of transport processes in controlled solidification, solute redistribution, and process stability reviewed in this chapter are based on results of pioneering studies during the last three decades. The analyses are rich in physical insight but are semiquantitative because of the limiting assumptions needed to permit closed-form analyses. The new possibilities for the large-scale numerical analysis of these transport processes are beginning to make practical the detailed simulation of melt growth. Examples of these results are included to give perspectives of the complexity involved in quantitative modeling of these systems.

Several examples of melt growth systems are described in the next section to facilitate the description of the transport processes in succeeding sections. The crystal growth systems are distinguished according to whether a surrounding solid ampoule or a melt–fluid meniscus shapes the crystal near the solidification interface. Methods with these features are classified as confined and meniscus-defined crystal growth techniques, respectively. Then, the fundamental transport processes associated with solidification of a dilute binary alloy in a temperature gradient are reviewed. The classical one-dimensional analysis of directional solidification is discussed for diffusion-controlled growth. Mechanisms for convection in the melt are reviewed, and models that account for convection in solute transport are presented.

Important development efforts have gone into any commercial crystal growth system to optimize the important aspects of each process with respect to the production of compositionally uniform cylindrical crystals with low densities of crystalline defects. Progress in analysis and optimization of real solidification systems is exemplified in later sections through descriptions of the vertical Bridgman–Stockbarger and Czochralski growth methods. The meniscus-defined growth of the Czochralski method leads to the issues of process stability and control for batchwise growth; these points are brought out in a subsection devoted to process stability and control.

Classification of Melt Crystal Growth Systems

The many technological innovations in melt crystal growth of semiconductor materials all build on the two basic concepts of confined and meniscus-defined crystal growth. Examples of these two systems are shown schematically in Figure 1. Typical semiconductor materials grown by these and other methods are listed in Table I. The discussion in this section focuses on some of the design variables for each of these methods that affect the quality of the product crystal. The remainder of the chapter addresses the relationship between these issues and the transport processes in crystal growth systems.

Several issues must be addressed. First, the heat-transfer environment must yield a well-controlled temperature field in the crystal and melt near the melt–crystal interface so that the crystallization rate, the shape of the solidification interface, and the thermoelastic stresses in the crystal can be controlled. Low dislocation and defect densities occur when the temperature gradients in the crystal are low. This point will become an underlying theme of this chapter and has manifestations in the analysis of many of the transport processes described here.

Second, the stoichiometry of the melt and of impurities introduced during processing must be controlled to the level demanded by application. Although these constraints vary with application, more control is clearly better in that the demands on purity and spatial uniformity of the material are becoming more stringent with the increasing miniaturization of electronic devices.

Confined Crystal Growth Systems. In confined growth geometries, such as the variations of the directional-solidification method (*1*), the material is loaded into an ampoule, melted, and resolidified by varying the temperature field either by translation of the ampoule through the furnace (the Bridgman–Stockbarger method; Figure 1a) or by time-dependent variation of the heater power (the gradient freeze method). After solidification, the material is removed from the ampoule.

Confined melt growth systems have been used primarily for laboratory preparation of exotic materials and for alloys with high vapor pressures. For

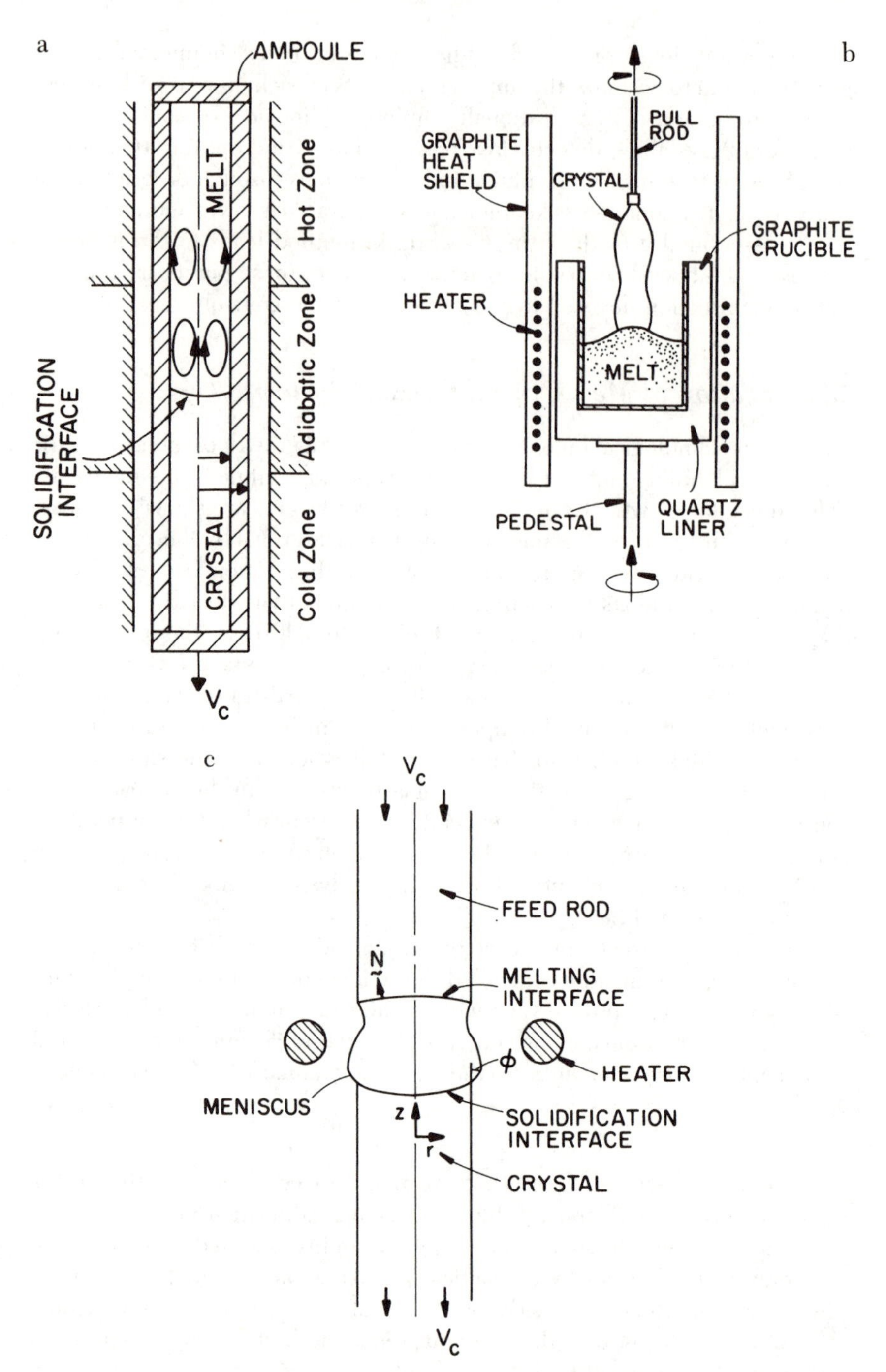

Figure 1. Schematic diagrams of several commonly used systems for melt crystal growth of electronic materials: (a) vertical Bridgman, (b) Czochralski, and (c) small-scale floating-zone systems.

Table I. Common Semiconductor Materials and Methods for Growing Them from the Melt

Material	*Application*	*Technique*
Single-crystal silicon	Integrated circuits	Czochralski
High-resistivity, single-crystal silicon	Power transistors	Floating zone
Polycrystalline silicon	Photovoltaic devices	Variety of capillary growth methods
Group III–V materials (e.g., GaAs, InP)	Optoelectronic devices, integrated circuits	Horizontal boat, horizontal Bridgman, liquid-encapsulated Czochralski, vertical gradient freeze
Group II–VI materials (e.g., CdTe, HgCdTe)	Detectors	Vertical Bridgman, horizontal Bridgman

such materials, the control of the stoichiometry in the melt is difficult without confinement (Table I). Directional solidification and the gradient freeze techniques are becoming popular for the growth of GaAs, InP, and other materials for which low axial temperature gradients are needed to produce crystals with low dislocation densities. Some of the renewed interest in these methods is a result of improvements in the control of the temperature gradients. Carefully designed heat–pipe furnaces have been successfully used in Bridgman–Stockbarger systems (*6*) but have been limited to temperatures below 1100 °C. Multizone resistance furnaces have found application in gradient freeze systems operating at higher temperatures (*7, 8*) for the growth of InP, GaP, and GaAs.

The horizontal Bridgman technique, or boat growth technique, is a variation in which the ampoule is laid horizontally with respect to gravity and the temperature gradient in the melt is changed by varying the temperature profile in the surrounding heater. The melt is contained in an open ampoule, or boat, so that the composition of the melt can be equilibrated with the surrounding ambient. For GaAs growth, a source of arsenic is placed in the system and the vapor pressure of arsenic is controlled independently to maintain the stoichiometry of the melt. This technique is a hybrid of the classical gradient-freezing technique, because of the method for varying the temperature profile, and of a meniscus-defined growth method, because of the presence of the melt–ambient surface.

Meniscus-Defined Crystal Growth Systems. In most conventional meniscus-defined growth systems, a seed crystal is dipped into a pool of melt, and the thermal environment is varied so that a crystal grows from the seed as it is pulled slowly out of the pool. Two examples of meniscus-defined growth are shown in Figure 1. The Czochralski (CZ) method (Figure 1b) and the closely related liquid-encapsulated Czochralski (LEC) method are batchwise processes in which the crystal is pulled from a crucible with decreasing melt volume. In the LEC system, an oxide material (usually B_2O_3) is layered over the melt to prevent the loss of volatile components in a high-vapor-pressure system, such as GaAs and InP. In these systems, the gas pressure is usually maintained significantly above the ambient pressure.

The details for implementation of crystal growth by the CZ method (*9–11*), the floating-zone method (*12, 13*), and meniscus-defined growth methods (*14, 15*) developed for producing thin sheets have been reviewed. The advantages and disadvantages of various confined and meniscus-defined growth methods determine the types of materials that are produced by each technique. For example, meniscus-defined methods have the advantage that the cooling crystal is free to expand and so is less likely to generate large thermoelastic stresses that lead to defect and dislocation generation. However, active control is needed to produce crystals of uniform cross section, because the shape of the crystal is constrained only by capillary action.

In the floating-zone (FZ) system, a molten pool is formed by a circumferential heat source that separates a melting polycrystalline feed rod and a solidifying cylindrical crystal. In small-scale, resistively heated zones, the pool of melt is held in shape by capillary forces and gravity and by hydrodynamic stresses caused by flow in the melt. The experimental techniques related to floating-zone systems are discussed in the pioneering book by Pfann (*12*). Floating-zone systems with conventional resistance heaters are limited on earth to the growth of crystals with diameters less than ~1 cm, because deformation of the meniscus by gravity causes loss of wetting of the crystal by the surrounding melt (*16*); floating-zone growth in outer space removes this restriction and is an area of active research (*17*). The shape of a small-scale floating zone is shown in Figure 1c.

Large-diameter industrial floating-zone systems have been developed with radio frequency (rf) induction heating elements shaped so that the induction coil has a smaller diameter than the growing crystal. These systems are used for the growth of high-purity semiconductor materials, such as high-resistivity silicon and germanium, and are described in the book by Muhlbauer and Keller (*13*). The explanation for the success of these systems has centered on the levitation of the melt caused by the Maxwell stresses induced by the coil; however, appeal to meniscus stabilization because of this additional levitation force may not be necessary because of the shape of the zone. The large melting interface of the polycrystalline feed rod is typically very concave, so that the distance between the two melt–solid interfaces is small, perhaps no more than a centimeter. The melting interface is covered with a thin film of melt, so that the meniscus surrounding the bulk fluid exists only adjacent to the crystal and the molten zone may be no larger in the system with rf heating than in the system with resistive heating, which is used for smaller diameter crystals. From this point of view, the rf coil is a clever method of locally heating a small melt pool that is bound by a meniscus with a shape that balances capillary force with gravity.

Modeling of Melt Crystal Growth Systems. Two key issues must be addressed before a detailed survey of modeling of melt crystal growth is presented. First, it must be clear why and on what levels modeling can play a role in optimization and control of systems for the growth of single crystals from the melt. The potential for modeling in the development of these processes will be obvious from a survey of the current state of the art in growth of different semiconductor materials by similar processes.

Variants of the CZ process make a good case study. During the last 25 years, the Czochralski process has been optimized for the growth of single-crystal silicon to the point that today dislocation-free crystals with 4–6-in. (10–15-cm) diameters are grown from 5–25-kg batches of melt. The diameter control is 0.1%, and impurity levels are controlled to be less than 5×10^{15} carbon atoms per cm^3 and 1×10^{18} oxygen atoms per cm^3. By contrast, GaAs crystals grown by the liquid-encapsulated Czochralski process are limited to less than a 3-in. (7.6-cm) diameter with poor diameter control and have 300–5000 dislocations per square centimeter, depending on the level of dopants added to the melt. Moreover, other semiconductor materials, such as CdTe, have not been grown successfully in Czochralski configurations. The difference in the difficulty of producing silicon and GaAs wafers is reflected in the difference by a factor of 100 in their cost per square centimeter!

The role of transport processes in setting the level of difficulty for the growth of each of these materials was described qualitatively by S. Motakeff at the Massachusetts Institute of Technology (MIT) in terms of a plot similar to Figure 2, in which the estimated thermal conductivities of several semiconductor materials are plotted against the approximate value of the critical resolved shear stress (CRSS). Materials with low conductivities and low values of CRSS are more difficult to grow because of the larger temperature gradients (proportional to the conductivity) that will occur during processing and the lower resistance of the crystal to the formation of dislocations (proportional to CRSS). Although this argument is extremely qualitative, the trend is clear. Analysis of linear thermoelastic stress in the crystal will be developed further in a later subsection on transport processes and defect formation. The newer materials used in the microelectronic industry can be produced only with better quantitative optimization and control of the growth systems. Modeling will play a substantial role in the development of the next generation of crystal growth systems.

Analysis of crystal growth systems transcends levels of detail ranging from thermal analysis of an entire crystal growth system to analysis of the dynamics of defects in the crystal lattice. These models are represented schematically in Figure 3 for a vertical Bridgman growth system. Several intermediate-length scales for modeling this system have been included that are not obvious without further discussion of the transport processes. These scales include an intermediate scale for analysis of melt flow and solute

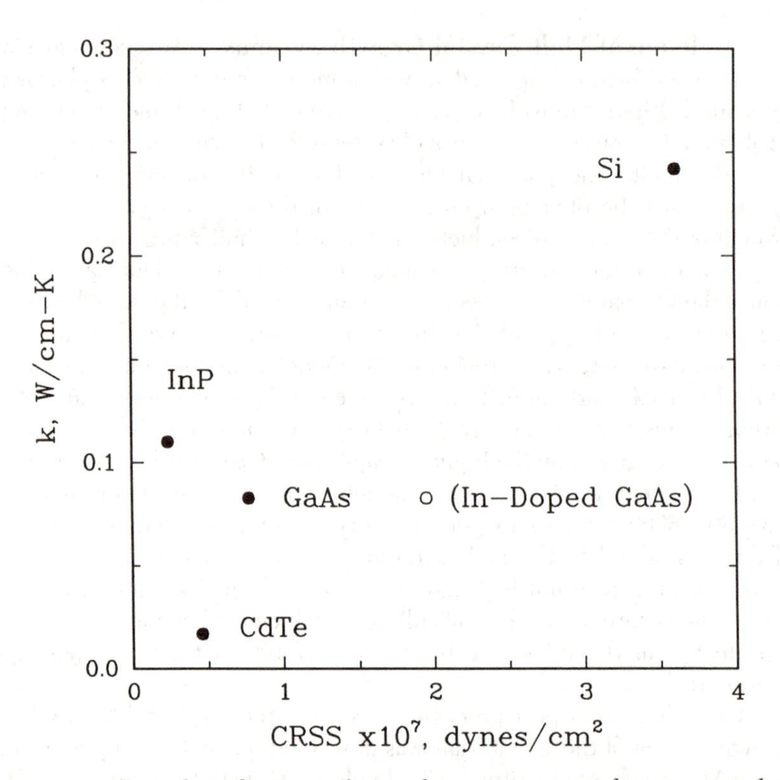

Figure 2. Thermal conductivity of several common semiconductor materials plotted against the best estimates for the critical resolved shear stress (CRSS) for the crystal. As explained in the text, materials with low thermal conductivity and low CRSS are hardest to grow.

transport in the melt with boundary conditions imposed by coupling between this level of detail and an entire system model and microscale models of melt–crystal interface morphology that account for surface forces and crystalline anisotropy. Analyses on these three length scales are based on continuum conservation equations and are described in this chapter.

Defect formation and dynamics in the crystal and at the melt–crystal interface are molecular-scale events that are only adequately simulated by lattice-scale models. A discussion of lattice-scale equilibrium and calculations of molecular dynamics is beyond the scope of this chapter.

An important question for any modeling effort, especially one aimed at a quantitative description of complex transport processes, is the level of accuracy of the model. As will become evident in the discussion of transport models and specific calculations, the values for thermophysical properties and transport coefficients must be known, as well as the dependence of these coefficients on temperature and pressure. Information is lacking for this data base. Critical material properties for semiconductor materials are not known

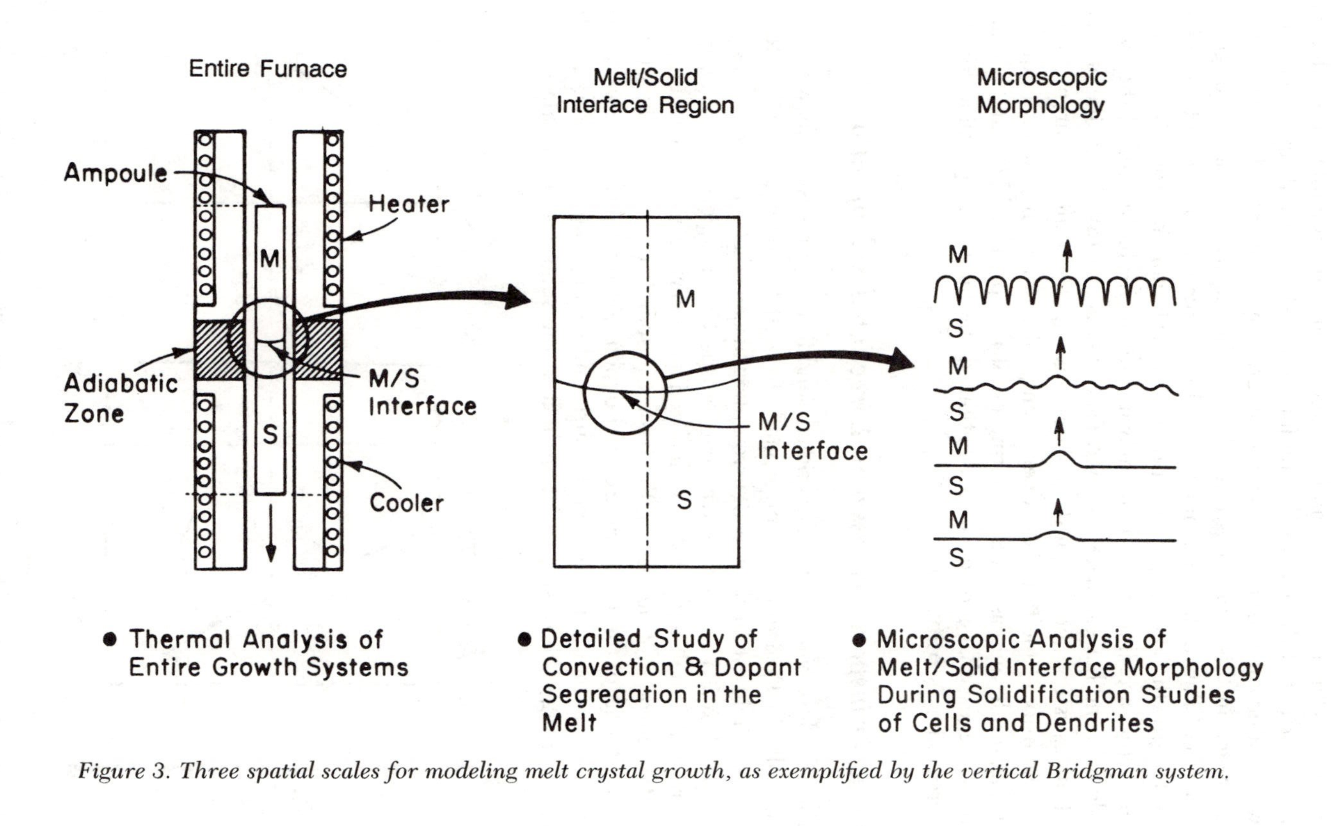

Figure 3. Three spatial scales for modeling melt crystal growth, as exemplified by the vertical Bridgman system.

with any accuracy, and little data are available on the sensitivity of the parameters to temperature and concentration. For example, the value of the coefficient of thermal expansion for molten silicon is only available from two isolated measurements, which disagree by a factor of three (*18*). The difference in this number is critical in predicting whether steady or time-dependent flow occurs in simulations of buoyancy-driven convection in CZ crystal growth. Thermophysical properties for important group III–V semiconductors are almost totally unknown; Jordan (*19*) has compiled the available data and has estimated properties for GaAs and InP. The value of such estimates is now being questioned. For example, recent measurements of the viscosity of molten GaAs over a range of temperatures show an order-of-magnitude increase as the melting point is approached (*20*), and the authors suggest that molecular association in the melt is responsible for the increase in viscosity.

Basic Transport Processes in Directional Solidification

The simplest picture of a directional solidification process is shown schematically in Figure 4 and corresponds to melt translating through a temperature field established by a surrounding furnace that brackets the melting temperature. When convective heat transport is unimportant, the temperature in the melt and crystal is idealized as a nearly constant axial gradient for the purpose of describing the basic transport mechanisms. The quantitative details of the temperature field depend on the thermophysical properties of the melt, crystal, and the surrounding ampoule and on the growth rate, through the release of latent heat at the solidification interface. In a

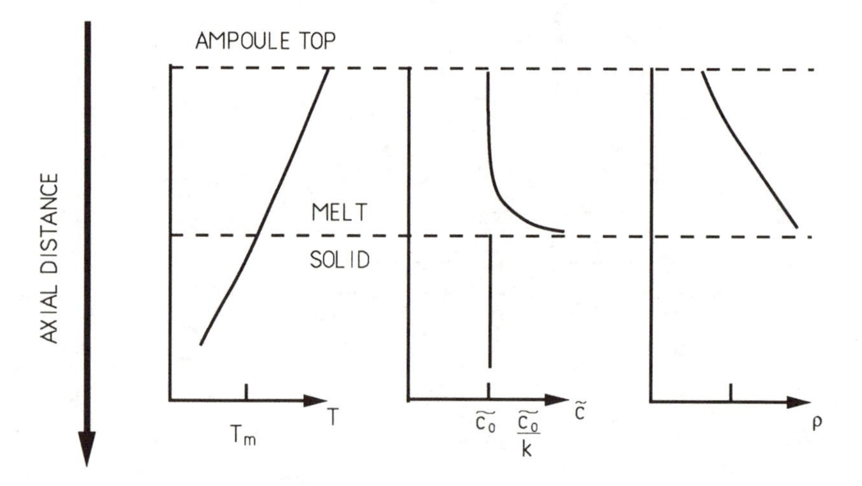

Figure 4. Profiles of temperature, solute concentration, and density in one-dimensional directional solidification of a binary alloy.

confined growth system, convection in the melt is caused by the translation of the ampoule and by buoyancy-driven convection due to temperature and concentration gradients. The details of the convection pattern will depend on heat transfer in the system and on the orientation of the melt and crystal with respect to gravity.

Heat transfer in melt crystal growth systems is by conduction in all phases; convection in the melt; and convective, conductive, and radiative exchanges between the various parts of the growth system. The goals of heat-transfer systems designed for any crystal growth configuration are to establish a nearly constant temperature gradient laterally along the melt–crystal interface and to control the cooling rate of the crystal. The details of implementing these conditions can be extremely complex because of the complicated geometries in systems like the CZ and LEC techniques and because of the influence of radiative heat exchange. Progress in understanding heat transport is discussed for specific systems in the later sections on vertical Bridgman–Stockbarger and Czochralski crystal growth. The discussion in this section focuses on species transport for solutes that form ideal solutions in the melt and single-phase crystalline solids. The shape of the phase diagram between melt and solid is described by the dependence of the liquidus $T_s(c)$ and solidus $T_l(c)$ temperatures on the solute concentration c. The liquidus and solidus curves are taken to be straight lines with slopes m and m/k, respectively, in which k is the equilibrium partition coefficient. Then the composition of melt ($\tilde{c}_m$) and solid ($\tilde{c}_s$) in equilibrium are related by

$$\tilde{c}_m = \tilde{c}_s/k \tag{1}$$

(The symbol ˜ is used throughout this chapter to denote dimensional variables; its absence corresponds to a dimensionless formulation.) The phase diagram corresponding to constant values of m and k is shown in Figure 5.

Taking the segregation coefficient to be independent of the local growth rate of the crystal is to assume that the solute concentration is in local equilibrium at the interface, that is, that interface kinetics plays no role in setting the composition of the crystal. Although this assumption is probably accurate for the slow growth rates (1–10 μm/s) on a microscopically rough crystal surface typical for the growth of many semiconductors, it is undoubtedly poor at higher solidification velocities or when the crystal grows along a facet so that the growth rate is governed by the movement of ledges across the surface.

One-Dimensional, Diffusion-Controlled Crystal Growth. Neglecting bulk convection leads to an idealized picture of diffusion-controlled solute transport of a dilute binary alloy with the solute composition c_0 far from an almost flat melt–crystal interface located at $\tilde{z} = 0$ (*1, 21*).

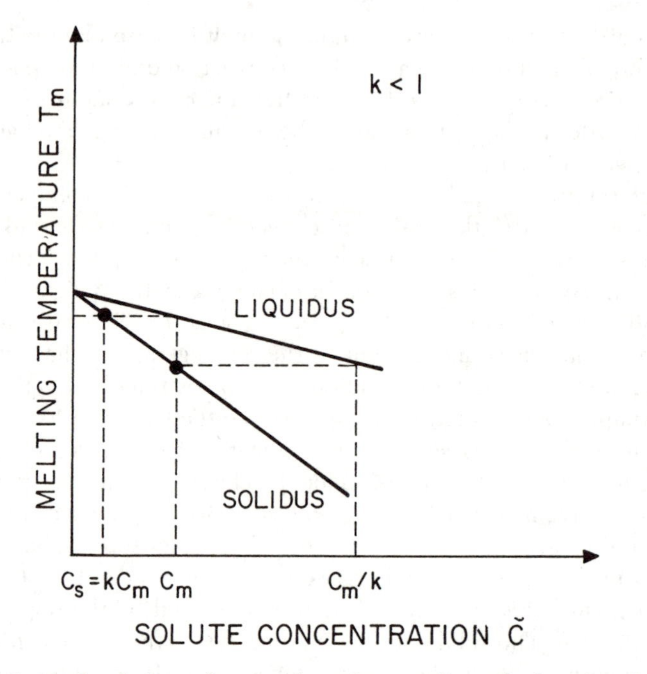

Figure 5. Idealized phase diagram for a binary alloy. The liquidus slope m *and equilibrium segregation coefficient* k *are defined in the text.*

When viewed from a reference frame that is stationary with respect to the solidification interface, the melt moves uniaxially toward the interface and the crystal moves away. Then solute transport in the melt is governed by the one-dimensional balance equation

$$D\frac{\partial^2 \tilde{c}}{\partial \tilde{z}^2} + V_g\frac{\partial \tilde{c}}{\partial \tilde{z}} = 0 \tag{2}$$

in which D is the solute diffusivity in the melt, $\tilde{c}$ is the dimensional concentration, $\tilde{z}$ is the dimensional axial coordinate, and V_g is the growth rate measured in terms of the melt velocity. Equation 2 is solved with the following condition of solute conservation at the melt–crystal interface

$$-D\frac{\partial \tilde{c}}{\partial \tilde{z}}\bigg|_{\tilde{z}=0} = (1 - k)\bigg|_{\tilde{z}=0} \tag{3}$$

in which k is the equilibrium partition coefficient defined by equation 1. Because the growth rate V_g is taken to be independent of the composition at the interface, the results are limited in practice to dilute alloy systems.

The solute field that decays to c_0 far from the interface ($\tilde{z} \rightarrow \infty$) is given by:

$$\tilde{c}(\tilde{z}) = c_0 \left[1 + \frac{1-k}{k} \exp\left(\frac{-V_g \tilde{z}}{D} \right) \right] \tag{4}$$

Equation 4 shows the existence of an exponential diffusion layer adjacent to the interface so that the variation in concentration caused by solute rejection ($k < 1$) or incorporation ($k > 1$) at the interface only persists for an e-folding distance in the order of D/V_g.

Analysis of transients in directional solidification is complicated by the coupling between heat transfer, the shape of the melt–crystal interface, and solute transport. Smith et al. (*22*) analyzed the idealized situation of one-dimensional, diffusion-controlled transport in which the diffusion of heat is much more rapid than that of solute, so that changes in the heat-transfer environment or the ampoule pull rate appear as a step change in the macroscopic solidification rate at the interface. Under these conditions, the transient solute profile is governed by equations 2 and 3, with an accumulation term ($\partial c/\partial t$) appearing on the right side of equation 2. Solution of this equation by separation of variables and the boundary condition (equation 4) gives an e-folding time of D/kV_g^2 for the solute profile in response to a step change in the pull rate.

When the alloy is nondilute, the melting temperature depends on the solute concentration adjacent to the interface through the shape of the liquidus curve. Then the transport of heat and solute and the location of the solidification interface do not decouple (*23*); Derby and Brown (*24*) presented a numerical algorithm for the analysis of this problem.

Convection in Melt Growth. Convection in the melt is pervasive in all terrestrial melt growth systems. Sources for flows include buoyancy-driven convection caused by the solute and temperature dependence of the density; surface tension gradients along melt–fluid menisci; forced convection introduced by the motion of solid surfaces, such as crucible and crystal rotation in the CZ and FZ systems; and the motion of the melt induced by the solidification of material. These flows are important causes of the convection of heat and species and can have a dominant influence on the temperature field in the system and on solute incorporation into the crystal. Moreover, flow transitions from steady laminar, to time-periodic, chaotic, and turbulent motions cause temporal nonuniformities at the growth interface. These fluctuations in temperature and concentration can cause the melt–crystal interface to melt and resolidify and can lead to solute striations (*25*) and to the formation of microdefects, which will be described later.

Equations of Motion and Driving Forces for Convection. The presentation in this section is limited to the case when the density variations in the melt caused by temperature and concentration gradients can be modeled by the Boussinesq approximation (26) in terms of the thermal and solutal expansion coefficients, $\beta_t \equiv -(1/\rho)(\partial\rho/\partial T)_{p,c}$ and $\beta_c \equiv (1/\rho)(\partial\rho/\partial c)_{p,T}$, respectively, in which p is pressure and T is temperature. Positive values of β_t correspond to a material with a density that decreases with increasing temperature; a positive value of β_c is appropriate for a solute with density that increases with increasing concentration.

The governing equations and boundary conditions for modeling melt crystal growth are described for the CZ growth geometry shown in Figure 6. The equations of motion, continuity, and transport of heat and of a dilute solute are as follows:

$$\rho_0 \left(\frac{\partial \tilde{\mathbf{v}}}{\partial \tilde{t}}\right) + \tilde{\mathbf{v}} \cdot \nabla \tilde{\mathbf{v}} = -\nabla \tilde{\mathbf{p}} + \mu \nabla^2 \tilde{\mathbf{v}} + \rho_0 g[1 - \beta_t(\tilde{T} - \tilde{T}_r) + \beta_c(\tilde{c} - c_0)]\mathbf{e}_z + \tilde{\mathbf{F}}_b(\tilde{\mathbf{V}}, \tilde{\mathbf{x}}) \tag{5}$$

$$\nabla \cdot \tilde{\mathbf{v}} = 0 \tag{6}$$

$$\left(\frac{\partial \tilde{T}}{\partial \tilde{t}} + \tilde{\mathbf{v}} \cdot \nabla \tilde{T}\right) = \alpha_m \nabla^2 \tilde{T} \tag{7}$$

$$\left(\frac{\partial \tilde{c}}{\partial \tilde{t}} + \tilde{\mathbf{v}} \cdot \nabla \tilde{c}\right) = D \nabla^2 \tilde{c} \tag{8}$$

In equations 5–8, the variables and symbols are defined as follows: ρ_0 is reference mass density, $\tilde{\mathbf{v}}$ is dimensional velocity field vector, $\tilde{\mathbf{p}}$ is dimensional pressure field vector, μ is Newtonian viscosity of the melt, g is acceleration due to gravity, $\tilde{T}$ is dimensional temperature, $\tilde{T}_r$ is the reference temperature, $\tilde{c}$ is dimensional concentration, c_0 is far-field level of concentration, $\mathbf{e}_z$ is a unit vector in the direction of the z axis, $\tilde{\mathbf{F}}_b$ is a dimensional applied body force field, ∇ is the gradient operator, $\mathbf{v}(\mathbf{x}, t)$ is the velocity vector field, $\mathbf{p}(\mathbf{x}, t)$ is the pressure field, μ is the fluid viscosity, α_m is the thermal diffusivity of the melt, and D is the solute diffusivity in the melt. The vector $\tilde{\mathbf{F}}_b$ is a body force imposed on the melt in addition to gravity. The body force caused by an imposed magnetic field $\mathbf{B}(\mathbf{x}, t)$ is the Lorentz force, $\tilde{\mathbf{F}}_b \equiv \sigma_c(\mathbf{v} \times \mathbf{v} \times \mathbf{B})$. The effect of this field on convection and segregation is discussed in a later section.

Other mechanisms for flows in melt crystal growth arise from surface stresses along or the relative motion of the boundaries of the melt. The no-slip boundary condition describes the relative motion of a rigid boundary ∂D_{b1}

$$\tilde{\mathbf{v}}(\mathbf{x}, t) = \tilde{\mathbf{V}}_s(\tilde{\mathbf{x}}, \tilde{t}) \text{ for } \tilde{\mathbf{x}} \in \partial D_{b1} \tag{9}$$

in which $\tilde{\mathbf{V}}_s(\tilde{\mathbf{x}}, \tilde{t})$ is the velocity of the portion of the boundary ∂D_{b1}. The melt is solidified normal to the melt–solid interface and obeys a no-slip tangential to the surface. These conditions are written as follows

$$(\mathbf{N} \cdot \tilde{\mathbf{v}}) = \tilde{V}_g(\mathbf{N} \cdot \mathbf{e}_g)$$
$$(\mathbf{T} \cdot \tilde{\mathbf{v}}) = \tilde{V}_g(\mathbf{T} \cdot \mathbf{e}_g), \ \tilde{\mathbf{x}} \in \partial D_{bI} \quad (10)$$

for a steadily solidifying solid, in which the shape of the melt–crystal surface is given by the unit normal **N** and tangent **T** vectors and the direction of steady-state solidification at the velocity V_g is given by the unit vector in the direction of gravity, $\mathbf{e}_g$.

Both the normal and tangential components of stress must balance, and the kinematics of the surface and flow field must be consistent along a melt–fluid interface. For the meniscus shown in Figure 6 and described by the normal and tangent vectors (**n**, **t**), these conditions dictate that

$$\mathbf{nn}:[-\tilde{\mathbf{p}}\mathbf{I} + \mu(\nabla\tilde{\mathbf{v}} + \nabla\tilde{\mathbf{v}}^T)] + p_0 = 2\sigma\tilde{H} + \mathbf{n} \cdot \tilde{\mathbf{F}}_s \quad (11a)$$

$$[\mu\mathbf{tn}:(\nabla\tilde{\mathbf{v}} + \nabla\tilde{\mathbf{v}}^T)] = \nabla_2\tilde{\sigma} + \mathbf{t} \cdot \tilde{\mathbf{F}}_s \quad (11b)$$

$$\frac{\partial \tilde{h}}{\partial \tilde{t}} = \mathbf{n}\tilde{\mathbf{v}}, \ \tilde{\mathbf{x}} \in \partial D_{bm} \quad (12)$$

in which $\tilde{H}$ is the mean curvature of the surface, σ is the interfacial tension, $\tilde{\mathbf{F}}_s$ is a dimensional surface force, ∇_2 is the gradient operator defined for the surface, and **I** is the identity tensor. The quantities in brackets signify the difference of the quantity evaluated from both phases bordering the meniscus. Equation 11a is the condition for balancing normal stress across the meniscus with the presence of the gas phase accounted for simply by a static pressure p_0. When viscous stresses (the term proportional to the viscosity μ) and the dynamic pressure in the melt are unimportant, this condition gives the Young–Laplace equation for the shape of a hydrostatic interface. Solutions for the Young–Laplace equation are available for most meniscus-defined growth configurations.

The term $\nabla_2\tilde{\sigma}$ in equation 11b represents the tangential stress caused by a spatial variation in the interfacial tension due to either concentration or temperature variation along the surface. The dependence of the interfacial tension on temperature and concentration is usually expressed by the following approximation

$$\tilde{\sigma} = \tilde{\sigma}_0 \left[1 + \left(\frac{\partial \tilde{\sigma}}{\partial \tilde{T}}\right)_c (\tilde{T} - T_0) + \left(\frac{\partial \tilde{\sigma}}{\partial \tilde{c}}\right)_T (\tilde{c} - c_0)\right] \quad (13)$$

in which $\tilde{\sigma}_0$, T_0, and c_0 are reference values. Unfortunately, the dependence of $\tilde{\sigma}$ on temperature even for pure melts of important electronic materials is not known. The recent measurements of Hardy (27) for silicon suggest that equation 13 is appropriate for an extended range of $\tilde{T}$. The additional surface traction $\tilde{\mathbf{F}}_s(\mathbf{x}, t)$ is included in equation 11a to account for imposed stresses, such as the direct coupling of an electromagnetic field generated by a radio frequency (rf) field, which leads to vigorous stirring of the melt. Inclusion of this effect as a surface force is a valid approximation for an electrically conducting melt and a high-frequency rf field (*28*).

A major complication in the analysis of convection and segregation in melt crystal growth is the need for simultaneous calculation of the melt–crystal interface shape with the temperature, velocity, and pressure fields. For low growth rates, for which the assumption of local thermal equilibrium is valid, the shape of the solidification interface ∂D_{bI} is given by the shape of the liquidus curve $T_m(c)$ for the binary phase diagram:

$$T(\tilde{\mathbf{x}}, t) = T_m(c)\ \tilde{\mathbf{x}} \in \partial D_{bI} \tag{14}$$

The influence of surface curvature and surface free energy on the melting temperature has been neglected in equation 14, because it is so small that it is unimportant in determining the macroscopic shape of the interface. However, microscopic details of the melt–crystal interface structure, such as the onset of cellular and dendritic growth, depend crucially on the contribution of the surface energy to the setting of the length scale of the pattern; this point is discussed further in the section on transport processes and solid microstructure. In addition to equation 14, latent heat that is released at the melt–crystal interface must be included in the interfacial energy balance.

Scaling Analysis. The complexity of practical crystal growth systems makes difficult the understanding of the roles of each of the driving forces for convection. As can be imagined from the expanded view of CZ growth shown in Figure 6, different driving forces—for example, crystal and crucible rotation, buoyancy-driven flows, and surface-tension-driven flows—dominate the flow in different parts of the melt. Clearly, full numerical solutions that account for all these effects are needed. However, scaling analysis that balances the effects of various driving forces in idealized geometries is useful in computing order-of-magnitude estimates of flow intensity and the scalings for changes in flow intensity with variations in boundary conditions and thermophysical properties.

In scaling analysis, the differential equations and boundary conditions are put in dimensionless form by introducing characteristic scales for the length, time, velocity, concentration, etc., that are chosen to reflect the dominant mechanisms for transport in each conservation law. Then the dimensionless groups in the equations supply estimates for the relative magnitudes of various driving forces. Some of the dimensionless groups for convection and segregation in the melt are listed in Table II.

The Schmidt (Sc) and Prandtl (Pr) numbers (Table II) are ratios of molecular diffusivities and thus are formed purely from the thermophysical

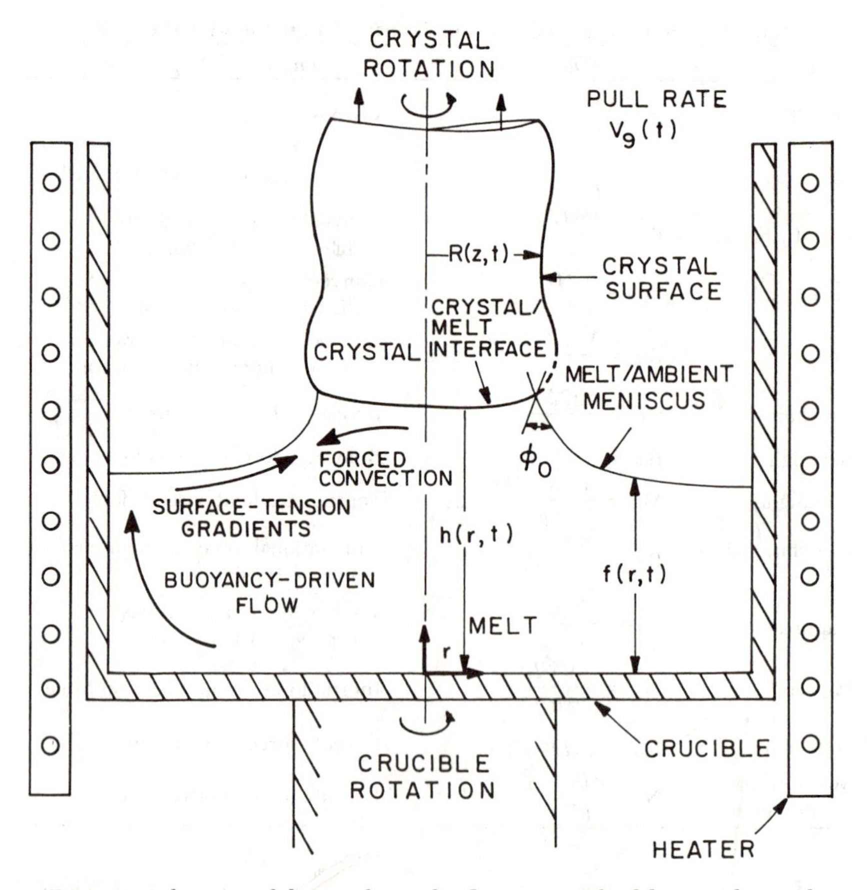

Figure 6. Schematic of driving forces for flows in Czochralski crystal growth system, which shows the regions where the driving forces will produce the strongest motions. The shape functions describing the unknown interface shapes are listed also.

properties of the melt. Prandtl numbers relevant to melt crystal growth vary between the large values (1–10) for oxide melts to the extremely low values (0.01–0.1) for semiconductor melts. The Schmidt numbers are large (10–100) because of the low solute diffusivities for typical melts.

Examples of dimensionless groups that specify ratios of transport mechanisms are listed next in Table II and depend on the size and shape of the domain. The Peclet numbers for heat (Pe_t) and solute (Pe_s) and momentum (Re) transport are ratios of scales for convective to diffusive transport and depend on the magnitudes of the velocity field and the length scale for the diffusion gradient. Boundary layers form at large Peclet numbers (Pe_t or Pe_s) or Reynolds numbers (Re). The formation of a boundary layer at a large Re is particularly important in crystal growth from the melt, because the low

Table II. Dimensionless Groups in Transport Equations for the Melt

Name	*Definition*	*Physical Interpretation*
Prandtl	$\mathrm{Pr} = \frac{\nu}{\alpha}$	Viscous diffusivity/heat diffusivity
Schmidt	$\mathrm{Sc} = \frac{\nu}{D}$	Viscous diffusivity/species diffusivity
Peclet (thermal)	$\mathrm{Pe} = \frac{V^*L^*}{\alpha}$	Convective heat transport/ diffusive heat transport
Peclet (solutal)	$\mathrm{Pe}_s = \frac{V^*L^*}{D}$	Convective species transport/ diffusive species transport
Reynolds	$\mathrm{Re} = \frac{V^*L^*}{\nu}$	Convective momentum transport/ viscous momentum transport
Grashof	$\mathrm{Gr} = \frac{\rho g \beta \Delta T L^3}{\nu^2}$	Buoyancy force/viscous force
Rayleigh	$\mathrm{Ra}_t = \mathrm{Gr}_t\mathrm{Pr}$	Buoyancy force/viscous force
Marangoni	$\mathrm{Ma}_t = \frac{\sigma(d\sigma/dT)L}{\rho\nu^2}\Delta T^*$	Tension gradient/viscous force
Gravitational bond	$\mathrm{Bo} = \frac{gL^{*2}}{\sigma_0}\Delta\rho$	Gravitational force on meniscus/ surface tension force
Capillary	$\mathrm{Ca} = \frac{V^*\mu}{\sigma_0}$	Viscous force on meniscus/ surface tension force
Weber	$\mathrm{We} = \frac{\rho V^{*2}L^*}{\sigma_0}$	Inertia force/surface tension force
Hartmann	$\mathrm{Ha} = B_0L^*(\sigma/\rho\nu)^{1/2}$	Lorentz force/viscous force
Magnetic interaction	$N = \frac{\sigma B_0^2 L^*}{\rho V^*}$	Lorentz force/inertia force

values of solute diffusivity lead to convectively dominated species transport, even at low fluid velocities. The choices for the velocity and gradient length scales in complex systems are not obvious, and different combinations are appropriate for different portions of the flow domain for real systems.

The next set of dimensionless groups listed in Table II scale the strength of a particular driving force for convection relative to the damping action of viscosity. The Rayleigh (Ra) and Grashof (Gr) numbers correspond to the scaling of the strength of buoyancy-driven convection relative to viscosity and arise when different scales are used for velocity and pressure in the equation of motion. The Marangoni number (Ma) scales the magnitude of the surface shear stress due to a surface tension gradient to viscosity. The scale of the surface tension gradient has been taken as $(d\tilde{\sigma}/d\tilde{T})(d\tilde{T}/d\tilde{x})$, and so only variations caused by temperature differences are taken into account.

The shape of the melt–fluid interface in a meniscus-defined crystal growth system is set by surface tension, gravitational force, and viscous and dynamic pressure forces on the surface. The Bond number (Bo; for hydro-

static pressure), capillary number (Ca; for viscous stress), and Weber number (We; for dynamic pressure) measure the magnitude of these forces scaled against surface tension. Most of the theoretical and numerical analyses of heat transfer and convection in meniscus-defined crystal growth that will be discussed later are for idealized geometries in which the meniscus is represented by a coordinate surface in a separable orthogonal coordinate system, such as a plane or a cylinder. The calculations assume that the underlying flow causes little deflection of the meniscus. This last assumption is mathematically equivalent to taking $Ca << 1$ and $We << 1$.

The Hartmann number (Ha) and the magnetic interaction parameter (N) in Table II scale the importance of an applied magnetic field to the action of viscosity and inertia, respectively. Treatment of the interaction of a magnetic field as a body force that introduces only the Lorentz force in equation 1 is a valid idealization when the melt is so electrically conductive that convection of charge by the flow is unimportant, that is, when the Peclet number appropriate for electrical charge transport is small. This assumption is justified in most semiconductor melts and has been used in all the numerical computations reported thus far. The electromagnetic force generated by a rf induction heater directly coupled to a floating zone leads to a surface Hartmann number that measures the effective surface force in the high-frequency limit. Muhlbauer et al. (*29*) studied the effect of the rf field in numerical simulations of flow in a large-scale floating zone.

Analysis of flows in which more than one driving force exists has been limited to several idealized cases of thermosolutal convection driven by vertical and constant temperature and concentration gradients. A discussion of thermosolutal convection is presented by Brown (*5*).

When the temperature and concentration gradients are perfectly vertical, thermosolutal convection begins at critical values of thermal or solutal Rayleigh numbers (Ra_t or Ra_s) as an instability from a static fluid. Increasing the driving force leads to finite-amplitude flows and to nonlinear transitions, as discussed in the section on transport processes and defect formation. Surface-tension-driven motions also begin at a critical value of the Marangoni number when the free surface is perpendicular to a perfectly vertical temperature gradient (*30*, *31*).

Imperfect alignment of the vertical gradient causes convection for any value of the driving force. The motion is weak for small values of the Grashof number (Gr) and represents a balance of viscous and buoyancy forces. Boundary layers form at higher values of the Grashof number, where viscous effects are confined to boundary and internal shear layers and the core flows are set by a balance of inertia and buoyancy. Increasing the driving force in these flows also leads to transitions to time-periodic and chaotic convection.

The complexity of the temperature field in even the most carefully designed crystal growth systems leads to both vertical and horizontal temperature gradients in the melt, so that buoyancy-driven and surface-tension-driven flows are always present. Analysis of the flows in a slender rectangular cavity heated at the ends has been used as a test problem for flows driven by lateral temperature variations. Use of the aspect ratio of the cavity as a perturbation parameter reduces the analysis to the solution of a sequence of problems in which the flow in the core of the cavity is dictated by an almost one-dimensional balance of buoyancy and viscous forces, and the turning flows at the ends of the cavity are two dimensional but inertialess. The rigorous perturbation theory for this problem was introduced by Cormack et al. (*32*) for two-dimensional buoyancy-driven motions and was extended by Hart (*33*, *34*) to include the stability of these motions for fluids with low Prandtl numbers. Hurle and co-workers (*35*) have observed three-dimensional oscillatory flows in a slender cavity of gallium heated from the ends.

Similar analyses are available for surface-tension-driven flows in a slender cavity with the additional assumption that the meniscus at the top of the cavity is also flat (*36*). Smith and Davis (*37–39*) have used this configuration to study the stability of the flow with respect to wavelike instabilities (*see* also reference 40). Homsy and co-workers (*41*, *42*) have analyzed the effect of a surface-active agent on the thermocapillary motion in a slender cavity.

Convection in the crystal growth systems discussed earlier cannot be characterized by analysis with either perfectly aligned vertical temperature gradients or slender cavities, because these systems have spatially varying temperature fields and nearly unit aspect ratios. Even when only one driving force is present, such as buoyancy-driven convection, the flow structure can be quite complex, and little insight into the nonlinear structure of the flow has been gained by asymptotic analysis.

Hjellming and Walker (*43*, *44*) and Langlois and Walker (*45*) discovered an important exception to this situation that arises when a strong axial magnetic field is imposed on the melt. They analyzed the motion in a prototype of the Czochralski crystal growth system for the case in which the magnetic interaction parameter (N) and Hartmann number (Ha) are so large that fluid inertia can be neglected everywhere and viscous forces are confined to Hartmann layers, which are needed to satisfy the no-slip and shear stress boundary conditions and to conserve mass between adjacent flow cells. The length scales for these Hartmann layers for rotational and buoyancy-driven flows are shown in Figure 7, which is taken from the analysis of Hjellming and Walker. The flow outside the boundary layers is determined by a balance of buoyancy and Lorentz forces. Moreover, the core flow can be written explicitly in terms of the temperature field. A self-consistent heat-transfer problem is developed in the limit where the thermal Peclet number is small enough that temperature boundary layers are much thicker than the Hartmann layers.

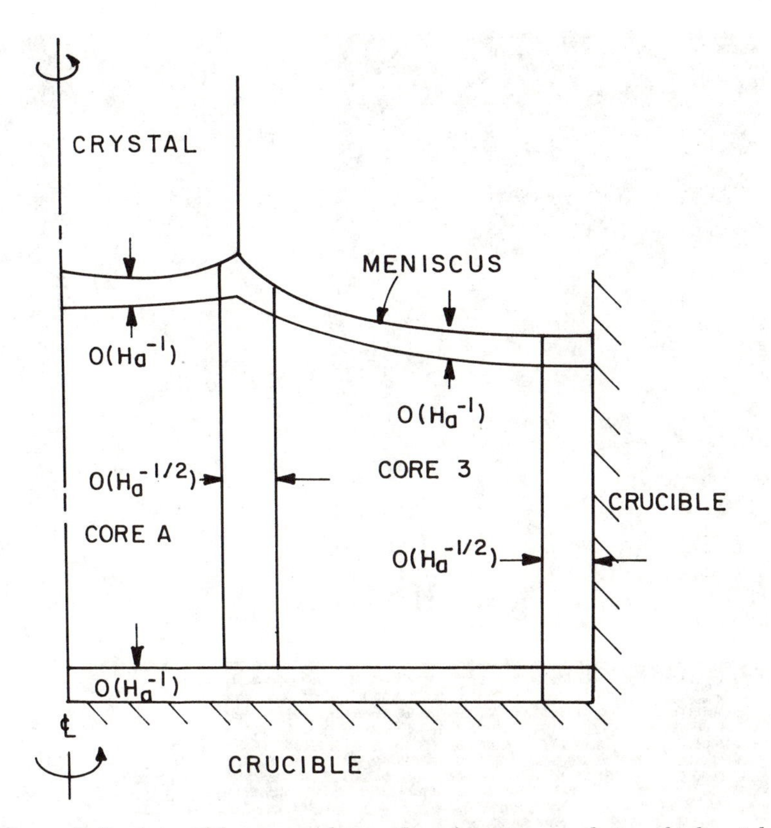

Figure 7. Regions of buoyancy-driven flow for CZ growth in a high axial magnetic field. The figure is based on the results of Hjellming and Walker (134).

Importance of Flow Transitions. Witt and his colleagues (*46, 47*) documented the microscopic changes in crystal growth rate and axial solute segregation caused by short-time-scale variations in the temperature and velocity fields. They did this study by perfecting the use of the Peltier interface demarcation (*48, 49*); a good description of this technique is given by Wargo and Witt (*50*). In this technique, pulses of current are passed through the melt and crystal at regular intervals, and the Peltier effect causes periodic local cooling of the melt–crystal interface and leads to rapid but small changes in the local freezing rate. These changes in the microscopic solidification rate result in layers of high solute incorporation, which can be detected optically in a polished and etched sample sliced along the growth axis. Pictures of the etching patterns taken from a vertical Bridgman experiment with melt below the crystal are shown in Figure 8 for the growth of gallium-doped germanium (*46*). The picture in Figure 8d is the etching pattern introduced solely by the periodic Peltier pulses used to impart a time record in the crystal.

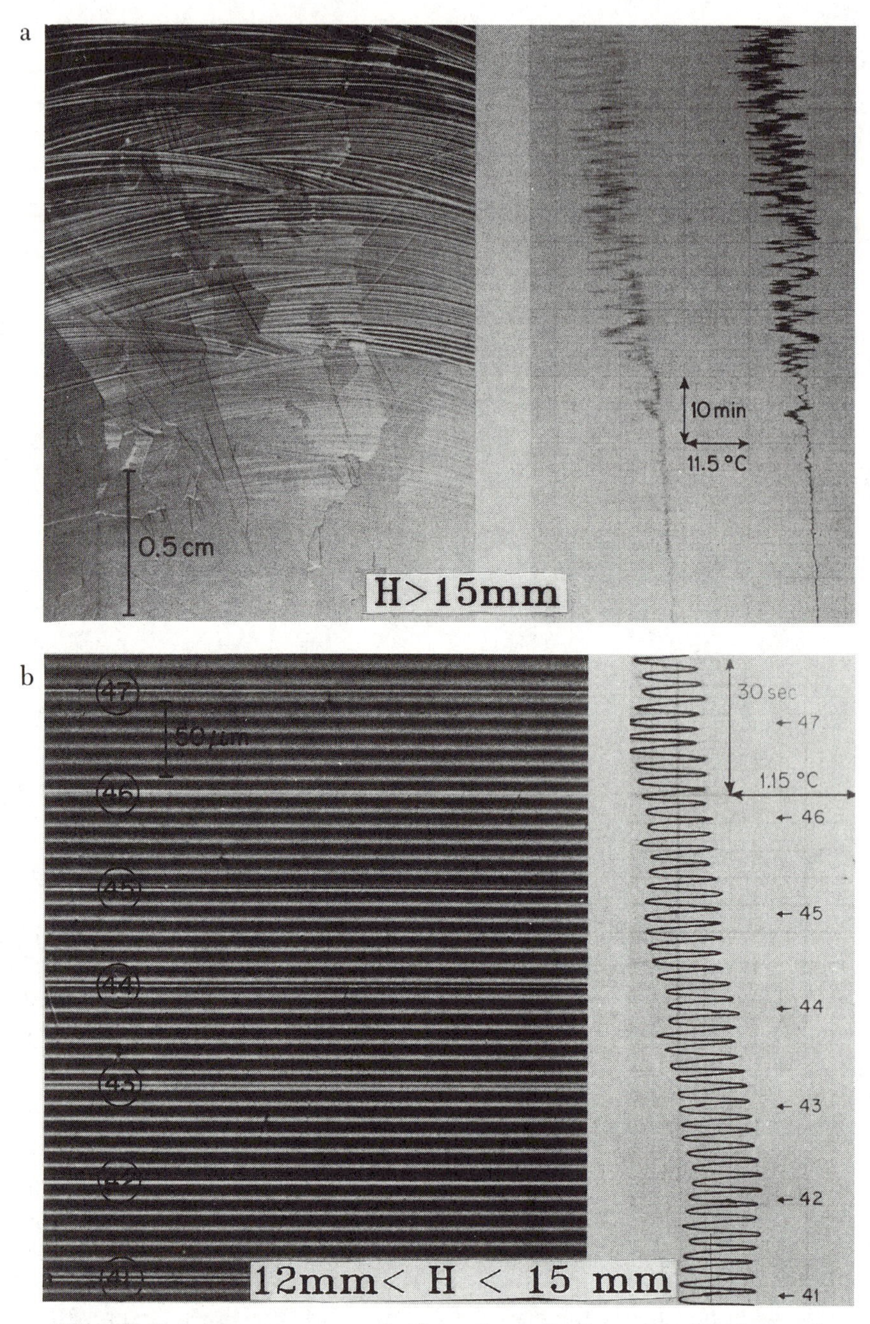

Figure 8. Striation patterns revealed by etching of gallium-doped InSb crystal grown in vertical unstable Bridgman system by Kim et al. (46). (a) Irregular pattern caused by chaotic convection. (b) Regular pattern caused by time-periodic convection.

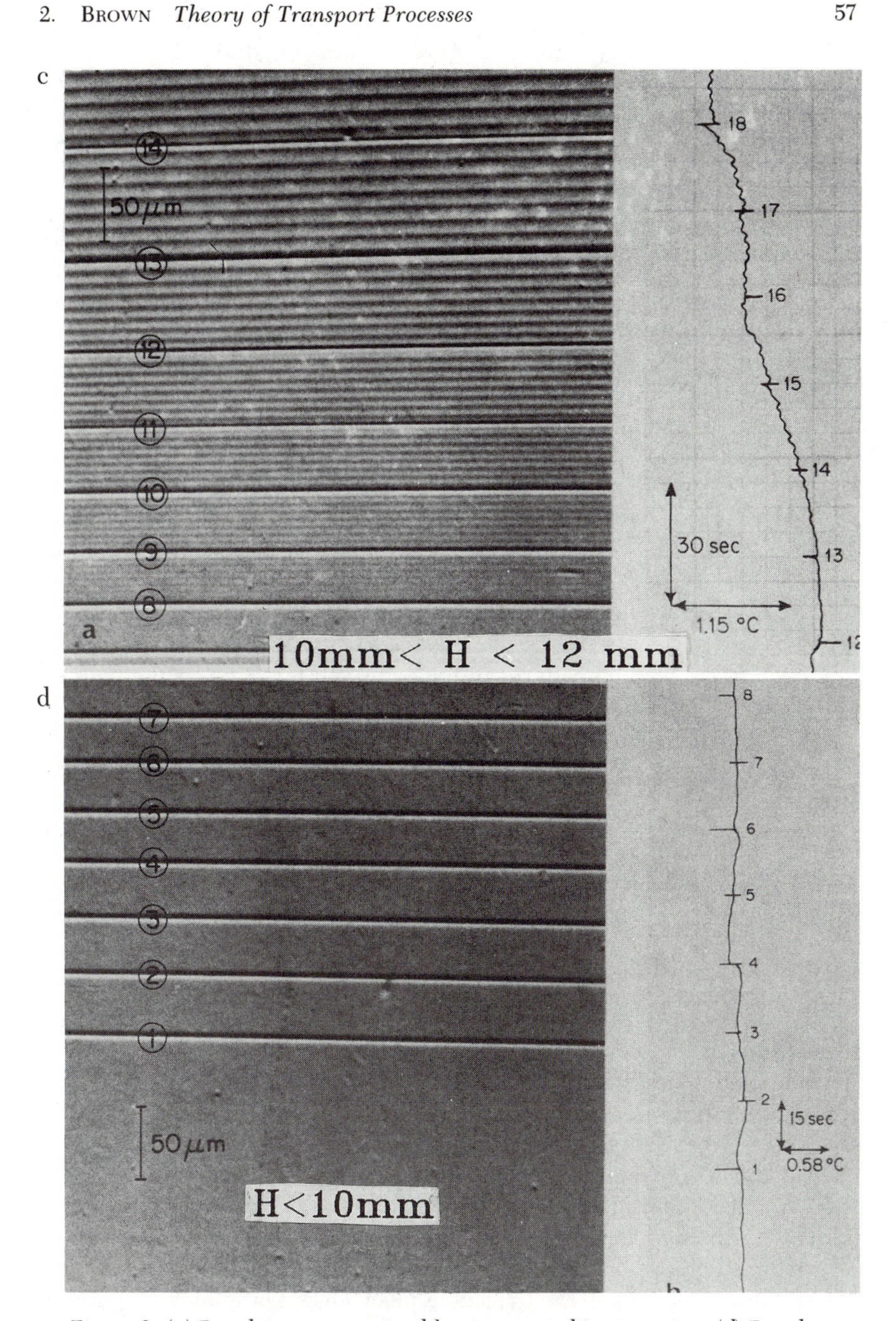

Figure 8. (c) Regular pattern caused by time-periodic convection. (d) Regular pattern caused by the Peltier pulses used to mark the growth rate of crystal. Transitions are caused by decreasing the length of melt during the growth of one crystal. Temperature measurements by a thermocouple in melt during growth of crystal are shown along the etchings.

The pictures are taken at various times during the translation of the ampoule through the furnace and thus correspond to different lengths of the melt. The intensity of the convection scales with the thermal Rayleigh number (Ra_t) for the vertical cylinder is given by

$$Ra_t \equiv \rho_0 \beta g L^3 \Delta T / \alpha \nu \tag{15}$$

in which ρ_0 is the reference density, β is the coefficient of thermal expansion, α is the thermal diffusivity, ν is the momentum diffusivity, ΔT is the temperature difference between the melting point and the hot end of the ampoule, and L is the length of the melt. The length of the melt and the value of Ra_t decrease from Figure 8a to Figure 8d. In Figure 8d, the etching pattern introduced solely by the periodic markers in Figure 8a indicates aperiodic or chaotic convection at the highest convection level. Decreasing Ra_t reduces the convection to simply periodic behavior, as shown by the set of periodic striations superimposed on the marking pattern in Figures 8b and 8c. The disappearance of all striations other than the current-induced markings in Figure 8d is indicative of laminar convection in the melt. The frequency of the convective oscillations is a few tenths of a hertz. The interpretation of the etching patterns just given is substantiated by the temperature measurements from a thermocouple in the melt during the same experiment; the transitions from chaotic, to periodic, and, finally, to steady or laminar temperature measurements are shown alongside the etching patterns in Figure 8.

Transitions from steady-state to time-dependent surface-tension-driven motions are well known also and are important in meniscus-defined crystal growth systems. For example, the experiments of Preisser et al. (*51*) indicate the development of an azimuthal traveling wave on the axisymmetric base flow in a small-scale floating zone.

Predicting these flow transitions and designing systems that suppress the onset of time-dependent motion are central to the controlled growth of compositionally uniform crystals. Although many theoretical and experimental studies treat such transitions in idealized flow problems, the relevance of these results to crystal growth systems is limited (*see* references 52 and 53 for comprehensive reviews). Direct prediction of the transitions in crystal growth systems is needed and is at the leading edge of current experimental and numerical analyses. Analysis of the transitions leading to chaotic convection in crystal growth experiments seems beyond present capabilities because of the three-dimensional nature of these flows and because of the sensitivity of the transitions following the onset of time-periodic motion to the details of the experimental system. Moreover, the oscillations in the melt lead to tremendously disparate time scales for transport, with time scales ranging from the short temporal wavelength for the convective oscillations (~1 s) to the long time scale (~1×10^4 s) to grow 1 cm of crystal at a rate of 10 μm/s.

Fortunately, the onset of oscillatory flow from a steady flow occurs as a Hopf bifurcation and is much more reproducible experimentally. Also, the existence of a Hopf bifurcation with increasing flow intensity can be detected in large-scale numerical calculations by solving the equation set for the linear stability of the underlying laminar flow. This approach has been used in investigations of buoyancy-driven convection in idealized geometries (*54*, *55*). Crochet and co-workers have studied the oscillations in the convection of a prototype of the horizontal Bridgman method for two-dimensional motion without (*56*, *57*) and including (*58*) the melt–crystal interface. Crochet et al. (*59*) computed three-dimensional motions by direct simulation of the nonlinear, time-dependent equations. Winters (*60*) has used computer-implemented perturbation methods to locate Hopf bifurcations to detect the onset of two-dimensional oscillations in the same system.

An important result of understanding the transitions in buoyancy-driven convection in melt growth is the quantitative development of mechanisms to suppress these transitions. Applied magnetic fields have been shown experimentally by many research groups to be effective in laminarizing the flows in metallic melts (*61*, *62*). Kim measured (*62*) the temperature as a function of time in a vertical Bridgman system with increasing intensity (**B**) of a transverse applied magnetic field (Figure 9). As **B** was increased, the temporal structure varied from an aperiodic recording without the field, to a time-periodic recording, and, finally, to a steady-state trace. These transitions have not been predicted theoretically. Only a few numerical calculations (*63*) give an indication of the effect of field intensity on the transition to time-periodic motion. Comprehensive theoretical and numerical analyses of the interaction of the applied fields with the flow have been limited to the large field strengths that lead to laminar flows.

Modeling the Influence of Melt Convection on Solute Segregation. Convection in the melt causes mixing of solutes and alters the diffusion layer adjacent to the melt–crystal interface. The spatial structure and intensity of the flow set the axial (along the growth direction) and lateral (perpendicular to the growth direction) profiles of solute concentration in the crystal. The different regimes for solute segregation are shown in Figure 10 in terms of the uniformity of the solute concentration across the crystal, Δc, which is defined as the maximum difference in the concentration across the crystal divided by the average value, $<\tilde{c}>_I$, and by the effective segregation coefficient k_{eff} defined as

$$k_{eff} \equiv \frac{k << \tilde{c} >>}{<\tilde{c}>_I} \tag{16}$$

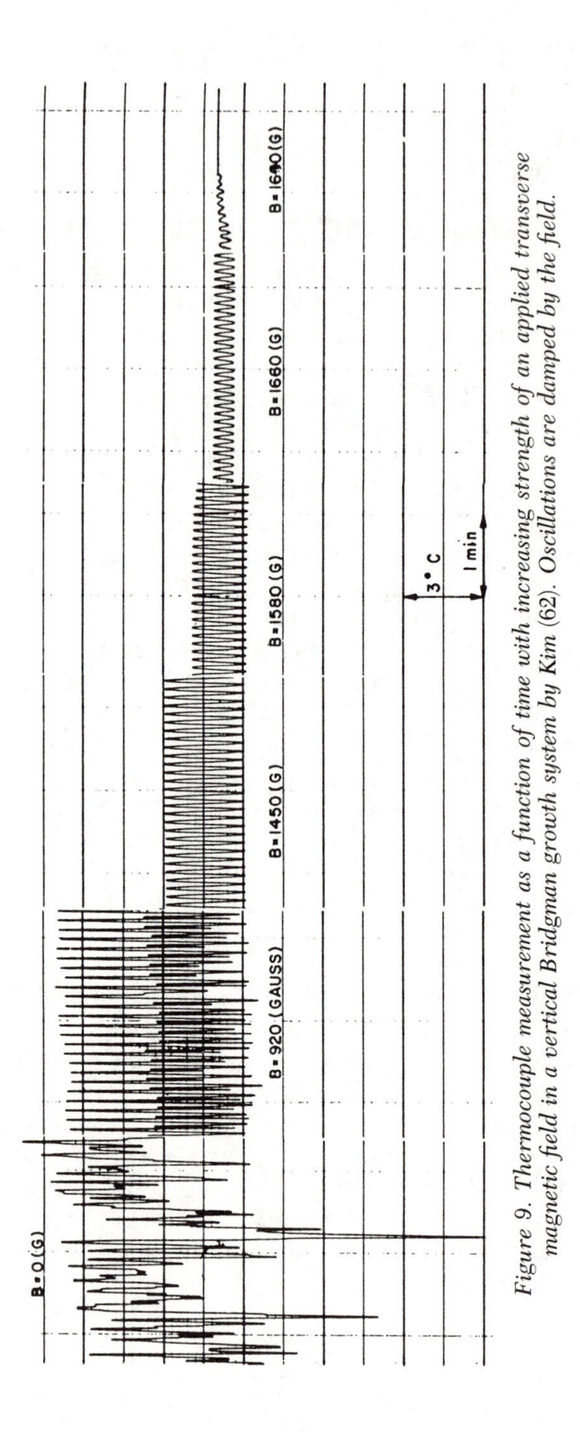

Figure 9. Thermocouple measurement as a function of time with increasing strength of an applied transverse magnetic field in a vertical Bridgman growth system by Kim (62). Oscillations are damped by the field.

in which $<<\tilde{c}>>$ is the volume-averaged concentration of solute in the melt.

These measures of solute segregation are closely related to the spatial and temporal patterns of the flow in the melt. Most of the theories that will be discussed are appropriate for laminar convection of varying strength and spatial structure. Intense laminar convection is rarely seen in the low-Prandtl-number melts typical of semiconductor materials. Instead, nonlinear flow transitions usually lead to time-periodic and chaotic fluctuations in the velocity and temperature fields and induce melting and accelerated crystal growth on the typically short time scale (order of 1 s) of the fluctuations.

In diffusion-controlled directional solidification, the only velocity present in the melt is due to solidification. When the melt–crystal interface is planar, this combination of diffusion and convection leads to a uniform radial distribution of solutes, that is, $\Delta c = 0$, and k_{eff} approaches unity, if the melt is sufficiently long that the diffusion layer next to the interface occupies only a small fraction of the length of the melt. Introduction of cellular convection in the melt distorts the diffusion-controlled profile (equation 4) independently of lateral concentration gradients due to the curvature of the melt–crystal interface (*64*, *65*). Harriott and Brown (*66*) demonstrated this mixing for flows driven by rotating the feed rod and crystal in small-scale floating zones in the limit at which the solutal Peclet number (Pe_s) based on the bulk velocity is small. The degree of radial nonuniformity is maximum for some intermediate level of convection. Increasing the intensity beyond this value homogenizes the melt and reduces the lateral segregation, that is, $\Delta c \rightarrow 0$. Convection also forces k_{eff} to approach k, because the bulk concentration $<<\tilde{c}>>$ is increased (if $k < 1$) because of the mixing of the diffusion layer with the melt in the remainder of the ampoule. In the limit of very intense convection, the variation in the concentration is confined to a boundary layer much thinner than D/V_g. Scheil (*67*) first described axial segregation in the limit of complete mixing, at which the boundary layer thickness approaches zero.

The fluctuations in the velocity and temperature fields caused by time-periodic or chaotic flows lead to fluctuations in the axial and radial segregations of solute, as indicated by the final region on Figure 10. The etching patterns in Figure 8 are the best experimental indication of the temporal characteristics of the convection and have been used as a guide to the analysis of the effect of magnetic fields on convection by Robertson and O'Connor (*68*) for large-scale floating-zone experiments. Relating the frequency response of the axial composition profile directly to the characteristics of the chaotic flow is complicated by the melting and accelerated growth of the crystal because of the concomitant fluctuations in the temperature field.

Solute segregation with bulk convection is given rigorously by solving the two-dimensional solute balance equation [for a two-dimensional velocity

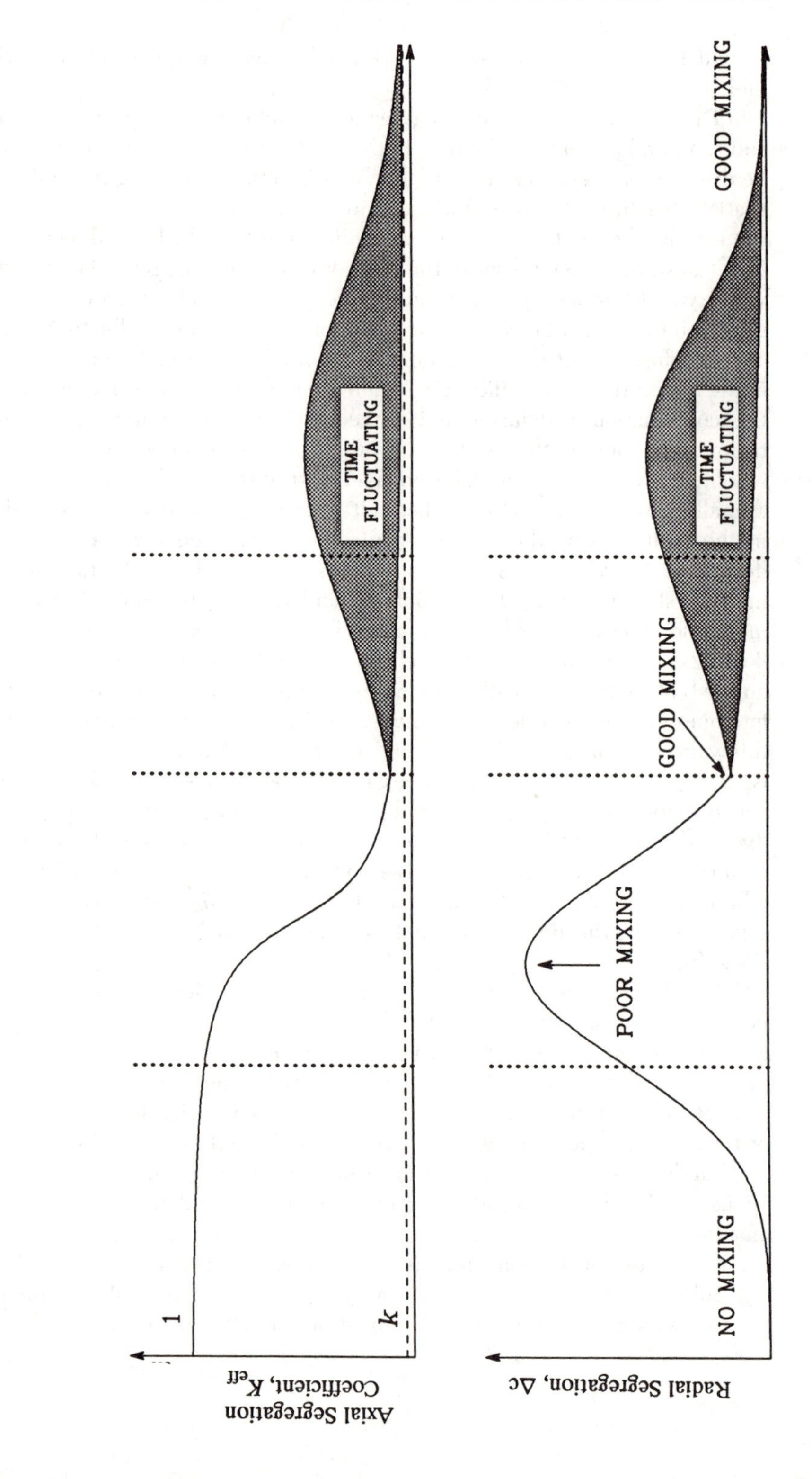
Axial Segregation Coefficient, K_{eff}
1
k
TIME FLUCTUATING
Radial Segregation, Δc
NO MIXING
POOR MIXING
GOOD MIXING
TIME FLUCTUATING
GOOD MIXING

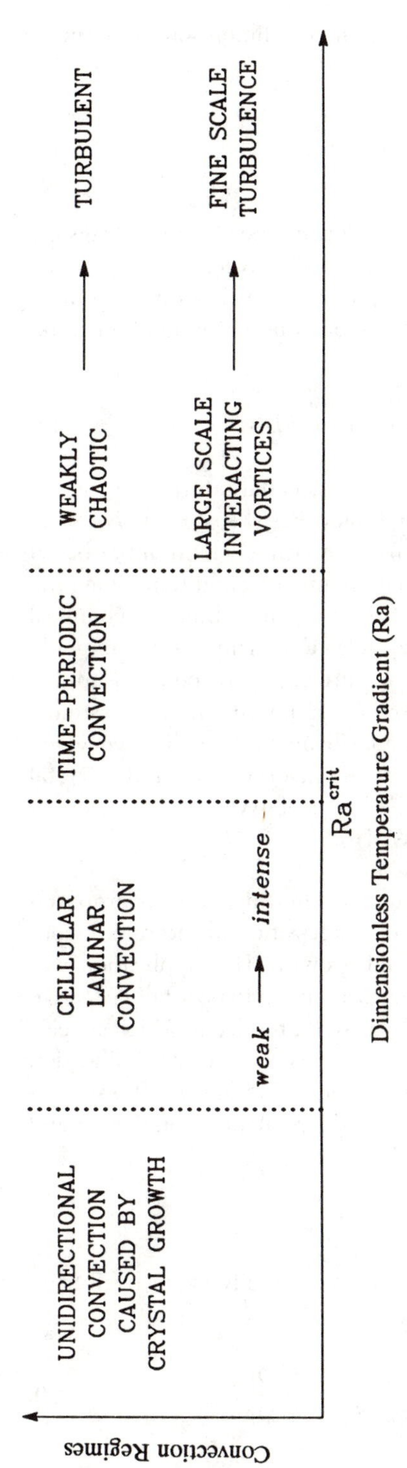

Figure 10. Diagrammatic representation of the regimes of solute transport in melt crystal growth in terms of the effects on radial segregation Δc *and the effective segregation coefficient* k_{eff}.

field $\mathbf{v}(r, z)$], which is given in equation 17 in the dimensionless form of equation 8:

$$\mathrm{Pe}_s \left(\frac{\partial c}{\partial t} + \mathbf{v} \cdot \nabla c \right) = \nabla^2 c \tag{17}$$

In equation 17, Pe_s ($\mathrm{Pe}_s \equiv V_0 L/D$) is the Peclet number for mass transfer, as defined in Table II, and scales the importance of convective solute transport by the velocity in the bulk (scaled by V_0) to diffusion. The dimensionless solute balance at the melt–crystal interface is the generalization of equation 3 for a steadily solidifying interface ∂D_I.

$$\mathbf{n} \cdot \nabla c = \mathrm{Pe}_g c(1 - k)\mathbf{n} \cdot \mathbf{e}_g \text{ at } \partial D_{bI} \tag{18}$$

In equation 18, $\mathbf{n}$ is the unit normal to the melt–crystal interface, $\mathbf{e}_g$ is the unit vector in the direction of crystal growth, and Pe_g ($\mathrm{Pe}_g \equiv V_g L/D$) is the dimensionless crystal growth rate or, alternatively, the Peclet number based on the solidification velocity. The appropriate boundary conditions along the other surfaces depend on the geometry of the system and on the chemical interactions of these boundaries with the melt. For many melts, ampoule and crucible materials are inert, and these surfaces correspond to no-flux boundaries for solutes. Chemical interactions are important in some systems. For example, the quartz crucibles used in silicon growth dissolve when contacted with the melt, and modeling of oxygen incorporation in the crystal must account for this flux, in addition to the losses of oxygen to the ambient that occur in meniscus-defined growth (*69*, *70*).

Stagnant Films. The concept of a stagnant-film thickness, as proposed by Nernst (*71*), is the most widely used characterization of the role of convection in solute segregation during crystal growth. This application was reviewed by Wilcox (*72*). Convective mixing is assumed to be totally effective outside of a thin layer adjacent to the melt–crystal interface in which species transport is by diffusion only and the melt motion is caused by solidification (Figure 11). If the composition of solute in the bulk ($\tilde{z} > \tilde{\delta}$) is c_0, the steady-state composition profile in the layer is derived by solving equations 2 and 3 with the following boundary condition:

$$c(\tilde{\delta}) = c_0 \tag{19}$$

This boundary condition yields equation 20, as originally given by Burton et al. (*73*):

$$\frac{\tilde{c}(\tilde{z})}{c_0} = \frac{k + (1 - k) \exp(-V_g \tilde{z}/D)}{k + (1 - k) \exp(-V_g \tilde{\delta}/D)} \tag{20}$$

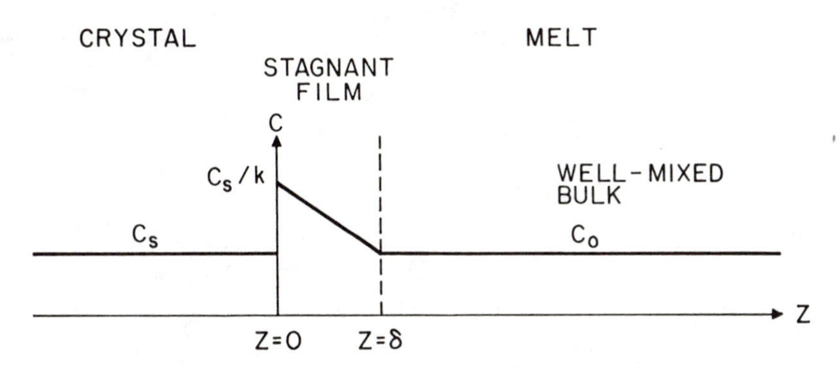

Figure 11. Representation of a stagnant film in solute transport in melt growth. The variables are defined in the text.

The effective segregation coefficient k_{eff} is defined in terms of the stagnant-film thickness as

$$k_{eff} \equiv \frac{k}{k + (1 - k) \exp(-V_g \tilde{\delta}/D)} \tag{21}$$

Equation 21, the most often used relation for fitting axial segregation data from experiment, is amazingly successful for this purpose. For a finite-length ampoule, equation 21 is rewritten in terms of the fraction (f) of the sample solidified to give the normal freezing expression for the composition of the crystal ($\tilde{c}_s = k\tilde{c}_m$) grown from a melt with initial composition c_0:

$$\frac{\tilde{c}_s}{c_0} = k_{eff}(1 - f)^{(k_{eff}-1)} \tag{22}$$

Figure 12 shows a sample profile from the growth of gallium-doped germanium by Wang (*6*). The profile fits equation 21, with an equilibrium segregation coefficient k of 0.087 and k_{eff} of 0.09–0.11.

Although the notion of a stagnant film is a useful way of thinking about the role of convection in axial solute segregation, it has well-known deficiencies (*72*) that prevent it from being predictive. The stagnant-film theory assumes that mixing is perfect outside of the stagnant film and that bulk convection does not modify the velocity field inside the layer. These assumptions prevent the prediction of changes in axial segregation (k_{eff}) caused by changes in the intensity of the flow without an additional empirical correlation for $\tilde{\delta}$ as a function of convection from either experimental (*73–75*) or computational (*76, 77*) data for the effective segregation coefficient and equation 21. Also, because the velocity field in the stagnant film is assumed to correspond only to the growth rate, the stagnant-film model gives no

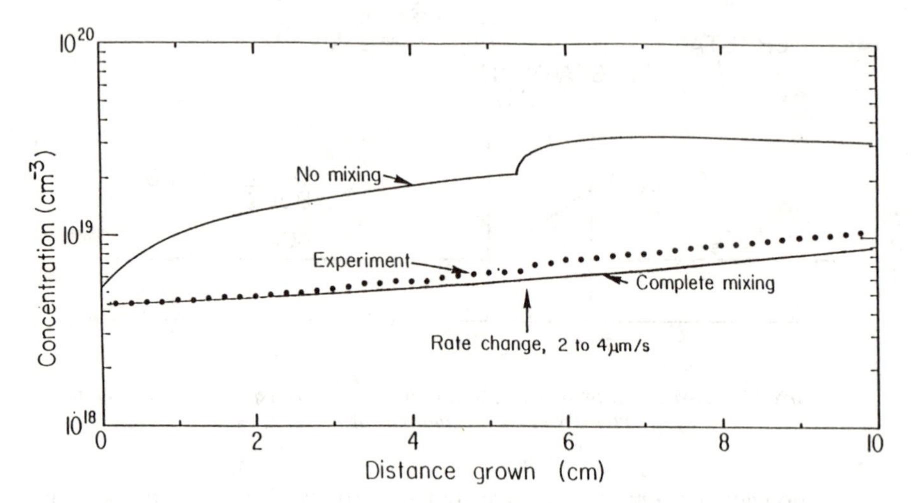

Figure 12. Axial segregation data for the growth of gallium-doped germanium in a vertical Bridgman furnace. The figure is taken from Wang (6).

prediction of the concentration variations along the crystal surface that are caused by lateral nonuniformities in the flow.

Analysis of Solute Boundary Layer. A closed-form solution of the species balance (equation 7) with the interfacial balance (equation 8) and the condition that $\tilde{c} \rightarrow c_0$ as $\tilde{z} \rightarrow \infty$ is only feasible when either convection contributes only weakly to transport, as was the case in the analysis of Harriott and Brown (*66*), or the variation of the concentration from unity is confined to a boundary layer adjacent to the growing crystal. Burton et al. (*74*) analyzed the specific case of solute segregation near a rotating crystal, as occurs in CZ growth. The velocity field near the crystal was taken to be the sum of a uniform crystal growth rate V_g and the self-similar axisymmetric velocity field due to rotation of an infinitely large crystal surface, as described by the asymptotic expansion of Cochran (*78*).

Because the axial velocity is independent of radial distance from the center of the crystal, the concentration field varies only axially and is given by the solution of the solute balance equation with the approximate velocity field of the form

$$\tilde{v}_z = -V_g - V_0(\tilde{z}^2/L^2) \tag{23}$$

in which L is a characteristic length scale for the bulk flow ($L^2 \equiv \nu/\Omega$, in which Ω is the angular velocity of the disk in the similarity solution of Cochran [*78*]) and $V_0 \equiv \Omega R$, in which R is the radius of the crystal. The closed-form solution of this problem is written in terms of exponential integrals, as described by Burton et al. (*74*). The effective segregation coefficient is defined by an equation analogous to equation 21:

$$k_{\text{eff}} = \frac{k}{k + (1 - k) \exp(-\Delta)} \tag{24}$$

The constant Δ in equation 24 is defined as

$$\Delta \equiv -\ln \left(1 - \int_0^\infty \exp[-(\xi + \lambda\xi^3)]d\xi\right) \tag{25}$$

in which λ is the ratio of the component of the velocity field due to the bulk flow to that due to solidification, $\lambda \equiv (V_0/3V_g)$, and ξ is the variable of integration.

Wilson (*79, 80*) pointed out that Δ is not the dimensionless thickness of the diffusion boundary layer scaled with D/V_g, as originally suggested by Burton et al. (*74*), except in the limit at which the velocity field in the layer is dominated by the bulk flow, that is, $\lambda >> 1$. In this case, the analysis reduces to the one first presented by Levich (*81*), and the integral in equation 25 is approximated as follows:

$$\Delta \approx \Gamma(1/3) \frac{\lambda^{-1/3}}{3}, \quad \lambda >> 1 \tag{26}$$

In equation 26, $\Gamma(\cdot)$ is the gamma function. When the ratio λ is rewritten in terms of the Peclet numbers ($Pe_s \equiv V_0L/D$ and $Pe_g \equiv V_gL/D$) as $\lambda = (Pe_s/3Pe_g)$, the scaling in equation 26 is recognized as the scaling for a boundary layer adjacent to a no-slip surface. In the limit at which the solidification rate dominates ($\lambda << 1$), the boundary layer is controlled by a balance of diffusion and convection caused by the solidification velocity. In this case, Δ scales as Pe_g^{-1}.

Wilson (*80*) defined a conventional boundary layer thickness $\tilde{\Delta}$ as follows:

$$\tilde{\Delta} \equiv \frac{\tilde{c}(0) - c_0}{[-d\tilde{c}/d\tilde{z}(0)]}\left(\frac{V_g}{D}\right) = 1 - \exp(\Delta) \tag{27}$$

The constant $\tilde{\Delta}$ is approximately Δ in the limit $\lambda \to \infty$, because Δ is small. Then the boundary layer thickness is expected to vary according to equation 26.

The boundary layer structure predicted by the analysis of Burton et al. (*74*) and by Wilson (*80*) is much more robust than just a description of the solute boundary layer caused by the rotational flow near a large crystal.

When a stagnation flow on the melt–crystal interface is created by any mechanism and is intense enough to lead to a thin solute boundary layer, the velocity field in the boundary layer can be described by

$$\tilde{v}_z(\tilde{r},\tilde{z}) = -V_g - V_0 \frac{F(\tilde{r})}{2} \tilde{z}^2 \tag{28}$$

in which V_0 sets the scale of the bulk convection, and the variation in the lateral direction, $F = F(\tilde{r})$, is slow compared with the rapid variation in the perpendicular direction $\tilde{z}$. With these conditions, the local concentration field, the effective segregation coefficient, and the lateral solute uniformity follow directly from the analysis just described. If $\mathrm{Pe}_s \equiv V_0 L/D >> 1$ and $\mathrm{Pe}_g \equiv V_g L/D = \mathbf{O}(1)$, then a dimensionless boundary layer equation is derived that is valid away from the edges of the crystal:

$$2\eta \frac{\partial c}{\partial r} \int_0^r F(\xi) d\xi - F(r)\eta^2 \frac{\partial c}{\partial \eta} = \frac{\partial^2 c}{\partial \eta^2} + \mathbf{O}(\mathrm{Pe}_s^{-2/3}) \tag{29}$$

In the preceding equations, $\eta \equiv \tilde{z}\mathrm{Pe}_s^{-1/3}$ is the stretched coordinate in the boundary layer and $(r, z) \equiv (\tilde{r}/L, \tilde{z}/L)$. The concentration is assumed to scale with a bulk value of c_0. The scaled form of the solute balance at the interface is given by:

$$\left. \frac{\partial c}{\partial n} \right|_{\eta=0} = -\mathrm{Pe}_g \mathrm{Pe}_s^{-1/3}(1 - k)c(r, 0) \tag{30}$$

Techniques from boundary layer analysis can be used to construct a series solution to equations 29 and 30 of the form

$$c(r, \eta) = 1 + C_1(r, \eta)\mathrm{Pe}_s^{-1/3} + \mathbf{O}(\mathrm{Pe}_s^{-2/3}) \tag{31}$$

in which C_1 is the functional dependence of the concentration computed at leading order in Pe_s. This result is the same as the structure of a concentration boundary layer adjacent to a no-slip boundary (82), which is implied by the form of equation 28 and the conservation of mass. This structure has been observed in concentration fields computed numerically for highly convected melts (66, 76, 77). Camel and Favier (83) predicted the same scaling for the radial segregation across the interface and the effective segregation coefficient from an order-of-magnitude analysis that corresponds to the scalings described here.

The multicellular structure of laminar convection in small-scale crystal growth systems complicates the interpretation of the boundary layer analysis, because the almost-constant level of the concentration field in each cell is unknown a priori and each cell communicates with its neighbors by diffusion through thin internal layers that separate them. These bulk concentrations also scale with Pe_s. Examples of multicellular structures are shown in the finite-element calculations of Harriott and Brown (*66*) for small-scale floating zones and the analysis of Adornato and Brown (*76, 77*) for directional solidification. The work of Adornato and Brown (*76, 77*) is reviewed later in the section on vertical Bridgman–Stockbarger crystal growth.

Several groups (*84–86*) have extended the similarity analysis of Burton et al. (*73*) to the case in which an axial magnetic field is imposed on the melt with sufficient strength such that $Ha >> 1$ and $N << 1$. With these limits, a closed-form asymptotic expression describes the variation in the flow field across the thin $\mathbf{O}(Ha^{-1/2})$ Hartmann layer adjacent to the disk. Axial solute segregation across this layer was analyzed by assuming that the melt outside of the Hartmann layers is well mixed. The effective segregation coefficient approaches 1 when the field strength is increased, as expected for any mechanism that damps convection near the crystal.

Hjellming and Walker (*44*) have presented a semiquantitative analysis of solute transfer with a strong magnetic field to an entire Czochralski system by coupling the asymptotic analysis of the flow and temperature field described previously with boundary layer models for solute transport across the Hartmann layers caused by the field. An important conclusion of this analysis is that the solute transfer for typical magnetic field strengths will be transient throughout the entire crystal growth because of the long diffusion time necessary for the species to traverse the Hartmann layers.

Connection between Transport Processes and Solid Microstructure. The formation of cellular and dendritic patterns in the microstructure of binary crystals grown by directional solidification results from interactions of the temperature and concentration fields with the shape of the melt–crystal interface. Tiller et al. (*21*) first described the mechanism for "constitutional supercooling" or the microscale instability of a planar melt–crystal interface toward the formation of cells and dendrites. They described a simple system with a constant-temperature gradient $\tilde{G}$ (in Kelvins per centimeter) and a melt that moves only to account for the solidification rate V_g. If the bulk composition of solute is c_0 and the solidification is at steady state, then the exponential diffusion layer forms in front of the interface. The elevated concentration (assuming $k < 1$) in this layer corresponds to the melt that solidifies at a lower temperature, which is given by the phase diagram (Figure 5) as

$$\tilde{T}_m = \tilde{T}_m{}^o + \tilde{m}\tilde{c} \tag{32}$$

in which $\tilde{m} < 0$ and $\tilde{T}_m{}^o$ is the melting temperature of the pure material.

The melt in the diffusion layer should solidify at a temperature higher than that at the interface ($\tilde{T}_m^o + m\tilde{c}_0/k$). If the temperature gradient $\tilde{G}$ is low enough, the melt in front of the interface may be supercooled so that any small protuberance of the solid will solidify.

Tiller et al. (*21*) derived a criterion for the onset of this instability by estimating the rate of change of the melting temperature of the melt in front of the interface as $\tilde{m}(d\tilde{c}/c\tilde{z})_I$ and evaluating the gradient from the solute balance at the interface equation 4. The criterion for the stability of the interface is that the actual temperature gradient must be larger than this value, or

$$\frac{\tilde{G}}{V_g} > -\tilde{m}\left(\frac{c_0}{Dk}\right)(1 - k) \qquad (33)$$

Decreasing the temperature gradient, increasing the concentration of solute, or increasing the growth rate leads to instability of the planar interface.

Although equation 33 gives a physical description of the mechanism of the instability that leads to microstructure formation during solidification, it is not rigorous because it does not consider the effects of the rates of heat and species transport on the evolution of the disturbance. Because of this deficiency, equation 33 cannot be used as a basis for further analysis of microstructure formation. This deficiency is shown clearly by the inability of equation 33 to predict the spatial wavelength of the microstructure formed along the interface.

The spatial microstructure of the interface is strongly influenced by its surface energy, which appears in the Gibbs–Thomson equation (*87*) for the melting temperature of a curved interface

$$\tilde{T}_m = \tilde{T}_m^o + \tilde{m}\tilde{c} + \tilde{\Gamma}(2\tilde{H}) \qquad (34)$$

in which $\tilde{\Gamma}$ ($\tilde{\Gamma} \equiv \gamma/\Delta H_f$) is the capillary length associated with the surface energy, γ, and the heat of fusion, ΔH_f; and $\tilde{H}$ is the local mean curvature of the solidification front. $\tilde{H}$ is defined so that $\tilde{H} > 0$ for an interface that is convex with respect to the crystal. Equation 34 implies that the melting point is increased by capillarity for short-wavelength disturbances ($\tilde{H} >> 1$). This increase in melting temperature compensates for the constitutional supercooling mechanism and implies a short-wavelength cutoff to the instability.

Mullins and Sekerka (*88*, *89*) analyzed the stability of a planar solidification interface to small disturbances by a rigorous solution of the equations for species and heat transport in melt and crystal and the constraint of equilibrium thermodynamics at the interface. For two-dimensional solidification samples in a constant-temperature gradient, the results predict the onset of a sinusoidal interfacial instability with a wavelength ($\tilde{\lambda}$) corresponding to the disturbance that is just marginally stable as either $\tilde{G}$ is decreased or V_g is increased. The following discussion focuses on the growth rate as the control parameter for transitions in the interface morphology.

The curves of neutral stability $[V_g = V_c(\tilde{\lambda})]$ correspond to disturbances that become unstable for any further increase in the growth rate. Figure 13 shows a sample set of neutral stability curves for the succinonitrile–acetone alloy for a range of concentrations of acetone. The combination of succinonitrile and acetone is a well-characterized organic alloy used extensively in experiments that simulate metal solidification (*90*). Several features of these curves are universal. First, the curves are closed; a disturbance with a particular wavelength $\tilde{\lambda}$ becomes unstable at a critical growth rate, $V_c(\tilde{\lambda})$, but is restabilized at the higher growth rate corresponding to the top section of the curve. Second, decreasing the concentration of the solute decreases the size of the region of unstable wavelengths, as expected from the criterion of constitutional supercooling.

Although the balance equations are linear, in the absence of bulk convection, the unknown shape of the melt–crystal interface and the dependence of the melting temperature on the energy and curvature of the surface make the model for microscopic interface shape rich in nonlinear structure. For a particular value of the spatial wavelength, a family of cellular interfaces evolves from the critical growth rate $V_c(\tilde{\lambda})$ when the velocity is increased.

The evolution of these cellular forms to deep cells and dendrites with increasing V_g is governed by the full nonlinear equations. Figure 14 shows

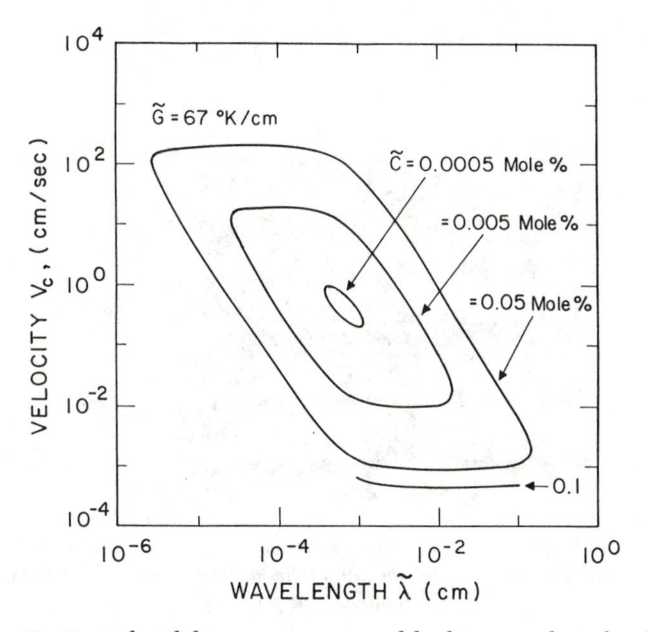

Figure 13. Neutral stability curves computed by linear analysis for the succinonitrile–acetone system as a function of acetone concentration for fixed temperature gradient of G = *67°/cm.*

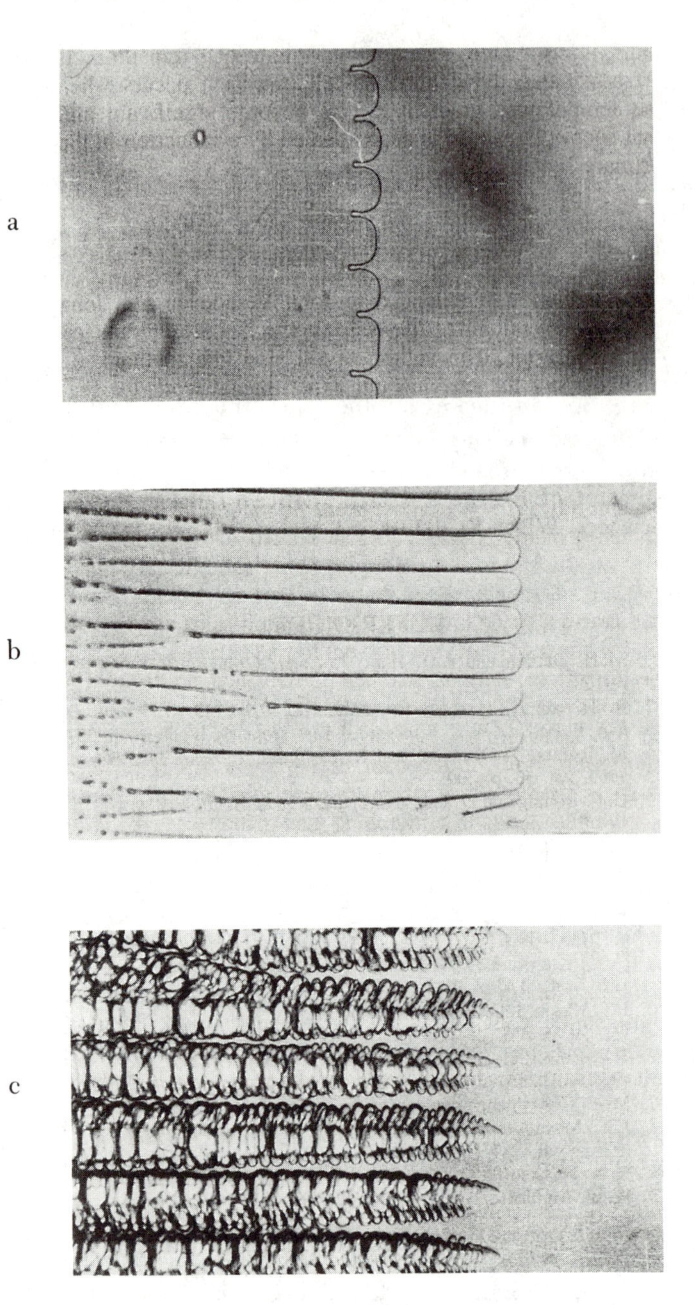

Figure 14. Photographs of cellular and dendritic structures in a thin-film solidification experiment using succinonitrile–acetone reported by Trivedi and Somboonsuk (91).

pictures of the cell-to-dendrite transition with increasing growth rate in a thin solidification sample for succinonitrile–acetone (*91*). Figure 14 shows shallow cells (Figure 14a) that develop deep grooves (Figure 14b) and, finally, dendritic side arms (Figure 14c) as the growth rate is increased. Even in the thin sample, the dendritic structures are not two dimensional, as indicated by the side arms that protrude out of the plane of the photograph.

Prediction of these nonlinear transitions, especially the spatial wavelength of the microstructures, is an area of active research. Full numerical solutions of nonlinear models for microscopic solidification (*92–94*) have shown the development of deep cells for ranges of spatial wavelengths within the unstable region of the neutral stability curve and have demonstrated mechanisms for splitting on individual cells (cell birth) and for merging of cell pairs (cell death). Dynamical calculations indicate that there may be no mechanism for wavelength selection in collections of shallow cells (*95*). Deep cells, like those shown in Figure 14b, have been computed as the growth rate is increased (*93*). These cells exist for ranges of wavelength and have rounded tips connected to slender side walls leading to a cell root with a smooth bottom. Inclusion of bulk convection in the models describing interface microstructure severely complicates the analysis. The current state of research in this area is reviewed by Glicksman et al. (*96*). Several analyses demonstrate that novel interactions between the local flow and the interface morphology are possible when the two phenomena are coupled (*97*).

Connection between Transport Processes and Defect Formation in Crystals. Analysis of the connection between macroscopic transport processes in the melt and crystal, such as the temperature field in both phases and solute transport near the melt–crystal interface, and defects in the crystal requires an understanding of the mechanisms for the generation (in the bulk crystal and at the interface), motion, combination, multiplication, and annihilation of crystallographic defects and dislocations. The description of this coupling for crystal growth has been confined to the simple picture from continuum thermoelasticity for a perfect crystal in which temperature gradients in the crystal during processing create stress in the solid that causes dislocations when its magnitude exceeds the critical resolved shear stress (CRSS), which is evaluated in the slip directions for the crystal. Billig (*98*) introduced this notion to explain the generation of dislocations in germanium crystals grown by the Czochralski method. Several authors have used this idea to developed qualitative estimates for the bounds on temperature gradients in CZ growth. Tsivinsky (*99*) estimated the maximum diameter crystal that can be grown before the CRSS is reached. Brice (*2*) found a relationship between the magnitude of the shear stress in the solid and crack formation.

Jordan et al. (*100*) refined the thermoelastic stress calculation for the analysis of the spatial distribution of dislocations in GaAs grown with the LEC method. Their analysis was based on a two-dimensional model for the temperature field throughout the grown crystal that included all of the critical parameters: pull rate, heat transfer between the crystal and the ambient, and thermophysical properties. In this and in other related works (*101–103*), the crystal was assumed to be an isotropic material with a constant coefficient of expansion, so that the strains in the material are only functions of the local axial and radial temperature gradients. Under these conditions, nonconstant displacements of the crystal, which lead to nonisotropic stresses in the solid, are caused by nonconstant temperature gradients.

Jordan calculated these strains in closed form by assuming that only deformations in the azimuthal plane of the crystal are important. Duseaux (*102*) and Kobayashi and Iwaki (*103*) considered both radial and axial displacements by numerically calculating the strains. Figure 15 shows a direct comparison between the pattern for the magnitude of the shear stress predicted by the analysis of Jordan et al. (*100*) and the pattern of dislocations in a GaAs crystal grown by the LEC technique. Although a quantitative comparison is not meaningful, the similar patterns for regions of high stress and high dislocation density near the edge of the crystal strongly suggest that thermoelastic stress plays an important role in dislocation formation.

The importance of the thermal conductivity of the crystal and the CRSS in determining the degree of difficulty of growing a specific material from the melt is understood in terms of the relationship between these parameters and the formation of dislocations in the crystal because of excess stress. Clearly, materials with lower values of the CRSS must be grown in systems with lower temperature gradients to prevent crystallographic slip. Low values of the conductivity make this difficult to achieve.

The isoelectronic doping of GaAs crystals with indium introduced by Mil'vidskii et al. (*104*) (*see* also references 105 and 106) was based on the idea that indium causes solid-solution hardening of the GaAs lattice and raises the CRSS. Crystals with low dislocation densities have been grown by many groups using this method. Recent measurements of the value of the CRSS as a function of temperature and indium concentration (*107*, *108*) reveal a twofold increase in CRSS in the temperature range just below the melting point. This difference is adequate to eliminate profuse dislocation multiplication in moderate-diameter crystals, but dislocation formation in crystals grown in a low-stress environment remain unexplained.

Although appealing from an engineering perspective, the analyses based on linear thermoelasticity do not address the action of defects and dislocations created by microscopic yield phenomena below the CRSS and of those that are incorporated in the crystal at the solidification front. In the previous works cited (*104–108*), the authors assume that no defects exist at the melt–crystal interface and that the stresses on this surface are zero. Constitutive equations incorporating models for plastic deformation in the crystal due to dislocation motion have been proposed by several authors (*109–111*) and have been used to describe dislocation motion in the initial stages of

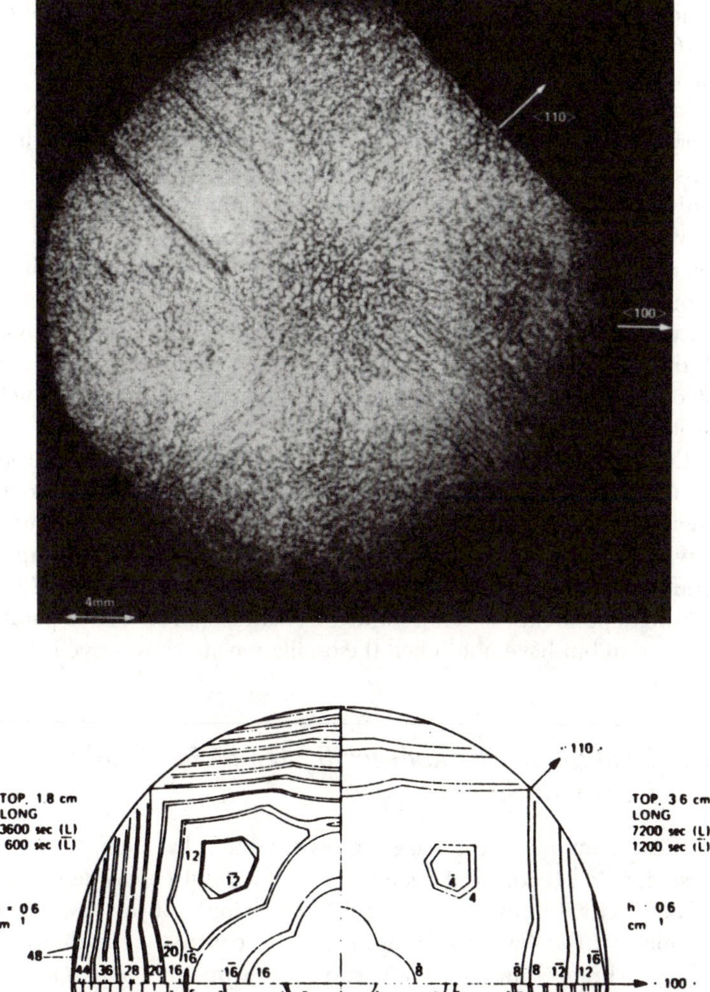

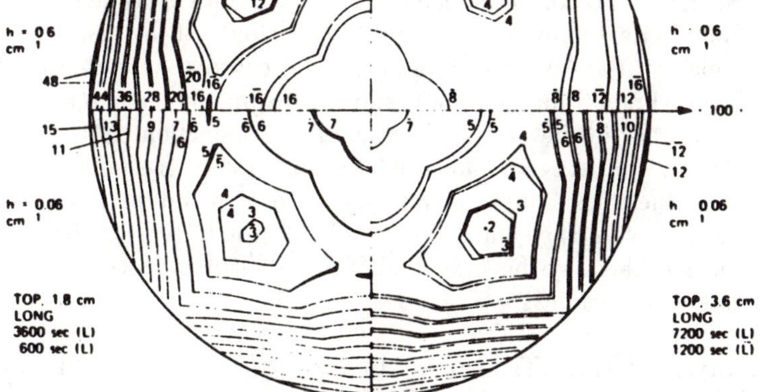

Figure 15. Comparison of measured dislocation density in a GaAs wafer grown by the LEC method (top) with the thermoelastic stress calculation by Jordan et al. (bottom) (100). The high dislocation density around the periphery is predicted by the calculations.

plastic deformation in CZ silicon crystals (*112–114*) and silicon sheet growth (*115, 116*).

An interesting feature of the model proposed by Haasen (*109*) is the explicit appearance of the dislocation density as a function of the stress level in the crystal. Then the dislocations grown into the crystal upon solidification are translated with the pulling rate of the crystal and multiply according to mechanisms associated with their mobility. Dillion et al. (*116*) have reported temperature profiles in silicon sheets in which explosive growth of the dislocation density is predicted by this model. The understanding of the effects of dopants on dislocation mobility is minimal, although experimental evidence exists that some solutes, such as O_2 in Si (*117*) and In in GaAs (*108*), inhibit dislocation mobility. The elastoplastic constitutive equations model this effect only to the extent that the parameters are known as a function of dopant level.

Little is known about the interactions between the transport properties in the melt and the production of defects at the melt–crystal interface. An exception is the swirl microdefect seen during processing of dislocation-free silicon wafers (*118*). The origins of this defect (*119*) are related to temperature oscillations and remelting of the interface. Kuroda and Kozuka (*120*) have studied the dependence of temperature oscillations on operating parameters in a CZ system but have not linked the oscillations to convective instabilities in the melt.

Vertical Bridgman–Stockbarger System: Case Study of Confined Growth System

The vertical Bridgman–Stockbarger growth system has been the subject of the most detailed theoretical analyses and of careful experiments, because the confined growth environment and the proximity of the ampoule to the furnace make it the easiest system for precise control of the thermal field. Also, for low-Prandtl-number fluids and small ampoule diameters (~1 cm), the heat transfer in the melt is dominated by conduction, so that thermal analysis of the furnace–ampoule–melt–crystal system is simplest. Investigations into the roles of furnace and ampoule design and the interactions of these features with convection in the melt and solute segregation have led to advances in these systems. Several of these developments are described in the following sections.

Analysis of Heat Transfer. In the vertical Bridgman–Stockbarger system shown in Figure 1a, the axial temperature gradient needed to induce solidification is created by separating hot and cold zones with a diabatic zone in which radial heat flow from the ampoule to the furnace is suppressed. Analyses of conductive heat transfer have focused on this geometry.

Chang and Wilcox (*121*) were the first to compute the temperature field in an idealized Bridgman–Stockbarger system and to identify the importance of a perfect diabatic zone in establishing a nearly flat melt–crystal interface. Naumann (*122*) and Jasinski et al. (*123*) used a one-dimensional analysis to define the minimum lengths of the hot and cold zones and the ampoule for which the temperature field in the melt and crystal reaches a steady-state profile, and the ampoule translation rate equals the macroscopic growth rate at the interface. Wang et al. (*124*) showed the validity of this analysis through Peltier demarcation measurements of the microscopic growth rate in a well-characterized vertical Bridgman–Stockbarger system.

Several investigators (*125*, *126*) have pointed out the importance of the thermal conductivity of the ampoule material with respect to those of the melt and crystal in setting the shape of the melt–crystal interface, especially when a thick ampoule is used, as is required in the growth of materials that have extremely high vapor pressures (e.g. HgCdTe). Simple design criteria for picking ampoule materials and thickness are given by Naumann and Lehoczky (*125*) and Jasinski and Witt (*126*). Besides setting the curvature of the melt–crystal interface, the choice of ampoule material strongly influences the radial temperature gradients in the melt and the magnitude and direction of buoyancy-driven flow in the melt (*76*, *77*, *127*).

Convection and Segregation. When the melt wets the ampoule walls (which may not be the case in low-gravity experiments; *128*), convection in the melt is driven only by density differences caused by temperature and concentration and by the solidification rate V_g. Chang and Brown (*127*) first computed the flows in an idealized Bridgman system in which the effects of the ampoule on heat transfer were neglected and the temperatures of the hot and cold zones were imposed directly on the melt and crystal with a perfect adiabatic zone separating the two regions. The analysis was based on the Boussinesq approximation for axisymmetric buoyancy-driven flow caused by temperature gradients and included an analysis of heat transfer in the crystal and calculation of the interface shape. The calculations used a finite-element–Newton algorithm to solve the coupled convection–solidification problem that has proved adaptable to the analysis of other melt growth systems. The calculations are formidable; the combination of analyses of intense convection on several length scales (for transport of momentum, heat, and solute) and the unknown interface location forces the use of supercomputers for exhaustive analyses.

These results verified that heat transfer in the melt was conduction dominated, except at intense convection levels because of the low-Prandtl-number characteristic of semiconductor melts. The shape of the melt–crystal interface changes with convection only at these higher convection levels. The flows are cellular, with the direction and magnitude of each cell determined by the radial temperature gradients induced by the thermal boundary conditions. In the idealized system studied, the mismatch in boundary conditions at the junction of the hot zone and the adiabatic region (Figure 16) causes the temperature to increase radially and drive a flow up along the

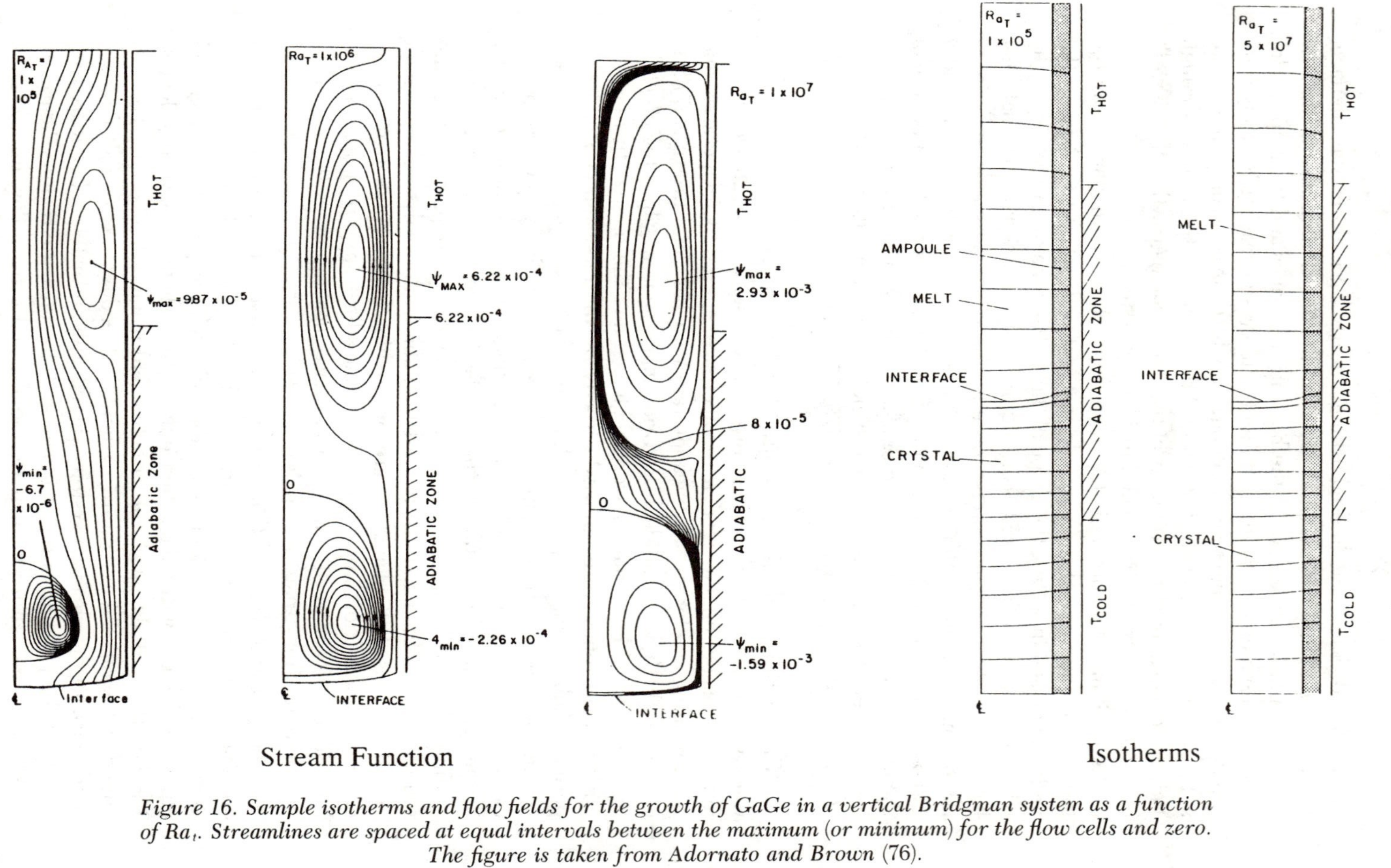

Figure 16. Sample isotherms and flow fields for the growth of GaGe in a vertical Bridgman system as a function of Ra_t. Streamlines are spaced at equal intervals between the maximum (or minimum) for the flow cells and zero. The figure is taken from Adornato and Brown (76).

wall and down along the center of the ampoule. Secondary flows develop adjacent to the melt–crystal interface when the convection is intense enough to alter the temperature field.

On the basis of these flows, calculations of solute transport for a dilute species were reported by Chang and Brown (*127*) for a pseudo-steady-state model of directional solidification in which a solute of given composition enters the ampoule at the end with the melt, and the crystal is pulled away at the growth rate V_g. The axial and radial segregations of solute were much more sensitive to changes in the convection level than was heat transfer because of the lower diffusivity of the species compared with the heat diffusivity ($Sc/Pr >> 1$) for the melt. The results showed the transition from diffusion-controlled growth to weak convection, with the associated maximum in radial segregation and finally the formation of a bulk-convection-controlled boundary layer along the solidification interface, as described qualitatively by Figure 10. The impression of convection adjacent to the melt–crystal interface on radial segregation Δc was accentuated by the changes caused by the formation of the secondary flow cell adjacent to it with increasing thermal Rayleigh number Ra_t.

A quantitative prediction of convection and solute segregation in any melt growth system requires careful control of the thermal boundary conditions that define the driving forces for the flow. Adornato and Brown (*76*, *77*) refined the finite-element–Newton method (*77*) and extended it to include calculation of heat transfer in the ampoule (*76*) and thermosolutal convection driven by both temperature and concentration gradients. Results are reported (*76*, *77*) for the growth of gallium-doped germanium in the Bridgman–Stockbarger furnace of Wang (*6*) and the growth of silicon–germanium alloys in the constant-gradient furnace designed by Rousaud et al. (*129*).

Samples of the isotherms and convection patterns computed for the Bridgman–Stockbarger system of Wang are shown in Figure 16. The results are shown with the frame of reference that the melt–crystal interface is stationary and that melt and crystal are translated downward at the growth rate V_g. The temperature field shows two distinct regions for the radial temperature gradients. Near the interface, the temperature is highest at the center of the ampoule because of the higher thermal conductivity of the melt and the intermediate conductivity of the boron nitride ampoule; the sign of the radial temperature gradient is reversed near the junction of the hot and adiabatic zones for the reason just given. Only the uniaxial convection caused by solidification is present for $Ra_t < 1 \times 10^5$; however, two toroidal flow cells driven in opposite directions by the temperature field are seen for higher values of Ra_t. The cells intensify with increasing Ra_t to the point that the streamlines corresponding to the growth velocity ($V_g = 4$ μm/s in this case) are confined to thin layers along the periphery of the cells.

Solute concentration fields are shown in Figure 17 for the flows in Figure 16. The diffusion-controlled profile for unidirectional solidification is un-

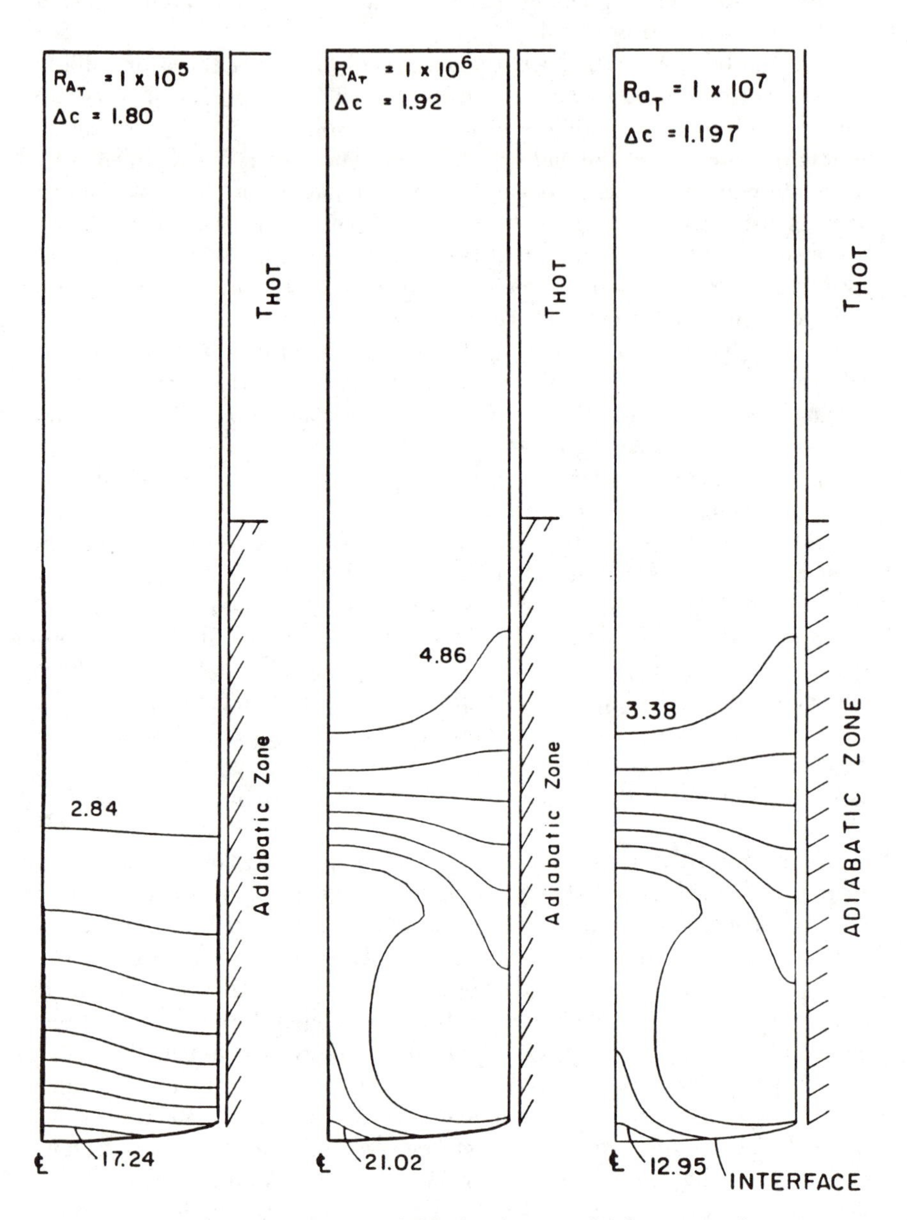

Figure 17. Solute concentration fields for the growth of GaGe in the system described in Figure 16. The growth rate (V_g) *is 4 μm/s.*

mistakable when $Ra_t = 1 \times 10^5$ (Figure 17, left). Radial segregation is present even without extensive bulk convection (the isoconcentration curves intersect with the solidification interface) because of the curvature of the melt–crystal interface. At higher Rayleigh numbers, the cellular flows cause mixing and increase the radial segregation (Figure 17, middle and right). At $Ra_t = 1 \times 10^7$, cores of constant solute composition form within the flow cells separated by the boundary and internal layers. Calculations of the solute field for convection levels corresponding to Ra_t values up to 2×10^8, which are values appropriate for Wang's furnace, are impossible with the finite-element mesh used in this analysis because of the underresolution of these layers.

The radial segregation predicted by these calculations is shown in Figure 18 for three growth rates and the range of Ra_t for which computations were possible. The transitions described by Figure 10 are apparent. The measurements of Wang (*6*) for radial segregation of gallium are shown also in this figure, and comparison of the radial segregations demonstrates that dopant redistribution in this system is controlled by intense laminar convection. Adornato and Brown (*76*) attempted to extrapolate the computational results for radial segregation to the Rayleigh number appropriate for the real furnace by using the power-law scaling suggested by the boundary layer analysis described previously. The results for Δc agree to within 4–30% for the three growth rates, without any attempt to adjust the thermophysical properties of the system beyond an initial calibration of the furnace temperature profile. More importantly, the exponent for the proportionality to Pe_s ranged from

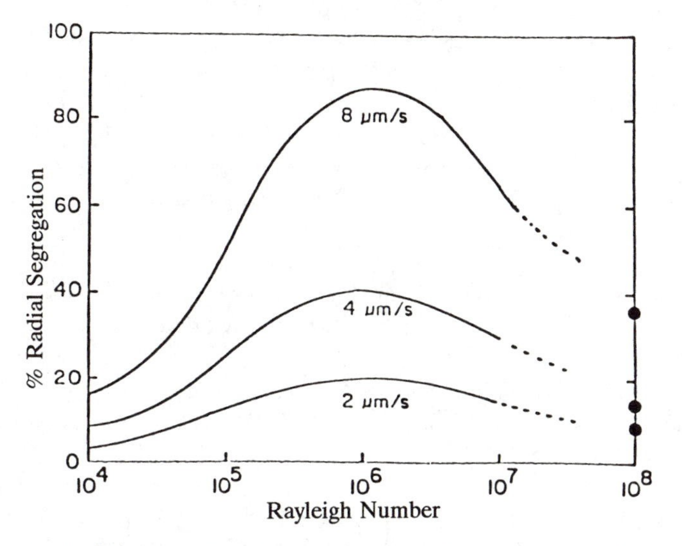

Figure 18. Radial segregation (Δc) as a function of Ra_t for the growth of GaGe in the furnace described by Figures 15 and 16. Solid dots correspond to the results of experiments by Wang (6).

0.28 to 0.30 when the velocity scale was taken as that for a vertical buoyant boundary layer (*130*), so that $V_0 \sim (g\beta R^2(\Delta T)_r)^{1/2}$, in which β is the coefficient of thermal expansion and $(\Delta T)_r$ is the characteristic radial temperature difference driving the flow. These values are in reasonable agreement with the simple asymptotic theory presented earlier.

Transitions to three-dimensional, time-periodic, and chaotic flows do occur in vertical Bridgman growth systems for higher convection levels with the melt above the solid or when the melt is below the crystal, so that the axial temperature gradient leads to an unstably stratified melt (*46*) (*see* the discussion in reference 5). The implementation of an imposed axial magnetic field removes these oscillations (*62*) and reduces the motion to laminar flow. Several investigators have analyzed the impact of the magnetic field on solute segregation in Bridgman growth numerically (*131–133*) and asymptotically (*133*). The structure of the asymptotic analysis follows the work of Hjellming and Walker (*134*), as discussed earlier.

The flows and the consequent solute segregation caused by thermosolutal convection in nondilute alloys is only beginning to be explored by experimental and computational analyses. Recent results are discussed by Brown (*5*).

Czochralski Crystal Growth: Case Study of Meniscus-Defined Growth System

The presence of the meniscus in Czochralski (CZ) growth brings new dimensions to the problems of optimization and control, as well as the analysis of transport processes. Because of its relatively enormous industrial significance, CZ growth has received the most attention of any melt growth method. Also, the distinctions made by using Figure 3 between different levels of modeling are clearest in the discussion of Czochralski crystal growth. The discussion in this section focuses on the analysis of macroscopic transport processes and their influence on the design and control of a CZ growth system.

Understanding the onset of microstructured crystal growth is very relevant also to CZ systems, especially for the growth of heavily doped crystals, as in the case of adding indium to GaAs to lower the dislocation density, which was discussed earlier in regard to defect formation in the crystal. In this case, segregation of indium into the melt during growth in a LEC system eventually leads to morphological instability, as documented by Ono et al. (*135*).

Meniscus Shape. Figure 6 shows a schematic representation of the inner portion of a CZ furnace formed by the melt, crystal, and crucible. The crystal is attached to the melt by the melt–ambient meniscus denoted by ∂D_m. If the traction caused by hydrodynamic forces is neglected (a good assumption for semiconductor melts with high surface tensions and low viscosities), the interface shape, shown in Figure 6 as $z = f(r, \theta, t)$, is set by a balance of capillary force with gravity, by the radius and height of the crystal, by the wetting of melt on the crucible, and by the wetting condition at the melt–crystal–ambient trijunction. The hypothesis for the wetting condition at this trijunction is that a constant angle ϕ_0 (Figure 6) is formed between the tangents to the local melt–ambient and crystal–ambient surfaces, which is independent of growth rate and other macroscopic parameters. The concept of a wetting angle was first proposed by Herring (*136*) to describe crystal sintering and then adapted by Bardsley et al. (*137*) for meniscus-defined crystal growth. The theoretical justification for such an angle hinges on the same arguments used to describe equilibrium contact angles formed at the trijunction of a liquid on an inert solid (*138*). The use of local equilibrium in the description of this point is more restrictive in the solidification process because of the dynamic phase transformation that must be proceeding even on a molecular level near the trijunction. The implications of nonequilibrium microscopic processes and bulk hydrodynamics on this angle are not understood.

Experimental justification for specification of the angle at the point of three-phase contact comes from the results of Surek and Chalmers (*139*), which verify that a particular value of ϕ_0 measured macroscopically can be associated with the crystal growth of a material in a specific crystallographic orientation and that ϕ_0 is roughly independent of growth rate.

The second justification for the angular condition is that this condition is necessary for the determination of the radius of the crystal at the trijunction as a function of heat-transfer conditions and pull rate. This argument is simple. The dimensionless Young–Laplace equation of capillary statics gives the shape of an axisymmetric melt–ambient meniscus as

$$\frac{\partial^2 f/dr^2}{[1 + (\partial f/\partial r)^2]^{3/2}} + \frac{\partial f/\partial r}{r[1 + (\partial f/\partial r)^2]^{1/2}} = \mathrm{Bo}[f(r, \tau) - \lambda_0] \qquad (35)$$

in which the radius of the crucible R_c has been taken as the length scale and Bo is the Bond number ($\mathrm{Bo} \equiv \Delta\rho g R_c^2/\sigma$) or the square of the ratio of this scale to the capillary length l_c [$l_c \equiv (\sigma/\rho g)^{1/2}$]. The left side of equation 35 is the local mean curvature of the meniscus. The constant λ_0 is a dimensionless reference value of the pressure difference across the meniscus if the curvature is zero and is determined by the constraint that the melt encloses a fixed volume. Equation 35 is a nonlinear second-order boundary-value problem, which requires two boundary conditions for solution. For a nonwetting crucible material these conditions are as follows:

$$h[R(f, t), t] = f(R, \tau), \frac{\partial f}{\partial r}(1, t) = 0 \qquad (36)$$

The first condition specifies the joining of the meniscus and the crystal, and the second condition specifies the wetting at the crucible wall. The radius of the crystal at the trijunction $R(f, \tau)$ is left undetermined and must be set by an auxiliary constraint, such as the wetting angle there.

For growth of a crystal with shape $R(z, \tau)$ evolving in time, the wetting-angle condition is written in dimensionless form as

$$\left.\frac{\partial R}{\partial t}\right|_{z=h(R,t)} = \left(V_g(\tau) - \left.\frac{\partial h}{\partial t}\right|_{r=R(h,t)}\right) \tan\left[\phi(\tau) - \phi_0\right] \tag{37}$$

Equation 37 describes the relationship between the rate of change of the crystal radius at the trijunction and the deviation of the local angle from the equilibrium value ϕ_0. In this expression, $\phi(\tau)$ is the dynamic angle formed between the local tangents to the melt–ambient and crystal–ambient surfaces, and $V_g(\tau)$ is the dimensionless pull rate of the crystal. For steady-state growth, equation 37 simply sets the angle with what must be a solid cylinder of constant radius. The importance of the dynamical form equation 37 is brought out in the next section.

When the crucible is large enough that its wetting properties have no influence on meniscus shape, that is, the meniscus becomes flat across the melt surface, the possible shapes come from the collection for unbounded asymmetric interfaces reported by Huh and Scriven (*140*) (*see* also reference 141). This case is a good approximation for Czochralski growth of large crystals in which the region of influence of surface tension, σ, measured by the capillary length, l_c [$l_c \equiv (\sigma/\rho g)^{1/2}$], is approximately 1 cm. Hurle (*142*) developed a closed-form approximation to these meniscus shapes, which are described by the shape function $f(r)$ that is valid in this limit.

Heat-Transfer Analysis: Thermal-Capillary Models. Numerous analyses of various aspects of heat transfer in the CZ system have been reported; many of these are cited by either Kobayashi (*143*) or Derby and Brown (*144*). The analyses vary in complexity and purpose, from the simple one-dimensional or "fin" approximations designed to give order-of-magnitude estimates for the axial temperature gradient in the crystal (98) to complex system-oriented calculations designed to optimize heater design and power requirements (*145, 146*). The system-oriented, large-scale calculations include radiation between components of the heater and the crucible assemblies, as well as conduction and convection.

The dynamics of the Czochralski system can be described only by heat-transfer models that include the interaction of the shape of the meniscus, which are referred to as thermal-capillary models (TCM), because only these models give self-consistent determination of the meniscus shape, crystal radius, and heat transfer in each phase.

Billig (*98*) realized this point and used a one-dimensional heat-transfer analysis for the crystal with assumptions about the temperature field in the melt to derive the the following relationship that has been used heavily in qualitative discussions of crystal growth dynamics:

$$R \approx V_g^2 \tag{38}$$

Kim and co-workers (*147*) presented the following empirical correlation for silicon growth

$$R = \frac{A}{V_g + B} \tag{39}$$

in which A and B are constants that depend on the details of the system and are determined from experiments.

Kobayashi (*143*) presented the first computer simulations that considered the determination of the crystal radius as part of the analysis but avoided the capillary problem by considering a flat melt–ambient surface, which is consistent with $\phi_0 = 90°$. Calculations were performed for a fixed crystal radius, and then the growth rate was adjusted to balance the heat flux into the crystal. Crowley (*148*) was the first to present numerical calculations of a conduction-dominated heat-transfer model for the simultaneous determination of the temperature fields in crystal and melt and of the shapes of the melt–crystal and melt–ambient surfaces for an idealized system with a melt pool so large that no interactions with the crucible are considered. She used a time-dependent formulation of the thermal-capillary model and computed the shape of an evolving crystal from a short initial configuration.

In a series of papers, Derby and Brown (*144*, *149–152*) developed a detailed TCM that included the calculation of the temperature field in the melt, crystal, and crucible; the location of the melt–crystal and melt–ambient surfaces; and the crystal shape. The analysis is based on a finite-element–Newton method, which has been described in detail (*152*). The heat-transfer model included conduction in each of the phases and an idealized model for radiation from the crystal, melt, and crucible surfaces without a systematic calculation of view factors and diffuse-gray radiative exchange (*153*).

Results from a quasi steady-state model (QSSM) valid for long crystals and a constant melt level (if some form of automatic replenishment of melt to the crucible exists) verified the correlation (equation 39) for the dependence of the radius on the growth rate (*144*) and predicted changes in the radius, the shape of the melt–crystal interface (which is a measure of radial temperature gradients in the crystal), and the axial temperature field with important control parameters like the heater temperature and the level of melt in the crucible. Processing strategies for holding the radius and solidification interface shape constant as the melt volume decreases through a typical batchwise growth cycle are presented by Derby and Brown (*149*), as computed by a strategy of augmenting the Newton method with the additional constraints for each control parameter.

Dudokovic and co-workers (*154, 155*) extended the analysis of a QSSM to include diffuse-gray radiation. They computed view factors by approximating the shapes of the crystal and melt by a few standard geometrical elements and incorporating analytical approximations to the view factors. Atherton et al. (*153*) developed a scheme for a self-consistent calculation of view factors and radiative fluxes within the finite-element framework and implemented this scheme in the QSSM.

Figure 19 shows sample isotherms and interface shapes predicted by the QSSM for calculations with decreasing melt volume in the crucible, as occurs in the batchwise process. Because the crystal pull rate and the heater temperature are maintained at constant values for this sequence, the crystal radius varies with the varying heat transfer in the system. Two effects are noticeable. First, decreasing the volume exposes the hot crucible wall to the crystal. The crucible wall heats the crystal and causes the decrease in

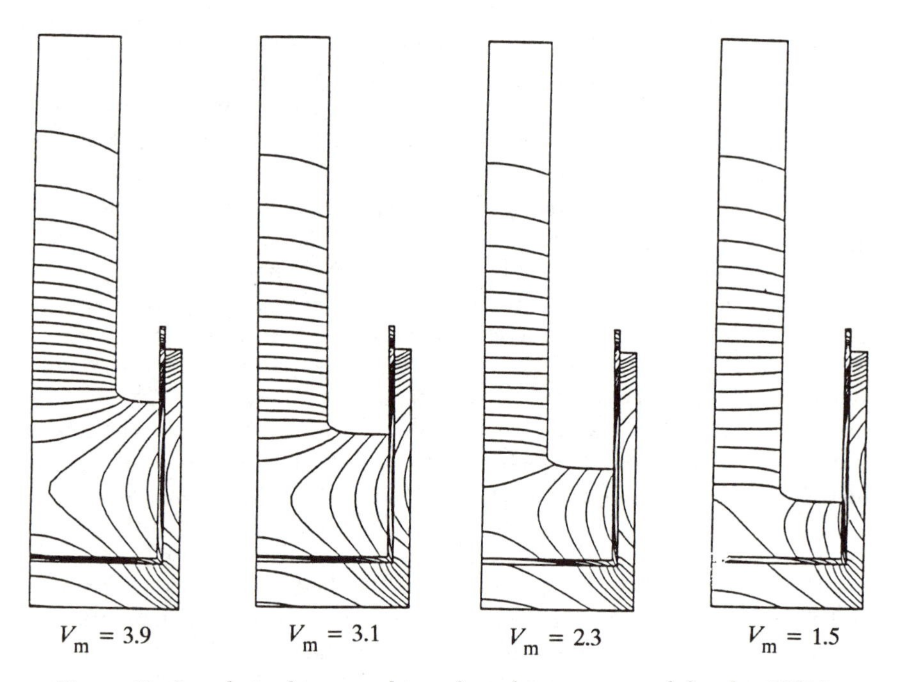

Figure 19. Sample isotherms and interface shapes computed for the QSSM for CZ crystal growth by Atherton et al. (153). *The model includes detailed radiation between the surfaces of the melt, crystal, and crucible. Isotherms are spaced at 10 K increments in the melt and 30 K increments in the other phases.*

radius seen for dimensionless melt volumes (V_m) of 3.1 and 2.3. Second, heat transfer from the crucible to the melt becomes ineffective for low volumes and causes a decrease in the temperature of the melt and an increase in the crystal radius. For the lowest volume, the melt temperature at the bottom of the crucible drops below the melting point, and a lump of solid forms at the center. The growth is terminated in the simulation by merging this mass of solid with the melt–crystal interface. Insulation or the introduction of heat along the bottom leads to a monotonic decrease in the radius to much lower melt volumes, with an almost linear relationship ($R \sim V_m$) when the bottom is heated.

The results of all the thermal-capillary models discussed so far have neglected the influence of convection in the melt in transporting heat to the solidification interface. The status of convection calculations that neglect the coupling to global heat transfer and capillary consideration is discussed later. The union of thermal-capillary analysis with detailed convection calculations is discussed in the subsection on melt flow.

Process Stability and Control. Operationally, automatic control of the crystal radius by varying either the input power to the heater or the crystal pull rate has been necessary for the reproducible growth of crystals with constant radius. Techniques for automatic diameter control have been used since the establishment of Czochralski growth. Optical imaging of the crystal or direct measurement of the crystal weight has been used to determine the instantaneous radius. Hurle (*156*) reviewed the techniques currently used for sensing the radius. Bardsley et al. (*157, 158*) described control based on the measurement of the crystal weight.

The details of the control strategy have received much less attention. The theoretical (*159*) and experimental (*160*) analyses of the transfer function for CZ growth are notable exceptions. Algorithms for model-based control are just being developed.

The implementation of a control algorithm using any sensing technology depends on the dominant dynamics of the Czochralski process, especially the stability of the process to perturbations in the thermal field and the interface shapes and the batchwise transient introduced by the decreasing melt volume. The control strategies differ substantially according to whether the process is stable or inherently unstable. This issue is discussed here in terms of the several notions for stability introduced by Hurle (*10*) that are relevant to meniscus-defined crystal growth. Three types of instabilities are distinguished according to the time scales for the disturbances: capillary instabilities of the meniscus, dynamic instability of the entire thermal-capillary system, and convective instabilities of the melt.

Capillary Instability. The meniscus shape may become unstable in the sense that small perturbations to it cause the melt to separate from the crystal in the time scale of a capillary-induced motion of the surface. This time scale is proportional to the capillary wave speed, which is a few seconds for the thermophysical properties of molten silicon. The capillary stability of meniscus shapes that are important in CZ growth have been analyzed by several authors (*161*, *162*) using arguments from the theory of energy stability for equilibrium configurations. Meniscus shapes that are stable to capillary instabilities are feasible for a range of contact angles and interface heights.

Dynamic Stability of Thermal Capillary System. Surek and co-workers (*163*, *164*) introduced the notion of *dynamic stability* for meniscus-defined crystal growth systems. They considered the question of whether a meniscus-defined growth system subjected to a perturbation to the crystal shape at fixed pull rate will return to the initial shape or diverge unstably away from it. The concept of dynamic stability does not include the effect of the batchwise drop in the melt volume on the evolution of the perturbation, even though the time scales for the decreasing melt volume and for the growth of the crystal are essentially the same, except when the crucible is much larger in diameter than the crystal being grown.

The initial analysis of Surek (*163*) considered the influence of the wetting condition at the trijunction (equation 39) and the Young–Laplace equation for meniscus shape but neglected the influence of heat transfer on the dynamics of the system. Without heat transport, the location of the melting-point isotherm is unspecified, and so the dynamics of the height of the trijunction $h(f, \tau)$ is unknown. On the basis of an isothermal analysis, Surek showed that the floating-zone techniques and the die-defined methods for growing thin solid sheets are stable to perturbations in the meniscus height but that the CZ method is inherently unstable. Refinements to the analysis that included simple descriptions of the interactions between heat transfer and the meniscus height showed that conditions exist for stable operation (*164*, *165*) without batchwise transients being considered.

Integration of a time-dependent thermal-capillary model for CZ growth (*150*, *152*) also has illuminated the idea of dynamic stability. Derby and Brown (*150*) first constructed a time-dependent TCM that included the transients associated with conduction in each phase, the evolution of the crystal shape in time, and the decrease in the melt level caused by the conservation of volume. However, the model idealized radiation to be to a uniform ambient. The technique for implicit numerical integration of the transient model was built around the finite-element–Newton method used for the QSSM. Linear and nonlinear stability calculations for the solutions of the QSSM (if the batchwise transient is neglected) showed that the CZ method is dynamically stable; small perturbations in the system at fixed operating parameters decayed with time, and changes in the parameters caused the process to evolve to the expected new solutions of the QSSM. The stability of the CZ process has been verified experimentally, at least partially, by the uncontrolled, stable growth of small-diameter germanium (*166*) and silicon crystals (M. Wargo, unpublished, Massachusetts Institute of Technology).

Atherton et al. (*153*) have extended the calculations to include diffuse-gray radiation between components of the enclosure and reached essentially the same conclusions regarding the stability of the process. However, they discovered a new mechanism for the damped oscillation of the crystal radius caused by the radiative interaction between the crystal surface just above the melt level and the hot crucible wall. These oscillations are especially apparent when the vertical temperature gradient in the crystal is low, so that radiative heat transport has a dominating influence.

The dynamic stability of the quasi steady-state process suggests that active control of the CZ system has to account only for random disturbances to the system about its set points and for the batchwise transient caused by the decreasing melt volume. Derby and Brown (*150*) implemented a simple proportional-integral (PI) controller that coupled the crystal radius to a set point temperature for the heater in an effort to control the dynamic CZ model with idealized radiation. Figure 20 shows the shapes of the crystal and melt predicted without control, with purely integral control, and with

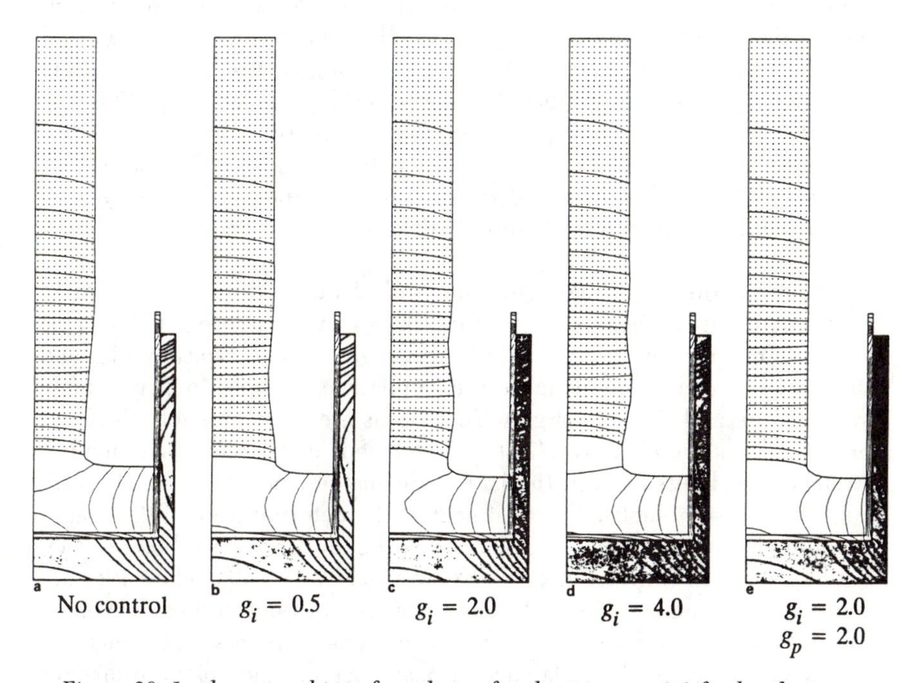

Figure 20. Isotherms and interface shapes for the time $\tau = 1.0$ for batchwise simulations of CZ growth. Results are shown for (a) uncontrolled, (b–d) integral control, and (e) proportional-integral control simulations. Isotherms are spaced as described for Figure 19. The figure is taken from Atherton et al. (153).

PI control. The oscillations in the radius about the set point predicted with integral control are not unexpected. Both the oscillations in the control of systems governed by linear models (*167*) and the occurrence of Hopf bifurcations in nonlinear models (*168*) have been predicted with integral control. Indeed, the oscillations shown for PI control are anticipated from the appearance of a Hopf bifurcation in the QSSM with increasing gain of the integral part of the controller (*150*). Simple PI control removes the oscillations and leads to a constant radius throughout a large portion of the growth run. Atherton et al. (*153*) have extended these calculations to include diffuse-gray radiation and have shown that PI control of the crystal radius is much more difficult there, because the increasing influence of radiation at the melt level drops makes a single set of gain values inappropriate for the whole growth run.

It remains to be shown that the dynamics predicted by thermal-capillary models adequately represents a real CZ system. A first step in this direction has been taken by Thomas et al. (*169*) by making a direct comparison between the predictions of a dynamic TCM for the liquid-encapsulated CZ growth of GaAs and the response of an experimental system using a large (2-kilogauss) axial magnetic field. The pull rate and heater temperature in the calculation were varied with time according to prescribed histories taken from experimental data. The shape of the crystal and the temperature field predicted by the simulation are shown in Figure 21 as a function of the time during the batch process. The predicted shape of the crystal agrees qualitatively with that produced in the experiment. The quantitative value of the radius at any time was within 15% of the experimental value; this difference is well within the error caused by uncertainties in the experiment and by poor knowledge of thermophysical property data used in the model.

Convection and Segregation. ***Melt Flow.*** Although analysis of convection and segregation in Czochralski growth has received the most attention of any melt growth system, the large size of a typical puller and the presence of a meniscus make the simulations the most difficult of any system. The real flows in large-scale systems are most probably three dimensional and temporally chaotic; a detailed analysis of such motions stretches the limits of even the largest calculations performed today. Calculations of axisymmetric flow that neglect the determination of the shapes of the melt–crystal and melt–ambient phase boundaries have been carried out by a number of investigators for some time. The results before 1985 are reviewed by Langlois (*170*) and focus on the interactions of flows driven by crucible and crystal rotation, buoyancy, and gradients in surface tensions. Starting with the initial calculations of Kobayashi (*171*), most of these analyses have focused on predicting the multicellular structure of the flows caused by combinations of various driving forces.

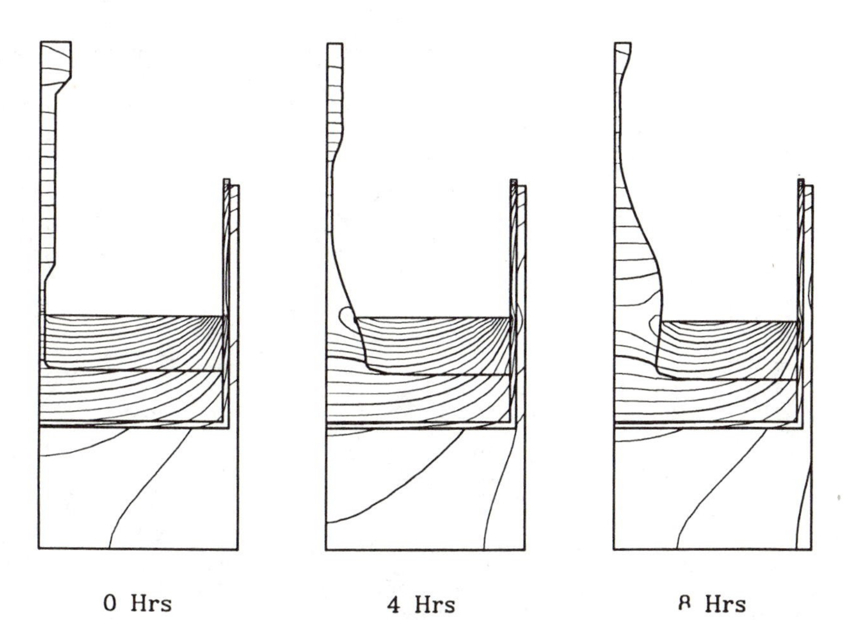

Figure 21. Isotherms and interface shapes for selected times during the batch-wise simulations of GaAs crystal in LEC growth. The figure is taken from Thomas et al. (169).

More-recent simulations concentrate on the onset of time-periodic convection, which marks the onset of a dominant mechanism for dopant striations in the crystal, and on the effect of an imposed magnetic field on this transition. Crochet et al. (56) first computed the onset of an axisymmetric time-periodic motion in CZ flow driven by heating of the side wall. The flow motion is caused by the competition between flow cells created by an instability that leads to the separation of the hydrodynamic boundary layer along the crucible side wall. Accurate calculations must resolve this boundary layer. A simulation reported by Crochet et al. (56) is shown in Figure 22, which is a plot of the kinetic energy in the flow as a function of time for $\mathrm{Gr} \equiv g\beta(T_{\mathrm{wall}} - T_{\mathrm{m}})R_{\mathrm{c}}^{3}/\nu^{2} = 1 \times 10^{7}$, in which T_{wall} is the constant temperature of the crucible wall. The oscillations develop from some initial state and tend toward constant values of the amplitude and frequency. The flow patterns and temperature fields reported by Crochet et al. (56) for several different times in this simulation are shown in Figure 23. The interaction between the two largest toroidal vortices is apparent.

Tangborn (*172*) and Patera (*173*) have reproduced the oscillations observed by Crochet at al. by using a spectral-element simulation with flat phase boundaries. They demonstrated that underresolution of the boundary layer leads to spurious oscillations at lower Grashof numbers. Others have

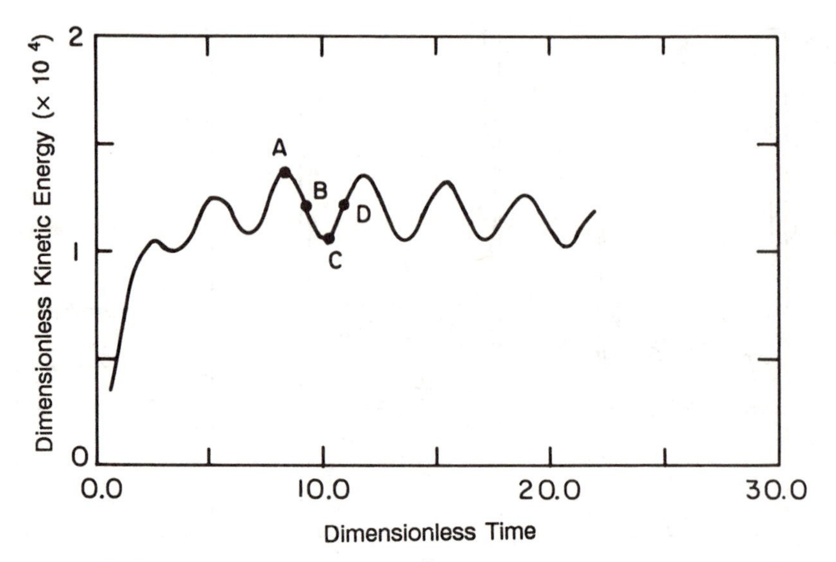

Figure 22. Kinetic energy as a function of time in a simulation of Czochralski bulk flow by Crochet et al. (55) *for Gr = 1 × 10^7. Points A and B correspond to flows shown in Figure 23.*

computed unsteady convection in a similar model for axisymmetric convection in CZ bulk flow.

Langlois (*44*, *63*) first demonstrated numerically the experimentally observed damping of convection caused by an imposed axial magnetic field on unsteady convection and the possibility of achieving steady, axisymmetric flows. Mihelcic and Winegrath (*174*) and Patera (*173*) have presented similar results with more emphasis on the transition between steady and oscillatory flows caused by increasing the magnetic field strength. Increasing the field beyond the level necessary to stabilize the flow decreases the mixing in the melt and leads to radial nonuniformity of dopants in the crystal, as would be the case for weak convective mixing shown in Figure 10. Oreper and Szekely (*131*) documented this effect, and Hjellming and Walker (*44*) presented an asymptotic analysis for the flow and temperature fields in this limit.

Numerical simulations that combine the details of the thermal-capillary models described previously with the calculation of convection in the melt should be able to predict heat transfer in the CZ system. Sackinger et al. (*175*) have added the calculation of steady-state, axisymmetric convection in the melt to the thermal-capillary model for quasi steady-state growth of a long cylindrical crystal. The calculations include melt motion driven by buoyancy, surface tension, and crucible and crystal rotation. Figure 24 shows sample calculations for growth of a 3-in. (7.6-cm)-diameter silicon crystal as a function of the depth of the melt in the crucible.

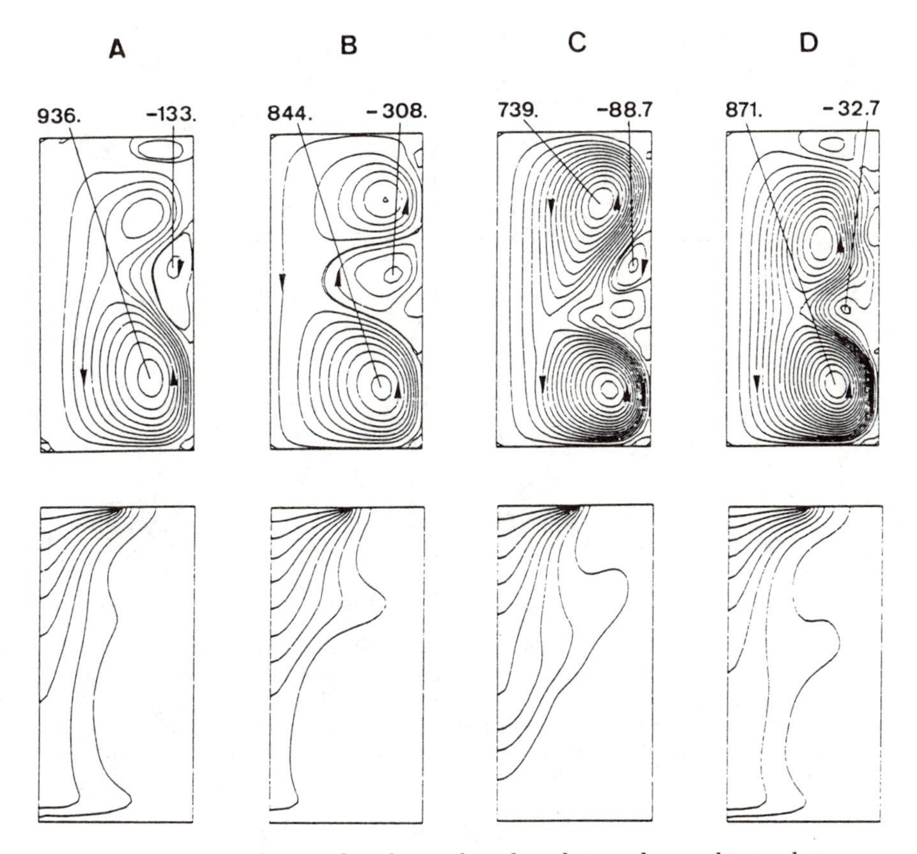

Figure 23. Streamlines and isotherms for selected times during the simulation shown in Figure 22. Flows correspond to points shown in Figure 22.

The radial temperature difference imposed by the heater arrangement drives a large toroidal flow cell that moves up the wall and down toward the center of the crucible. For the deepest melt volumes, the flow is separated along the wall. This steady axisymmetric motion is probably not stable. Time-periodic oscillations caused by the instability of this separation have been reported at values of the Grashof number below the values that correspond to the calculations in Figure 24. Adding an axial magnetic field leads to the reattachment of the flow cell to the wall of the crucible and to the flow structure predicted by Hjellming and Walker (43) at high field levels.

The flow separation along the crucible side wall disappears, and a separation along the bottom appears as the melt level is lowered (Figure 24). The melt–crystal interface deforms to be convex at the center of the crystal as it senses the cold crucible bottom with the decreasing melt depth. This interface flipping is well documented in experimental systems and is also seen in calculations based solely on conduction in the melt.

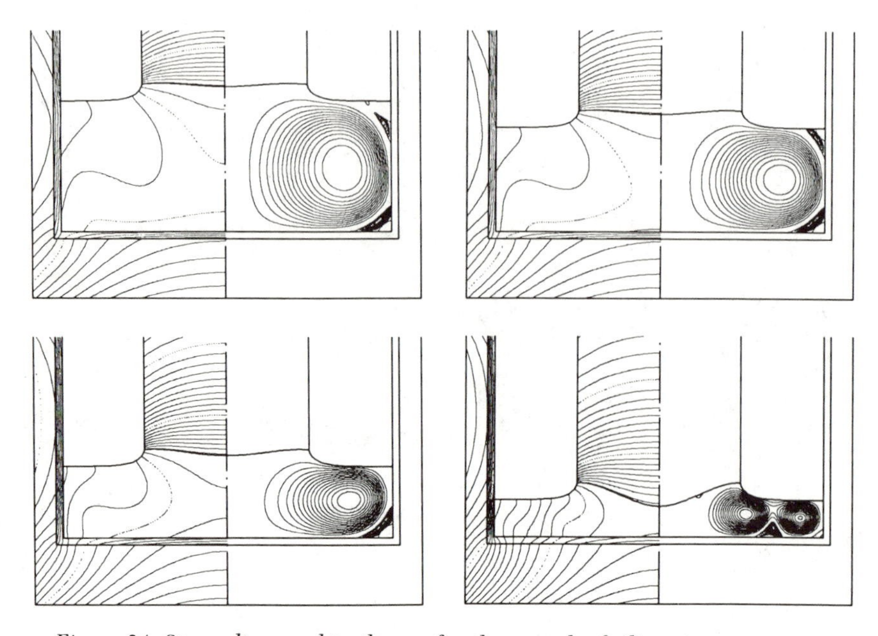

Figure 24. Streamlines and isotherms for the growth of silicon in a prototype Czochralski system with self-consistent calculation of interface and crystal shapes by using the quasi steady-state thermal-capillary model and the condition that the crystal radius remains constant. Calculations are for decreasing melt volume. The Grashof number (scaled with the maximum temperature difference in the melt) varies between 1.0×10^7 and 2.0×10^7 with decreasing melt volume.

Few calculations of three-dimensional convection in CZ melts (or other systems) have been presented because of the prohibitive expense of such simulations. Mihelcic et al. (*176*) have computed the effect of asymmetries in the heater temperature on the flow pattern and showed that crystal rotation will eliminate three-dimensional convection driven by this mechanism. Tangborn (*172*) and Patera (*173*) have used a spectral-element method combined with linear stability analysis to compute the stability of axisymmetric flows to three-dimensional instabilities. Such a stability calculation is the most essential part of a three-dimensional analysis, because nonaxisymmetric flows are undesirable.

Solute Transport. Burton et al. (*72*) originated the boundary layer analyses for solute transport in CZ crystal growth described earlier. The analysis of the boundary layer has been the starting point for studying axial solute segregation in CZ growth. Wilson (*80*) has refined this analysis by including a full numerical calculation of the self-similar velocity field due to crystal rotation instead of using the asymptotic approximation of Cochran

(*78*). Wilson (*177, 178*) also used the transient, self-similar velocity field caused by crystal rotation to analyze the time-dependent axial dopant striations caused by a fluctuating local growth rate. By simultaneous calculation of the flow and the solute field without (*177*) and with (*178*) accounting for the possible back melting of the solid, she demonstrated periodic axial striations in the concentration field that were extremely similar to ones reported in CZ growth for systems with poor thermal symmetry that resulted in growth rate fluctuations proportional to the crystal rotation rate (*25*). Earlier, Hurle et al. (*179*) made the connection between dopant striations and temperature oscillations by using a simpler model.

Analysis of solute transport in the presence of a magnetic field has received considerable attention. Hurle and Series (*85*) and Cartwright and Hurle (*180*) used the self-similar form of the velocity field near a rotating crystal as a framework for examining the role of an axial magnetic field in modifying axial segregation in the crystal.

Lee et al. (*181*) reported the only detailed calculations of solute transport in Czochralski growth for oxygen transport in the presence of an axially aligned magnetic field. Their results indicated that oxygen transport is reduced by the damping of convection and that the structure of the flow cells has a significant effect on the transfer rate, if the flow is steady. Hoshikawa et al. (*182*) first demonstrated the control of oxygen concentration in silicon grown by the CZ method by using an applied magnetic field. More recent analysis by Kim and Langlois (*183*) of boron transport in CZ growth with an axial magnetic field shows large radial variations in boron concentration when bulk convection is weak. Such large variations have been reported in several as yet unpublished experimental studies.

Summary and Outlook

Pioneering efforts in the understanding of transport processes in melt crystal growth have elucidated the important fundamental mechanisms in these systems and are the beginning of comprehensive understanding of property–processing relationships for electronic materials. The segregation analysis of Burton et al. (*74*), the analysis of the onset of morphological instabilities by Mullins and Sekerka (*88, 89*), and the documentation of the transitions to chaotic convection in melt crystal growth by Witt and his collaborators (*46–50*) are examples of such seminal investigations.

Each of these efforts has spawned nearly a generation of scientific exploration and engineering application of these concepts. As this chapter indicates, the level of generic understanding of transport processes and macroscopic properties of the crystal is good. Even so, quantitative predictions of the performance of specific crystal growth systems and the design and optimization of these systems based solely on theoretical understanding are not yet feasible. The design of systems for specific materials still involves a great deal of experimental input and empiricisms. The reasons for this limitation are many.

First, the role of system design on the details of convection and solute segregation in industrial-scale crystal growth systems has not been adequately studied. This deficiency is mostly because numerical simulations of the three-dimensional, weakly turbulent convection present in these systems are at the very limit of what is computationally feasible today. New developments in computational power may lift this limitation. Also, the extensive use of applied magnetic fields to control the intensity of the convection actually makes the calculations much more feasible.

Even if the extensive calculations needed to define the dependence of segregation and heat transfer on operating conditions can be performed, they may not be justified in the sense that the results cannot be supported either by the level of quantitative characterization of a typical crystal growth system or by the accuracy of the data base for the thermophysical properties of exotic semiconductor alloys, especially under high-temperature processing conditions. The first deficiency requires extensive developments in the technology of sensors for monitoring variables during the crystal growth process, such as the temperature gradients in the crystal that are directly responsible for defect generation in the solid. Characterization of the thermophysical properties of these materials is clearly a prerequisite for the accurate understanding of the processing conditions for any melt crystal growth system.

The most conspicuous shortcomings of modern theory of crystal growth appear when the links between macroscopic processing conditions and the formation of crystalline defects are considered. Other than the intuition derived by applying the theory of linear thermoelasticity to the calculation of stress fields in the cooling solids, no guidelines exist for designing growth systems. Elucidating the mechanisms of defect formation and the importance of the chemistry of the melt and hence transport processes in these mechanisms are important frontiers in the processing of electronic materials.

Abbreviations and Symbols

$\mathbf{B}$	magnetic field
B_0	magnitude of magnetic field
c^*	characteristic concentration scale
$\tilde{c}$	dimensional concentration field
c_0	reference value of c (equation 4)
c_o	far-field level of concentration
$<\tilde{c}>_I$	interfacial average concentration (equation 16)
$c(\mathbf{x}, t)$	dimensionless concentration field
Δc^*	characteristic scale for concentration difference
D	molecular diffusivity of solute in melt

∂D_b	boundary of computational domain
∂D_{bI}	fraction of boundary adjoining melt–solid interface
∂D_{bl}	fraction of boundary adjoining solid surfaces
$\mathbf{e}_g$	unit vector in direction of gravity
$\mathbf{e}_z$	unit vector in z direction (equation 5)
$f(r)$	dimensionless meniscus shape function
$\mathbf{F}_b(\tilde{\mathbf{v}}, \tilde{\mathbf{x}})$	dimensionless applied body force field
$\mathbf{F}_s(\tilde{\mathbf{x}})$	dimensionless surface force
g	acceleration of gravity
G	dimensionless axial temperature gradient
$h(r)$	dimensionless melt–crystal interface shape function
H	dimensionless mean curvature of interface
$\tilde{H}$	dimensional mean curvature of interface
ΔH_f	latent heat of fusion
I	refers to the melt–crystal interface when used as a subscript
$\mathbf{I}$	identity tensor
k	equilibrium partition coefficient (equation 1)
k_{eff}	effective segregation coefficient
l_c	capillary length, $(\sigma/\rho_m g)^{1/2}$
L^*, L	characteristic length scale
m	refers to the melt phase when used as a subscript
$\tilde{m}$	slope of liquidus curve
$\mathbf{n}$	unit normal vector to melt–ambient meniscus
$\mathbf{N}$	unit normal vector to melt–crystal interface
$\mathbf{O}$	order of magnitude
$\mathbf{p}$	dimensionless pressure field
p_0	reference pressure
$\tilde{\mathbf{p}}$	dimensional pressure field
r	dimensionless radial coordinate
R	dimensionless crystal radius
s	refers to the solid of crystal phase when used as a subscript
t	dimensionless time
$\tilde{t}$	dimensional time
$\mathbf{t}$	unit tangent vector to melt–ambient meniscus
$\mathbf{T}$	unit tangent vector to melt–crystal interface
$\tilde{T}_r$	dimensional value of T (equation 5)
T^*	characteristic temperature scale
T_0	reference value of T
T_m	melting temperature
T_m^o	melting temperature of pure material
$T(\mathbf{x}, t)$	dimensionless temperature field
ΔT	maximum temperature difference
ΔT^*	characteristic temperature difference
$\tilde{\mathbf{v}}$	dimensional velocity vector

$\mathbf{v}(\mathbf{x}, t)$	dimensionless velocity field
V^*	characteristic velocity scale
V_g	growth rate
V_o	velocity scale for bulk motion
$\mathbf{V}_s(\mathbf{x}, t)$	dimensionless velocity of solid surfaces
$\mathbf{x}$	vector of dimensionless coordinates
x_i	thermal conductivity of phase i
$\tilde{\mathbf{x}}$	dimensional vector of coordinates
$\tilde{z}$	dimensional axial distance
α_i	thermal diffusivity of phase i
β_c	coefficient of solutal expansion
β_t	coefficient of thermal expansion
γ	melt–crystal interfacial energy
$\bar{\Gamma}$	melt–crystal capillary length, $\gamma/\Delta H_f$
δ	dimensionless stagnant-film thickness
Δ	dimensionless boundary layer thickness
η	boundary layer coordinate
λ	$\mathrm{Pe}_s/3\mathrm{Pe}_g$ (equation 26)
λ_0	reference value of λ
ξ	variable of integration
ν	momentum diffusivity, μ/ρ_m
μ	Newtonian viscosity of melt
Ω	characteristic rotation rate
ρ_0	reference density of melt
ρ_i	mass density of phase i
σ	interfacial tension of melt–ambient meniscus
σ^*	characteristic value of σ
σ_0	reference value of σ
τ	deviatoric stress tensor
ϕ	angle between melt–ambient and crystal–melt surfaces
ϕ_0	reference value of ϕ
0	refers to reference values when used as a subscript
~	refers to dimensional values of parameter when used over a variable

Acknowledgments

The writing of this paper and the research contributions of the crystal growth group at the Massachusetts Institute of Technology described within it were made possible by grants from the Microgravity Sciences Program of the National Aeronautics and Space Administration, the Defense Advanced Research Project Agency, and the Mobil Foundation and by a Teacher–Scholar Award from the Camille and Henry Dreyfus Foundation to me. I am deeply indebted to these sponsors. I am also grateful to the many colleagues who have contributed to this paper through their research and discussions with me; these include L. J. Atherton, S. R. Coriell, J. J. Derby, D.–H. Kim, G. B. McFadden, P. H. Sackinger, P. D. Thomas, C. A. Wang, M. J. Wargo, and A. F. Witt. I am is especially grateful to E. D. Bourret for her critical reading of the manuscript and to D. Maroudas who helped with the references.

References

1. Flemings, M. C. *Solidification Processing;* McGraw–Hill: New York, 1974.
2. Brice, J. C. *The Growth of Crystals from Liquids;* North-Holland: New York, 1976.
3. Rosenberger, F. *Fundamentals of Crystal Growth;* Springer–Verlag: New York, 1979.
4. Pimputkar, S. M.; Ostrach, S. *J. Cryst. Growth* **1981,** *55*, 614–640.
5. Brown, R. A. In *Advanced Crystal Growth;* Dryburgh, P., Ed.; Prentice Hall: Englewood Cliffs, NJ, 1987; pp 41–95.
6. Wang, C. A.; Ph. D. Thesis, Massachusetts Institute of Technology; Cambridge, 1984.
7. Parsey, J. M.; Thiel, G. A. *J. Cryst. Growth* **1985,** *73*, pp 211–220.
8. Gault, W. A.; Monberg, E. M.; Clemans, J. E. *J. Cryst. Growth* **1986,** *74*, pp 491–506.
9. Zulehner, W.; Huber, D. In *Crystals: Growth, Properties, and Applications;* Grabmaier J., Ed.; Springer–Verlag: New York, 1982; Vol. 8 pp 1–144.
10. Hurle, D. T. J. In *Crystal Growth of Electronic Materials;* E. Kaldis Ed.; Elsevier Science: New York, 1985a.
11. Lin, W.; Benson, K. E. *Ann. Rev. Mater. Sci.* **1987,** *17*, 293–298.
12. Pfann, W. G. *Zone Melting;* R. F. Krueger: Huntington, 1978.
13. Muhlbauer, A.; Keller, W. *Floating Zone Silicon;* Dekker: New York, 1981.
14. Ciszek, T. F. *J. Cryst. Growth* **1984,** *66*, 655–672.
15. Dietl, J. J.; Helmreich, D.; Sirtl, E. In *Crystals: Growth, Properties, and Applications;* Grabmaier, J., Ed.; Springer: Berlin, 1981; Vol. 5; pp 43–108.
16. Duranceau, J. L.; Brown R. A. *J. Cryst. Growth* **1986,** *75*, 367–389.
17. Brown, R. A. In *Materials Sciences in Space;* Feuerbacher, B.; Hamacher, H.; Naumann, R. J., Eds.; Springer–Verlag: Berlin, 1986: pp 55–94.
18. Langlois, W. E. *J. Cryst. Growth* **1982,** *56*, 15–19.
19. Jordan, A. S. *J. Appl. Phys.* **1985,** *71*, 551–558.
20. Kakimoto, K.; Hibiya T. *Appl. Phys. Lett.* **1987,** *50*, 1249–1250.
21. Tiller, W. A.; Jackson, K. A.; Rutter, J. W.; Chalmers, B. *Acta Metall.* **1953,** *1*, 428–437.
22. Smith, V. G.; Tiller, W. A.; Chalmers, B. *Can. J. Phys.* **1955,** *33*, 723–745.
23. Bourret, E. D.; Derby, J. J.; Brown, R. A. *J. Cryst. Growth* **1985,** *71*, 587–596.
24. Derby, J. J.; Brown, R. A. *Chem. Eng. Sci.* **1986,** *41*, 37–46.
25. Carruthers, J. R.; Witt, A. F. In *Crystal Growth and Characterization;* Ueda, R.; Mullins, J. B., Eds.; North Holland: Amsterdam, 1975.
26. Turner, J. S. *Buoyancy Effects in Fluids*; Cambridge University Press: Cambridge, 1973.
27. Hardy, S. *J. Cryst. Growth* **1984,** *69*, 456.
28. Sneyd, A. D. *J. Fluid Mech.* **1979,** *92*, 35–51.
29. Muhlbauer, A.; Erdmann, W.; Keller, W. *J. Cryst. Growth* **1983,** *64*, 529–545.
30. Pearson, J. R. A. *J. Fluid Mech.* **1958,** *4*, 489–500.

31. Scriven, L. E.; Sterling C. V. *J. Fluid Mech.* **1964,** *19,* 321–340.
32. Cormack, D. E.; Leal L. G., Imberger, J. *J. Fluid Mech.* **1974,** *65,* 209–229.
33. Hart, J. E. *Int. J. Heat Mass Transfer* **1983,** *26,* 1069–1074.
34. Hart, J. E. *J. Fluid Mech.* **1983,** *132,* 271–281.
35. Hurle, D. T. J.; Jakeman, E.; Johnson, C. P. *J. Fluid Mech.* **1974,** *64,* 565–576.
36. Sen, A. K.; Davis, S. H. *J. Fluid Mech.* **1982,** *121,* 163–186.
37. Smith, M. C.; Davis, S. H. *J. Fluid Mech.* **1982,** *121,* 187–206.
38. Smith, M. C.; Davis, S. H. *J. Fluid Mech.* **1983,** *132,* 119–144.
39. Smith, M. C.; Davis, S. H. *J. Fluid Mech.*, **1983,** *132,* 145–162.
40. Davis, S. H. *Annu. Rev. Fluid Mech.* **1987,** *19,* 403–435.
41. Homsy, G. M.; Meiburg, E. *J. Fluid Mech.* **1984,** *139,* 443–459.
42. Carpenter, B.; Homsy, G. M. *J. Fluid Mech.* **1985,** *155,* 429–439.
43. Hjellming, L. N.; Walker, J. S. *J. Cryst. Growth* **1988,** *87,* 18–32.
44. Hjellming, L. N.; Walker, J. S. *J. Cryst. Growth* **1987,** *85,* 25–31.
45. Langlois, W. E; Walker, J. S. In *Computational and Asymptotic Methods for Boundary and Interior Layers;* Miller, J. J. H., Ed.; 1982; 299–304.
46. Kim, K. M.; Witt, A. F.; Gatos, H. *J. Electrochem. Soc.* **1972,** *119,* 1218–1226.
47. Kim, K. M.; Witt, A. F.; Lichtensteiger, M.; Gatos, H. *J. Electrochem. Soc.* **1978,** *125,* 475–480.
48. Singh, R.; Witt A. F.; Gatos, H. C. *J. Electrochem. Soc.* **1968,** *115,* 112–113.
49. Witt, A. F.; Lichtensteiger, N.; Gatos, H. C. *J. Electrochem. Soc.* **1983,** *120,* 1119–1123.
50. Wargo, M. J.; Witt, A. F. *J. Cryst. Growth* **1984,** *66,* 289–298.
51. Preisser, G.; Schwabe, D.; Scharmann, A. *J. Fluid Mech.* **1982,** *126,* 545–567.
52. Carruthers, J. R. *J. Cryst. Growth* **1976,** *32,* 13–26.
53. Carruthers, J. R. In *Preparation and Properties of Solid State Materials;* Wilcox, W. R.; Lefever, R. A., Eds.; Dekker: New York, 1977; Vol. 3.
54. McLaughlin, J.; Orszag, S. A. *J. Fluid Mech.* **1982,** *123,* 123–142.
55. Curry, J. H.; Herring, J. R.; Loncaric, J.; Orszag, S. A. *J. Fluid Mech.* **1984,** *147,* 1–38.
56. Crochet, M. J.; Wouters, P. J.; Geyling, F. T.; Jordan, A. S. *J. Cryst. Growth* **1983,** *65,* 153–165.
57. Crochet, M. J.; Geyling, F. T.; Van Schaftinger, J. J. *Int. J. Numer. Methods Fluids* **1987,** *7,* 29–48.
58. Wouters, P.; Van Schaftingen, J. J.; Crochet, M. J.; Geyling, F. T. *Int. J. Numer. Methods Fluids* **1987,** *7,* 131–153.
59. Crochet, M. J.; Geyling, F. T.; Van Schaftingen J. J.; *Int. J. Numer. Methods Fluids* **1987,** *7,* 49–67.
60. Winters, K. H. In *Proc. Fifth Internat. Conf. Numer. Methods in Thermal Problems;* Pineridge: 1987.
61. Utech, H.; Flemings, M. C. *J. Appl. Phys.* **1966,** *37,* 2021–2024.
62. Kim, K. M. *J. Electrochem. Soc.* **1982,** *129,* 427–429.
63. Langlois, W. E. *J. Cryst. Growth* **1984,** *70,* 73–77.
64. Coriell, S. R; Sekerka, R. F. *J. Cryst. Growth* **1979,** *46,* 479–482.
65. Coriell, S. R; Boisvert, R. F.; Rehm, R. G.; Sekerka, R. F. *J. Cryst. Growth* **1981,** *54,* 167–175.
66. Harriott, G. M.; Brown, R. A.; *J. Cryst. Growth* **1984,** *69,* 589–604.
67. Scheil, E. Z. *Metall.* **1942,** *34,* 70.
68. Robertson, G. D., Jr.; O'Connor, D. J. *J. Cryst. Growth* **1986,** *76,* 111–122.
69. Carlberg, T.; King, T. B.; Witt, A. F. *J. Electrochem. Soc.* **1982,** *129,* 189–193.
70. Murgai, A. In *Crystal Growth of Electronic Materials;* Kaldis, E., Ed.; Elsevier: New York, 1985; pp 211–226.
71. Nernst, W. *Z. Phys. Chem.* **1904,** *47,* 52–55.
72. Wilcox, W. R. *Mater. Res. Bull.* **1969,** *4,* 265–274.

73. Burton, J. A.; Kolb, J. A.; Slichter, W. P.; Struthers, J. D. *J. Chem. Phys.* **1953,** *21*, 1981–1986.
74. Burton, J. A.; Prim, R. C.; Slichter, W. P. *J. Chem. Phys.* **1953,** *21*, 1987–1991.
75. Zief, M.; Wilcox, W. R. *Fractional Solidification;* Dekker: New York, 1967.
76. Adornato, P. M.; Brown, R. A. *J. Cryst. Growth* **1987,** *80*, 155–190.
77. Adornato, P. M.; Brown, R. A. *Inter. J. Numer. Methods Fluids* **1987,** *7*, 761–791.
78. Cochran, W. G.; *Proc. Cambridge Philos. Soc.* **1934,** *30*, 365–375.
79. Wilson, L. O.; *J. Cryst. Growth* **1978,** *44*, 247–250.
80. Wilson, L. O. *J. Cryst. Growth* **1978,** *44*, 371–376.
81. Levich, V. G. *Physicochemical Hydrodynamics;* Prentice Hall: Englewood Cliffs, NJ, 1962.
82. Pan, Y. F.; Acrivos, A. *Int. J. Heat Mass Trans.* **1968,** *11*, 439–444.
83. Camel, D.; Favier, J. J. *J. Phys.* **1986,** *47*, 1001–1014.
84. Hurle, D. T. J.; Series, R. W. *J. Cryst. Growth* **1985,** *73*, 1–9.
85. Series, R. W.; Hurle, D. T. J.; Barraclough, K. G. *IMA J. Appl. Math.* **1985,** *35*, 195–203.
86. Cartwright, R. A.; El-Kaddah, N.; Szekely, J. *IMA J. Appl. Math.* **1985,** *35*, 175–194.
87. Woodruff, D. P. *The Solid–Liquid Interface;* Cambridge Press: Cambridge, 1973.
88. Mullins, W. W.; Sekerka, R. F. *J. Appl. Phys.* **1964,** *34*, 323–329.
89. Mullins, W. W.; Sekerka, R. F. *J. Appl. Phys.* **1964,** *35*, 444–451.
90. Huang, S.-C.; Glicksman, M. E. *Acta Metall.* **1981,** *29*, 701–715.
91. Trivedi, R.; Somboonsuk, S. In *Proceedings of the Flat-Plate Solar Array Research Forum on the High Speed Growth and Characterization of Crystals for Solar Cells;* JPL Publication 84–23, 1984.
92. Ungar, L. H.; Brown, R. A. *Phys. Rev. B* **1984,** *31*, 1367–1380.
93. Ungar, L. H.; Brown, R. A. *Phys. Rev. B* **1985,** *31*, 5931–5940.
94. Bennett, M. J.; Brown, R. A.; Ungar, L. H. In *Proc. 1986 Inter. Symp. Phys. Structure Formation;* Springer–Verlag: Berlin, 1988; pp 180–190.
95. Brown, R. A. In *Supercomputer Research in Chemistry and Chemical Engineering;* Jensen, K. F.; Truhlar, D. G., Eds.; ACS Symposium Series 353; American Chemical Society: Washington, DC, 1987, pp 295–333.
96. Glicksman, M. E.; Coriell, S. R.; McFadden, G. B. *Annu. Rev. Fluid Mech.* **1986,** *18*, 307–335.
97. Coriell, S. R.; Cordes, M. R.; Boettinger, W. J.; Sekerka, R. F. *J. Cryst. Growth* **1980,** *49*, 13–28.
98. Billig, E. *Proc. Roy. Soc. (London)* **1956,** *A235*, 37–55.
99. Tsivinsky, S. V. *Kristall. Tech.* **1979,** *10*, 5–35.
100. Jordan, A. S.; Caruso, A. R.; Von Nieda, A. R. *Bell System Tech. J.* **1980,** *59*, 593–637.
101. Jordan, A. S.; Caruso, A. R.; Von Nieda, A. R.; Nielsen, J. W. *J. Appl. Phys.* **1981,** *52*, 3331–3336.
102. Duseaux, M. *J. Cryst. Growth* **1983,** *61*, 576–590.
103. Kobayashi, N.; Iwaki, T. *J. Cryst. Growth* **1985,** *73*, 96–110.
104. Mil'vidskii, M. G.; Osvenskii, V. B.; Shirfrin, S. S. *J. Cryst. Growth* **1981,** *52*, 396.
105. Fornari, R.; Paorici, C.; Zanotti, L.; Zuccalli, G. *J. Cryst. Growth* **1983,** *63*, 415–418.
106. McGuigan, S.; Thomas, R. N.; Barnett, D. L.; Hobgood, H. M.; Swanson, B. W. *Appl. Phys. Lett.* **1986,** *48*, 1377–1379.
107. Tabache, M. G.; Bourret, E. D.; Elliot, A. G. *Appl. Phys. Letts.* **1986,** *49*, 289–291.

108. Guruswamy, S.; Rai, R. S.; Faber, K. T.; Hirth, J. P. *J. Appl. Phys.* **1987,** *62,* 4130–4134.
109. Haasen, P. In *Dislocation Dynamics;* Rosenfield, A. R.; Hahn, G. T.; Bement, A. L.; Jaffee, I., Eds.; Battelle Institute Materials Colloquia: Columbus, 1967.
110. Alexander, H.; Haasen, P. *Solid State Phys.* **1968,** *22,* 28.
111. Myshlyaev, M.; Nikitenko, V. I.; Nestenenko, V. I. *Phys. Status Solidi* **1969,** *36,* 89–96.
112. Suezawa, M.; Sumino, K.; Yonenaga, I. *Phys. Status Solidi* **1979,** *51,* 217–226.
113. Schroter, W.; Brion, H. G.; Siethoff, H. *J. Appl. Phys.* **1983,** *54,* 1816–1820.
114. Lambropoulos, J. C. *J. Cryst. Growth* **1987,** *80,* 245–256.
115. Lambropoulos, J. C.; Hutchinson, J. W.; Bell, R. O; Chalmers, B.; Kalejs, J. P. *J. Cryst. Growth* **1983,** *65,* 324–330.
116. Dillion, O. W.; Tsai, C. T.; Angelis, R. J; *J. Cryst. Growth* **1987,** *82,* 50–59.
117. Sumino, K.; Harada, H.; Yonenaga, I. *Jpn. J. Appl. Phys.* **1980,** *19,* L49–52.
118. Ravi, K. V. *Imperfections and Impurities in Semiconductor Silicon;* Wiley: New York, 1981.
119. Chikawa, J.; Shirai, S. *J. Cryst. Growth* **1977,** *39,* 328–340.
120. Kuroda, E.; Kozuka, H. *J. Cryst. Growth* **1983,** *63,* 276–284.
121. Chang, C. E.; Wilcox, W. R. *J. Cryst. Growth* **1974,** *21,* 135–140.
122. Naumann, R. J. *J. Cryst. Growth* **1983,** *58,* 554–568.
123. Jasinski, T. J.; Rohsensow, W. M.; Witt, A. F. *J. Cryst. Growth* **1983,** *61,* 339–354.
124. Wang, C. A.; Witt, A. F.; Carruthers, J. R. *J. Cryst. Growth* **1984,** *66,* 299–308.
125. Naumann, R. J.; Lehoczky, S. L. *J. Cryst. Growth* **1983,** *61,* 707–710.
126. Jasinski, T. J.; Witt, A. F. *J. Cryst. Growth* **1985,** *71,* 295–304.
127. Chang, C. J.; Brown, R. A. *J. Cryst. Growth* **1983,** *63,* 343–364.
128. Haynes, J. M. In *Materials Sciences in Space;* Feuerbacher, B.; Hamacher, H.; Naumann, R. J., Eds.; Springer–Verlag: Berlin, 1986; pp 129–146.
129. Rousaud, A.; Camel, D.; Favier, J. J. *J. Cryst. Growth* **1985,** *73,* 149–166.
130. Acrivos, A. *Chem. Eng. Sci.* **1966,** *21,* 343–352.
131. Oreper, G. M.; Szekely, J. *J. Cryst. Growth* **1983,** *63,* 505–515.
132. Oreper, G. M.; Szekely, J. *J. Cryst. Growth* **1984,** *67,* 405–419.
133. Kim, D.-H.; Adornato, P. M.; Brown, R. A *J. Cryst. Growth* **1988,** *89,* 339–356.
134. Hjellming, L. N.; Walker, J. S. *J. Fluid Mech.* **1986,** *164,* 237–273.
135. Ono, H.; Watanabe, H.; Kamejima, T.; Matsui, J. *J. Cryst. Growth* **1986,** *74,* 446–452.
136. Herring, C. In *The Physics of Powder Metallurgy;* Kingston, W. E., Ed.; McGraw–Hill: New York, 1951.
137. Bardsley, W.; Frank, F. C.; Green, G. W.; Hurle, D. T. J. *J. Cryst. Growth* **1974,** *23,* 341–344.
138. Rowlinson, J. S.; Widom, B. *Molecular Theory of Capillarity;* Clarendon: Oxford, 1982.
139. Surek, T.; Chalmers, B. *J. Cryst. Growth* **1975,** *29,* 1–11.
140. Huh, C.; Scriven, L. E. *J. Colloid. Inter. Sci.* **1969,** *30,* 323–337.
141. Padday, J. F. *Philos. Trans. R. Soc. Lond.* **1971,** *269,* 265–293.
142. Hurle, D. T. J. *J. Cryst. Growth* **1983,** *63,* 13–17.
143. Kobayashi, N. In *Preparation and Properties of Solid State Materials;* Wilcox, W. R., Ed.; Dekker: New York, 1981; Vol. 6.
144. Derby, J. J.; Brown, R. A. *J. Cryst. Growth* **1986,** *74,* 605–624.
145. Williams, G.; Reusser, W. E. *J. Cryst. Growth* **1983,** *64,* 448–460.
146. Dupret, F.; Ryckmans, Y.; Wouters, P.; Crochet, M. J. In *Crystal Growth;* Cockayne, B.; Hogg, J. H. C; Lunn, B.; Wright, P. J., Eds.; North Holland: New York, 1986; pp 84–91.

147. Kim, K. M.; Kran, A.; Smetana, P.; Schwuttke, G. H. *J. Electrochem. Soc.* **1983,** *130*, 1156–1160.
148. Crowley, A. B. *IMA J. Appl. Math.* **1983,** *147*, 173–189.
149. Derby, J. J.; Brown, R. A. *J. Cryst. Growth* **1986,** *75*, 227–240.
150. Derby, J. J.; Brown, R. A. *J. Cryst. Growth* **1987,** *83*, 137–151.
151. Derby, J. J.; Brown, R. A. *J. Cryst. Growth* **1988,** *87*, 251–260.
152. Derby, J. J.; Atherton, L. J.; Thomas, P. D.; Brown, R. A. *J. Sci. Comput.* **1988,** *2*, 297–343.
153. Atherton, L. J.; Derby, J. J., Brown, R. A. *J. Cryst. Growth* **1987,** *84*, 57–78.
154. Sristava, R. K.; Ramachandran, P. A.; Dudokovic, M. P. *J. Cryst. Growth* **1985,** *73*, 487–504.
155. Ramachandran, P. A.; Dudokovic, M. P. *J. Cryst. Growth* **1985,** *71*, 399–408.
156. Hurle, D. T. J. *J. Cryst. Growth* **1977,** *42*, 473–482.
157. Bardsley, W.; Hurle, D. T. J.; Joyce, G. C. *J. Cryst. Growth* **1974,** *40*, 13–20.
158. Bardsley, W.; Hurle, D. T. J; Joyce, G. C. *J. Cryst. Growth* **1977,** *40*, 21–28.
159. Steel, G. K.; Hill, M. J. *J. Cryst. Growth* **1975,** *30*, 45–53.
160. Hurle, D. T. J.; Joyce, G. C.; Wilson, G. C.; Ghassempoory, M.; Morgan, C. *J. Cryst. Growth* **1986,** *74*, 480–490.
161. Mika, K.; Uehoff, W. *J. Cryst. Growth* **1975,** *30*, 9–20.
162. Boucher, E. A.; Kent, H. J. *Proc. R. Soc. London* A**1977,** *356*, 61–75.
163. Surek, T. *J. Appl. Phys.* **1976,** *47*, 4384–4393.
164. Surek, T.; Coriell, S. R.; Chalmers, B. *J. Cryst. Growth* **1980,** *50*, 21–31.
165. Tatarchenko, V. A.; Brener, E. A. *J. Cryst. Growth* **1980,** *50*, 33–44.
166. Robertson, D. S.; Young, I. M. *J. Phys. D: Appl. Phys.* **1975,** *8*, L59–L61.
167. Stephanopoulos, G. *Chemical Process Control* Prentice–Hall: Englewood Cliffs, NJ, 1984.
168. Chang, H.; Chen, L. *Chem. Eng. Sci.* **1984,** *39*, 1127–1142.
169. Thomas, P. J.; Brown, R. A.; Derby, J. J.; Atherton, L. J.; Wargo, M. J. *J. Cryst. Growth*, in press, 1989.
170. Langlois, W. E. *Annu. Rev. Fluid Mech* **1985,** *17*, 191–215.
171. Kobayashi, N. *J. Cryst. Growth* **1978,** *43*, 357–363.
172. Tangborn, Ph.D. Thesis, Massachusetts Institute of Technology; 1987, Cambridge.
173. Patera, A. T., Fall Meeting of the Society of Industrial and Applied Mathematics, Tempe, 1985.
174. Mihelcic, M.; Winegrath, K. *J. Cryst. Growth* **1985,** *71*, 163.
175. Sackinger, P. A.; Derby, J. J.; Brown, R. A. *Inter. J. Numer. Methods Fluids*, in press.
176. Mihelcic, M.; Winegrath, K.; Pirron, C. *J. Cryst. Growth* **1984,** *69*, 473–488.
177. Wilson, L. O. *J. Cryst. Growth* **1980,** *48*, 435–450.
178. Wilson, L. O. *J. Cryst. Growth* **1980,** *48*, 451–458.
179. Hurle, D. T. J.; Jakeman, E.; Pike, E. R. *J. Cryst. Growth* **1968,** *3*, 4, 633–640.
180. Cartwright, R. A.; Hurle, D. T. J.; Series, R. W.; Szekely, J. *J. Cryst. Growth* **1987,** *82*, 327–334.
181. Lee, K. J.; Langlois, W. E.; Kim, K. M. *Physicochem. Hydrodyn.* **1984,** *5*, 135–141.
182. Hoshikawa, I.; Kohda, H.; Nakanishi, H.; Ikuta, K. *Jpn. J. Appl. Phys.* **1980,** *19*, L33–L36.
183. Kim, K. M.; Langlois, W. E. *J. Electrochem. Soc.* **1986,** *133*, 2586–2590.

RECEIVED for review October 14, 1988. ACCEPTED revised manuscript February 6, 1989.

3

Liquid-Phase Epitaxy and Phase Diagrams of Compound Semiconductors

Timothy J. Anderson

Department of Chemical Engineering, University of Florida, Gainesville, FL 32611

The fundamentals of liquid-phase epitaxy of compound semiconductors are presented. Selected topics associated with the chemical processing of semiconductors by this technique are discussed. These topics include compositional variations, growth mechanisms, and the initial stages of epitaxy. A formalism for the treatment of solid–liquid phase equilibrium in multicomponent compound-semiconductor systems is given. Procedures for interpolating, extrapolating, and predicting multicomponent phase diagrams are suggested.

EPITAXY IS THE PROCESS OF GROWING regularly oriented, thin films on a substrate. The goals of epitaxy are to produce perfectly crystalline films with control of composition, thickness, and purity and to reproducibly repeat the process on either a new substrate or a previously deposited film. Several processes for epitaxial film growth are available, including liquid-phase epitaxy (LPE), chemical vapor deposition (CVD), molecular-beam epitaxy (MBE), plasma deposition, and ion-cluster-beam deposition. These deposition technologies are at different stages of development and offer certain or potential advantages for the fabrication of solid-state devices.

Advantages and Limitations of Liquid-Phase Epitaxy

Liquid-phase epitaxy is the process of growing films from a liquid solution and is an attractive method of film growth for several reasons. LPE was the

0065–2393/89/0221–0105$15.90/0

first process broadly developed for the growth of compound-semiconductor materials and has proven to be a reliable process. Compared with other methods, LPE is a simple process and thus requires a short downtime and demands less operator skill. The required investment in capital equipment is modest, and the operating expenses are minimal relative to other growth technologies.

In LPE, film growth occurs very near the equilibrium state, and thus the technique is reproducible and gives films with low concentrations of growth-induced defects. The starting materials are generally in elemental form, and their availability in high purity translates into low levels of unintentional doping. Impurity levels are further decreased by the tendency of most contaminants to segregate into the melt. Furthermore, the liquid-metal solvents used in LPE can dissolve most elements to give a large selection of dopants.

The high growth rate that can be achieved by LPE reduces growth time for thick-film applications. In addition, processing from the melt tends to equalize the distribution coefficients of the liquid constituents. In contrast, large differences in distribution coefficients can exist between the transportable vapor species used in other processes. For example, solid solutions containing Al mixed with Ga or In are easy to grow by LPE but are difficult to grow with near-equilibrium CVD techniques. Relative to some alternative growth methods, safety considerations for LPE are less restrictive, with hydrogen handling being the most hazardous operation. The success of LPE is evident from its prominence in the production of commercial devices, particularly for optoelectronic applications. For many device structures, the highest device performance characteristics have been achieved in devices produced by LPE.

Although the advantages of LPE are considerable, this process has several serious drawbacks. LPE is inherently a batch process and is usually scaled up by the simple duplication of systems. Single-wafer batch processing and long baking times allow low wafer throughput compared with the throughput realized with Si CVD technology. Great care must be taken to yield acceptable surface morphologies, because the process is often plagued by surface defects such as incomplete melt removal, terrace formation, pin holes, and meniscus lines. Thickness uniformity can be poor as a result of natural and forced convection in the melt and variations in substrate temperature. Because of high initial growth rates, film thicknesses of less than several hundred angstroms or within tight thickness tolerances are extremely difficult to achieve.

During the growth of solid solutions (e.g., pseudobinary alloys and doped compounds), the melt composition at the interface can be a function of time, a situation that results in a natural composition gradient in the growth direction. Heteroepitaxy, the growth of an epitaxial film with a composition different from that of the substrate, is more difficult compared with other techniques because of meltback problems. For example, it is difficult to grow a layer of InP on $In_xGa_{1-x}As$ with an abrupt interface because of the limited transport of As and Ga across the interface. Despite these limitations, liquid-phase epitaxy is still commonly used to grow epitaxial films of compound semiconductors.

Several excellent reviews of liquid-phase epitaxy have appeared in the literature over the past 15 years (*1–12*). The discussion in this chapter will be limited in scope but will supplement the material discussed in previous reviews. In particular, issues that can be analyzed by traditional methods of chemical engineering are addressed for this chemical process. Because the growing solid–liquid interface is near equilibrium, the calculation of multicomponent compound-semiconductor phase diagrams will be emphasized.

LPE Growth Techniques

The LPE growth system must provide a driving force for nucleation of the solid film, a physical means of contacting a liquid solution to the substrate, an efficient way of removing the solution from the substrate after growth, and, for most device applications, a method of repeating the growth process from a solution of different composition.

Nucleation. The driving force must decrease the Gibbs energy of the liquid solution near the substrate to a value sufficiently low to produce nucleation on the substrate but not so low that homogeneous or heterogeneous nucleation on nonsubstrate solid surfaces occurs. Thus a small window in the values of the solution Gibbs energy exists in which successful growth can take place.

The upper limit of the Gibbs energy is determined primarily by substrate properties (i.e., orientation, impurity content, crystallinity, and composition), whereas the lower limit is fixed by the properties of the liquid solution and the materials of construction. The solution Gibbs energy is a function of temperature, of composition, and, weakly, of pressure. Most LPE techniques are driven by temperature decreases caused by precooling of the substrate, continuous or step cooling of the solution, Peltier cooling, or the establishment of a temperature gradient between the substrate and a constant-composition source. Electroepitaxy, in which a large current is passed normal to the substrate, represents an example in which the liquid composition near the solid–liquid interface is changed by electromigration, in addition to Peltier cooling. In response to a driving force, the system adjusts the value of the liquid Gibbs energy by changing the composition through deposition. In heteroepitaxy, the required driving force during the initial stages of epitaxy is clearly different from that required during subsequent growth, which is essentially homoepitaxial deposition.

Contact between Solution and Substrate. The physical processes used to contact the liquid solution to the substrate can be classified broadly into tipping, dipping, or sliding methods (Figure 1).

A tipping procedure was first suggested by Nelson (*13*). In this method, a liquid solution is successfully brought into and removed from contact with the substrate by either tipping a horizontal boat (Figure 1a) or rotating a cylinder (Figure 1b) containing a fixed substrate and the solution. Tipping techniques are simple, but melt removal is difficult if the technique relies on gravity. The early versions of these designs could not grow multiple layers and gave poor thickness uniformity.

Both horizontal (Figure 1c) and vertical (Figure 1d) dipping processes have been investigated, but again, these techniques rely on gravity for melt removal. To improve thickness uniformity, dipping techniques are often accompanied by mechanical stirring.

In modern LPE systems, film growth is almost exclusively performed in a linear sliding-boat design (Figure 1e). The sliding-boat technique has evolved over the past 20 years, with many variations existing in different laboratories. The basic design consists of two main pieces (Figure 1e): a

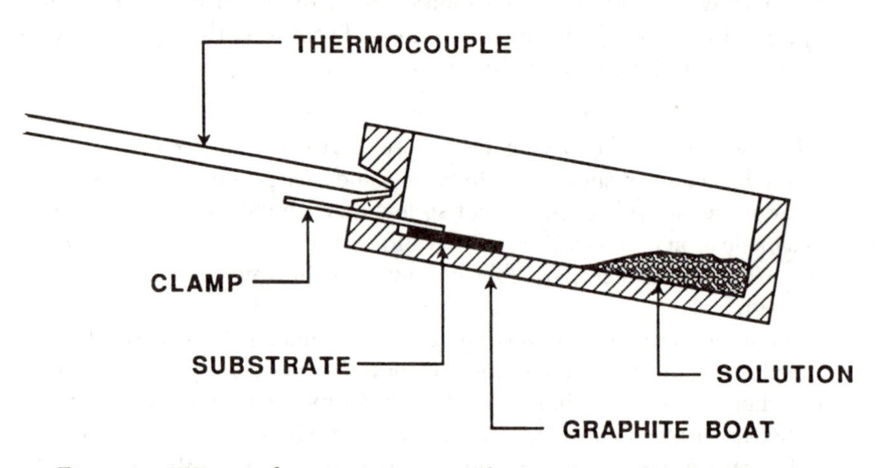

Figure 1a. LPE growth systems: tipping technique. (Reproduced with permission from reference 13. Copyright 1963 RCA Research and Engineering.)

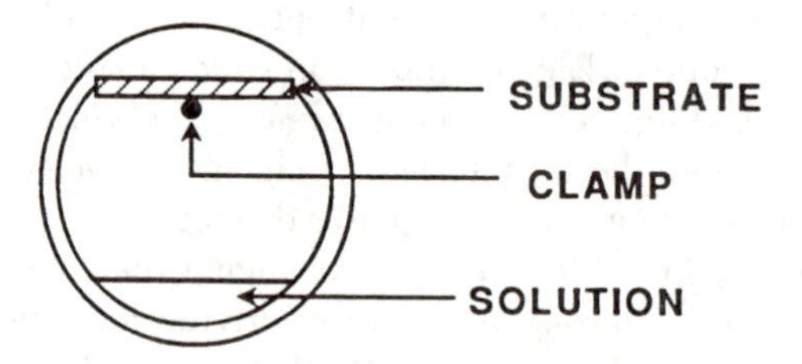

Figure 1b. LPE growth systems: rotary method.

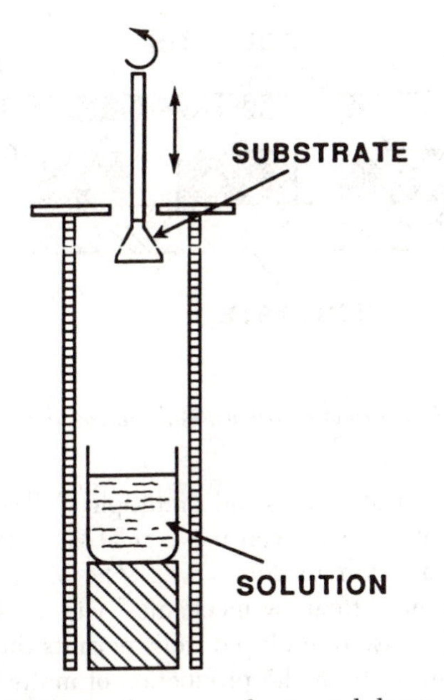

Figure 1c. LPE growth systems: horizontal dipping method.

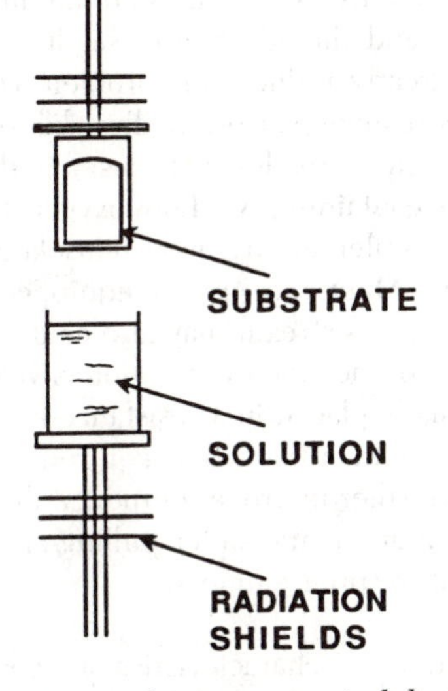

Figure 1d. LPE growth systems: vertical dipping method.

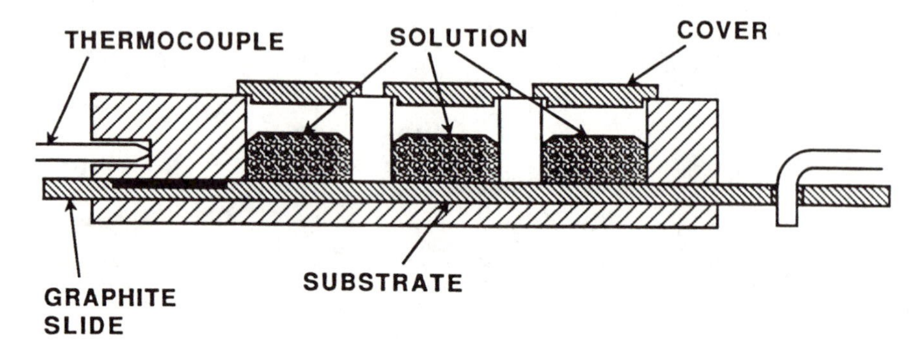

Figure 1e. LPE growth systems: multibin sliding-boat method.

slider plate with a substrate recess and a multiple-well assembly. A substrate fitted in the slider plate recess can move relative to the well assembly to place the substrate beneath a solution well. The pieces and recess are machined to fine tolerances to allow mechanical wiping of the melt from the film after growth. The use of multiple wells permits the growth of multiple layers on a single substrate or the production of multiple substrates.

The slider boat is usually constructed of graphite, although alternative materials such as boron nitride and silica (*14*) have been used. Graphite is readily obtained in high purity, easily machined, relatively inert with respect to the solution, and nearly frictionless in operation. Graphite boats do have a limited lifetime, and care must be exercised in desorbing gases contained in this relatively porous material.

The slider boat assembly is normally surrounded by a fused silica tube. Reduction of the silica tube by hydrogen can unintentionally dope the semiconductor with Si (*15*), and the addition of small amounts of Cl (as metal chloride or HCl) can greatly reduce this problem (*16*). The slider boat is blanketed in a flowing stream of palladium-alloy-diffused hydrogen. Extreme care must be used to minimize the level of oxygen in the system, particularly for the growth of films containing Al. Low oxygen levels are achieved by prebaking the system, implementing good housekeeping procedures, and using purified hydrogen. Many systems are equipped with a hygrometer to measure H_2O levels. The gas stream may also contain desired dopants (*17*) or major species used to reduce the evaporation of volatile components (*18*). The arrangement is usually placed in a resistance-heated furnace, and temperature control to better than 0.1 °C is required to achieve reproducible growth thicknesses. To further improve reproducibility, modern systems are generally equipped with automatic slider pulling-and-pushing mechanisms and automatic temperature programmers.

Solvent Selection. The characteristics of a good solvent include low vapor pressure (<1.0 Pa), low cost, availability in high purity, low solubility in the film, compatibility with the materials of construction, and low toxicity. A small slope of the liquidus line is desirable for improved thickness control. A moderate solubility of the solute (film material) allows the economical use of the materials. For effective melt removal, low values of surface tension, kinematic viscosity, and eutectic temperatures or other solid-transformation temperatures are helpful.

For the growth of compound semiconductors, the solvent is usually the constituent element of the compound with the lowest vapor pressure and distribution coefficient. For example, Ga is usually the solvent for the growth of $Al_xGa_{1-x}As$. Although both Al and Ga satisfy the low-vapor-pressure requirement, the distribution coefficient of Ga is considerably smaller than that of Al. Other solvents can be used to alter the properties of the melt. For example, Panek et al. (*19*) performed the growth of GaAs from a solution containing the isoelectronic element Bi. They reported a significant increase in growth rate, probably because of a change in the slope of the liquidus line.

The growth of group III–V semiconductors from solutions containing large amounts of group IV elements has also been investigated (*20–22*). These elements are amphoteric dopants in group III–V compounds, and relatively large doping levels can be achieved at high solvent concentrations. The use of a group IV element as a solvent strongly influences the shape of the phase diagram. For example, changes in the slope of the liquidus altered the growth rate of GaAs (*20*) and increased P solubility, which in turn increased the wettability of the difficult-to-grow (111) InP surface during InP growth (*22*). However, the use of alternative solvents has received minimal attention. The optimal solvent composition should depend on the desired film characteristics.

Growth Procedures. The exact procedures for the growth of device-quality epitaxial films by LPE vary with the deposited material, the design of the growth equipment, and the structure to be fabricated. The general procedure involves preparation of the substrate, loading of the melt constituents and pretreatment of the system, and growth.

Substrate Preparation. Substrates are commercially available for most binary compound semiconductors, principally the group III–V gallium and indium compounds and tellurides. The materials GaAs, InP, and CdTe are produced in the largest volume. In recent years, compound semiconductors have been grown successfully on noncompound-semiconductor substrates, particularly Si substrates. The successful deposition techniques are those that operate at conditions that allow a large departure from equilibrium at the growing interface (i.e., MBE and metallorganic chemical vapor deposition [MOCVD]). LPE is a near-equilibrium process, and this fact makes heteroepitaxy difficult.

The growth of large single crystals of compound semiconductors is accomplished by many techniques, including the Czochralski process and the horizontal and vertical Bridgman processes (*23*). After crystal growth, the orientation of the boule is obtained by chemical etching or X-ray measurement (*24*). Internal saws are used to slice into wafers an ingot that is attached to a mounting block with wax. The wafers are then ground to the same thickness. The slicing and grinding procedures produce considerable waste and result in work damage to a depth of 50–80 μm.

This work damage is usually removed by fast chemical etching. The etching step is accomplished by the reaction of the semiconductor with the etchant to produce soluble oxides. The etching rate may be limited by either a surface reaction or mass transfer. For surface-reaction-limited etchants, the etching rate is a strong function of temperature and orientation. Preferential etching along dislocation lines can give rise to etch pits. For mass-transfer-limited etchants, the stirring conditions in the liquid etchant must be controlled to avoid nonuniform etching rates.

The final step is a chemical or chemomechanical polishing procedure. Orange-peel textures that are generally present on chemically polished surfaces have been eliminated successfully by using a hydroplaning chemomechanical polishing step (*25*). Particularly important to LPE is control of the wafer thickness to a tolerance of <25 μm. Several excellent review articles have been published on the etching of semiconductor surfaces (*26–28*). Commercial wafers are generally at this stage of processing. The quality of compound-semiconductor wafers in terms of defect density, radial homogeneity, and maximum diameter is not as high as that available for Si wafers, although the quality has improved tremendously during the last few years.

One of the most important steps in the growth of device quality layers by LPE is substrate preparation. The exact substrate preparation procedure varies with material and user, but usually includes a degreasing step (e.g., sequential rinsing with trichloroethylene, acetone, and methanol in an ultrasonic bath, followed by drying in an inert gas stream) and a short isotropic etch to remove surface oxides. The recent literature on MBE sample preparation should be consulted regularly, because the effectiveness of substrate preparation procedures can be monitored by surface analytical tools installed in MBE systems. A large body of literature is available on the polishing of GaAs (*29–44*), InP (*44–49*), and other compound semiconductors (*50–52*).

Predeposition Procedures. When the substrate surface has been prepared, the substrate is loaded into the recessed area of the slider. At this time, the melt constituents are loaded into the bins in the slider boat. Dopants, volatile components, and elements with large distribution coefficients are required in small quantities and must be carefully weighed.

Effect of Baking. Baking the system at elevated temperatures with flowing purified hydrogen greatly reduces the concentration of impurities incorporated in the grown layers (53–57). The baking process supposedly removes impurities absorbed on the walls of the materials of construction and purifies the liquid solution by simple evaporation of volatile species or reduction by hydrogen.

Shown in Figure 2 are the electron donor and acceptor concentrations as a function of baking time at 800 °C for the growth of GaAs by LPE (57). The level of unintentional impurity dramatically decreases with increasing baking time. The residual acceptor and donor were attributed to C and Si, respectively.

Similar results are shown in Figure 3 for the growth of $In_xGa_{1-x}As$ layers on InP (55). The measured carrier concentration at room temperature decreases with increased baking time for baking times up to 50 h.

Purification of Solution. An approximate model for the purification of the solution can be developed by assuming that a stagnant melt initially contains an impurity at a uniform concentration, C_i^o, and loss of dopant occurs by evaporation at the top surface. The rate of evaporation is assumed to be directly proportional to the difference in concentrations at the top surface and at equilibrium. If the proportionality constant is z, the diffusion coefficient of the impurity in the melt is D_i, and the depth of the melt is l, then the following expression for the impurity concentration in the melt,

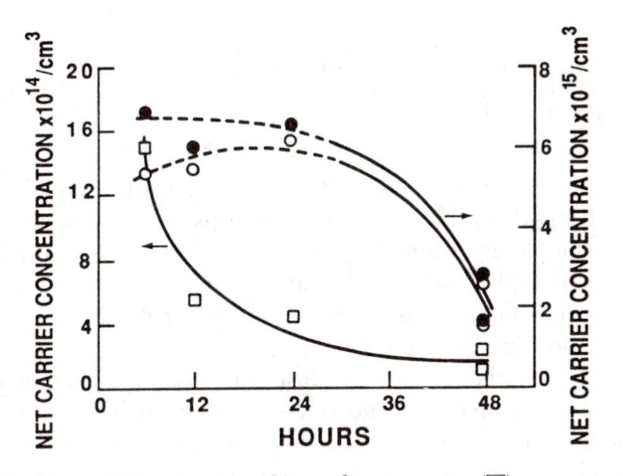

Figure 2. Donor (●), acceptor (○), and net carrier (□) concentrations as a function of baking time at 800 °C for growth of GaAs. (Reproduced with permission from reference 57. Copyright 1984 American Institute of Physics.)

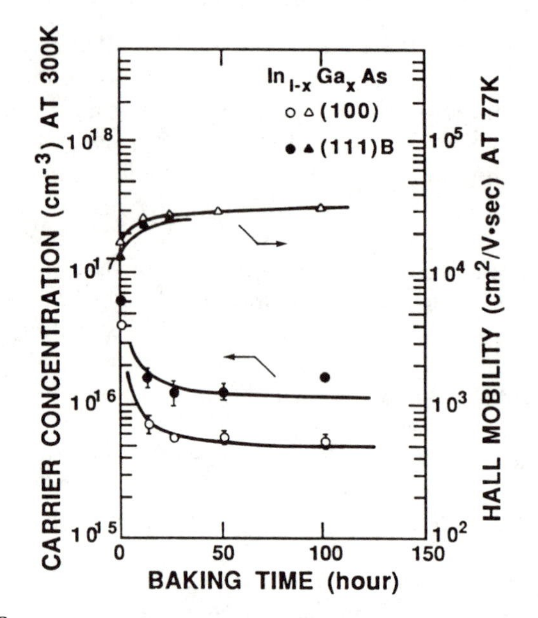

Figure 3. Room-temperature carrier concentration and Hall mobility at 77 K of (100) and (111) B $In_{1-x}Ga_xAs$ LPE layers grown from solutions baked at 670 °C (Reproduced with permission from reference 55. Copyright 1982 Elsevier.)

C_i^l, as a function of time (t) at the solid–liquid interface is obtained (58)

$$\frac{C_i^l}{C_i^o} = \operatorname{erf}\left(\frac{l}{2\sqrt{D_i t}}\right) - \exp\left(\frac{z(l + zt)}{D_i}\right) \operatorname{erfc}\left(\frac{l}{2\sqrt{D_i t}} + z\sqrt{\frac{t}{D_i}}\right) \quad (1)$$

in which erf is error function and erfc is error function complement. If the rate of the first-order surface process is high (large value of z), equation 1 reduces to a simple error function distribution with a characteristic time of $l^2/4D_i$. By using typical values for l and D_i, the time required for 84.3% of the impurity to be depleted is about 7 h, a value that is consistent with the results shown in Figures 2 and 3.

This model is rather simple, because it neglects possible mixing effects caused by natural convection and convection forced by H_2 flow or slider motion and the dependence of impurity diffusion coefficients on the concentrations of other impurities present in the melt. The exact mechanism by which baking influences the concentration of trace impurities is not well understood. However, the use of a prebaking step is considered necessary to achieve high-purity film growth by LPE.

The time that the substrate and melt are at an elevated temperature before growth is initiated can be sufficient to allow pitting of the substrate by incongruent evaporation of the substrate. Incongruent evaporation of phosphorus from InP substrates begins at 300 °C in vacuum and adversely affects the defect density, surface morphology, and local alloy composition of epitaxial films (*59–61*). Above the incongruent-evaporation temperature, the flux of phosphorus is greater than that of In. After sufficient excess phosphorus has evaporated, nucleation of P-saturated In droplets takes place on the surface. The P evaporation rate from these droplets is higher than that from the solid surface and results in the formation of pits.

Chu et al. (*62*) examined the morphological evolution of InP surfaces in flowing H_2 under a microscope with a hot stage. The temperature required to nucleate In droplets on (100) InP is shown in Figure 4 as end points of heating curves. Nucleation of decomposition pits did not occur until the temperature exceeded 600 °C in these short-time studies. In these experiments, the rate of thermal decomposition was limited by mass transfer, as indicated by the dependence on hydrogen flow rate of the time to nucleation. These results suggest that phosphorus evaporation can be reduced by locating the substrate in an enclosed portion of the boat, in which convective transport of phosphorus in the flowing H_2 ambient cannot occur. Another effective technique to reduce this problem is to use a GaAs or InP cover piece to shield the substrate (*63*). A localized P ambient generated from

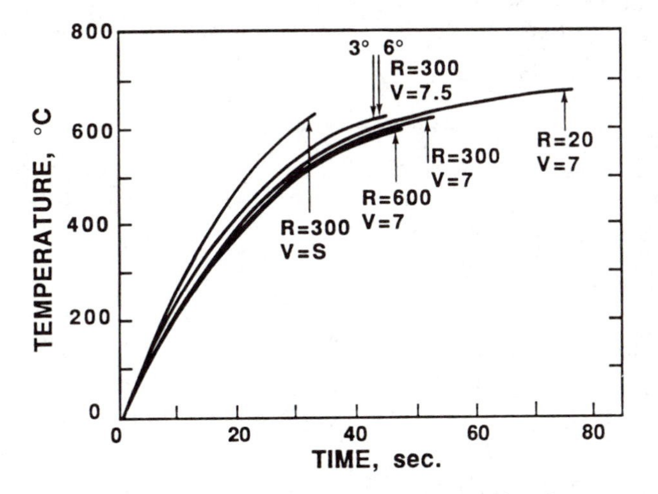

Figure 4. Heating curves for (100) and 3°-misoriented and 6°-misoriented (100) InP surfaces at different hydrogen flow rates and heating rates. The time and temperature required for the nucleation of saturated indium droplets is indicated by arrows. R *is the* H_2 *flow rate (expressed in standard cubic centimeters per minute [sccm]), and* V *is the heating voltage (expressed in volts; direct current). (Reproduced with permission from reference 62. Copyright 1983 The Electrochemical Society.)*

elemental phosphorus (*64*), phosphine (*65, 66*), or Sn–InP porous container over the substrate can also minimize evaporative losses (*67–69*). The Sn–InP solution can establish a sufficient P overpressure to preserve an InP substrate and will not require an intentional meltback of the substrate. Intentional meltback procedures can lead to poor surface quality because of morphological instabilities occurring during dissolution (*70*).

In addition to thermal decomposition of the substrate, holding the system at elevated temperatures for long times can result in volatilization of desired constituents. As an example, Figure 5 shows the amount of phosphorus loss from In–P solutions in terms of a decrease in the liquidus temperature as a function of baking time at 670 °C (*55*). This level of evaporative loss is significant and must be accurately accounted for to control subsequent growth. In addition, the evaporating species can be transported downstream to other bins and alter the composition of these melts.

Deposition Procedures. A schematic of a typical growth sequence is shown in Figure 6. After baking the system (step a), the system temperature is adjusted for growth and the substrate is brought into contact with the melt in the first bin (step b). In normal LPE practice, supersaturation of the liquid solution is achieved by temperature adjustment. The initial temperature can be set above, at, or below the liquidus value. If meltback is desired, a small increase in temperature above saturation is fixed first. The subsequent rate of etching decreases with time as the composition locally adjusts towards the equilibrium value.

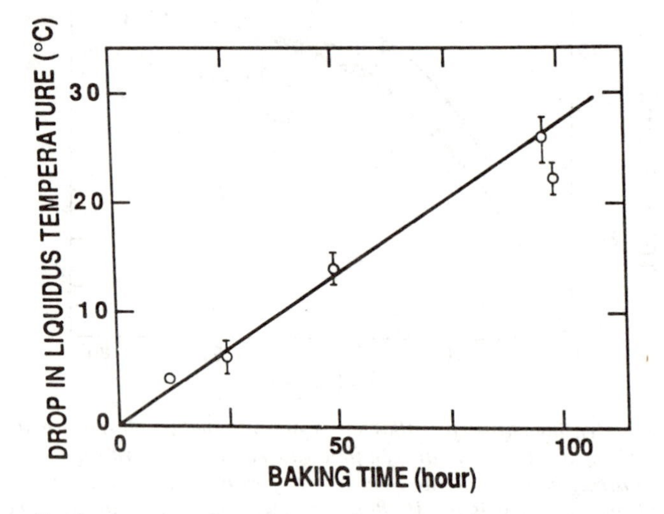

Figure 5. The amount of P evaporated, expressed as a drop in saturation temperature, from In–P solutions after a baking process at 670 °C. (Reproduced with permission from reference 55. Copyright 1982 Elsevier.)

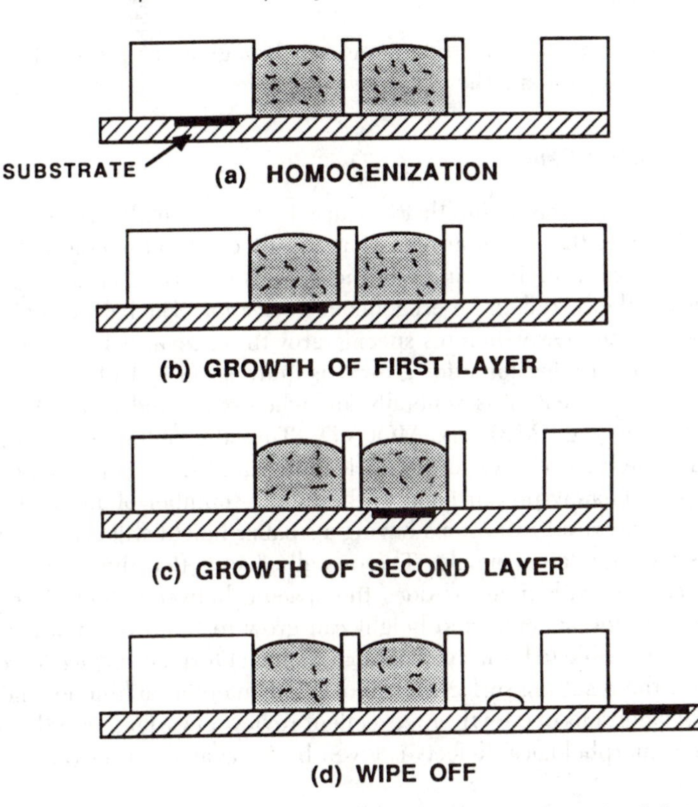

Figure 6. Schematic of a typical LPE growth sequence using the sliding-boat method.

If the temperature is initially fixed at the value of the liquidus temperature, cooling of the system must be programmed (~0.1–0.5 °C/min) to effect growth. This process is termed *equilibrium cooling*. If the initial temperature is sufficiently below the equilibrium value to produce heterogeneous nucleation on the single-crystal substrate, no further cooling is required to achieve finite growth. An isothermal growth process results and is termed *step cooling*. Application of an initial step cooling followed by ramp or equilibrium cooling is termed *supercooling*.

Growth is terminated by sliding the substrate from underneath the melt and using the well wall to wipe off excess melt. If another film of a different doping level or bulk composition is desired, the substrate is brought into contact with an adjacent well containing the next solution (step c). The system temperature can now be several degrees below the initial homogenization temperature, and to prevent homogeneous nucleation in these subsequent melts, the composition of the solution should have been previously adjusted to produce a liquidus temperature in the vicinity of the actual temperature and a solidus composition appropriate for the device application. After the final layer is grown, the substrate is moved through an empty bin for final melt removal (step d), and the system is cooled slowly to room temperature.

Growth Mechanisms

The growth of an epitaxial film that is nearly lattice matched to the substrate typically involves the lateral growth of atoms attached to the edges of steps on the crystal surface. The source of these steps may be dislocations, misorientation of the crystal, or heterogeneous nucleation. The balance between these different sources generates specific growth mechanisms. Because the thermodynamic driving force for growth is quite small in LPE, the rate of heterogeneous nucleation is generally low when compared with other epitaxial processes (e.g., MBE and MOCVD). Thus the lateral growth rate is expected to be high relative to the nucleation rate. The lateral velocity of a step is approximately proportional to the height (number of atomic layers) of the step. As a result, the steps having a smaller height will catch up with the steps having a larger height. The overall effect is that the average step height increases with time, as does the spacing between steps. For long growth times, the average step height can grow to the level at which the steps become visible to form a morphological defect termed *surface terraces*. Obviously, the resulting surface morphology of an epitaxial film depends on the initial conditions of growth. An excellent review of various growth mechanisms and morphological defects is given by Benz and Bauser (*71*).

Homogeneous Nucleation. Homogeneous nucleation in the melt must be avoided, because the growth of nuclei produces local compositional variations in the melt, which translate into local variations in growth rate. According to classical nucleation theory, spontaneous fluctuations of atomic configurations serve to form nuclei of the solid phase. The difference between the molar Gibbs energy of the bulk solid and that of the liquid solution, ΔG_B, is negative if the melt temperature is below the equilibrium value and provides a driving force for continued growth. However, most of the atoms in a small embryo are not as fully bonded as they might be in the bulk crystal, and these atoms make a bigger contribution to the Gibbs energy of the system. The total Gibbs energy of formation for a nucleus, ΔG_f, is the sum of the contributions from the bulk and surface atoms and is a function of the particle volume and shape:

$$\Delta G_f = \frac{V\rho\Delta G_B}{M} + A\sigma \tag{2}$$

In equation 2, V is the volume of the nucleus, ρ is the density of the solid, M is the molecular weight of the solid, A is the surface area of the nucleus, and σ is the surface Gibbs energy per unit area.

The value of ΔG_f shows a positive maximum as the size of the nucleus increases. For a spherical nucleus, the maximum value of ΔG_f (ΔG_f^*) is given by:

$$\Delta G_f^* = \frac{16\pi\sigma^3 M^2}{3(\rho \Delta G_B)^2} \tag{3}$$

At ΔG_f^*, the critical nucleus radius is given by:

$$r^* = \frac{-2\sigma M}{\rho \Delta G_B} \tag{4}$$

When a nucleus has reached a radius of r^*, further growth lowers the Gibbs energy of the total system. To a first-order approximation, ΔG_B depends linearly on temperature. As the temperature decreases below the equilibrium value, the absolute value of ΔG_B increases and the critical radius decreases.

On the basis of the development of Kishi et al. (72), the rate of formation of stable nuclei, N, for diffusional growth is proportional to exp ($-\Delta G_f^*/RT$), in which R is the gas constant and T is the temperature, according to

$$N = K \exp\left(\frac{-\Delta G_f^*}{RT}\right) \tag{5}$$

with

$$K = 2C_i^l(C_i^l - C_e) \ln\left(\frac{C_i^l}{C_e}\right) D_i \left(\frac{RT}{\sigma}\right)^{1/2} \tag{6}$$

and

$$\Delta G_f^* = \frac{4\pi\sigma^3}{3}\left[\frac{M}{\rho RT \ln\left(\frac{C_i^l}{C_e}\right)}\right]^2 \tag{7}$$

in which C_i^l is the actual melt concentration of the limiting component i, C_e is the liquidus concentration at the temperature of interest, and D_i is the diffusion coefficient of the limiting species.

The temperature dependence of the nucleation rate allows many critical nuclei to be formed on a time scale that is small relative to the growth time when the difference between the actual solution and equilibrium temperatures is greater than a critical value, ΔT_c. If the temperature variations of liquid density are neglected, the critical supersaturation, ΔT_c, will vary with the average growth temperature through variations in the shape of the phase diagram (C_i^1/C_e), surface tension (σ), and diffusion coefficient (D_i).

As the average growth temperature is increased, the rate of homogeneous nucleation for a constant temperature decrease below saturation will increase when the shape of the phase diagram and the temperature dependence of the surface tension (which decreases with temperature) and diffusion coefficient are considered. These relationships imply that a smaller supersaturation temperature window is available to avoid homogeneous nucleation as the average growth temperature is increased. This expectation is supported by the experimental results shown in Figure 7 (72). In Figure 7, experimental values of the critical supersaturation, ΔT_c, are plotted against the liquidus temperature. Similar studies with GaAs melts have shown that critical supercooling in homogeneous solutions can be greater than 10 °C (73–75). On the other hand, the critical supersaturation required for heterogeneous nucleation on low-surface-energy single-crystal substrates is less than 0.25 °C (76, 77). Apparently the high surface tension of the solid graphite–melt prevents heterogeneous nucleation on the graphite boat.

Initial Stages of Epitaxy. The physical processes that take place during the initial stages of epitaxy (less than a few seconds after the substrate is contacted to the melt) are not well understood (79–90).

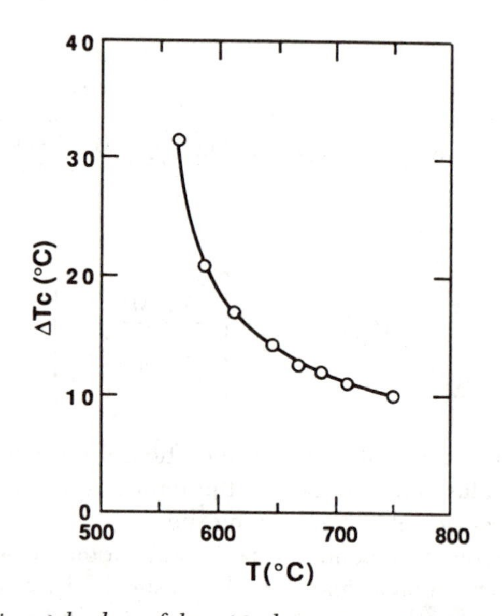

Figure 7. Experimental values of the critical supersaturation, ΔT_c, as a function of saturation temperature in In-rich In–P solutions. (Reproduced with permission from reference 72. Copyright 1986 Elsevier.)

Homoepitaxy. The simplest situation is *homoepitaxy*, with the solution initially saturated with respect to the solid. Microscopic variations in the Gibbs energy of the substrate surface are expected from the presence of dislocations and any roughness exposing different crystal orientations. Presumably, the Gibbs energy of the liquid is constant, and local differences between the Gibbs energy of the substrate and that of the liquid will drive mass-transfer processes (dissolution or deposition) across the interface. Thus, significant dissolution can occur along dislocation lines and at the edges of the substrate at long exposure times.

Growth is initiated by supersaturating the melt to a critical level. At the beginning of growth, all elements are present in sufficient quantities, so that a kinetics-limited situation (atom attachment or nucleation) must surely exist at very short times ($<<1$ s). The rate of growth becomes very fast when a critical supersaturation is reached, and the high growth rate results in rapid initial growth (thicknesses of a few hundred angstroms). This rapid growth will be slowed by a local depletion of solutes and a local increase in the temperature because of the energy of crystallization released at the interface. The thermal diffusivity of liquid metals at elevated temperature is approximately four orders of magnitude greater than the mass diffusivity, and thus the limitation due to solute transport is expected to be the dominant limitation.

Heteroepitaxy. *Heteroepitaxy* (e.g., deposition of $Al_xGa_{1-x}As$ on GaAs) is somewhat different, because the solid and liquid cannot initially be in equilibrium, that is, a chemical potential difference exists across the solid–liquid interface. In compound semiconductors, the chemical potential of each element is constrained by compound stoichiometry. For example, for a ternary solid ($A_xB_{1-x}C$) in equilibrium with a ternary liquid, the conditions of equilibrium are given by equations 8 and 9:

$$\mu_A^s + \mu_C^s \equiv \mu_{AC}^s = \mu_A^l + \mu_C^l \quad (8)$$

$$\mu_B^s + \mu_C^s \equiv \mu_{BC}^s = \mu_B^l + \mu_C^l \quad (9)$$

In these equations, μ_A^s, μ_B^s, and μ_C^s are the chemical potentials of A, B, and C, respectively, in the solid phase and μ_A^l, μ_B^l, and μ_C^l are the chemical potentials of A, B, and C, respectively, in the liquid phase. Equations 8 and 9 are not necessarily satisfied when a ternary A–B–C liquid is contacted with an arbitrary $A_xB_{1-x}C$ solid solution. Furthermore, a driving force for the formation of a new phase may exist in a single phase (e.g., a liquid may be supersaturated).

A driving force (e.g., a difference in chemical potential) causes a physical process (manifested as deposition or dissolution) to occur. The physical process restores the balance of the component chemical potentials. The direction of the process is from the region of higher chemical potential to the region of lower chemical potential. The chemical potentials of each component do not adjust independently, because they are coupled through their dependence on concentration, temperature, and pressure (negligibly). If the rates of surface processes are assumed to be fast relative to the rates of chemical-potential-difference-driven mass-transfer processes, then the surface chemical potentials rapidly satisfy equations 8 and 9. Although shifts in component chemical potential have occurred, net differences between the chemical potentials of the two bulk phases still exist with the same sign, as in the initial situation. The signs of these differences for each component can either be the same or opposite. The gradient in chemical potential will drive mass-transfer processes in both phases, with transport in the liquid being more rapid than that in the solid. When the gradients are opposing, growth rate reversal, that is, initial dissolution followed by growth or the same processes occurring in the reverse order, may occur. The first process to occur will be determined by the liquid phase, because transport is more rapid in this phase.

As an example, Small and Ghez (*87*) calculated growth thicknesses after a slightly supersaturated Ga–Al–As liquid solution was contacted with an $Al_xGa_{1-x}As$ alloy, with compositions adjusted such that the liquidus temperature of the liquid is lower than the solidus temperature of the alloy. Essentially, the liquid phase wants to form a solid because it is initially supersaturated, but at long times, the solid should dissolve. Figure 8 shows the calculated nondimensional thickness as a function of nondimensional time (*87*) of a case demonstrating growth reversal. The chemical potential of AlAs is higher in the liquid phase and adjusts rapidly by depositing an AlAs-rich solid. The GaAs chemical potential is higher in the bulk solid phase and is driven towards dissolution, but transport through the initial AlAs-rich surface layer is slow. The initial adjustment of the liquid-phase chemical potentials by deposition reduces the driving force in the liquid phase and permits transport in the solid to increase relative to that in the liquid. Eventually, dissolution occurs, and at long times, a pseudosteady state is reached, but a thin layer of AlAs-rich solid remains on the surface.

Forced Convection. An additional complication arises from convection in the melt forced by the motion of the slider and only marginally assisted by the gas flow above the melt. Forced convection will transport solute across the substrate from the back edge. Moving a solid horizontal boundary across the bottom of an initially stagnant and semiinfinite liquid is a classical problem of unsteady viscous flow (*91*). The ratio of the velocity of the fluid in the direction of motion, $v(y)$, to the solid-boundary velocity, V, is given by

$$\frac{v(y)}{V} = \text{erfc}\left(\frac{x}{\sqrt{4\nu t}}\right) \tag{10}$$

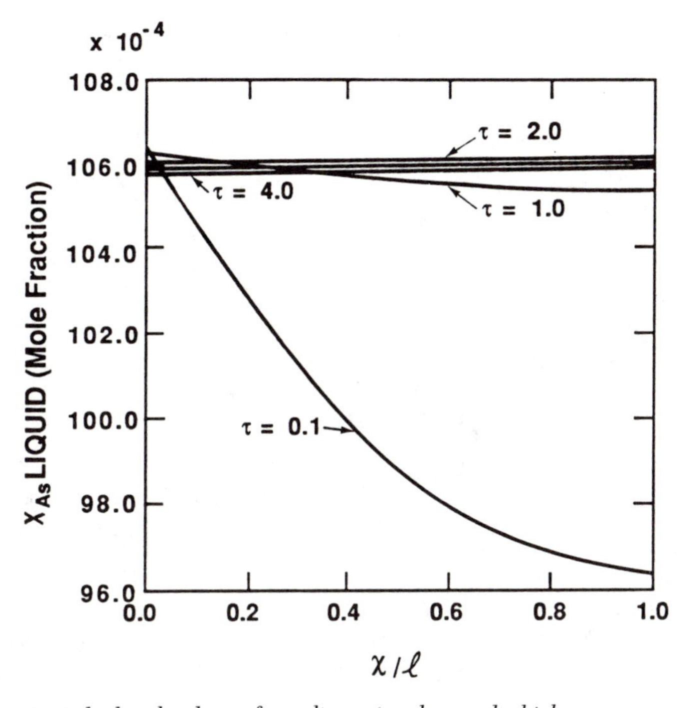

Figure 8. Calculated values of nondimensional growth thickness versus nondimensional time ($\tau = D_i t/l^2$) *demonstrating growth reversal for the growth of* $Ga_xAl_{1-x}As$ *on a substrate having a melting temperature higher than the equilibrium liquidus temperature. (Reproduced with permission from reference 87. Copyright 1984 American Institute of Physics.)*

in which x is the distance above the moving boundary and ν is the kinematic viscosity of the melt. A characteristic penetration depth of the disturbance is in the order of $4\sqrt{\nu t}$. For liquid Ga at 1000 K after 1 s of motion, the penetration depth is ~0.1 cm (92).

The actual situation during LPE growth is complicated by the presence of the solid well walls, which produce a three-dimensional rotating-flow pattern. The time necessary for the fluid to cease motion by viscous dissipation should be characterized by the quantity l^2/ν, in which l is a characteristic length. For typical LPE conditions, this time should be in the order of 1–100 s. Decreasing the aspect ratio (ratio of melt height to melt width) and placing a solid boundary at the top surface should decrease the time during which forced convection is important. Shown in Figure 9 are measured film thicknesses as a function of time for the growth of GaAs and $Al_{0.4}Ga_{0.6}As$ by LPE at 700 °C (84). The growth temperature program included an ~0.5 °C initial supersaturation, followed by a cooling rate of 0.1 °C/min. The measured thicknesses are greater than expected for diffusion-limited growth by 0.11 μm for GaAs and 0.06 μm for the alloy. Presumably,

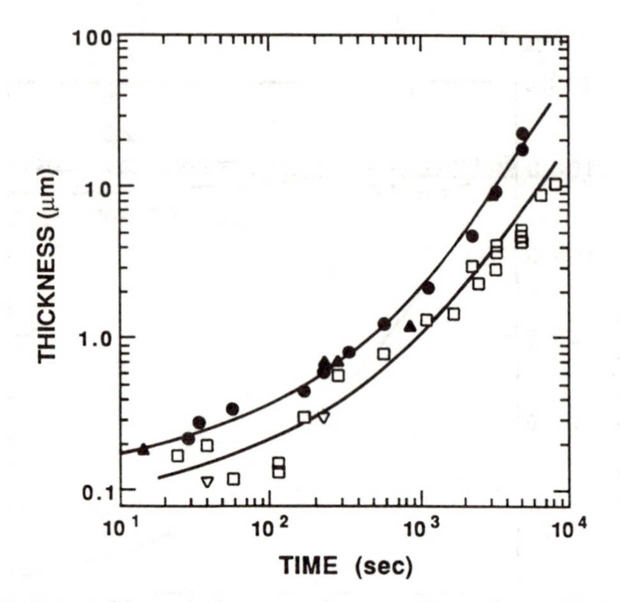

Figure 9. Epitaxial layer thickness as a function of growth time for the growth of GaAs (●, ▲) and $Al_{0.4}Ga_{0.6}As$ (□, ▽). (Reproduced with permission from reference 84. Copyright 1982 Elsevier.)

the enhanced mass-transfer rate created by forced convection produces the increased growth thickness.

Compositional Variations

Figure 10 shows a binary A–AB system in which the only compound that exists is AB. The system forms a simple eutectic-type phase diagram with element A. A liquid solution with [B] = C_o is saturated with respect to the formation of compound AB at the liquidus temperature T_o. The concentration of B in the solid compound AB is C_s and is related to the density of the compound, ρ_{AB}, and the molecular weight, M_{AB}, by $\rho_{AB}/2M_{AB}$. According to the phase diagram, cooling to temperature T_e results in the formation of the solid compound and a shift in the concentration of the remaining liquid to a composition C_e. The film thickness, d, after equilibrium cooling at a rate r for a time t is given by a simple material balance on element B as:

$$d \simeq \frac{lrt}{m(C_s - C_o) + rt} \tag{11}$$

In equation 11, l is the height of the liquid and m is the slope of the liquidus ($m \simeq [T_o - T_e]/[C_o - C_e]$).

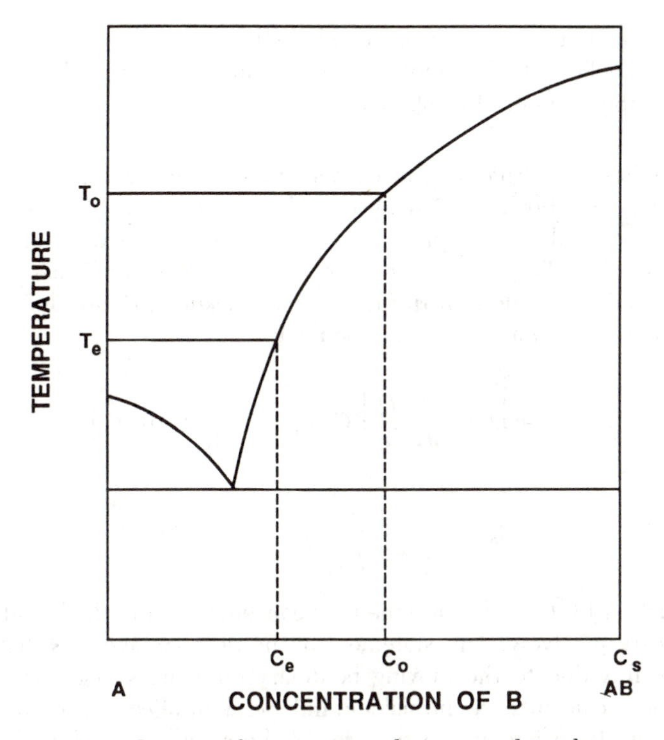

Figure 10. Schematic of an A–AB simple eutectic phase diagram.

According to this equilibrium model, the thickness of the grown film should vary linearly with both the height of the liquid and the growth times typically encountered in LPE. Experimental thicknesses from growth by equilibrium cooling are independent of the solution height and proportional to $t^{3/2}$ for typical liquid levels and growth times. Furthermore, the measured growth rates are about an order of magnitude less than that predicted by equation 11. These results and the facts that the growth rate is independent of surface orientation for most material systems (*93*) and that a small supercooling is needed for nucleation suggest that the growth process is diffusion limited.

Models of Diffusion during LPE. An exact formulation of the LPE diffusion problem constitutes a nontrivial Stefan problem (*80*). An analytical first-order model capable of representing long-time experimental thickness data to within experimental error considers the following one-dimensional conservation equation:

$$\frac{\partial C_i^l(x, t)}{\partial t} = D_i \frac{\partial^2 C_i^l(x, t)}{\partial x^2} \quad i = 1, 2, \ldots n \qquad (12)$$

In this equation, x is the distance above the growing interface, D_i is the diffusion coefficient of component i in the multicomponent liquid solution, and n is the number of components.

Semiinfinite Extent. The conservation equations are subject to the boundary conditions implied by equilibrium at the solid–liquid interface with a specified cooling program and a semiinfinite extent of the liquid ($l >> [D_i t]^{1/2}$). The concentration of the liquid solution is assumed to be uniform initially. The growth rate, v, and thickness, d, are determined by mass conservation at the moving boundary:

$$v = D_i \frac{\partial C_i^l(0, t)}{\partial x} \left[C_i^s(0, t) - \frac{C^l}{C^s} C_i^l(0, t) \right]^{-1} \tag{13}$$

$$d = \int_0^t v \, dt \tag{14}$$

In equation 13, C^l and C^s are the total concentrations in the liquid and solid phases, respectively. This statement of the problem assumes that the convective flux due to the moving boundary (growing surface) is small, the diffusion coefficients are mutual and independent of concentration, the area of the substrate is equal to the area of the solution, the liquid density is constant, and no transport occurs in the solid phase. Further, the conservation equations are uncoupled from the equations for the conservation of energy and momentum. Mass flows resulting from other forces (e.g., thermal diffusion and Marangoni or slider-motion-induced convective flow) are neglected.

The solution to equations 12–14 for the growth of a pseudobinary compound, $A_xB_{1-x}C$, can be expressed in analytical form (*94*) for the three common temperature programs: ramp or equilibrium cooling, step cooling, and supercooling. The expressions for thickness as a function of time are given by:

$$\text{ramp or equilibrium cooling:} \quad d = \frac{2Krt^{3/2}}{3} \tag{15}$$

$$\text{step cooling:} \quad d = K\Delta T t^{1/2} \tag{16}$$

$$\text{supercooling:} \quad d = K\left(\Delta T t^{1/2} + \frac{2rt^{3/2}}{3}\right) \tag{17}$$

In equations 16 and 17, ΔT is the difference between the saturation and actual temperatures. The constant K in equations 15–17 is defined by equation 18:

$$K = \frac{2C^l}{\sqrt{\pi}}\left[\sum_i \frac{m_i}{\sqrt{D_i}}\left(C_i^s - C_i^l\frac{C^s}{C^l}\right)\right]^{-1} \quad i = 1, 2 \tag{18}$$

In equation 18, m_i is the slope of the liquidus curve for species i, that is, $m_i = (\partial T/\partial C_i^l)_{C_j^l=C_e}$.

Figure 11 illustrates the differences between the various temperature schedules when the value of K is taken as 4.0×10^{-6} cm/Ks$^{1/2}$. The growth rate for step cooling is very high during the early portion of epitaxy but decreases with time. On the other hand, the growth rate for equilibrium cooling is initially zero and increases with time. The growth rate expression for supercooling represents a superposition of the equations for step cooling and equilibrium cooling, with the step-cooling solution dominating at short times and the equilibrium-cooling solution dominating at long times.

Finite Extent. As the growth time approaches $l^2/2D_i$, the assumption of a semiinfinite growth solution is no longer valid. The concentration of solutes at the top of the solution will begin to decrease, and this decrease

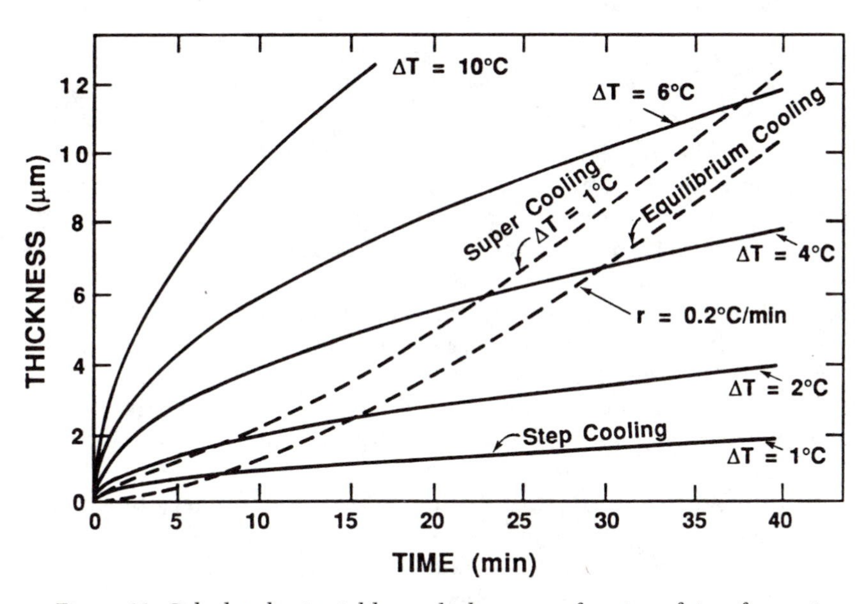

Figure 11. Calculated epitaxial layer thickness as a function of time for equilibrium, step-cooling, and supercooling temperature programs. (Reproduced with permission from reference 94. Copyright 1984 American Institute of Physics.)

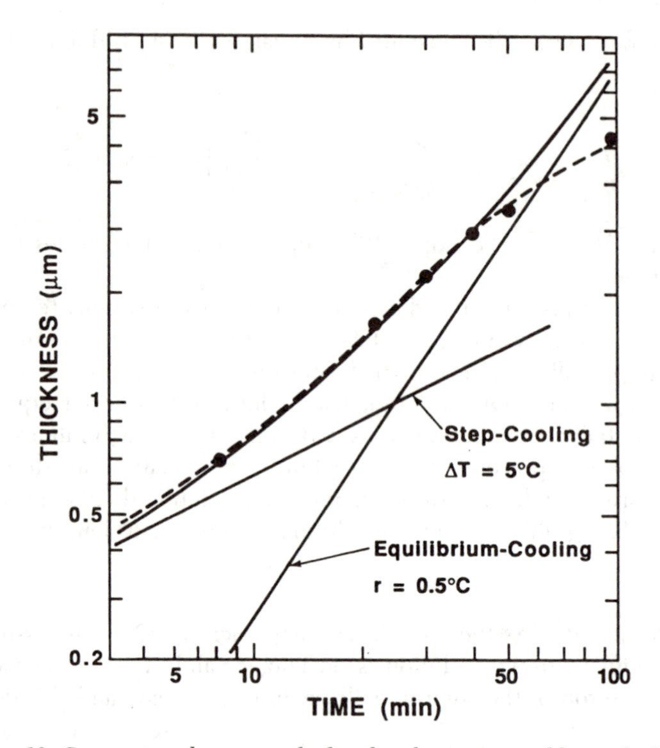

Figure 12. Comparison between calculated and experimental layer thicknesses as a function time for the growth of GaP by supercooling. (Reproduced with permission from reference 95. Copyright 1983 American Institute of Physics.)

will lower the flux of solute to the growing surface. Such a decrease in growth rate is illustrated in Figure 12 for the growth of GaP by the supercooling technique (95). A Ga-rich solution with a liquidus temperature of 848 °C was subjected to an initial supercooling of 5 °C, and a cooling schedule of 0.5 °C/min was started. The growth expression given by equation 17 gives an excellent fit to the experimental thickness data for $t < 40$ min. However, significant deviations from this relation are seen at longer times. At the growth conditions used to obtain the thickness data in Figure 12, the value of $l^2/2D_i$ is 61 min, a value in reasonable agreement with the experimental value. Similar results for the growth of GaAs were determined by Hsieh (74).

Two other factors could contribute to the observed departure of the growth thickness at long times from that predicted by equation 17. First, homogeneous nucleation in the melt and the subsequent growth of precipitates will act as sinks for solute atoms, just as film growth does. The exact location in the melt where homogeneous nucleation will occur depends on the melt compositional and thermal profiles (96). The precipitates are less dense than the melt and will rise to the top of the melt. Such precipitates were formed during long-time LPE growth experiments in which ramp cooling was programmed (*74*), with the amount of precipitation generally increasing with increased cooling rate. The level of supersaturation that exists at long times can be significant. For example, a 25 °C decrease in temperature below the original solution liquidus temperature is in effect after 40 min for the experimental conditions used to generate the results depicted in Figure 12.

The second reason for expecting the growth thickness to be reduced below the value given by equation 17 is associated with solute evaporation at the upper surface of the solution. For the growth of compound semiconductors, a solvent with low vapor pressure is chosen generally. The preferential evaporation of solute in a supersaturated solution shifts the composition towards the liquidus line and decreases the degree of supersaturation. The maximum rate of evaporation is given by the Langmuir–Hertz equation

$$J_i = \frac{P_i}{2\pi M_i RT} \tag{19}$$

in which P_i is the partial pressure of the solute, M_i is the molecular weight of the evaporating species, R is the gas constant, and T is temperature.

For ideal solutions, the partial pressure of a component is directly proportional to the mole fraction of that component in solution and depends on the temperature and the vapor pressure of the pure component. The situation with group III–V systems is somewhat more complicated because of polymerization reactions in the gas phase (e.g., the formation of P_2 or P_4). Maximum evaporation rates can become comparable with deposition rates (0.01–0.1 μm/min) when the partial pressure is in the order of 0.01–1.0 Pa, a situation sometimes encountered in LPE. This problem is analogous to the problem of solute loss during bakeout, and the concentration variation in the melt is given by equation 1, with l replaced by the distance below the gas–liquid interface and z taken from equation 19. The concentration variation will penetrate the liquid solution from the top surface to a depth that is nearly independent of z/D_i and comparable with the penetration depth produced by film growth. As result of solute loss at each boundary, the variation in solute concentration will show a maximum located in the melt. The density will show an extremum, and the system could be unstable with respect to natural convection.

Comparison between Temperature Programs. All three techniques (ramp cooling, step cooling, and supercooling) have been used to achieve device-quality epitaxial film growth. An advantage of ramp cooling is the lower growth rate at short periods that permit better control of the thickness of thin films. Because a small amount of supercooling is required for heterogeneous nucleation, ramp cooling is actually the supercooling procedure with a small initial ΔT.

The use of ramp cooling in the growth of multiple layers after the first layer of growth can present problems associated with meltback. For both ramp-cooling and supercooling methods, the previously discussed problem of homogeneous nucleation exists at long growth times. In addition, the growth rate increases as growth proceeds, and the thickness reproducibility decreases. Because the variation of the growth rate with time for step cooling is opposite to that of ramp cooling, thickness control improves with growth time. From a thickness control viewpoint, supercooling has the negative attributes of both limiting cases: high initial growth rates and an increasing growth rate at long periods. The presence of an initial supersaturation (step cooling and supercooling) seems to improve the surface morphology (*74*, *97*). The supersaturation increases the driving force for nucleation and causes more uniform nucleation over the substrate.

Pure step cooling is an isothermal operation with simple temperature control. De Crémoux (*98*) showed that step-cooling growth from a multicomponent solution should produce epitaxial layers with a constant composition. This prediction has been experimentally verified (*99*, *100*). The variation of thickness with growth time for the growth of lattice-matched InGaAsP layers on (100) InP by the step-cooling technique is shown in Figure 13 (*100*). The thickness data are consistent with diffusion-limited growth kinetics for $t < 30$ min. After this time, the thickness is less than that predicted by equation 16, and growth finally stops after 60 min. The composition of the quaternary layer was constant only when the process was diffusion limited ($t < 30$ min).

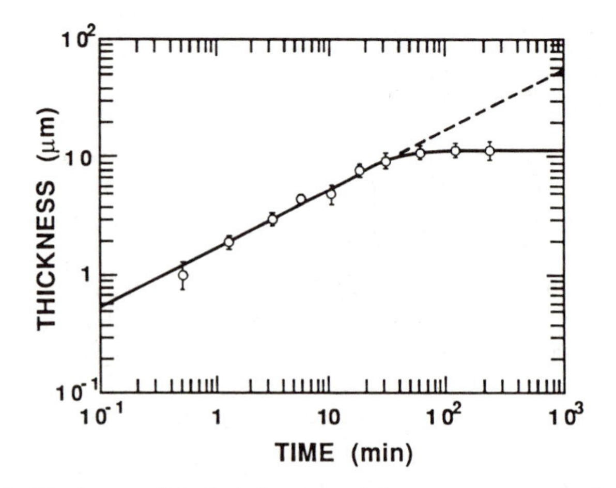

Figure 13. Thickness of InGaAsP epitaxial layers versus growth time for growth by the step-cooling technique with 10 °C supersaturation. (Reproduced with permission from reference 100. Copyright 1980 American Institute of Physics.)

Growth Limited by Kinetics. A diffusion-limited growth model adequately describes the experimental growth rate data for most group III–V systems (*77, 93, 95, 100*). The growth of InGaAsP, InGaAs, and InGaP on (111) InP planes appears to be controlled by surface kinetics (*93, 101–103*).

The variation of layer thickness as a function of supersaturation for the growth of InGaAsP on various InP substrate orientations is shown in Figure 14 (*93*). The step-cooling method was used, and the thickness data displayed in this figure were for a growth time of 10 min. According to equation 16, the thickness should be directly proportional to ΔT for a constant growth time. The results shown in Figure 14 for growth on (001) InP are consistent with this relationship. For growth on (111) faces, initial growth does not occur until a critical supersaturation of about 4 °C for nucleation is achieved.

When nucleation was accomplished for growth on the (111) A plane, the thickness data were consistent with the diffusion-limited model with a slightly increased effective growth rate. In contrast, the data for growth on the (111) B face deviate from a linear relationship. In addition, the compositions of the quaternary films grown under identical conditions were nearly the same and uniform for growth on the (001) and (111) A planes, but the composition for growth on the (111) B face was nonuniform and significantly different from the composition deposited on the other faces.

The results just described have been explained in terms of two phe-

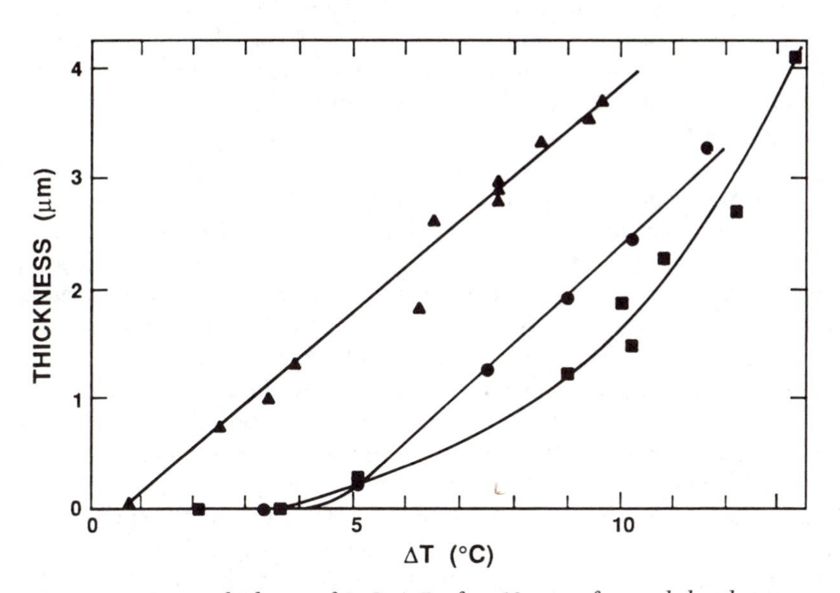

Figure 14. Layer thickness of InGaAsP after 10 min of growth by the step-cooling method as a function of supersaturation. The liquid solution mole fractions are as follows: $x_{Ga} = 5.54 \times 10^{-3}$, $x_{As} = 4.39 \times 10^{-2}$, *and* $x_P = 3.22 \times 10^{-3}$. *The curves are for (001) (▲), (111) A (●), and (111) B (■). (Reproduced with permission from reference 93. Copyright 1986 Elsevier.)*

nomena: (1) orientation dependence of the liquidus temperature and (2) limitation by surface reaction. The variation of the liquidus temperature with orientation was determined by seed dissolution studies (i.e., measurement of weight loss after ternary or quaternary solutions are contacted with binary seeds). Clearly these results are not representative of true equilibrium, because the remaining seed is primarily InP. However, the experiments are representative of the process that occurs during heteroepitaxy. As previously discussed for the initial stages of epitaxy, chemical potential gradients exist (i.e., for the contact of InGaP liquid with an InP seed, μ_{GaP} is higher in the solution and μ_{InP} is higher in the bulk seed), in addition to possible differences with respect to the initial saturation condition of the liquid and the hypothetical solid in equilibrium with it.

The difference in the saturation temperature measured by seed dissolution between (001) and (111) faces is related possibly to anisotropy in the solid-state diffusion coefficient. In the work of Thijs et al. (93), the saturation temperature measure by InP seed dissolution experiments was higher by the same amount on both (111) A and (111) B surfaces than on the (001) plane. The variation of composition with time for the growth of InGaAsP on (111) B InP (93) is consistent with a variation of the growth velocity from a $t^{-1/2}$ dependency as a result of a limiting surface reaction for short times. If the equilibrium boundary condition of the diffusion equation is changed to one in which the flux at the solid–liquid interface is equated to the rate of a first-order reaction, the concentration variation is functionally similar to that given by equation 1, with l replaced by distance above the surface, z replaced by the first-order rate constant k, and concentration relative to the equilibrium value. The growth thickness will exhibit a $t^{1/2}$ dependency at times longer than D_i/k^2.

Impurity Incorporation. Useful solid-state devices are products of the precisely controlled incorporation of electrically active elements. As an example, GaAs can be grown with a conductivity that is n type, p type, or semi-insulating; the range of electrical conductivity spans more than 10 orders of magnitude. The elements can be added intentionally (dopant) or unintentionally (impurity). In growth by LPE, doping is accomplished usually by adding dopants (elemental or compounds) during solution formulation, but dopants can be introduced into the solution from the gaseous ambient also. The impurities can come from the solution starting materials, the gas stream, and the materials of construction (primarily impurities in graphite and from the reduction of SiO_2 by H_2). In addition, impurities or dopants present in the substrate or previously deposited layers can redistribute during growth. In the growth of compound semiconductors by LPE, most impurities tend to segregate in the liquid phase, and thus growth involves a single-stage fractionation. One of the advantages of LPE is the capability to grow high-purity films; carrier concentrations in the $10^{14}/cm^3$ range are possible on a routine basis.

The equations used to describe dopant incorporation are identical to those used to describe the deposition of the semiconductor. Thus equations 12–14 are applicable to a diffusion-limited model, with the number of components, n, increased by the number of dopants added. The equilibrium distribution coefficient, k_i, is defined as

$$k_i = \frac{C_i^s}{C_i^l} \tag{20}$$

and indicates the distribution of the impurity between the solid and liquid phases. With this definition, the expression for the incorporation rate of the impurity (equation 13) becomes

$$D_i \frac{\partial C_i^l(0, t)}{\partial x} = v \left(k_i - \frac{C^l}{C^s} \right) C_i^l(0, t) \tag{21}$$

The growth velocity, v, is largely determined by the processes that control the deposition of the semiconductor. For growth under isothermal conditions (step cooling), the composition of the impurity in the solid does not vary, similar to growth of multicomponent alloys. An approximate expression for the effective distribution coefficient, k_{eff}, for the growth of a binary compound by step cooling in the unbounded case is given by (*4*)

$$k_{eff} \equiv \frac{C_i^s(0, t)}{C_i^l(\infty)} = k_i \left[1 + \frac{\Delta T m_j (k_i - 1)}{C^s} \left(\frac{D_j}{D_i} \right)^{1/2} \right]^{-1} \tag{22}$$

in which $C_i^l(\infty)$ is the initial dopant concentration and the subscript j represents a species in the compound. For step cooling, the effective distribution coefficient is independent of time and is closer to unity than the equilibrium distribution coefficient, k_i. The accumulation ($k_i < 1$) or depletion ($k_i > 1$) of impurities at the growing interface changes the impurity concentration in the solid to higher ($k_i < 1$) or lower ($k_i > 1$) values. The situation for a nonisothermal boundary condition (ramp cooling or supercooling) is more complex and requires a numerical solution of the diffusion problem. Because the boundary condition at the interface varies with time, the incorporation rate of impurities also varies with time and can lead to an extremum in the concentration profile of the dopant.

Solid–Liquid Equilibrium

Analysis of the growth process by LPE usually stipulates an equilibrium boundary condition at the solid–liquid interface. The solid–liquid phase diagrams of interest to LPE are those for the pure semiconductor and the semiconductor–impurity systems. Most solid alloys exhibit complete miscibility for mixing on the same sublattice (e.g., $Al_xGa_{1-x}As$) and extremely limited miscibility for mixing between the two sublattices or a sublattice and the interstitial lattice (i.e., the point defect structure). To calculate growth rates and variations in bulk composition, only the mixing on the same sublattice is usually considered. When transport processes in the solid solution are treated or when the relation of dopant concentration to the electrical properties is of interest, the point defect structure must be considered. The procedures for calculating phase diagrams in which deviations from stoichiometry are neglected are discussed in the following sections.

The temperature–composition phase diagram of a generalized ternary A–B–C system is shown in Figure 15, in which triangular coordinates are used to represent composition. The binary A–C or B–C limits are similar and show the formation of a single compound AC or BC. The AC–C or BC–C portions of the phase diagram are simple eutectic types, whereas the AC–A or BC–B limits shown in Figure 15 are nearly degenerate eutectic types. The elements A and B are from the same column of the periodic table and exhibit a variety of types of phase diagrams. For simplicity, an isomorphous phase diagram is shown in Figure 15. Also shown in this figure is the

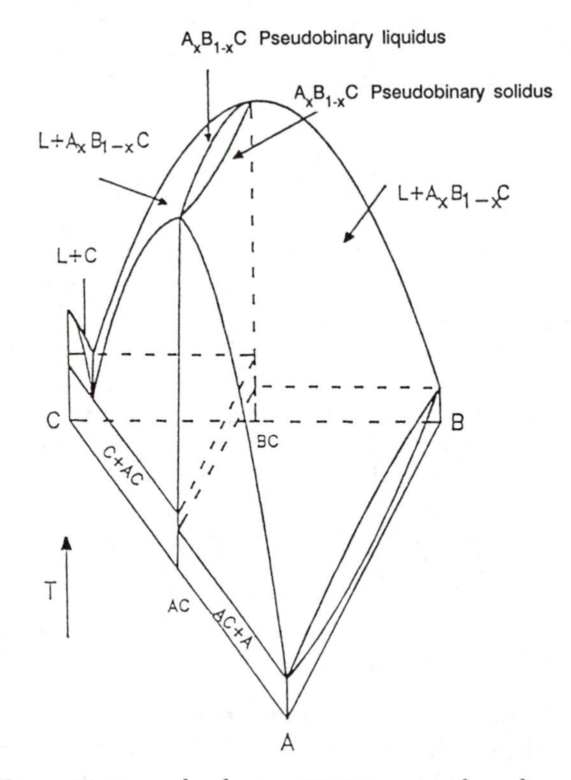

Figure 15. Generalized group III–V ternary phase diagram.

pseudobinary phase diagram ($x_C = 0.5$), which is isomorphous for most group III–V systems. A single liquidus surface covers nearly the entire phase diagram, and a eutectic valley is formed in C-rich solutions. The solidus sheet of $(A_xB_{1-x})_{1+\delta}C_{1-\delta}$ is simplified to zero thickness ($\delta = 0$) and would be slightly inclined if the maximum melting temperatures of the alloys were located at nonstoichiometric compositions. The liquidus surface in regions of low vapor pressure and the corresponding solidus tie lines are of interest to LPE.

Problem Formulation. The conditions of equilibrium require the equivalence in each phase of temperature, pressure, and chemical potential for each component that is transferable between the phases and are subject to constraints of stoichiometry. A statement of the equivalence of chemical potential is identical to equations 8 and 9. An example is the $A_xB_{1-x}C_yD_{1-y}$ quaternary system. This system contains four binary compounds, AC, BC, AD, and BD, and the conditions of equilibrium allow three equations (of the type given by equation 8) to be written. The fourth possible equation is redundant as a result of the stoichiometric constraint (i.e., equal number of atoms on each sublattice).

For solid–liquid equilibrium in a quaternary system, the Gibbs phase rule allows four degrees of freedom. If T, P, x_C, and x_D (in which x_i is the mole fraction of component i in liquid solution) are specified, then x_A, x, y, and x_{AC} (in which x_{ij} is the mole fraction of component ij in solid solution) are determined, and the system is invariant. These variables are defined by the following equations:

$$x_A = \frac{1 - (1 - x_C - x_D)(P_{BC} + P_{BD})}{(P_{AC} + P_{AD}) - (P_{BC} + P_{BD})} \tag{23}$$

$$x = \frac{(P_{AC} + P_{AD})[1 - (1 - x_C - x_D)(P_{BC} + P_{BD})]}{(P_{AC} + P_{AD}) - (P_{BC} + P_{BD})} \tag{24}$$

$$y = \frac{(P_{AC} - P_{BC}) - (1 - x_C - x_D)(P_{AC}P_{BD} - P_{AD}P_{BC})}{(P_{AC} + P_{AD}) - (P_{BC} + P_{BD})} \tag{25}$$

$$x_{AC} = \frac{\Gamma_{AD}\Gamma_{BC}}{\Gamma_{AC}\Gamma_{BD}} - \frac{x_{AD}x_{BC}}{x_{BD}} \exp\left(\theta_{AD} + \theta_{BC} - \theta_{AC} - \theta_{BD}\right) \tag{26}$$

in which

$$P_{ij} \equiv \Gamma_{ij}x_j \exp\left(-\theta_{ij}\right) \quad ij = AC, AD, BC, \text{and } BD \tag{27}$$

$$\theta_{ij} \equiv \frac{\mu_{ij} - \mu_i^{o,l} - \mu_j^{o,l}}{RT} \tag{28}$$

$$\Gamma_{ij} \equiv \frac{\gamma_i^{l} \gamma_j^{l}}{\gamma_{ij}^{s}} \tag{29}$$

In equations 27–29, P_{ij} is the partial distribution coefficient of component ij, Γ_{ij} is the ratio of activity coefficients, θ_{ij} is the reduced standard-state chemical potential difference, $\mu_i^{o,p}$ is the standard-state chemical potential of component i in phase p, and γ_i^{p} and γ_{ij}^{p} are the activity coefficients of components i and ij, respectively, in phase p. The working equations (equations 23–26) describing phase equilibria, along with the equation defining a mole fraction, are implicitly complex relations for T, P, x, y, x_{AC}, x_A, x_C, and x_D but involve only two thermodynamic quantities, θ_{ij} and Γ_{ij}. Equations 23–25 are implicit in composition only through the Γ_{ij} term, which is itself only a weak function of composition.

The thermodynamic quantity θ_{ij} is a reduced standard-state chemical potential difference and is a function only of T, P, and the choice of standard state. The principal temperature dependence of the liquidus and solidus surfaces is contained in θ_{ij}. The term Γ_{ij} is the ratio of the deviation from ideal-solution behavior in the liquid phase to that in the solid phase. This term is consistent with the notion that only the difference between the values of the Gibbs energy for the solid and liquid phases determines which equilibrium phases are present. Expressions for the limits of the quaternary phase diagram are easily obtained (e.g., for a ternary $A_xB_{1-x}C$ system, $y = 1$ and $x_D = 0$; for a pseudobinary section, $y = 1$, $x_D = 0$, and $x_C = 1/2$; and for a binary AC system, $x = y = x_{AC} = 1$ and $x_B = x_D = 0$).

This system of equations (equations 23–26) involves the three solid-phase-composition variables x, y, and x_{AC}. The other three solid-solution mole fractions are related to x, y, and x_{AC} by the following relations:

$$x_{AD} = x - x_{AC} \tag{30}$$

$$x_{BC} = y - x_{AC} \tag{31}$$

$$x_{BD} = (1 - x - y) + x_{AC} \tag{32}$$

This formulation implies that the alloy is a pseudoquaternary system composed of the four compounds, AC, AD, BC, and BD. For a property that is a function of composition, three composition variables must be specified to uniquely determine that property. For example, Sonomura et al. (*104*) have pointed out that the lattice parameter of an $A_xB_{1-x}C_yD_{1-y}$ alloy is not single valued for specified values of x and y. Variation in the actual crystal structure (i.e., distribution of bonding) is permitted for a given overall alloy composition. Similarly, the Gibbs energy of the alloy can take on different values for different bonding distributions in an alloy of the same bulk composition, $A_xB_{1-x}C_yD_{1-y}$.

Reduced Standard-State Chemical Potential Difference. The choice of reference state is entirely arbitrary and normally dictated by the available experimental data base. The standard state usually selected to describe the phase diagrams of compound semiconductors is the pure component of the same phase at the temperature and pressure of interest. The pressure dependence of the θ_{ij} term is negligible for all pressures of practical interest.

Method I. With the reference state chosen to be the pure component at the temperature of interest, θ_{ij} is simply the molar Gibbs energy of the solid compound ij minus the molar Gibbs energy of each pure liquid element i and j, all reduced by RT. This difference can be determined from a variety of thermodynamic sequences for which thermodynamic information is available. One such sequence, which was first suggested by Wagner (*105*) and later applied to group III–V binary systems by Vieland (*106*), is shown in Figure 16a. The solid compound ij at the temperature of interest, T, is first raised to the melting temperature, T_m^{ij}, and then melted. The stoichiometric liquid is then subcooled to T and finally separated into the pure liquid elements. This sequence (termed method I) results in the following relation:

$$\theta_{ij}^{I} = \ln\left[a_i^{sl}(T)a_j^{sl}(T)\right] - \frac{\Delta H_m^{ij}}{R} \times \left(\frac{1}{T} - \frac{1}{T_m^{ij}}\right) + \frac{1}{RT}\int_T^{T_m^{ij}}\int_T^{T_m^{ij}} \frac{\Delta C_p[ij]}{T}\, dT^2 \qquad (33)$$

In equation 33, the superscript I refers to the use of method I, $a_i^{sl}(T)$ is the activity of component i in the stoichiometric liquid (sl) at the temperature of interest, ΔH_m^{ij} is the molar enthalpy of fusion of the compound ij, and $\Delta C_p[ij]$ is the difference between the molar heat capacities of the stoichiometric liquid and the compound ij. This representation requires values of the Gibbs energy of mixing and heat capacity for the stoichiometric liquid mixture as a function of temperature in a range for which the mixture is not stable and thus generally not observable. When equation 33 is combined with equations 23 and 24 in the limit of the AC binary system, it is termed the *fusion equation* for the liquidus (*107–111*).

In applying equation 33, C_p^{sl} (the constant-pressure molar heat capacity of the stoichiometric liquid) is usually extrapolated from high-temperature measurements or assumed to be equal to C_p^{ij} of the compound, whereas the activity product, $a_i^{sl}(T)a_j^{sl}(T)$, is estimated by interjection of a solution model with the parameters estimated from phase-equilibrium data involving the liquid phase (e.g., solid–liquid or vapor–liquid equilibrium systems). To relate equation 33 to an available data base, the activity product is expressed

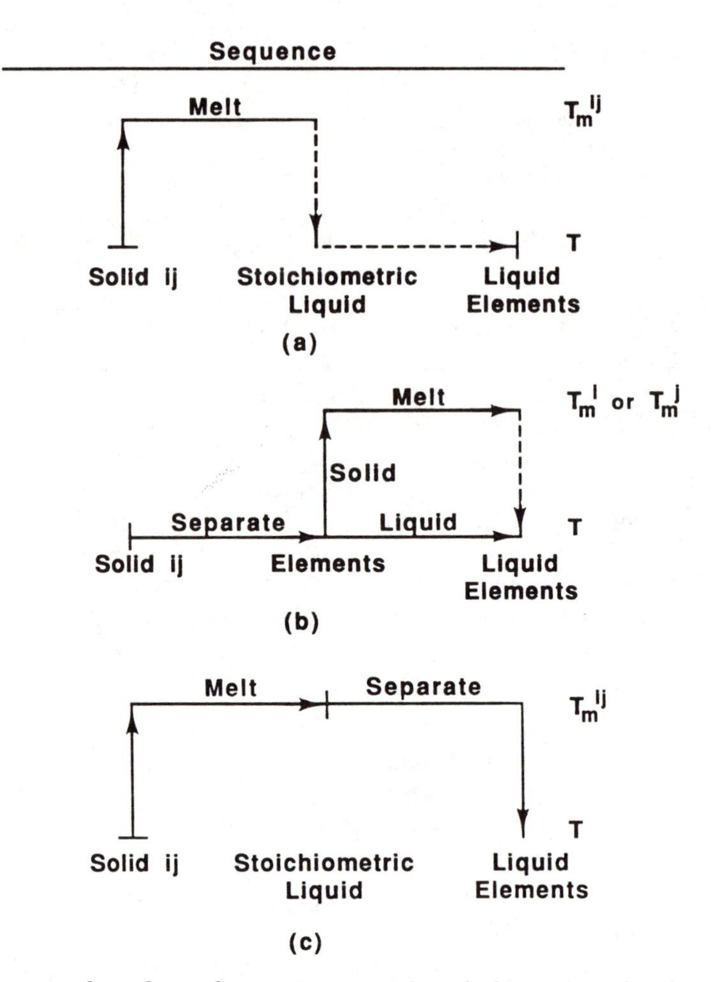

Figure 16. Three thermodynamic sequences for calculating the reduced standard-state chemical potential difference: (a) method I, (b) method II, and (c) method III.

in terms of the measurable activity product at some temperature, T^*, greater than or equal to T_m^{ij} and corrected to the temperature of interest with the relatively temperature-insensitive enthalpy of mixing of the stoichiometric liquid, ΔH_{mix}^{sl}. Equation 33 then becomes

$$\theta_{ij}^{I} = \ln [a_i^{sl}(T^*)a_j^{sl}(T^*)] - \int_T^{T^*} \frac{\Delta H_{mix}^{sl}}{RT^2}\, dT - \frac{\Delta H_m^{ij}}{R} \times \left(\frac{1}{T} - \frac{1}{T_m^{ij}}\right) + \frac{1}{RT}\int_T^{T_m^{ij}} \int_T^{T_m^{ij}} \frac{\Delta C_p[ij]}{T}\, dT^2 \quad (34)$$

The data base for method I consists of the measurable quantities $a_i^{sl}(T^*)$, $a_j^{sl}(T^*)$, ΔH_m^{ij}, T_m^{ij}, and $\Delta C_p[ij]$ and the extrapolated quantities ΔH_{mix}^{sl} and C_p^{sl}. Equation 34 contains liquid-mixture information.

Method II. A second thermodynamic sequence that is useful for the calculation of θ_{ij} (Figure 16b) consists of performing the reverse direct-formation reaction at the temperature of interest. If the elements are stable liquids at this temperature, the Gibbs energy of formation, $\Delta G_f^o[ij, T]$, is proportional to θ_{ij}. However, if one or both of the elements are solids at these conditions, the following sequence, which is similar to that applied to the compound in method I, is performed: The solid element is raised from T to the element melting temperature, T_m^i or T_m^j; the element is melted; and finally, the element is subcooled to the original temperature of interest. The resulting expression for θ_{ij} by method II (θ_{ij}^{II}) is

$$\theta_{ij}^{II} = \frac{\Delta G_f^o(ij, T)}{RT} - \sum_{n=i \text{ or } j} \frac{K_n \Delta H_m^n}{RT} \times \left(1 - \frac{T}{T_m^n}\right) - \int_T^{T_m^n} \int_T^{T_m^n} \frac{\Delta C_p[n]}{T} \, dT^2 \quad (35)$$

in which $K_n = 0$ if component n is a liquid at T and $K_n = 1$ if component n is a solid at T, ΔH_m^n is the enthalpy of fusion of component n, and $\Delta C_p[n]$ is the difference between the heat capacities of the pure liquid and the solid element n. If the pure element n is a vapor at T, then the ΔH_m^n and $\Delta C_p[n]$ terms indicate the vapor–liquid transition. For many compounds, a range of temperatures exists for which $K_n = 0$ for both elements (e.g., GaSb and AlSb), and therefore θ_{ij}^{II} is directly measurable. Substitution of equation 35 in equations 23 and 24 results in the *formation expression* for the binary liquidus (*107*, *110–114*).

The data required for method II are the measurable quantities $\Delta G_f^o[ij]$ and, if K_n is not 0, ΔH_m^n, T_m^n, and $C_p^s[n]$ and an extrapolated quantity $C_p^l[n]$. To make use of available data, the standard Gibbs energy of formation can be expressed in terms of the standard enthalpy and entropy of formation. The advantage of this formulation is that thermodynamic information for unstable or metastable systems is required only for the elements (i.e., $C_p^l[n]$), for which a better estimate can usually be made. This data base contains no explicit liquid-solution properties.

Method III. A third sequence (Figure 16c) is different from method I in that the pure liquid elements are separated from the stoichiometric mixture at the melting temperature and not at the temperature of interest. The pure liquid elements are then cooled back to the original temperature and, possibly, become unstable or metastable. The expression for θ_{ij} by method

III (θ_{ij}^{III}) is

$$\theta_{ij}^{III} = \frac{T_m^{ij}}{T} \ln [a_i^{sl}(T_m^{ij})a_j^{sl}(T_m^{ij})] + \frac{\Delta S_f^o[ij, T_m^{ij}]}{R} \times \left(\frac{T_m^{ij}}{T} - 1\right) + \frac{1}{RT}\int_T^{T_m^{ij}}\int_T^{T_m^{ij}} \frac{\Delta C_p}{T}\, dT^2 \quad (36)$$

in which $\Delta S_f^o[ij, T_m^{ij}]$ is the standard entropy of formation of the compound ij from the pure liquid elements, and ΔC_p is the difference between the heat capacities of the pure liquids and the solid compound:

$$\Delta C_p = C_p^l[i] + C_p^l[j] - C_p^s[ij] \quad (37)$$

If the elements are liquid in the standard state specified by the formation reaction, the Neumann–Kopp rule (*115*) suggests that $\Delta C_p = 0$. For the standard formation reaction in which $K_n = 0$ at the compound melting temperature, the standard entropy of formation can be obtained from the temperature derivative of $\Delta G_f^o[ij]$ or, often more accurately, from a combination of values of $\Delta G_f^o[ij, T_m^{ij}]$ and $\Delta H_f^o[ij, T_m^{ij}]$. When $K_n = 1$ at T_m^{ij}, the value of the natural standard entropy of formation, $\Delta' S_f^o[ij, T_m^{ij}]$, must be corrected as follows:

$$\Delta S_f^o[ij, T_m^{ij}] = \Delta' S_f^o[ij, T_m^{ij}] - \sum_{n=i \text{ or } j} K_n \left(\frac{\Delta H_m^n}{T_m^n} + \int_{T_m^n}^{T_m^{ij}} \frac{\Delta C_p[n]}{T}\, dT\right) \quad (38)$$

Equation 38 shows that $\Delta S_f^o[ij, T_m^{ij}]$ is independent of temperature and requires extrapolation of $C_p^l[n]$ only over the limited temperature range from T_m^n to T_m^{ij}, and the extrapolated $C_p^l[n]$ contribution would be partially cancelled by the last term in equation 36 for $T_m^n < T_m^{ij}$. The data base for the application of equation 37 is the measurable activity product at the melting temperature, T_m^{ij}, $\Delta S_f^o[ij, T_m^{ij}]$ (as discussed earlier), $C_p^s[ij]$, C_p^l[i or j], C_p^s[i or j], and $\Delta H_m^{i \text{ or } j}$. The advantages of method III are the explicit statement of the principal temperature dependence and the extrapolation of, at most, only the heat capacity of the liquid element.

Method IV. A final process by which θ_{ij} can be directly determined from experimental results is through application of equations 23 and 24. In the simplest form of the binary limit, the result is

$$\begin{aligned}\theta_{ij}^{IV} &= \ln [a_i(T^l, x_i^l)a_j(T^l, x_i^l)] \\ &= \ln [a_i(T^*, x_i^l)a_j(T^*, x_{jl})] \\ &\quad - \int_{T^*}^{T^l} \frac{\Delta\overline{H}(i, x_i^l) + \Delta\overline{H}(j, x_i^l)}{RT^2}\, dT \end{aligned} \quad (39)$$

in which $a_i(T^*, x_i^l)a_j(T^*, x_i^l)$ is the activity product at some measurement temperature, T^*, and liquidus composition, x_i^l, and $\Delta\overline{H}(\mathrm{i}, x_i^l)$ is the relative partial molar enthalpy for component i at the liquidus composition. The data base required for this final procedure is the measurable phase diagram and the activity product at the liquidus temperature and composition (or an isothermal activity product and the liquid-phase enthalpy of mixing). This procedure requires binary-mixture information and will produce two values of θ_{ij} at each liquidus temperature, one from each side of the phase diagram of the compound. Both values of θ_{ij} should be identical, given a consistent set of data. In the less likely event that multicomponent data are available, equations 23–26 can be solved for θ_{ij} simultaneously, with the solution requiring the activity and relative partial molar enthalpy of each component in the liquid and solid solutions.

Estimation of θ_{ij}. ***From Experimental Data.*** If the following forms for the temperature dependence of the heat capacity and the enthalpy of mixing, respectively, are assumed

$$C_p = d_1 + d_2T + \frac{d_3}{T^2} \tag{40}$$

$$\Delta H_{mix}^{sl} = A + BT \tag{41}$$

in which d_1, d_2, d_3, A, and B are empirical constants, then the expression for θ_{ij} on the basis of methods I–III has the following functionality

$$\theta_{ij} = C_0 + C_1T + \frac{C_2}{T} + \frac{C_3}{T^2} + C_4 \ln T \tag{42}$$

in which C_0, C_1, C_2, C_3, and C_4 are constants. The dominant terms in this expression are the first and third terms, with the remaining terms reflecting the magnitude of the various ΔC_ps. Listed in Table I are values of the constants C_0–C_4 suggested by analysis of the available data for several binary systems.

The four procedures are thermodynamically equivalent and are therefore useful in performing consistency tests on available sets of data. As an ex-

Table I. Suggested Values of Constants for the Calculation of θ_{ij}

Constant	*AlSb*	*GaSb*	*InSb*	*CdTe*	*HgTe*
C_0	−5.481	5.37	−5.724	−6.334	−12.34
C_1	-0.513×10^{-3}	0	-0.706×10^{-3}	-1.142×10^{-3}	-1.510×10^{-3}
C_2	−9177	−7950	−5902	−14,930	−5534
C_3	0	0	0	44,290	0
C_4	1.522	0.1	1.656	1.787	2.436

NOTE: These values are intended to be used in conjunction with equation 42.

ample, the consistency of In (*116*) and Ga (*117*) activity measurements with the reported phase diagram and the enthalpy of mixing in liquid solution with Sb (method IV) were compared with the available data sets required by methods I and II. In particular, the comparison showed that method II is more sensitive than method I, because the partial molar solution properties are required relative to the pure components rather than to the stoichiometric liquid.

As another example, methods I–IV were applied to the In–Sb system by using only experimental values of the required experimental properties. Figure 17a shows the recommended value of θ_{InSb} as a function of $1/T$. Also shown in this figure are the upper and lower experimental limits on θ_{InSb}^{I}. Figure 17b shows experimental values for θ_{InSb} with method II. The upper and lower experimental limits on θ_{InSb} with method III are shown in Figure 17c, along with the recommended value for θ_{InSb}. The data base used in calculating these limits were taken from the literature (*116*, *118–120* for θ_{InSb}^{I} [upper bound]; *121–126* for θ_{InSb}^{I} [lower bound]; *116*, *120*, *127*, *128* for θ_{InSb}^{III} [upper bound]; *124*, *125*, *129–131* for θ_{InSb}^{III} [lower bound]; and *132–134* for the elements).

The experimental bounds for θ_{InSb}^{I} shown in Figure 17a are narrow at elevated temperatures, but a wide range exists at lower temperature. This situation is not unexpected, because the last two terms in equation 33 are 0 at T_m^{InSb}, a fact indicating a consistent set of activity data. As temperature is decreased below T_m^{InSb}, the experimentally inaccessible properties of the stoichiometric liquid increase in importance and are sensitive to the extrapolation procedure. Such a wide range in θ_{ij} can produce significant differences in the calculated phase diagram, because θ_{ij} appears as the argument of an exponential in equation 27. The upper and lower limits for method III are narrower than those for method I and reflect the explicit temperature dependence found in equation 34 and the smaller uncertainty in the required difference in heat capacity.

Method II for calculating θ_{ij} is particularly useful in the temperature range accessible with galvanic cells (for the measurement of ΔG_f^o) or calorimetry (for the measurement of ΔH_f^o). The values of θ_{InSb}^{II} shown in Figure 17b were obtained by using direct high-temperature electromotive force determinations of ΔG_f^o[InSb, T] (*129*, *135*, *136*). Also shown in this figure are low-temperature values of θ_{InSb}^{II} calculated from calorimetric determinations of ΔH_f^o[InSb] (*123*, *137*, *138*) and the absolute entropy of the elements (*132*) and InSb (*139*). Room-temperature calorimetric measurements permit the low-temperature value of θ_{ij} to be pinned down. Finally, the change in the reduced standard-state chemical potential was calculated by method IV from the reported phase diagram (*140*) and enthalpy of mixing (*118*) at the compositions for the available activity measurements (*116*, *125*, *141*). The results are summarized in Table II and are consistent with the recommended value of θ_{InSb}.

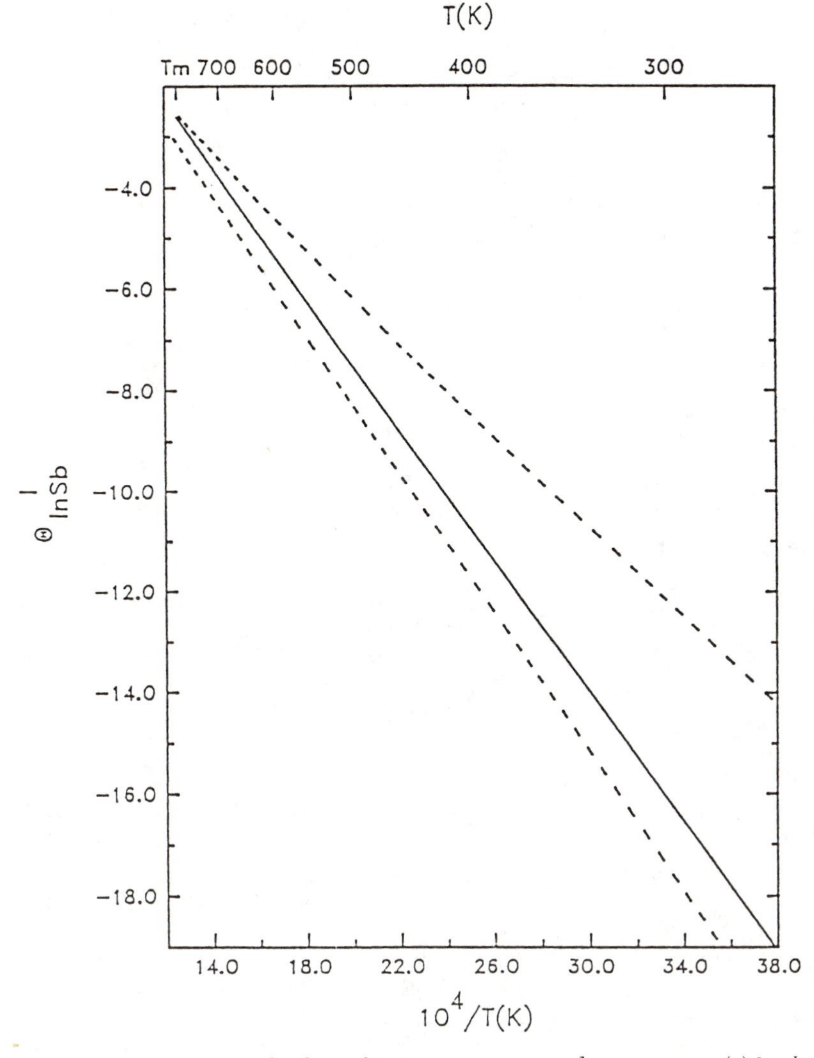

Figure 17. Experimental values of θ_{InSb} versus reciprocal temperature. (a) $\theta_{InSb}{}^{l}$: ---, upper and lower limits; and —, recommended value. Continued on next page.

From a Solution Model. Calculation of the difference in reduced standard-state chemical potentials by methods I or III in the absence of experimental thermodynamic properties for the liquid phase necessitates the imposition of a solution model to represent the activity coefficients of the stoichiometric liquid. Method I is equivalent to the equation of Vieland (*106*) and has been used almost exclusively in the literature. The principal difference between methods I and III is in the evaluation of the activity coefficients

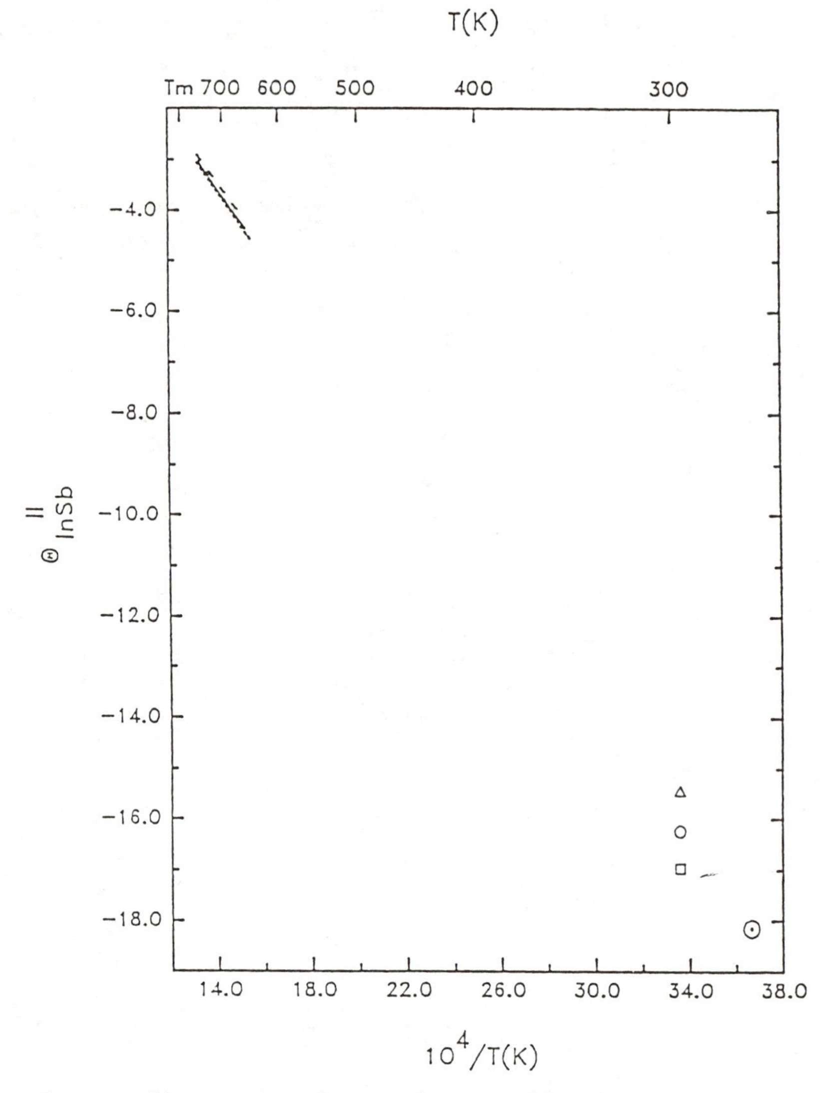

Figure 17. (b) θ_{InSb}^{II}: ΔG_f^o reference values were obtained from references 129 (---), 136 (—), and 135 (· · ·), and ΔH_f^o reference values were obtained from references 123 (△ and ○), 137 (⊙), and 138 (□).

of the stoichiometric liquid. With method I, the activity coefficients must be calculated at each liquidus temperature T^l, whereas with method III, only the activity coefficients at the melting point of the compound are required.

The most commonly used solution model to represent the liquid-phase behavior is the simple-solution model. Figure 18 shows a comparison between the recommended value of θ_{AlSb} and the value calculated from the

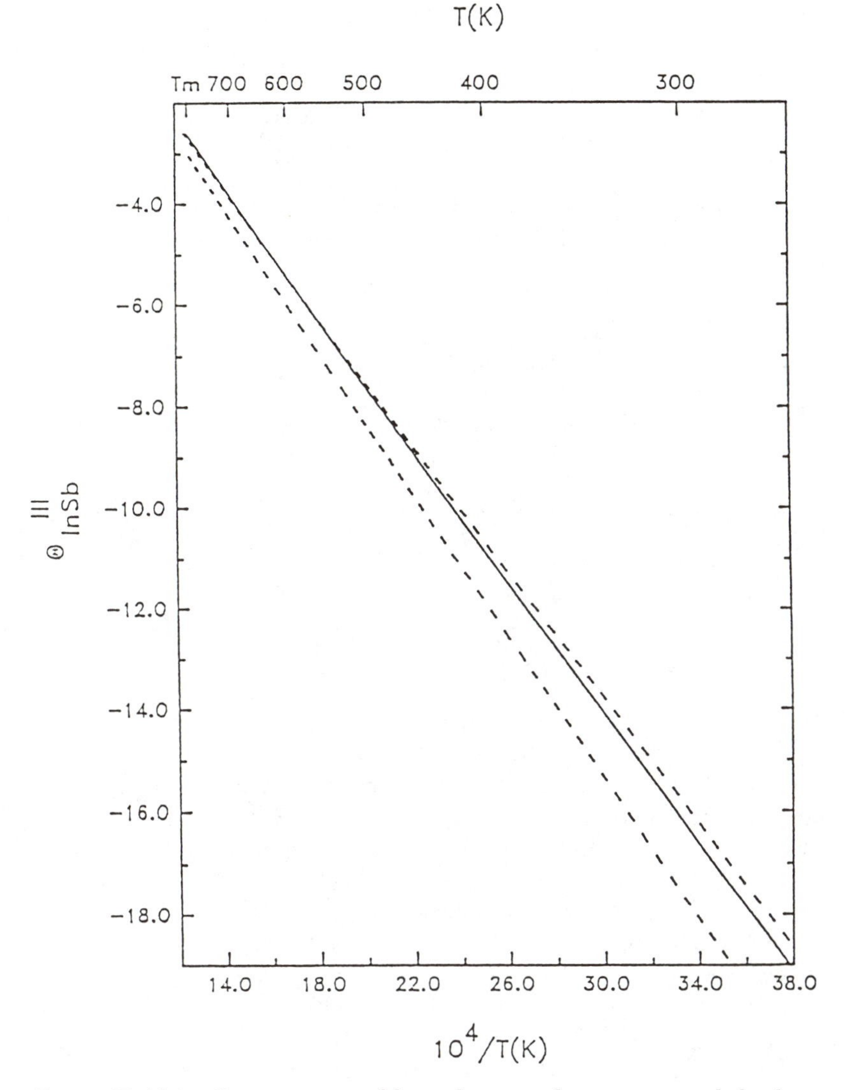

Figure 17. (*c*) θ_{InSb}^{III}: *---, upper and lower limits; and —, recommended value.*

parameters of the simple-solution model (*142–145*). These parameters were estimated from a fit to the liquidus measurements. In all cases, a reasonable fit of the liquidus data was reported, but the calculated values of θ_{AlSb} show significant variation (Figure 18). Obviously variations in the liquid-solution activities must exist to yield a reasonable fit to the phase diagram, and these variations become more pronounced as temperature is decreased. Thus, an accurate description of the liquid-solution behavior is difficult to obtain by

Table II. Values of θ_{InSb}^{IV}

x_{In}	T (*K*)	θ_{InSb}^{IV}	*Reference*
0.3	774.3	−3.41	125
0.4	788.6	−3.12	125
0.5	798.2	−3.01	125
0.6	787.6	−3.14	125
0.7	756.9	−3.56	125
0.8	699.3	−4.42	125
0.9	607.2	−6.24	125
0.3	774.3	−2.87	141
0.4	788.6	−2.69	141
0.5	798.2	−2.55	141
0.6	787.6	−2.66	141
0.7	756.9	−3.04	141
0.8	699.3	−3.87	141
0.9	607.2	−5.61	141
0.398	788.3	−2.90	116
0.500	798.2	−2.60	116
0.600	787.6	−2.75	116
0.656	772.1	−2.78	116
0.703	755.8	−3.16	116
0.787	707.8	−3.96	116
0.890	616.7	−5.81	116

using method I combined with phase diagram data to estimate liquid-solution-model parameters.

To test the effect of using a solution model in the calculation of θ_{ij} the simple-solution model was used in conjunction with either method I or method III to fit sets of data consisting of the liquidus temperature alone (*146*), liquidus temperature and enthalpy of mixing (*147*), liquidus temperature and activity (*147*), and all three types of data combined. With the parameters determined from the fit, values of θ_{ij} as a function of temperature were calculated and compared with the recommended value.

Figure 19 shows the results obtained with method I for the Al–Sb system. Examination of the results for the fit of the liquidus temperature alone (the usual procedure) indicates that both the activity product at the melting temperature and the temperature dependence of θ_{AlSb} are represented rather poorly. The results for the data base consisting of the liquidus and the enthalpy of mixing are significantly better than the results obtained with the liquidus data set alone. Only when activity data are also included in the data set does the value of θ_{AlSb} agree with the recommended value.

The results of similar calculations using the same solution data base and method III are shown in Figure 20. In all cases, the agreement between the calculated value of θ_{AlSb} and the recommended value is improved. A similar improvement by using method III has also been shown for the Ga–Sb system (*148*). For those systems in which no solution measurements have

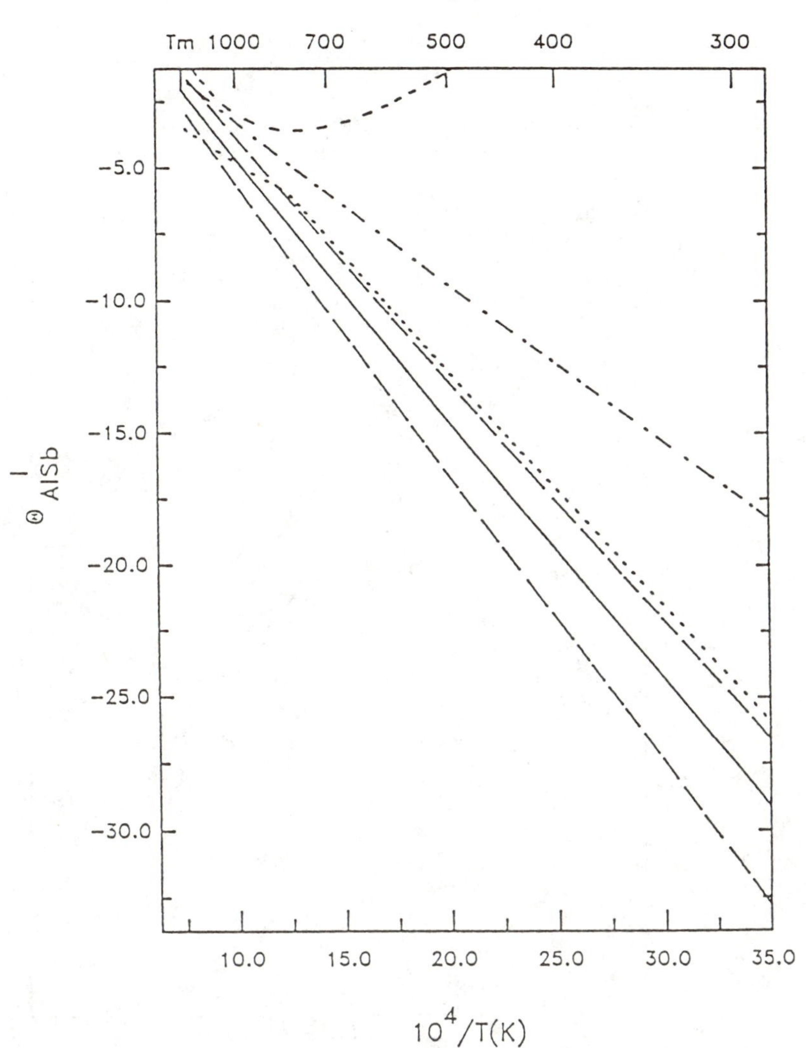

Figure 18. Values of $\theta_{AlSb}{}^{l}$ calculated from solution model parameters estimated by fitting the phase diagram. Data are from references 144 (— —), 143 (· · ·), 142 (— · —), and 145 (---). The recommended values are indicated by —.

been reported (e.g., phosphides), method III is a superior procedure for estimating θ_{ij} and solution model parameters.

By using the procedures just outlined, the reduced standard-state chemical potential can be estimated for all compounds. This value of θ_{ij} is valid for any solid–liquid phase equilibrium problem that contains the compound

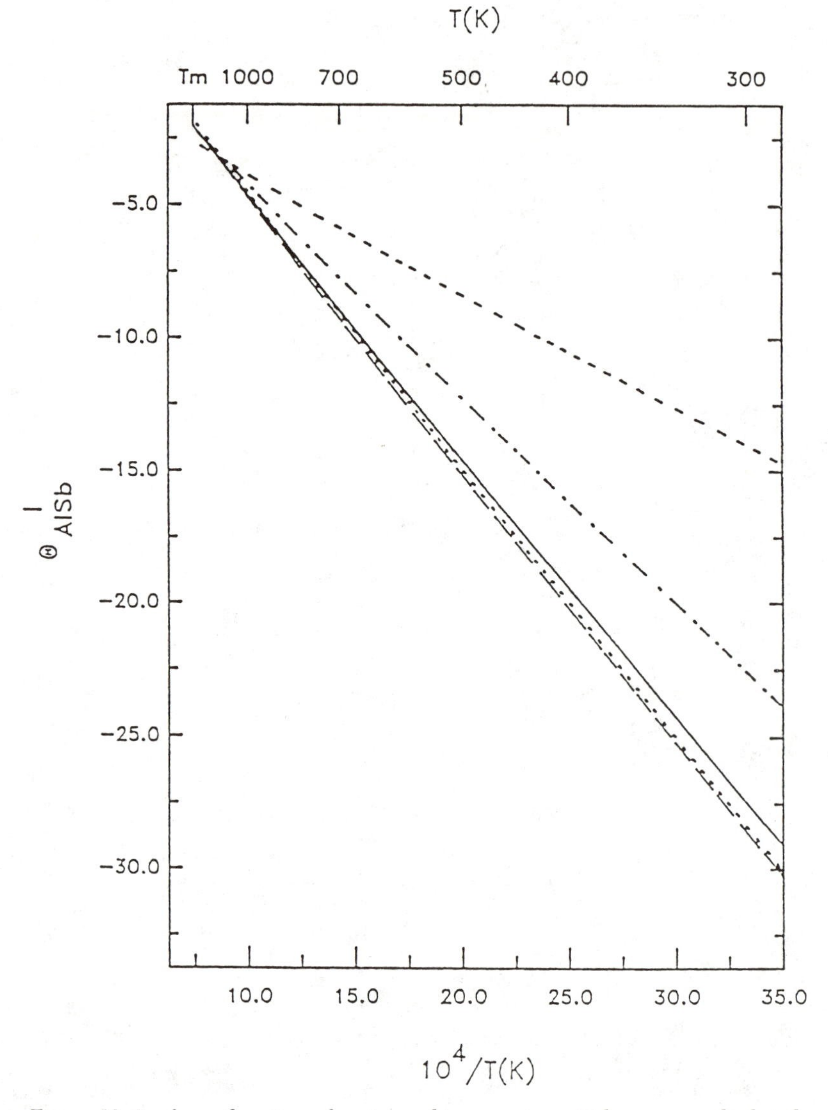

Figure 19. $\theta_{AlSb}{}^{l}$ *as a function of reciprocal temperature. Values were calculated by the simple-solution model, with parameters estimated from a fit of the combined data set (· · ·), liquidus data only (---), liquidus and activity data (— —), and liquidus and enthalpy of mixing data (— · —). The recommended values are indicated by —.*

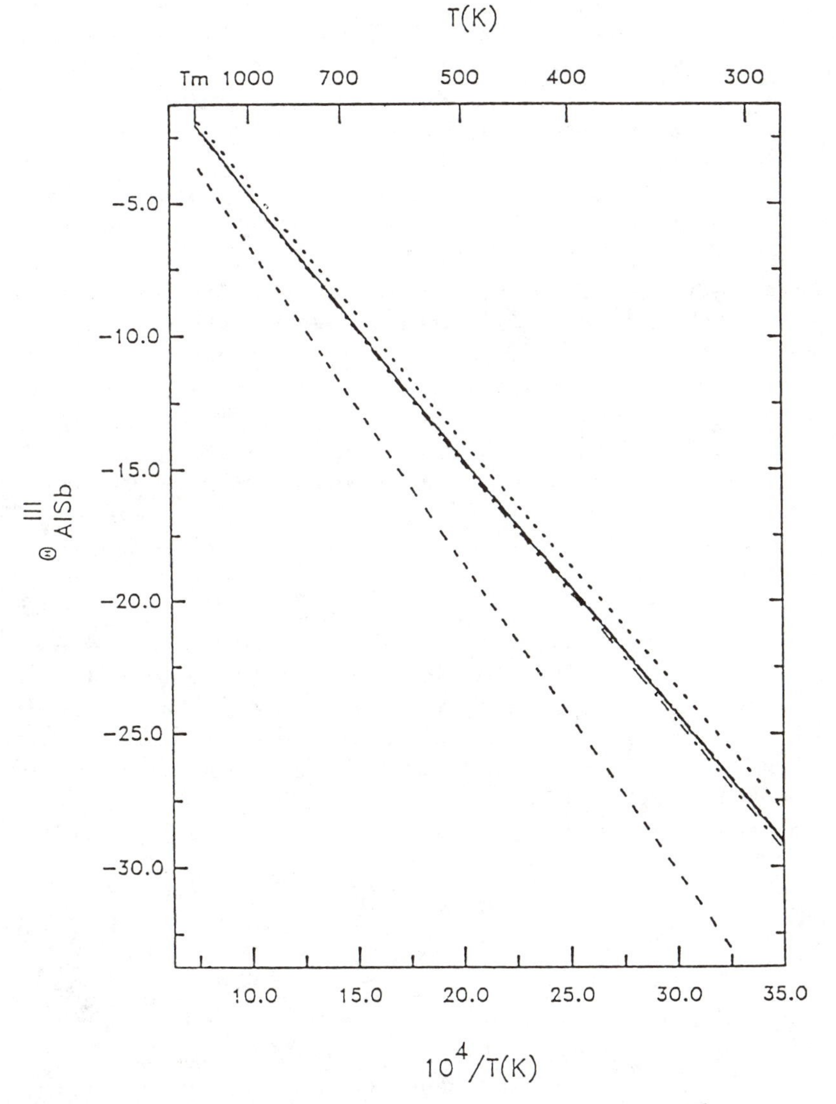

Figure 20. Values of θ_{AlSb}^{III} *versus reciprocal temperature. Values were calculated with simple-solution model parameters estimated by a fit to the combined data set (— —) and nearly coincident with the recommended values, liquidus data only (---), liquidus and activity data (— · —), and liquidus and enthalpy of mixing data (· · ·). The recommended values are indicated by —.*

(pure or in solution). However, the available data base for some systems is not complete (e.g., arsenides and phosphides). The lack of a comprehensive data base is primarily a result of difficulties in performing the required experiments (e.g., the high vapor pressures of the arsenides and phosphides). Because the standard state is entirely arbitrary, an infinite-dilution standard state would avoid the experimental difficulties encountered with high-vapor-pressure species. Unfortunately, the infinite-dilution properties of these species have received little experimental attention.

Determination of Γ_{ij}. In the formulation of the phase equilibrium problem presented earlier, component chemical potentials were separated into three terms: (1) θ_{ij}, which expresses the primary temperature dependence, (2) solution mole fractions, which represent the primary composition dependence (ideal entropic contribution), and (3) Γ_{ij}, which accounts for relative mixture nonidealities. Because little data about the experimental properties of solutions exist, Γ_{ij} is usually evaluated by imposing a model to describe the behavior of the liquid and solid mixtures and estimating model parameters by semiempirical methods or fitting limited segments of the phase diagram. Various solution models used to describe the liquid and solid mixtures are discussed in the following sections, and the behavior of Γ_{ij} is presented.

Liquid-Solution Models. The simple-solution model has been used most extensively to describe the dependence of the excess integral molar Gibbs energy, G^{xs}, on temperature and composition in binary (*142–144, 149–155*), quasi binary (*156–160*), ternary (*156, 160–174*), and quaternary (*175–181*) compound-semiconductor phase diagram calculations. For a simple multicomponent system, the excess integral molar Gibbs energy of solution is expressed by

$$G^{xs} = \frac{1}{2} \sum_{k=1}^{n} \sum_{j=1; j \neq k}^{n} w_{kj} x_k x_j \tag{43}$$

in which the interchange energies are equal ($w_{kj} = w_{jk}$), are functions of temperature and pressure, and are independent of composition. For condensed phases, the pressure dependence of w_{kj} can be neglected and w_{kj} is usually permitted a linear temperature dependence, $a + bT$ (in which a and b are constants). A strictly regular solution (with random mixing and excess entropy of mixing [S^{xs}] is 0) and an athermal solution (enthalpy of mixing [ΔH_{mix}] is 0) are two limiting cases of the simple solution. For strictly regular solutions, $w_{kj} = a$, and deviations from ideal-solution behavior arise from heat effects, whereas for athermal solutions, $w_{kj} = bT$, and deviations from ideality arise from entropy rather than heat effects.

In general, the simple-solution model performs quite well for symmetric binary systems (e.g., Ga–Sb). Inspection of equation 43 with $n = 2$ shows that the excess integral molar Gibbs energy is a symmetrical function of composition for binary systems. For a highly asymmetric system (e.g., Al–Sb), Anderson et al. (*143*) and Joullie and Gautier (*144*) used different values of the interaction parameter on either side of the compound melting point, whereas Cheng et al. (*182*) added a concentration-dependent term to the interaction parameter w_{kj}. Similarly, a composition-dependent w_{ij} has been used to describe the Ga–In (*183*), Ga–As, and Ga–P systems (*157*). The use of two different values of w_{kj} for a binary compound in the prediction of a ternary phase diagram will give a discontinuity in the liquidus at the quasi binary composition. In addition, the use of this extra term to calculate a multicomponent phase diagram will lead to a thermodynamic inconsistency in expressions of ternary activity, because the Gibbs–Duhem equation is not satisfied.

Although the simple-solution model provides a good analytical representation of the binary phase diagrams, good values for the thermodynamic properties of the liquid solution are not obtained when parameters determined from a fit of the binary liquidus are used. For example, values of the enthalpy of mixing predicted from these liquidus fits are always positive, whereas available experimental data show negative values. Indeed, the enthalpy of mixing is expected to be always negative because of the strong attractive interactions. This expectation is also expressed in the phase diagrams as a negative deviation from ideality and a tendency toward compound formation.

Similarly, Thurmond (*150*) and Arthur (*151*) found that the interaction coefficients obtained from a fit of the experimental liquidus or vapor pressure in the arsenide and phosphide systems did not produce the same temperature dependence. Panish et al. (*142*, *154*) pointed out that these discrepancies may be due to (1) errors resulting from the assumed values for ΔH_f^{ij} and the approximation $\Delta C_p[ij] = 0$ in θ_{ij}^{I}, (2) deviations from simple-solution behavior, or (3) uncertainties in the interpretation of the vapor pressure data, because some of the quantities necessary in the calculations are not accurately known (e.g., reference-state vapor pressures for pure liquid As and P). Knobloch et al. (*184*, *185*) and Peuschel et al. (*186*, *187*) have obtained excellent agreement between calculated and experimental activities and vapor pressures with the use of Krupkowski's asymmetrical formalism for activity coefficients, whereas Ilegems et al. (*111*) demonstrated that satisfactory agreement between liquidus and vapor pressure measurements exists when an accurate expression for the liquidus is used.

In addition, several other models have been used with method I to calculate binary or ternary phase diagrams (*183*, *188–201*). Among these models are the quasi chemical equilibrium model (*188*, *190*), truncated Margules expansions (*183*, *191*, *192*), Gaussian formalism (*193*), orthogonal series expansions (*194*), Darken's formalism (*195*), and various chemical theories (*196–201*).

The most popular chemical theory postulates stoichiometric chemical species that interact accordingly as a regular solution (regular associated-solution model). The associated-solution model is based usually on the following assumptions:

1. the moleculelike stoichiometric species of unlike atoms called "clusters", "associates", or "complexes" exist in the liquid state,
2. the associated complexes are in a dynamic equilibrium with the nonassociated atoms that can be described by a mass action law,
3. the associated complexes behave as independent particles,
4. all species are statistically distributed, and
5. the excess thermodynamic properties consist of contributions from both the physical interactions and the chemical reactions.

This model has been widely used for strongly interacting systems that exhibit asymmetric properties. However, the problem becomes complicated when this model is generalized to multicomponent systems because of the possibility of new species being formed. The formation of new species results in a large number of parameters involved in the calculation.

The case of binary solid–liquid equilibrium permits one to focus on liquid-phase nonidealities because the activity coefficient of solid component ij, γ_{ij}, equals unity. Aselage et al. (*148*) investigated the liquid-solution behavior in the well-characterized Ga–Sb and In–Sb systems. The availability of a thermodynamically consistent data base (measurements of liquidus, component activity, and enthalpy of mixing) provided the opportunity to examine a variety of solution models. Little difference was found among seven models in their ability to fit the combined data base, although asymmetric models are expected to perform better in some systems.

Often, a complete data base is unavailable for a particular system. In general, the results of attempted cross predictions (e.g., prediction of component activities from a fit to liquidus data) are unsatisfactory (*190*). When model parameters are allowed to vary with temperature, a significant correlation between the parameters is introduced. As an example, 87%-confidence ellipses for simple-solution parameters determined from fits to the combined Ga–Sb data set and liquidus alone are shown in Figure 21. The importance of the ellipses is in the shape rather than the relative magnitude. The ellipse resulting from the liquidus fit in Figure 21 is quite narrow and elongated. This ellipse shape indicates a high degree of correlation between the parameters. Thus a wide range of values exists for the parameter a, and

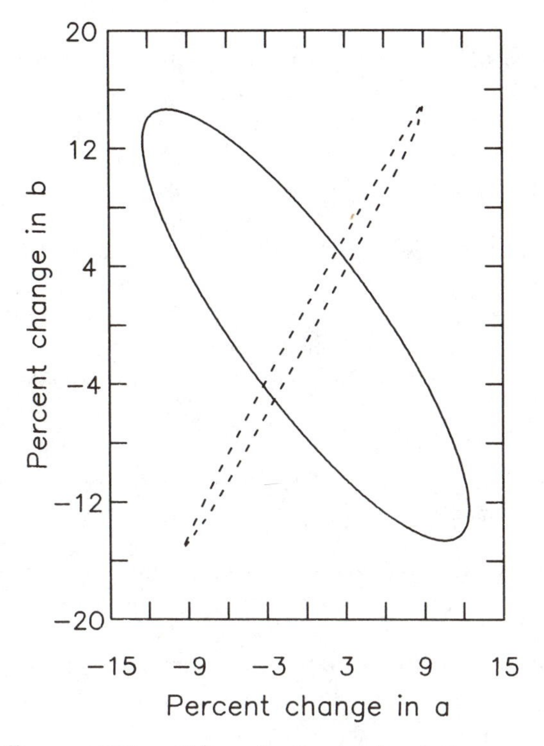

Figure 21. Ellipses at 87% confidence for the simple-solution parameters a *and* b *determined by a fit of the Ga–Sb data base: —, combined data set; ---, liquidus data only. (Reproduced with permission from reference 190. Copyright 1985 Pergamon.)*

associated with each value of a is a narrow range for the parameter b. These parameters would give nearly the same description of the liquidus. As long as the property to be calculated requires the sum $a + bT$ in the model equation (as in the case of isothermal activity coefficient), this correlation is not a major problem. This result helps explain the difficulty in extracting reliable information on the enthalpy of mixing from a fit to the liquidus; solution models are imperfect in their ability to represent the complex solution chemistry.

Solid-Solution Models. Compared with the liquid phase, very few direct experimental determinations of the thermochemical properties of compound-semiconductor solid solutions have been reported. Rather, procedures for calculating phase diagrams have relied on two methods for estimating solid-solution model parameters. The first method uses semiempirical relationships to describe the enthalpy of mixing on the basis of the known physical properties of the binary compounds (*202*, *203*). This approach does not provide an estimate for the excess entropy of mixing and thus neglects clustering that has been suggested in some systems (*204*). The second method estimates solution model parameters from a fit to experimental values of the multicomponent solid–liquid phase diagram (*142*). This procedure is subject to uncertainties in the liquid-solution thermodynamic properties, the experimental solidus and liquidus temperature values, and the appropriateness of the solution model.

Semiempirical Models. Attempts have been made to calculate the solid- and liquid-interaction parameters from the physical properties of the constituents. Ilegems and Pearson (*163*), Foster (*205*), and Panish et al. (*142*) have suggested that the the solid-phase interaction parameter can be determined approximately from the magnitude of the mismatch between the lattice parameters of the two end compounds, although no analytical expressions were presented to quantify the contribution to the excess Gibbs energy. For example, the solid-interaction parameters might be zero for the Al_xGa_{1-x}–V ternary systems, because the lattice parameters of the binary compounds are nearly identical.

Using the Phillips–Van Vechten theory (*206*) of chemical bonding to calculate the solid-interaction parameters and the molar volumes, Hildebrand's solubility parameters (*207*), and electronegativities (*208*, *209*) of the constituent elements to calculate the liquid-interaction parameters, Stringfellow (*203*, *210*) calculated the binary and ternary phase diagrams of group III–V systems. However, the agreement with the experimentally determined phase boundaries was poor in several cases. Stringfellow (*203*, *211*) presented a simpler, more accurate semiempirical model, called the delta–lattice–parameter (DLP) model, which is based on the Phillips theory (*212*) of predicting the solid-interaction parameters for group III–V systems. The results of the calculations are in good agreement with those determined by fitting the experimental phase diagram. This model, however, is not always appropriate for group II–VI and other systems, because the DLP model neglects mismatches in the ionicities and dehybridization factors of the two binary compounds (*212*).

Fedders and Muller (*213*) have derived an estimate of the solid-interaction parameter from another point of view, which ascribes the mixing enthalpy to bond distortions associated with the alloy formation and relates these distortions to the macroscopic elastic properties of the crystal. They concluded that the results based on elastic-crystal parameters yield a similar form for the thermodynamic properties as those estimated by DLP model based on optical-crystal parameters.

These semiempirical models postulate that local strain associated with different atomic sizes of the elements is the major contribution to the solid-solution enthalpy of mixing. An estimate of lattice strain energy has been compared to fitted values of the enthalpy of mixing for several group III–V systems (*156*). The results led to a calculated enthalpy of mixing that was a factor of 5 too high. In this elastic model (*156*), the strain energy was assumed to be that required to increase the bond distance of one group III–V pair and decrease the bond distance of the other pair to achieve the average bond distance suggested by the alloy lattice constant. Extended X-ray absorption-fine-structure measurements of group III–V systems (*214–216*) show that a constant difference between bond lengths always exists at any given alloy composition. An expression for the strain energy was derived that was consistent with the regular-solution formulation, the DLP model, and the fitted values of the enthalpy of mixing. Further support for the DLP model and estimates developed from an assessment of the phase diagram are found in recent dissolution calorimetric measurements of the enthalpy of mixing of $Ga_xIn_{1-x}P$ (Figure 22) (*202, 203, 217–219*) However, some evidence indicates that the excess entropy of mixing is not zero (*217*).

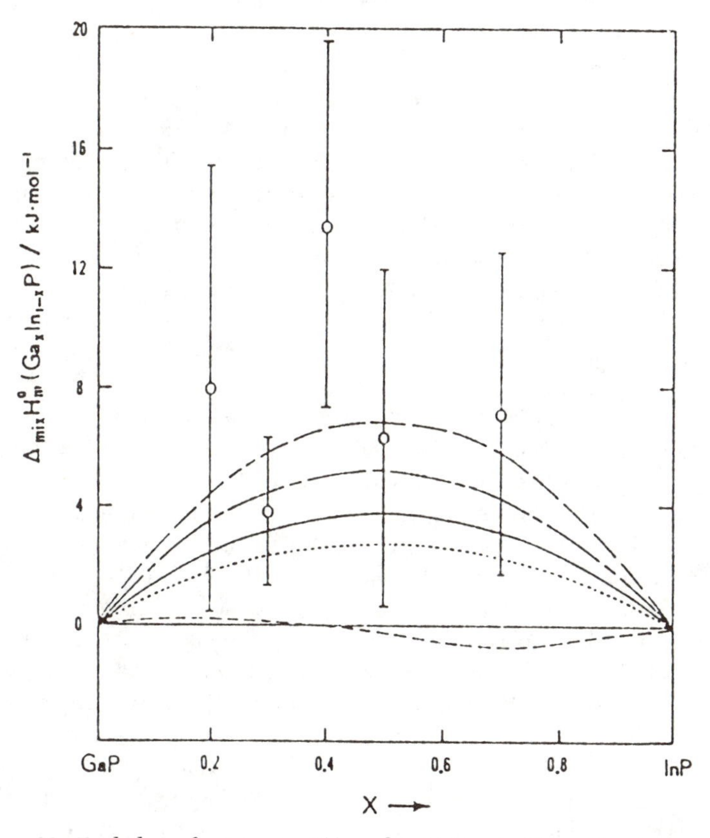

Figure 22. Enthalpy of mixing GaP(s) and InP(s) at 1048 K. The data were determined by dissolution calorimetry (— —; 217) and from references 202 (----), 203 (—), 218 (— - —), and 219 (---). (Reproduced with permission from reference 217. Copyright 1987 Elsevier.)

Parameter Estimation from Phase Diagram Assessment. The second approach used to determine solid-solution behavior is to estimate solution model parameters from a fit to the measured phase diagram. Many of the solution models used to describe the liquid solution have been used to model the solid mixture. The simple-solution expression and its special cases have been used most extensively.

The data base generally used is either the pseudobinary phase diagram or group III-rich portions of the ternary phase diagram. Foster and coworkers (*156*, *220–222*) reported that six pseudobinary sections can be satisfactorily fitted on the assumptions that the liquid phase is ideal and the solid phase is athermal. Panish and Ilegems (*142*) obtained somewhat poorer fits on the assumption that both liquid and solid solutions are strictly regular. Brebrick and Panlener (*159*) investigated the ideal, strictly regular, athermal, and quasi regular models for each phase and concluded that the strictly regular liquid with the simple solid is the simplest formulation giving satisfactory fits for each of seven systems examined.

A general result of predicting multicomponent phase diagrams from a fit to binary and pseudobinary or ternary phase diagrams is that liquidus isotherms are in fair or good agreement with experimental data and are insensitive to the choice of model. The calculated solidus isotherms, however, often give fair agreement only with experimental data and appear to be more sensitive to the values of the interaction energies. As an example, Gratton and Wolley measured solidus isotherms in the Ga–In–Sb system (*160*). Although the parameter set chosen by these authors, as well as by several others (*142*, *161*, *162*, *223*), showed reasonable agreement with experimental liquidus isotherms, the agreement with the solidus isotherms was poor in each case.

One of the main difficulties with using the pseudobinary phase diagram as a data base for estimating the solid solution properties is that the phase diagram represents the high-temperature behavior only. For most applications, the lower temperature portion of the phase diagram is important (e.g., for LPE and the prediction of miscibility gaps in the solid solution). The temperature dependence of the solid-solution Gibbs excess energy is sensitive to the solution model and the method of data reduction used.

As an example, Chang et al. (*224*) studied the $Ga_xIn_{1-x}Sb$ pseudobinary system. The $Ga_xIn_{1-x}Sb$ liquid mixture was treated either as a ternary mixture of Ga, In, and Sb, with the thermodynamic properties estimated with binary parameters, or as a binary mixture of GaSb and InSb, with the thermodynamic properties calculated from the simple-solution model by using the parameters estimated from a fit of the pseudobinary phase diagram. For both descriptions of the liquid mixture, the simple-solution equation was used to model the solid-solution behavior, and parameters were estimated from a fit of the pseudobinary phase diagram. Both treatments of the liquid phase gave standard deviations in the liquidus and solidus temperatures within the experimental uncertainty. The variations of the Gibbs excess energy of the solid solution with temperature, however, were in opposite directions for the two different treatments of the liquid phase. Thus the low-temperature segment of the ternary phase diagram is predicted differently by each assumption.

Experimental Values of Γ_{ij}. An experimental value for Γ_{ij} can be assigned by solving the equilibrium equations for Γ_{ij}. In the case of a ternary equilibrium, the result is given by

$$\Gamma_{ij} = \frac{x_{ij}}{x_i x_j} \exp(\theta_{ij}) \tag{44}$$

By using data from experimental phase diagram to obtain values of x_{ij}, x_i, and x_j and methods I–IV to calculate the quantity θ_{ij}, values of Γ_{GaSb} were calculated for the Al–Ga–Sb and Ga–In–Sb systems. Values of $1 - \Gamma_{GaSb}$ are shown as a function of composition along several isotherms in Figures 23 and 24. These two systems represent two extremes in the solid-solution behavior, with the $Al_xGa_{1-x}Sb$ system deviating only slightly from ideal behavior and the $Ga_xIn_{1-x}Sb$ system showing strong positive deviations from Raoult's law.

As shown in Figure 23, the value of Γ_{GaSb} in the Al–Ga–Sb system is significantly different from unity as a result of nonidealities in the liquid solution. The value of Γ_{GaSb} in this system is also nearly independent of both composition and temperature. Thus this system can be modeled as an ideal solution in both phases if the value of θ_{GaSb} was suitably adjusted.

Shown in Figure 24 (*160*, *225*) is a similar plot of $1 - \Gamma_{GaSb}$ versus x_{Sb} along six isotherms for the Ga–In–Sb system. Γ_{GaSb} is seen to be a more complex function of both temperature and composition, although this quantity appears to be independent of composition along several isotherms. A closer inspection of equation 44 reveals that the quantity $\exp(\theta_{ij})$ is constant along an isotherm and that the quantities x_{ij}/x_i and $1/x_j$ are equal to twice the distribution coefficients for elements i and j, respectively. Thus, not surprisingly, the variation of Γ_{ij} with composition and temperature is not as pronounced as that of the individual activity coefficients. This partial cancellation of the composition and temperature dependence of the liquid-solution activity coefficient product by the dependence of the solid solution activity coefficient is largely responsible for the ability of rather simple solution models to represent phase diagrams, and makes extraction of meaningful estimates of the individual mixture properties from fitted phase diagrams difficult.

Binodal and Spinodal Decompositions. As the temperature of a solid solution is lowered, the entropy contribution to the solid Gibbs energy

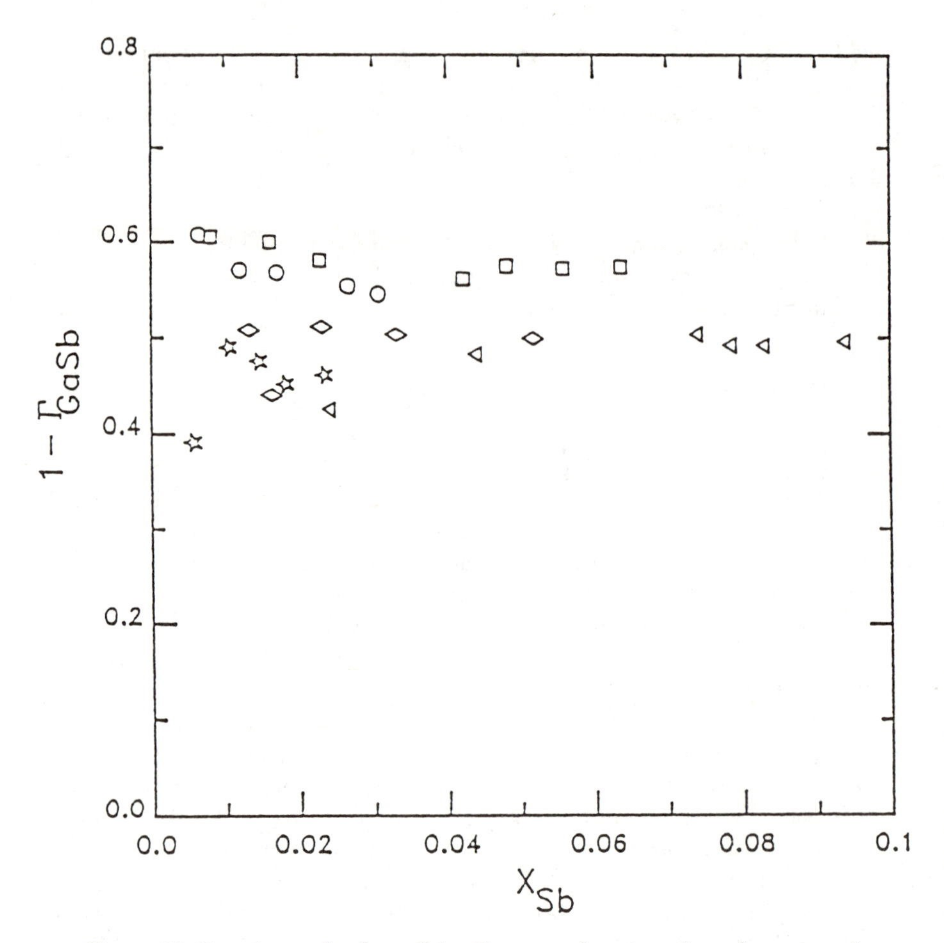

Figure 23. Experimental values of 1 − Γ_{GaSb} as a function of x_{Sb} along several isotherms in the Al–Ga–Sb system. The calculation used the recommended value of θ_{GaSb} in Table I and the phase diagram measurements of Dedegkaev et al. (145) *(☆, 778 K; ◇, 825 K; △, 873 K) and of Cheng and Pearson* (182) *(○, 773 K; □, 823 K).*

is decreased. This decrease in Gibbs energy can permit more-ordered phases to be stable and be insoluble in compound-semiconductor solid solutions. A typical pseudobinary phase diagram exhibiting a miscibility gap is shown in Figure 25. Two different two-phase regions are depicted. At a high temperature, the normal liquid–solid (L + S) two-phase field is formed, but below a critical temperature, T_c, a two-solid field is formed. The solid-phase boundary line represents the miscibility or binodal line and is determined by the condition of equal chemical potentials in each solid phase. For regular solid-solution behavior, this line is symmetric with respect to the axis x =

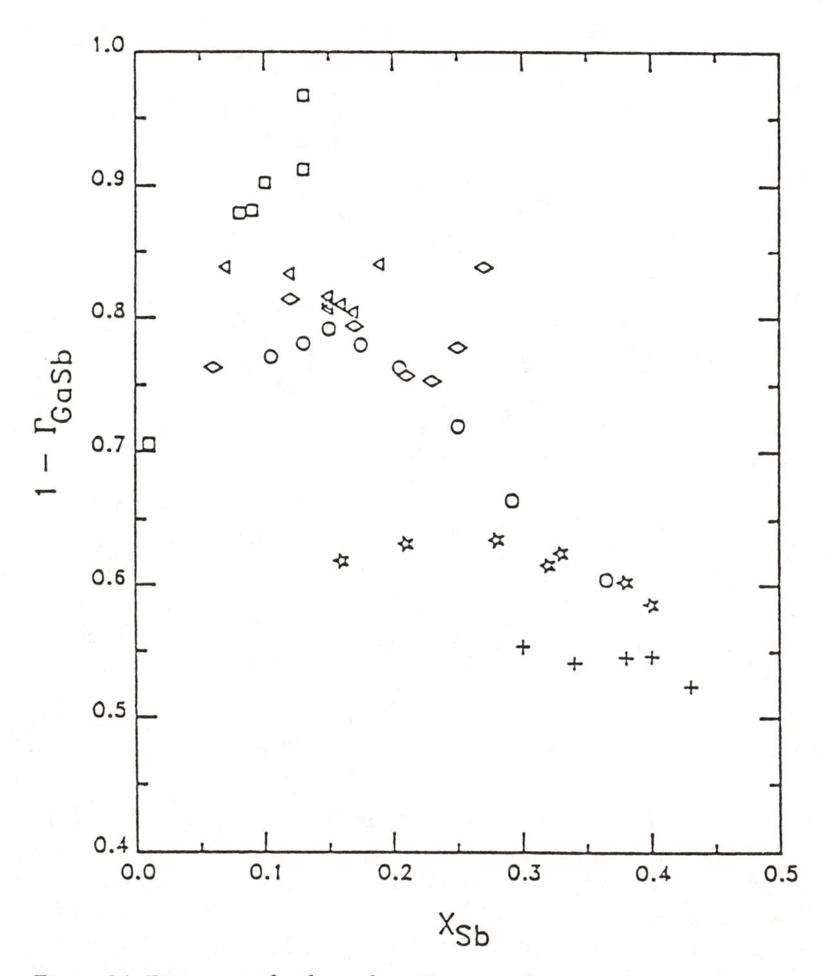

Figure 24. Experimental values of $1 - \Gamma_{GaSb}$ as a function of x_{Sb} along several isotherms in the Ga–In–Sb system. The calculation used the recommended values of θ_{GaSb} listed in Table I and the phase diagram measurements of Antypas (225) *(○, 773 K) and of Gratton and Wolley* (160) *(□, 653 K; △, 703 K; ◇, 748 K; ☆, 873 K; +, 923 K).*

½ and is represented by the equation

$$\ln\left(\frac{1}{1 - x}\right) = \frac{w_{kj}(2x - 1)}{R} T = \frac{2T_c(2x - 1)}{T} \quad (45)$$

The spinodal line is determined by the state at which the second derivative of the Gibbs energy with respect to composition is equal to zero. For a

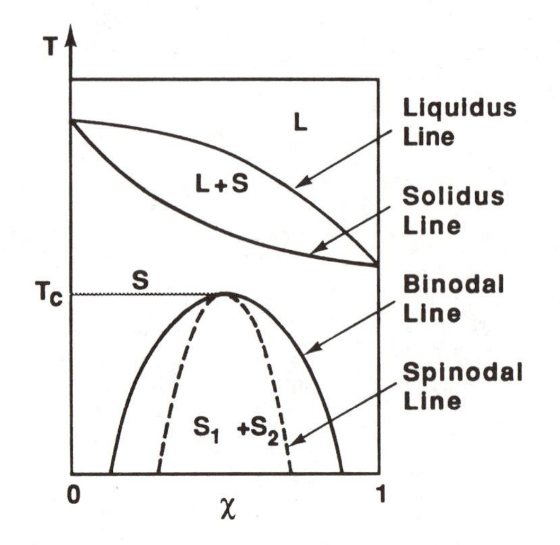

Figure 25. Schematic of a pseudobinary phase diagram exhibiting a miscibility gap.

regular solution, the spinodal line is determined by

$$x(1 - x) = \frac{RT}{2w_{kj}} \tag{46}$$

The solutions in the region inside the spinodal domain are unstable, whereas the solid solutions in the region between the binodal and spinodal lines are metastable. The presence of a miscibility gap limits the potential usefulness of these materials in device applications. Solutions with compositions lying inside the spinodal domain cannot be grown by LPE, whereas metastable solid solutions have a tendency toward phase separation and, eventually, device degradation.

Three types of behavior are exhibited by compound-semiconductor systems (*226*). First, some systems do not involve a miscibility gap (e.g., $Al_xGa_{1-x}As$). Second, a class of pseudobinary phase diagrams show a miscibility gap with T_c below the solidus (e.g., $Ga_xIn_{1-x}As$, $Ga_xIn_{1-x}P$, and GaP_xAs_{1-x}; Figure 25). Third, in immiscible systems, the binodal curve can penetrate the solidus line to produce a peritectic-type phase diagram (e.g., $Al_xP_xSb_{1-x}$ and GaP_xSb_{1-x}). Several attempts have been made to predict binodal and spinodal curves for compound semiconductors from semiempirical models and analysis of the phase diagrams (*227–231*). By using these methods, the values of T_c reported by different authors for a particular system show considerable scatter when fitting phase diagrams to determine the solid-solution behavior. The value of T_c is sensitive to the model and data base used. An experimental determination of the binodal decomposition curves is difficult because of slow transformation kinetics in the solid phase.

Summary

Liquid-phase epitaxy is a mature process that is receiving intense competition from alternative processes such as MBE and MOCVD. Although LPE is a small-scale operation, its ability to produce high-purity, low-defect-density films has made this process a useful deposition technique for many optoelectronic-device applications. The future of LPE as a viable commercial process is uncertain and will depend on the progress toward improving the film quality in alternative processes. LPE is a method that is rich in chemical engineering process phenomena, including interfacial and transport phenomena and phase equilibria. Many of the chemical processes are not fully understood, although first-order models and a vast amount of laboratory experience have rendered this method reliable for the growth of a variety of alloys.

A general formulation of the problem of solid–liquid phase equilibrium in quaternary systems was presented and required the evaluation of two thermodynamic quantities, θ_{ij} and Γ_{ij}. Four methods for calculating θ_{ij} from experimental data were suggested. With these methods, reliable values of θ_{ij} for most compound semiconductors could be determined. The term Γ_{ij} involves the deviation of the liquid solution from ideal behavior relative to that in the solid. This term is less important than the individual activity coefficients because of a partial cancellation of the composition and temperature dependence of the individual activity coefficients. The thermodynamic data base available for liquid mixtures is far more extensive than that for solid solutions. Future work aimed at measurement of solid-mixture properties would be helpful in identifying miscibility limits and their relation to LPE as a problem of constrained equilibrium.

Abbreviations and Symbols

a	activity
a_i^{sl}, a_j^{sl}	activities of components i and j in the stoichiometric liquid
A	area
c	refers to a critical point when used as a subscript
C	concentration
C_e	equilibrium concentration
C_i^l, C_i^s	concentrations of component i in the liquid and solid, respectively

C_i^o	reference concentration of component i
C^l	concentration in the liquid
C^s or C_s	concentration in the solid
C_o	initial concentration
C_p	constant-pressure molar heat capacity
$C_p^r[i]$, $C_p^r[j]$, $C_p^r[ij]$	constant-pressure molar heat capacity of components i, j, and ij in phase r
ΔC_p	difference between the molar heat capacities of the pure liquids and the solid compound
$\Delta C_p[ij]$	difference between the molar heat capacities of the stoichiometric liquid and the compound ij
$\Delta C_p[n]$	difference between the heat capacities of the pure liquid and the solid element n
d	film thickness
d_1, d_2, d_3	empirical constants in expression for molar heat capacity
D_i, D_j	diffusion coefficients of components i and j, respectively
e	refers to equilibrium when used as a subscript
f	refers to formation when used as a subscript
G	molar Gibbs energy
ΔG_f	molar Gibbs energy of formation
$\Delta G_f^o(ij, T)$	standard molar Gibbs energy of formation of compound ij at temperature T
ΔG_B	difference between the molar Gibbs energy of the bulk solid and that of the liquid solution
G^{xs}	excess integral molar Gibbs energy of solution
H	molar enthalpy
ΔH_{mix}^{sl}	molar enthalpy of mixing of stoichiometric liquid mixture
ΔH_m^n	molar enthalpy of fusion of pure element n
$\Delta\overline{H}(i, x_i^l)$, $\Delta\overline{H}(j, x_i^l)$	relative partial molar enthalpy of component i in liquid solution at composition x_i^l
ij	refers to solid compound ij when used a superscript or subscript
i, j, k	refer to mixture components when used as superscripts or subscripts
J	flux
J_i	evaporation rate of component i
k	first-order rate constant
k_i	equilibrium distribution coefficient, C_i^s/C_i^l
k_{eff}	effective distribution coefficient
K	constant defined by equation 18

K_n	constant with value 0 if element n is liquid at T and 1 if element n is solid at T
l	refers to the liquid when used as a superscript
l	liquid height
m	refers to melting when used as a subscript
m	slope of liquidus
$m_{i \text{ or } j}$	slope of liquidus for component i or j
M	molecular weight
M_i	molecular weight of component i
mix	refers to a mixing property when used as a subscript
n	number of components in liquid phase
N	rate of formation of nuclei
o	refers to an initial or reference value when used as a subscript
p	refers to the phase when used as a superscript
P_{ij}	partial distribution coefficient of component ij
r	cooling rate
r^*	critical nucleus radius
R	gas constant
s	refers to solid when used as a superscript or subscript
S	molar entropy
$\Delta S_f^o[ij, T_m^{ij}]$	standard molar entropy of formation of solid compound ij from the pure liquid elements at the compound melting temperature, T_m^{ij}
$\Delta' S_f^o[ij, T_m^{ij}]$	standard molar entropy of formation of solid compound ij from the elements in their natural state at the compound melting temperature, T_m^{ij}
sl	refers to the stoichiometric liquid when used as a superscript
t	time
T	temperature
T^*	measurement temperature
T_c	critical temperature
T^l	liquidus temperature
T_m^{ij}	melting temperature of compound ij
T_m^n	melting temperature of element n
ΔT	difference between saturation and actual temperatures
v	growth velocity
$v(y)$	velocity in the y direction
V	volume

w_{ki}, w_{kj}	interchange energy
x_i	mole fraction of component i in liquid solution
x_{ij}	mole fraction of component ij in solid solution
xs	refers to an excess function when used as a superscript
x, y	stoichiometry of solid solution $A_xB_{1-x}C_yD_{1-y}$
x	distance above growth interface
z	proportionality constant (equation 1)
δ	deviation from stoichiometry
γ	activity coefficient
γ_i^l, γ_j^l	activity coefficients of components i and j in the liquid
γ_{ij}^s	activity coefficient of component ij in the solid
Γ_{ij}	activity coefficient ratio (defined by equation 29)
θ_{ij}	reduced standard-state chemical potential difference (defined by equation 28)
θ_{ij}^I, θ_{ij}^{II}, θ_{ij}^{III}, θ_{ij}^{IV}	reduced standard-state chemical potential differences calculated by methods I–IV
μ	chemical potential
μ_A^l, μ_B^l, μ_C^l	chemical potentials of A, B, and C in the liquid
μ_A^s, μ_B^s, μ_C^s, μ_{AC}^s, μ_{BC}^s	chemical potentials of A, B, C, AC, and BC in the solid
μ_{ij}	chemical potential of component ij
$\mu_i^{o,l}$, $\mu_j^{o,l}$	standard-state chemical potential of component i or j in the liquid
ν	kinematic viscosity
ρ	density
σ	surface Gibbs energy per unit area

Acknowledgments

I appreciate the diligent efforts of former graduate students Terry Aselage and Kow Ming Chang. I also acknowledge the support of the Department of Energy under Grant DE–FG05–86ER45276.

References

1. Dawson, L. R. In *Progress in Solid State Chemistry;* Reiss, H.; McCaldin, J. O.; Eds.; Pergamon: Oxford, 1972; Vol. 7, Chapter 4.
2. Geiss, E. A.; Ghez, R. In *Epitaxial Growth, Part A;* Mathews, J. W., Ed.; Academic: London, 1975; p 183.
3. Elwell, D.; Scheel, H. J. In *Crystal Growth from High Temperature Solutions;* Academic: London, 1975.
4. Zschauer, K. H. *Festkörperprobleme* **1975,** *15,* 1.

5. Brice, J. C. *Curr. Topics Mater. Sci.* **1977,** *2*, 571.
6. Stringfellow, G. B. In *Crystal Growth: A Tutorial Approach;* Bardsley, W.; Hurle, D. T. J.; Mullin, J. B., Eds.; North Holland: Amsterdam, 1979; p 217.
7. Brice, J. C. *Curr. Topics Mater. Sci.* **1979,** *4*, 237.
8. Hollan, L.; Hallais, J. P.; Brice, J. C.; *Curr. Topics Mater. Sci.* **1980,** *5*, 1.
9. Nakajima, K. In *GaInAsP Alloy Semiconductors;* Pearsall, T. P.; Ed.; Wiley: New York, 1982; p 43.
10. Nakajima, K. *Semicond. Semimetals* **1985,** *22*, 1.
11. Bauser, E. In *Crystal Growth of Electronic Materials;* Kaldis, E., Ed.; Elsevier: Amsterdam, 1985; p 41.
12. Nakajima, K. *Prog. Cryst. Growth Charact.* **1986,** *12*, 97.
13. Nelson, H. *RCA Rev.* **1963,** *24*, 603.
14. Aylett, M. R.; Faktor, M. M.; Haigh, J.; White, E. A. D. *J. Cryst. Growth* **1981,** *54*, 604.
15. Chatillon, C.; Bernard, C. *J. Phys.* **1982,** *43*, 357.
16. Horikoshi, Y. *Rev. Electr. Commun. Lab.* **1978,** *26*, 748.
17. Bugajski, M.; Kazmierski, K. *Proc. Inst. Tech. Elek.* **1982,** *8*, 39.
18. Novotny, J.; Srobar, F.; Zelinka, J. *Cryst. Res. Technol.* **1983,** *18*, 651.
19. Panek, M.; Ratuszek, M.; Tlaczala, M. *Cryst. Res. Technol.* **1985,** *20*, 1577.
20. Dutartre, D.; Gavand, M. *Rev. Phys. Appl.* **1984,** *19*, 21.
21. Fiedler, F.; Wehmann, H. H.; Schlachetzki, A. *J. Cryst. Growth* **1986,** *74*, 27.
22. Logan, R. A.; Temkin, H. *J. Cryst. Growth* **1986,** *76*, 17.
23. Duncan, W. M.; Westphal, G. H. *VLSI Electron. Microstruct. Sci.* **1985,** *11*, 41.
24. Schiller, C. *C. R. Acad. Sci.* **1971,** *272*, 764.
25. Gormley, J. V.; Manfra, M. J.; Calawa, A. R. *Rev. Sci. Instrum.* **1981,** *52*, 1256.
26. Kern, W.; Deckert, C. A. In *Thin Film Processes;* Academic: London, 1978; p 401.
27. Kern, W. *RCA Rev.* **1978,** *39*, 278.
28. Heimann, R. B. In *Crystals: Growth, Properties, and Applications;* Grabmaier J., Ed.; Springer–Verlag: New York, 1982; Vol. 8, p 173.
29. Sullivan, M. V.; Kolb, G. A. *J. Electrochem. Soc.* **1963,** *110*, 585.
30. Reisman, A.; Rohr, R. *J. Electrochem. Soc.* **1964,** *111*, 1425.
31. Packard, R. D. *J. Electrochem. Soc.* **1965,** *112*, 871.
32. Dyment, J. C.; Rozgonyi, G. A. *J. Electrochem. Soc.* **1971,** *118*, 1346.
33. Rideout, V. L. *J. Electrochem. Soc.* **1972,** *119*, 1778.
34. Wood, D. R.; Morgan, D. V. *J. Electrochem. Soc.* **1975,** *122*, 773.
35. Chang, C. C.; Citrin, P. H.; Schwartz, B. *J. Vac. Sci. Technol.* **1977,** *14*, 943.
36. Laurence, G.; Simondet, F.; Saget, P. *Appl. Phys.* **1979,** *19*, 63.
37. Munoz-Yague, A.; Piqueras, J.; Fabre, N. *J. Electrochem. Soc.* **1981,** *128*, 149.
38. Zilko, J. L.; Williams, R. S. *J. Electrochem. Soc.* **1982,** *129*, 406.
39. Vasquez, R. P.; Lewis, B. F.; Grunthaner, F. J. *Appl. Phys. Lett.* **1983,** *42*, 293.
40. Adachi, S.; Oe, K. *J. Electrochem. Soc.* **1983,** *130*, 2427.
41. Adachi, S.; Oe, K. *J. Electrochem. Soc.* **1984,** *131*, 126.
42. Chen, J. A.; Lee, S. C. *J. Electrochem. Soc.* **1985,** *132*, 3016.
43. Ven, J.; Meerakker, J. E. A. M.; Kelly, J. J. *J. Electrochem. Soc.* **1985,** *132*, 3020.
44. Contour, J. P.; Massies, J.; Saletes, A. *Jpn. J. Appl. Phys.* **1985,** *24*, L563.
45. Tuck, B.; Baker, A. J. *J. Mater. Sci.* **1973,** *8*, 1559.
46. Nishitani, Y.; Kotani, T. *J. Electrochem. Soc.* **1979,** *126*, 2269.
47. Singh, S.; Williams, R. S.; Uitert, L. G.; Schlierr, A.; Camlibel, I.; Bonner, W. A. *J. Electrochem. Soc.* **1982,** *129*, 447.

48. Adachi, S. *J. Electrochem. Soc.* **1982,** *129*, 609.
49. Krowczyk, S. K.; Garrigues, M.; Bouredoucen, H. *J. Appl. Phys.* **1986,** *60*, 392.
50. Vasquez, R. P.; Lewis, B. F.; Grunthaner, F. J. *J. Vac. Sci. Technol. B* **1983,** *1*, 791.
51. Jensen, E. W. *Solid State Technol.* **1973,** *August*, 49.
52. Vasquez, R. P.; Lewis, B. F.; Grunthaner, F. J. *J. Appl. Phys.* **1983,** *54*, 1365.
53. Munoz, E.; Snyder, W. L.; Moll, J. L. *Appl. Phys. Lett.* **1970,** *16*, 262.
54. Mike, H.; Otsubo, M. *Jpn. J. Appl. Phys.* **1971,** *10*, 509.
55. Nakajima, K.; Yamazaki, S.; Takanohashi, T.; Akita, K. *J. Cryst. Growth* **1982,** *59*, 572.
56. Lin, L.; Fang, Z.; Zhou, B.; Zhu, S.; Xiang, X.; Wu, R. *J. Cryst. Growth* **1982,** *56*, 533.
57. Garrido, J.; Castano, J. L.; Piqueras, J. *J. Appl. Phys.* **1984,** *56*, 569.
58. Tuck, B. In *Introduction to Diffusion in Semiconductors,* IEE Monograph Series 16; Peter Peregrinus: Stevenage, England, 1974.
59. Brown, K. E. *J. Cryst. Growth* **1973,** *20*, 161.
60. Pak, K.; Nishinaga, T.; Uchiyama, S. *Jpn. J. Appl. Phys.* **1975,** *14*, 1613.
61. Temkin, H.; Keramidas, G.; Mahajan, S. *J. Electrochem. Soc.* **1981,** *128*, 1088.
62. Chu, S. N. G.; Jodlauk, C. M.; Johnston, W. D. *J. Electrochem. Soc.* **1983,** *130*, 2398.
63. Kinoshita, J.; Okuda, H.; Ulmatsu, Y. *Electron. Lett.* **1983,** *19*, 215.
64. DiGiuseppe, M. A.; Temkin, H.; Bonner, W. A. *J. Cryst. Growth* **1982,** *58*, 279.
65. Clawson, A. R.; Lum, W. Y.; McWilliams, G. E. *J. Cryst. Growth* **1979,** *46*, 300.
66. Lum, W. Y.; Clawson, A. R. *J. Appl. Phys.* **1979,** *50*, 5296.
67. Antypas, G. *Appl. Phys. Lett.* **1980,** *37*, 64.
68. Besomi, P.; Wilson, R. B.; Wagner, W. R.; Nelson, R. J. *J. Appl. Phys.* **1983,** *54*, 535.
69. Chin, B. H.; Schwartz, G. P.; Dautremont-Smith, W. C.; Dick, J. R. *J. Electrochem. Soc.* **1986,** *133*, 2161.
70. Nelson, A. W.; Westbrook, L. D.; White, E. A. D. *J. Cryst. Growth* **1982,** *58*, 236.
71. Benz, K. W.; Bauser, E. *Crystals: Growth, Properties, and Applications;* Grabmaier, J., Ed.; Springer: Berlin, 1980; Vol. 3, p 1.
72. Kishi, Y.; Yamazaki, S.; Nakajima, K.; Akita, K. *J. Cryst. Growth* 1986, *74*, 135.
73. Crossley, I.; Small, M. B. *J. Cryst. Growth* **1971,** *11*, 157.
74. Hsieh, J. J. *J. Cryst. Growth* **1974,** *27*, 49.
75. Panish, M. B. *J. Chem. Thermodyn.* **1970,** *2*, 319.
76. Crossley, I.; Small, M. B. *J. Cryst. Growth* **1972,** *15*, 275.
77. Rode, D. L. *J. Cryst. Growth* **1973,** *20*, 13.
78. Woodall, J. M.; Hovel, H. J. *Appl. Phys. Lett.* **1977,** *30*, 492.
79. Kordos, P.; Pearson, G. L.; Panish, M. B. *J. Appl. Phys.* **1979,** *50*, 6902.
80. Small, M. D.; Ghez, R. *J. Appl. Phys.* **1979,** *50*, 5322.
81. Rezek, E. A.; Vojak, B. A.; Chin, R.; Holonyak, N.; Sammaan, E. A. *J. Electron. Mater.* **1981,** *10*, 255.
82. Ghez, R.; Small, M. B. *J. Cryst. Growth* **1981,** *52*, 699.
83. Bolkhovityanov, Yu. B. *J. Cryst. Growth* **1982,** *57*, 84.
84. Reynolds, C. L.; Tamargo, M. C.; Anthony, P. J.; Zilko, J. L. *J. Cryst. Growth* **1982,** *57*, 109.
85. Bolkhovityanov, Yu. B. *J. Cryst. Growth* **1981,** *55*, 591.

86. Bolkhovityanov, Yu. B.; Vaulin, Yu. D. *Thin Solid Films* **1982,** *98,* 41.
87. Small, M. B.; Ghez, R. *J. Appl. Phys.* **1984,** *55,* 926.
88. Hiramatsu, K.; Tanaka, S.; Sawaki, N.; Akasaki, I. *Jpn. J. Appl. Phys.* **1985,** *24,* 1030.
89. Hiramatsu, K.; Tanaka, S.; Sawaki, N.; Akasaki, I. *Jpn. J. Appl. Phys.* **1985,** *24,* 822.
90. Bolkhovityanov, Yu. B.; Yudaev, V. I.; Gutakovsky, A. K. *Thin Solid Films* **1986,** *137,* 111.
91. Fahien, R. W. In *Fundamentals of Transport Phenomena;* McGraw–Hill: New York, 1983; p 273.
92. *Liquid Metals;* Beer, S. Z., Ed.; Dekker: New York, 1972.
93. Thijs, P. J. A.; Nijman, W.; Metselaar, R. *J. Cryst. Growth* **1986,** *74,* 625.
94. Vassilieff, G.; Saint-Cricq, B. *J. Appl. Phys.* **1984,** *55,* 743.
95. Kao, Y. C.; Eknoyan, O. *J. Appl. Phys.* **1983,** *54,* 1865.
96. Tiller, W. A. *J. Cryst. Growth* **1968,** *2,* 345.
97. Crossley, I.; Small, M. B. *J. Cryst. Growth* **1973,** *19,* 160.
98. de Crémoux, B. *Inst. Phys. Conf. Ser.* **1979,** *45,* 52.
99. Feng, M.; Cook, L. W.; Tashima, M. M.; Windhorn, T. A.; Stillman, G. E. *Appl. Phys. Lett.* **1979,** *34,* 292.
100. Cook, L. W.; Tashima, M. M.; Stillman, G. E. *Appl. Phys. Lett.* **1980,** *36,* 904.
101. Yamazaki, S.; Nakajima, K.; Kishi, Y. *Fujitsu Sci. Tech. J.* **1984,** *20,* 329.
102. Kume, M.; Ohta, J.; Ogasawara, N.; Ito, R. *Jpn. J. Appl. Phys.* **1982,** *21,* L424.
103. Nakajima, K.; Akita, K. *J. Electrochem. Soc.* **1982,** *129,* 2603.
104. Sonomura, H.; Horinaka, H.; Miyauchi, T. *Jpn. J. Appl. Phys.* **1983,** *22,* L689.
105. Wagner, C. *Acta Metall.* **1958,** *6,* 309.
106. Vieland, L. J. *Acta Metall.* **1963,** *11,* 137.
107. Brebrick, R. F. *Metall. Trans.* **1971,** *2,* 1657.
108. Jordan, A. S. *Metall. Trans.* **1971,** *2,* 1559.
109. Brebrick, R. F. *Metall. Trans.* **1971,** *2,* 3377.
110. Jordan, A. S.; Weiner, M. E. *J. Phys. Chem. Solids* **1975,** *36,* 1335.
111. Ilegems, M.; Panish, M. B.; Arthur, J. R. *J. Chem. Thermodyn.* **1974,** *6,* 157.
112. Furukawa, Y.; Thurmond, C. D. *J. Phys. Chem. Solids* **1965,** *26,* 1535.
113. Panish, M. B.; Sumski, S. *J. Phys. Chem. Solids* **1969,** *30,* 129.
114. Tung, T.; Liao, S.; Brebrick, R. F. *J. Vac. Sci. Technol.* **1982,** *21,* 117.
115. Kubaschewski, O.; Alcock, C. B. In *Metallurgical Thermochemistry;* Pergamon: Oxford, 1979; 5th ed.
116. Anderson, T. J.; Donaghey, L. F. *J. Electrochem. Soc.* **1984,** *131,* 3006.
117. Anderson, T. J.; Aselage, T.; Donaghey, L. F. *J. Chem. Thermodyn.* **1983,** *15,* 927.
118. Rosa, C. J.; Rupf-Bolz, N.; Sommer, F.; Predel, B. Z. *Metallkde* **1980,** *71,* 320.
119. Richman, D.; Hockings, E. F. *J. Electrochem. Soc.* **1965,** *112,* 461.
120. Lichter, L. D.; Sommelet, P. *Trans. Metall. Soc. AIME* **1969,** *245,* 99.
121. Predel, B.; Oehme, G. *Z. Metallkde.* **1976,** *67,* 826.
122. Mechkovskii, L. A.; Savitskii, A. A.; Skums, V. F.; Vecher, A. A. *Russ. J. Phys. Chem.* **1971,** *45,* 1143.
123. Schottky, W. F.; Bever, M. B. *Acta Metall.* **1958,** *6,* 320.
124. Glazov, V. M.; Petrov, D. A. *Izv. Akad. Nauk SSSR, Otd. Tekhn. Nauk* **1958,** *4,* 125.
125. Terpilowski, J. *Arch. Hutn.* **1959,** *4,* 355.
126. Cox, R. H.; Pool, M. J. *J. Chem. Eng. Data* **1967,** *12,* 247.
127. Lundin, C. E.; Pool, M. J.; Sullivan, R. W. *Denver Research Institute Final Report No. AFORL–63–156,* 1963.
128. Drowart, J.; Goldfinger, P. *J. Chim. Phys.* **1958,** *55,* 721.

129. Abbasov, A. S.; Mamedov, K. N. *Isv. Akad. Nauk, Azerb. SSR, Ser. Fiz. Matall. Tekhn. Nauk* **1970,** *3,* 86.
130. Kleppa, O. J. *J. Am. Chem. Soc.* **1955,** *77,* 897.
131. Maslov, P. G.; Maslov, Y. P. *Chem. Bonds Semicond. Solids* **1972,** *3,* 191.
132. Hultgren, R.; Desai, P. D.; Hawkins, D. T.; Gleizer, M.; Kelley, K. K. In *Selected Values of the Thermodynamic Properties of the Elements,* American Society for Metals: Metals Park, OH, 1973.
133. Vecher, A. A.; Geiderikh, V. A.; Gerasimov, Y. I. *Russ. J. Phys. Chem.* **1965,** *39,* 144.
134. Garbato, L.; Ledda, F. *Thermochim. Acta* **1977,** *14,* 267.
135. Terpilowski, J.; Trzebiatawski, W. *Bull. Acad. Polon. Sci., Ser. Sci. Chim.* **1960,** *8,* 95.
136. Sirota, N. N.; Yushkevich, N. N. *Chem. Bonds Semicond. Solids* **1967,** *1,* 95.
137. Jena, A. K.; Bever, M. B.; Banus, M. D. *Trans. Metall. Soc. AIME* **1967,** *239,* 725.
138. Gadzhiev, S. N.; Sharifov, K. A. *Dolk. Akad. Nauk SSSR,* **1961,** *136,* 1339.
139. Hultgren, R.; Desai, P. D.; Hawkins, D. T.; Gleizer, M.; Kelley, K. K. In *Selected Values of the Thermodynamic Properties of Binary Alloys,* American Society for Metals: Metals Park, OH, 1973.
140. Liu, T. S.; Peretti, E. A. *Trans. Amer. Soc. Metals* **1952,** *44,* 539.
141. Hoshino, H.; Nakamura, Y.; Shimoji, M.; Niwa, K. *Ber. Bunsenges. Phys. Chem.* **1965, 69.**
142. Panish, M. B.; Ilegems, M. In *Progress in Solid State Chemistry;* McCaldin, J. O., Ed.; Pergamon: New York, 1972; Vol. 7.
143. Anderson, S. J.; School, F.; Harris, J. S. In *Proc. 6th Int. Symp. GaAs Related Compounds,* 1976, p 346.
144. Joullie, A.; Gautier, P. *J. Cryst. Growth* **1979,** *47,* 100.
145. Dedegkaev, T. T.; Kryukov, I. I.; Lideikis, T. P.; Tsarenkov, B. V.; Yakovlev, Y. P. *Sov. Phys. Tech. Phys.* **1978,** *23,* 350.
146. Urasov, G. G. *Izv. Inst. Fiz., Khim. Anal.* **1921,** *1,* 461.
147. Predel, B.; Schallner, U. *Mater. Sci. Eng.* **1970,** *5,* 210.
148. Aselage, T. L.; Chang, K. M.; Anderson, T. J. In *Integrated Circuits: Chemical and Physical Processing;* Stroeve, P., Ed.; ACS Symposium Series 290; American Chemical Society: Washington, DC, 1985; p 276.
149. Hall, R. N. *J. Electrochem. Soc.* **1963,** *110,* 385.
150. Thurmond, C. D. *J. Phys. Chem. Solids* **1965,** *26,* 785.
151. Arthur, J. R. *J. Phys. Chem. Solids* **1967,** *28,* 2257.
152. Nougaret, P.; Potier, A. *J. Chim. Phys.* **1969,** *66,* 764.
153. Ilegems, M.; Panish, M. B. *J. Chem. Thermodyn.* **1973,** *5,* 291.
154. Panish, M. B. *J. Cryst. Growth* **1974,** *27,* 6.
155. Nguyen Van Mau, A.; Ance, C.; Bougnot, G. *J. Cryst. Growth* **1976,** *36,* 273.
156. Foster, L. M.; Woods, J. F. *J. Electrochem. Soc.* **1971,** *118,* 1175.
157. Wu, T. Y.; Pearson, G. L. *J. Phys. Chem. Solids* **1972,** *33,* 409.
158. Gorelenok, A. T.; Mdivani, V. N.; Moskvin, P. P.; Sorokin, V. S.; Usikov, A. S. *Russ. J. Phys. Chem.* **1982,** *56,* 1481.
159. Brebrick, R. F.; Panlener, R. J. *J. Electrochem. Soc.* **1974,** *121,* 932.
160. Gratton, M. F.; Wolley, J. C. *J. Electrochem. Soc.* **1978,** *125,* 657.
161. Blom, G. M.; Plaskett, T. S. *J. Electrochem. Soc.* **1971,** *118,* 1831.
162. Joullie, A.; Dedies, R.; Chevrier, J.; Bougnot, G. *Rev. Phys. Appl.* **1974,** 9, 455.
163. Ilegems, M.; Pearson, G. L. *Proc. Symp. GaAs* **1968,** 3.
164. Mabbitt, A. W. *J. Mat. Sci.* **1970,** *5,* 1043.
165. Blom, G. M. *J. Electrochem. Soc.* **1971,** *118,* 1834.
166. Panish, M. B.; Ilegems, M. *Proc. Symp. GaAs* **1970,** 67.

167. Kajiyama, K. *Jpn. J. Appl. Phys.* **1971,** *10,* 561.
168. Sugiura, T.; Sugiura, H.; Tanaka, A.; Sukegawa, T. *J. Cryst. Growth* **1980,** *49,* 559.
169. Antypas, G. A. *J. Electrochem. Soc.* **1970,** *117,* 1393.
170. Tomlinson, J. L. *Naval Ocean Systems Center, Technical Note 386,* San Deigo, CA, 1978.
171. Tanaka, A.; Sugiura, T.; Sukegawa, T. *J. Cryst. Growth* **1982,** *60,* 120.
172. Gratton, M. F.; Wolley, J. C. *J. Electrochem. Soc.* **1980,** *127,* 55.
173. Ilegems, M.; Panish, M. B. *J. Cryst. Growth* **1973,** *20,* 77.
174. Doi, A.; Hirao, M.; Ito, R. *Jpn. J. Appl. Phys.* **1978,** *17,* 503.
175. Onabe, K. *J. Phys. Chem. Solids* **1982,** *43,* 1071.
176. Nakajima, K.; Kusunoki, T.; Akita, K.; Kotani, T. *J. Electrochem. Soc.* **1978,** *125,* 123.
177. Ilegems, M.; Panish, M. B. *J. Phys. Chem. Solids* **1974,** *35,* 409.
178. Gorelenok, A. T.; Mdivani, V. N.; Moskvin, P. P.; Sorokin, V. S.; Usikov, A. S. *J. Cryst. Growth* **1982,** *60,* 355.
179. Alavi, K. T.; Perea, E. H.; Fonstad, C. G. *J. Electron. Mater.* **1981,** *10,* 591.
180. Perea, E. H.; Fonstad, C. G. *J. Appl. Phys.* **1980,** *51,* 331.
181. Henry, Y.; Moulin, M. *J. Cryst. Growth* **1981,** *51,* 387.
182. Cheng, K. Y.; Pearson, G. L. *J. Electrochem. Soc.* **1977,** *124,* 753.
183. Rao, M. V.; Tiller, W. A. *J. Phys. Chem. Solids* **1970,** *31,* 191.
184. Knoblock, G.; Butter, E.; Reiffarth, S. *Z. Phys. Chemie* **1978,** *259,* 667.
185. Knobloch, G. *Krist. Tech.* **1975,** *10,* 605.
186. Peuschel, G. P.; Apelt, R.; Knobloch, G.; Butter, E. *Krist. Tech.* **1979,** *14,* 409.
187. Peuschel, G. P.; Knobloch, G.; Butter, E.; Apelt, R. *Cryst. Res. Tech.* **1981,** *16,* 13.
188. Stringfellow, G. B.; Greene, P. E. *J. Phys. Solids* **1969,** *30,* 1779.
189. Stringfellow, G. B.; Greene, P. E. *J. Electrochem. Soc.* **1970,** *117,* 1075.
190. Aselage, T. L.; Chang, K. M.; Anderson, T. J. *CALPHAD: Comput. Coupling Phase Diagrams Thermochem.* **1985,** *9,* 227.
191. Brebrick, R. F. *Metall. Trans.* **1977,** *8,* 403.
192. Kaufman, L.; Nell, J.; Taylor, K.; Hayes, F. *CALPHAD: Comput. Coupling Phase Diagrams Thermochem.* **1981,** *5,* 185.
193. Esdaile, J. D. *Metall. Soc. AIME* **1982,** *13*(B), 213.
194. Bale, C. W.; Pelton, A. D. *Metall. Trans.* **1974,** *5,* 2323.
195. Darken, L. S. *Trans. TMS–AIME* 1967, *239,* 80.
196. Eckert, C. A.; Smith, J. S.; Irwin, R. B.; Cox, K. R. *AIChE J.* **1982,** *28,* 325.
197. Jordan, A. S. *Metall. Trans.* **1970,** *1,* 239.
198. Osamura, K.; Predel, B. *Trans. Jpn. Inst. Met.* **1977,** *18,* 765.
199. Osamura, K.; Murakami, Y. *J. Phys. Chem. Solids* **1975,** *36,* 931.
200. Kharif, Y. L.; Kovtunenko, P. V.; Maier, A. A. *Russ. J. Phys. Chem.* **1982,** *56,* 34.
201. Liao, P. K.; Su, C. H.; Tung, T.; Brebrick, R. F. *CALPHAD: Comput. Coupling Phase Diagrams Thermochem.* **1982,***6,* 141.
202. Stringfellow, G. B. *J. Phys. Chem. Solids* **1972,** *33,* 665.
203. Stringfellow, G. B. *J. Phys. Chem. Solids* **1973,** *34,* 1749.
204. Marbeuf, A.; Guillaume, J. C. *Rev. Phys. Appl.* **1984,** *19,* 311.
205. Foster, L. M. In *Electrochemical Society Extended Abstracts,* Houston Meeting; Electrochemical Society: Pennington, NJ, 1972; p 147.
206. Phillips, J. C.; Van Vechten, J. A. *Phys. Rev.* **1970,** *B2,* 2147.
207. Hildebrand, J. H.; Scott, R. L. In *The Solubility of Non-Electrolytes,* Dover: New York, 1964.
208. Gordy, W. *Phys. Rev.* **1946,** *69,* 604.

209. Gordy, W.; Thomas, J. O. *J. Chem. Phys.* **1956,** *25*, 439.
210. Stringfellow, G. B. *Mater. Res. Bull.* **1971,** *6*, 371.
211. Stringfellow, G. B. *J. Cryst. Growth* **1974,** *27*, 21.
212. Phillips, J. C. *Phys. Rev. Lett.* **1968,** *20*, 550.
213. Fedders, P. A.; Muller, M. W. *J. Phys. Chem. Solids* **1984,** *45*, 685.
214. Mikkelsen, J. C.; Boyce, J. B. *Phys. Rev. Lett.* **1982,** *49*, 1412.
215. Mikkelsen, J. C.; Boyce, J. B. In *Proc. 17th Int. Conf. Phys. Semicond.;* Chadi, J. D.; Harrison, W. A.; Bachrach, R. Z., Eds.; Springer–Verlag: New York, 1985.
216. Mikkelsen, J. C. *J. Electrochem. Soc.* **1985,** *132*, 500.
217. Anderson, T. J.; Colinet, C.; Chatillon, C.; Tmar, M. *J. Cryst. Growth* **1987,** *83*, 252.
218. Foxon, C. T.; Joyce, B. A.; Farrow, R. F. C.; Griffiths, R. M. *J. Phys. D* **1974,** *7*, 2422.
219. Ufimtsev, V. B.; Shumilin, V. P.; Krestovnikov, A. N.; Vigdorovich, V. N. *Russ. J. Phys. Chem.* **1970,** *44*, 624.
220. Foster, L. M.; Woods, J. F. *J. Electrochem. Soc.* **1972,** *119*, 504.
221. Foster, L. M.; Scardefield, J. E.; Woods, J. F. *J. Electrochem. Soc.* **1972,** *119*, 765.
222. Foster, L. M.; Scardefield, J. E.; Woods, J. F. *J. Electrochem. Soc.* **1972,** *119*, 1426.
223. Aselage, T. L.; Anderson, T. J. *High Temp. Sci.* **1985,** *20*, 207.
224. Chang, K. M.; Coughanowr, C. A.; Anderson, T. J. *Chem. Eng. Comm.* **1985,** *38*, 275.
225. Antypas, G. A. *J. Cryst. Growth* **1972,** *16*, 181.
226. Onda, T.; Ito, R. *J. Cryst. Growth* **1986,** *78*, 479.
227. Onabe, K. *Jpn. J. Appl. Phys.* **1983,** *22*, 201.
228. Kuphal, E. *J. Cryst. Growth* **1984,** *67*, 441.
229. Marbeuf, A.; Guillaume, J. C. *J. Phys.* **1982,** *43*, 45.
230. de Crémoux, B.; Hirtz, P.; Ricciardi, J. *Inst. Phys. Conf. Ser.* **1981,** *56*, 115.
231. Stringfellow, G. B. *J. Appl. Phys.* **1983,** *54*, 404.

RECEIVED for review December 30, 1987. ACCEPTED revised manuscript February 6, 1989.

4

Physical Vapor Deposition Reactors

T. W. Fraser Russell, Bill N. Baron, Scott C. Jackson,[1] and Richard E. Rocheleau[2]

Institute of Energy Conversion and the Department of Chemical Engineering, University of Delaware, Newark, DE 19716

Physical vapor deposition (PVD) is used for the deposition of semiconductor, insulator, and metal layers in the fabrication of a variety of electronic devices. Reactors for PVD are characterized by direct line-of-sight transport of molecular species from the gas phase to the desired substrate, where they react to form solid films with the desired properties. A reactor-and-reaction analysis of PVD quantitatively examines the generation of gas-phase species, spatial distribution of species arriving on the substrate, and surface reactions leading to film growth.

THE SEMICONDUCTOR, INSULATOR, or conductor layers in microscale or larger scale electronic devices such as a photovoltaic cell are created in a reactor. The reactor needs to be designed and operated to produce materials that have the desired optical and electronic properties. The design of reactors is a nontrivial research and design problem. In this chapter, some of the theoretical and experimental framework for this research and for more-effective designs of physical-vapor-deposition-type reactors will be developed.

A wide variety of names are used in the electronic industry for various types of reactors, but two broad classifications of reactors based on the means by which the molecular species are delivered to the substrate are useful: direct line-of-sight impingement and diffusive–convective mass transfer. The term "chemical vapor deposition" (CVD) has been used generally to describe

Current addresses:
[1]Engineering Department, E. I. du Pont de Nemours and Company, Wilmington, DE 19898
[2]Hawaii Natural Energy Institute, University of Hawaii, Honolulu, HI 96822

0065–2393/89/0221–0171$08.00/0

the diffusive–convective mass transfer. The critical issues in CVD are the following:

- chemical reaction in fluid phase
- transport to substrate (momentum, mass transfer, and heat transfer)
- film growth

The term "physical vapor deposition" (PVD) has been used for line-of-sight transport. The critical issues in PVD are the following:

- film growth
- spatial distribution on substrate
- molecular beam from source

Growth of the film is a primary concern for both reactor types, but the transport phenomena in a CVD reactor are more difficult to analyze. Knowledge of the fluid mechanics and heat and mass transfers, often for a very complex geometry, is required. In a line-of-sight PVD reactor, the transport of molecular species to the substrate can be analyzed more easily.

A PVD-type reactor can be one in which molecules reach the surface directly in a molecular beam from some source or sources in which raw materials are vaporized. At the pressures commonly used ($<10^{-6}$ Pa), the vaporized material encounters few intermolecular collisions while traveling to the substrate. Historically, higher pressure processes, such as sputtering and close-spaced vapor transport, have been classified as PVD (*1*). These processes also use physical means to generate the gas-phase species. However, the transport phenomena that need to be modeled for such higher pressure processes are more similar to CVD than PVD because of the diffusive–convective nature of transfer from the gas phase to the substrate.

PVD reactors may use a solid, liquid, or vapor raw material in a variety of source configurations. The energy required to evaporate liquid or solid sources can be supplied in various ways. Resistive heating is common, induction heating of the source bottle is sometimes used, and electron beams are also employed. Molecular-beam-epitaxy (MBE) systems are PVD-type reactors that operate at ultrahigh vacuum. Very low growth rates are used ($\sim$1 μm/h), and considerable attention is devoted to in situ material characterization to obtain high-purity epitaxial layers (*2*).

A reactor-and-reaction analysis quantitatively examines the molecular and transport phenomena in a reacting system. In PVD-type reactors, the most interesting molecular phenomena are those connected with the growing film, although chemical reactions that govern generation of gas-phase species in the source may also have to be examined. The analysis of a PVD-type reactor requires the quantification of the rate of effusion from the source and the prediction of the spatial distribution of the fluxes of molecular species arriving at the substrate.

When a molecule has reached the surface, the required reaction analysis is the same regardless of reactor type. The molecular phenomena that must be considered are the following:

- surface adsorption
- surface diffusion
- surface reaction
- film or crystal growth

The tools needed to analyze adsorption, surface diffusion, and surface reaction to form a product are the same as those used to analyze reactions on catalytic surfaces, the only difference being that in catalytic systems the product leaves the surface and desorbs into the fluid phase. In the processing of electronic materials, the product is the thin film that is formed on the surface.

In this chapter, the rate of effusion from a source will be quantified first. Then, the analysis of the spatial distribution of species arriving at the substrate will be developed. The final section will consider the surface reaction zone where the semiconductor film is formed.

Rate of Effusion from a Source

A typical line-of-sight PVD-type reactor is shown in Figure 1. Raw material is placed in the source bottle and heated to produce a molecular beam that impinges on the substrate. Power to the heater can be controlled by measuring the temperature of the source bottle, and it must be adjusted as the material is depleted. CdS or ZnS films can be made from a single-source (*3*) bottle containing CdS or ZnS powder. A ternary compound such as $CuInSe_2$ requires three source bottles: one for Cu, one for In, and one for Se (*3*). GaAs films typically require two sources, one for Ga and one for As (*4*).

The rate of effusion can be predicted for any source with a defined geometry and for any material as a function of the source bottle temperature. Figure 2 shows a typical source bottle and a sketch showing how the initial charge behaves. The nozzle and orifice allow one to design for proper placement of the molecular species on the substrate independently of the design for the rate-limiting output. This aspect will be discussed quantitatively in the next section.

An orifice with pressure drop $P_c - P_i$, in which P_c is the vapor pressure in the evaporation chamber and P_i is the vapor pressure in the intermediate chamber, is sometimes referred to as a rate-controlling orifice. In a properly

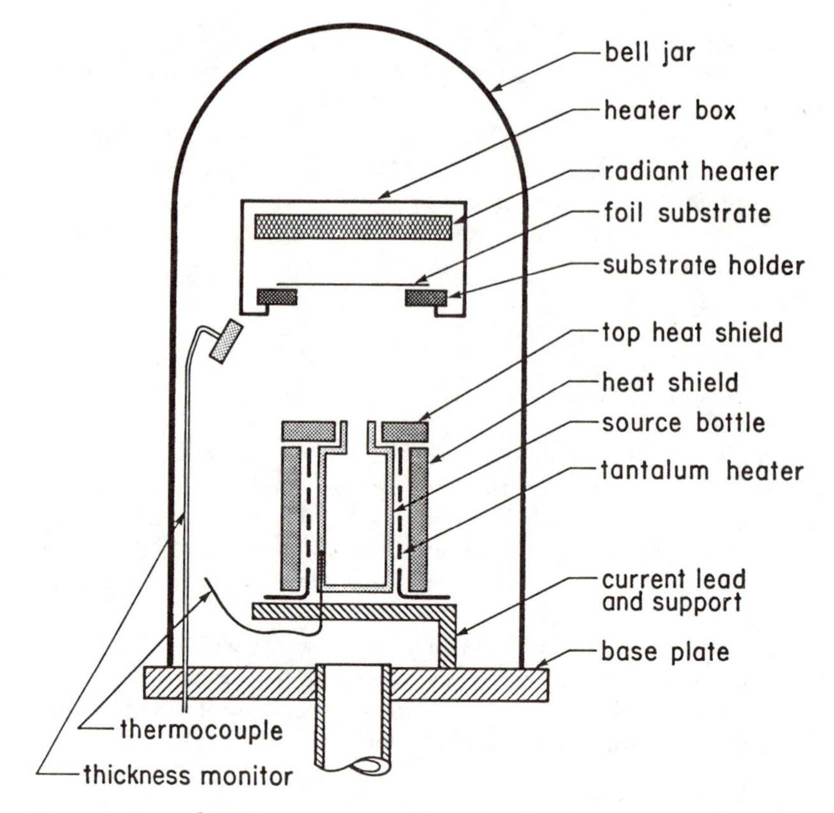

Figure 1. Typical PVD-type reactor. (Reproduced with permission from reference 5. Copyright 1982 American Institute of Chemical Engineers.)

designed system, the orifice area is small compared with the charge area to ensure that P_c is the equilibrium vapor pressure of the charge material.

Model Equations. If the material in the bottle is taken as the control volume (Figure 1), application of the law of conservation of mass yields

$$\partial \rho V/\partial t = -\rho_g q \tag{1}$$

in which ρ is the bulk density of the source material charge, V is the volume of the source material charge, t is time, ρ_g is the mass density of the gas phase, and q is the volumetric flow rate of the exit stream.

The change in density with time can be significant for subliming powder materials that sinter during evaporation. This dependence can be determined with a separate experiment and analyzed by using equation 2.

$$V\partial \rho/\partial t + \rho \partial V/\partial t = \rho_g q \tag{2}$$

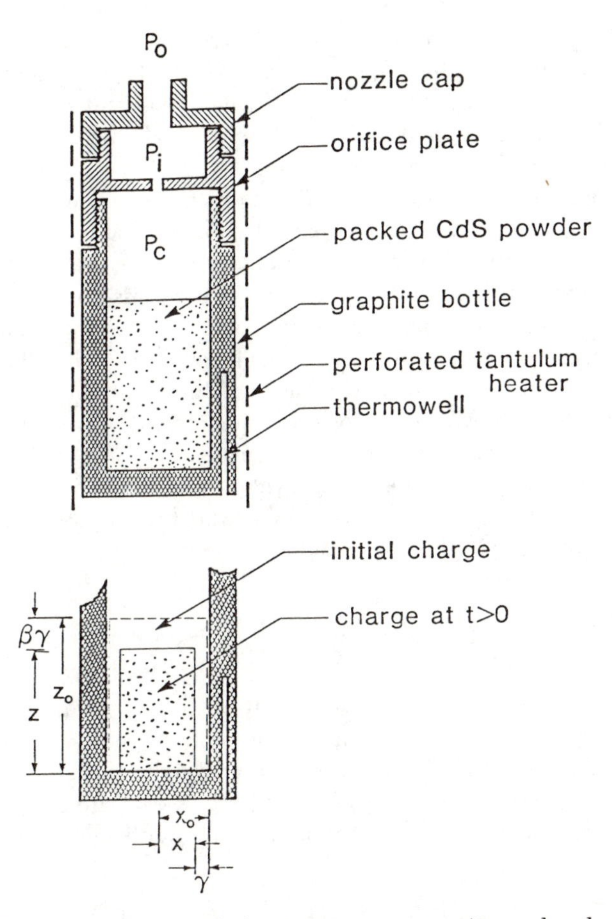

Figure 2. Typical source bottle for PVD-type reactor. (Reproduced with permission from reference 5. Copyright 1982 American Institute of Chemical Engineers.)

For the evaporation of CdS powder, $\partial\rho/\partial t$ is constant at about 0.17×10^{-3} g/cm^3-s (5). Neglecting the density change in this example would introduce an error of only about 10% in the calculated effusion rates.

The volumetric flow rate from the source depends upon the pressure difference across the orifice, $P_c - P_i$. The vapor pressure in the chamber containing the raw material, P_c, is calculated from the charge temperature. The charge temperature, T_1, is determined from an energy balance that considers the heat transfer from the wall of the source to the charge and the loss due to the heat of vaporization of the source material. If the heat transfer is by radiation, as is the case for a solid source, and if the charge temperature T_1 is uniform, then the energy balance takes the following form:

$$VC_p\partial T_1/\partial t = -\rho_g\Delta H_R + F_vF_e\sigma(T_2^4 - DT_1^4)A_s \tag{3}$$

in which C_p is the heat capacity of the source material, ΔH_R is the latent heat of vaporization of the source material, F_v is the view factor, F_e is the effective emissivity, σ is the Stefan–Boltzmann constant, T_1 is the temperature of the charge, T_2 is the temperature of the bottle wall, and A_s is the surface area of the source material charge.

Usually, the wall temperature, T_2, can be measured and used to control power to the source. A more complete treatment that includes the energy balance for the source bottle was given by Rocheleau (6).

Equations 1 and 3 are coupled through the mass flow, $\rho_g q$, which depends in a complicated manner on the charge temperature and the vapor pressure of the source material. The flow through the exit orifice can range from free molecular flow to viscous flow. If the Knudsen number (ratio of the mean free path, λ_m, to the orifice diameter, D) is greater than 1 (i.e., $\lambda_m/D > 1$), then the flow is free molecular. If $\lambda_m/D < 0.01$, the flow is viscous. A transition region exists between these limits.

The mean free path λ_m is related to the pressure by simple kinetic molecular theory according to equation 4

$$\lambda_m = (\pi R T_1/2M_w)^{1/2}\mu/p_m \tag{4}$$

in which p_m is the pressure at the point of interest (usually taken as the average of the pressures upstream and downstream of the restriction), R is the ideal gas constant, M_w is the molecular weight of the material flowing from the source, and μ is the gas viscosity. μ is evaluated at p_m and T_1.

For a compound AB that evaporates congruently according to the following reaction

$$(AB)_s \leftrightarrows A_g + 1/n(B_n)_g \tag{5}$$

or for the evaporation of an alloy, the properties used to determine the gas-phase properties are those of the mixture. For the binary dissociative evaporation of compound AB, the average molecular weight is given by

$$M_w = [nM_{AB}/(M_{B_n}{}^{1/2}) + (nM_A{}^{1/2})]^2 \tag{6}$$

in which n is the stoichiometric coefficient. The equilibrium vapor pressure is given by

$$K_p(T_1) = P_A P_{B_n}{}^{1/n} \tag{7}$$

$$P_e = K_p{}^{n/(n+1)}[(1 + n)n^{-n/(1+n)}] \tag{8}$$

in which K_p is the equilibrium constant for dissociative evaporation and P_e is the equilibrium vapor pressure. K_p is a function only of temperature T_1 and can be found in appropriate handbooks or the literature.

Equations 1 and 3 are solved at the same time by using an appropriate numerical algorithm for simultaneous first-order differential equations. Calculation of the term $\rho_g q$ may require an iteration within the integration routine, depending on the number of flow restrictions in series and the flow regime that is encountered. For two restrictions in series, $\rho_g q$ depends upon the pressures P_c, P_i, and P_o.

When the area of the charge is sufficiently large relative to the area of the orifice (10 times or greater), the chamber pressure, P_c, can be set equal to the equilibrium pressure, P_e, at T_1. P_o is the pressure in the vacuum chamber and is usually orders of magnitude below P_c. Therefore, little error is made if $P_o = 0$ is assumed. An iteration for systems with an orifice but no nozzle proceeds as follows:

1. λ_m is computed by using $\frac{1}{2}p_e$ for p_m (equation 4).
2. The Knudsen number, λ_m/D, and the aspect ratio, L/Γ, are obtained by using the mean free path (λ_m), orifice radius, Γ, and orifice length, L. L/Γ is fixed for a given geometry, but λ_m/D is a function of pressure.
3. When λ_m/D is greater than 1, free molecular flow exists, and a factor, K, is used to multiply the equation for an orifice for free molecular flow (Table I). K aids in dealing with the transitions between orifice and pipe flows (5). For an ideal orifice, $K = 1$, and for a long pipe, $K = \pi/2[(2/f - 1)\Gamma/L$. For intermediate situations in which $L/\Gamma > 1.5$

$$K = (1 + 0.4L/\Gamma)/[1 + 0.95L/\Gamma + 0.15(L/\Gamma)^2] \qquad (9)$$

Table I. Constitutive Flow Equations

Flow Regime	*Orifice*	*Pipe*
Free molecular		
$(\lambda_m/D > 1)$	$\rho_g q = \Gamma^2(\pi M_w/2RT)^{1/2} \times (p_1 - p_2)$	$\rho_g q = \dfrac{\pi\Gamma^4 M_w}{16\mu LRT}(p_1^2 - p_2^2) \times \left[4\left(\dfrac{2}{f} - 1\right)\dfrac{\lambda_m}{\Gamma}\right]$
Viscous		
$(\lambda_m/D < 0.01)$	$\rho_g q = \pi\Gamma^2 C_0 \times [2\rho_g(p_1 - p_2)]^{1/2}$	$\rho_g q = \dfrac{\pi\Gamma^4 M_w}{16\mu LRT}(p_1^2 - p_2^2)$

NOTE: The variables are defined in the text and at the end of the chapter.

4. For λ_m/D between 1 and 0.01, a fairly common condition, the equation for free molecular orifice flow is modified by multiplying with a constant C. The total pressure drop through the orifice is obtained by using the following equations

$$\rho_g q = C(M_w/2\pi R T_1)^{1/2}(p_1 - p') \quad (10a)$$

or

$$\rho_g q = (\Gamma^2/16\mu L)(p'^2 - p_2^2)\left[1 + \frac{4\lambda_m}{\Gamma}\left(\frac{2}{f} - 1\right)\right](M_w/RT_1) \quad (10b)$$

in which p_1 is the pressure upstream of the pipe or orifice, p' is the pressure of the pipe entrance after contraction, and f is the fraction of gas molecules diffusively reflected from the wall. p' is an intermediate pressure and must be obtained by an iteration using both forms of equation 10 with a fixed value of C.

5. For purely viscous flow, the equations in Table I can be used directly. If necessary an iteration using p' can be carried out.

For the CdS system investigated by Rocheleau et al. (5), the CdS charge was typically at temperatures of ~1220 K. The vapor pressure was given by:

$$P_e = 1.572 \times 10^{12} \exp(-2.633 \times 10^4/T_1) \quad (11)$$

For orifice areas ranging from 2×10^{-2} to 26×10^{-2} cm^2 and orifice aspect ratios (L/Γ) between 1 and 4, the flow was almost always in the transition regime ($0.01 < \lambda_m/D < 1.0$), and equations for the short pipe were required. The value of C (equation 10) was taken to be 20, but a C value between 10 and 30 had a small effect on flow. At $C = 20$, 5–25% of the total pressure drop was attributable to the entrance region of the orifices.

Behavior of Model Equations. Equations 1 and 3 can be solved numerically if initial charge mass and dimensions and wall chamber temperature, T_2, are provided. The numerical solution must incorporate the iteration schemes for $\rho_g q$ outlined in the previous section. Complete details are provided by Rocheleau (6) and Rocheleau et al. (5).

The model predictions have been compared with experimental results for the evaporation of CdS. A typical result showing CdS mass loss from the bottle as a function of time is shown in Figure 3. Figure 4 shows the CdS

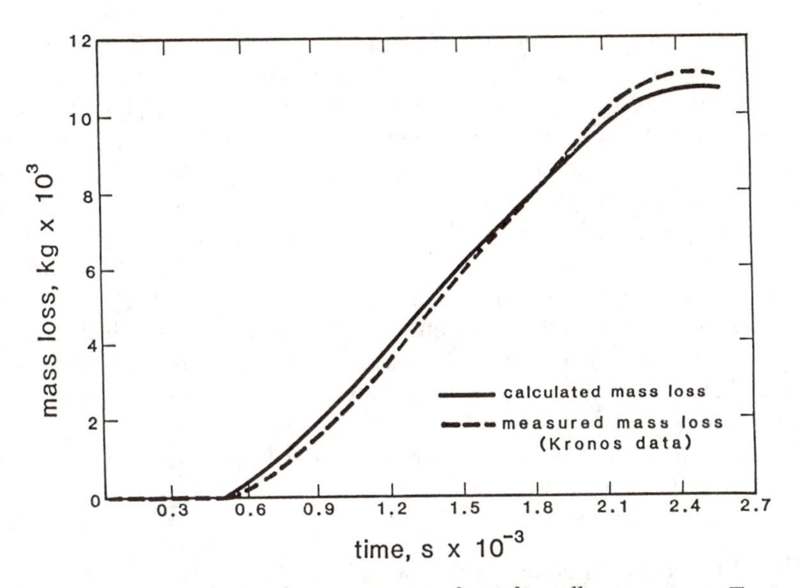

Figure 3. Measured mass loss versus time for CdS wall temperature, T_2 = *1227 K. (Reproduced with permission from reference 5. Copyright 1982 American Institute of Chemical Engineers.)*

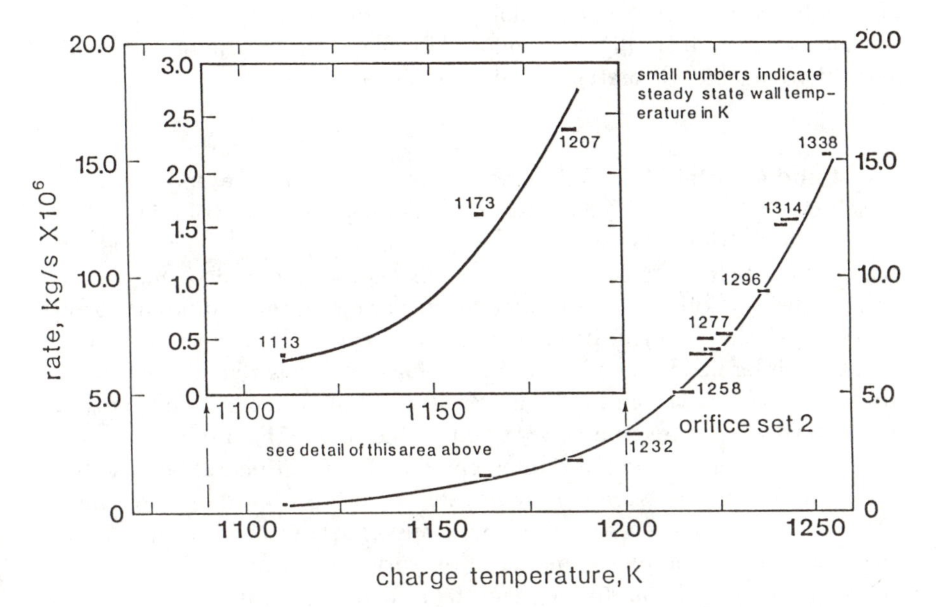

Figure 4. CdS effusion rate versus charge temperature. (Reproduced with permission from reference 5. Copyright 1982 American Institute of Chemical Engineers.)

effusion rate, $\rho_g q$, as a function of charge temperature, T_1. The wall temperature, T_2, is shown beside each point representing the experimental data. The solid curves are model predictions resulting from equations 1 and 3.

Figure 4 shows that the wall temperature, T_2, can be quite different from the charge temperature, T_1. For example, for the charge temperature to be about 1250 K, the wall temperature must be about 1338 K. If the charge temperature is erroneously assumed to equal the wall temperature, this 88 K difference leads to an overprediction of more than 100%. At low rates, the difference between T_1 and T_2 is smaller. Some investigators generally assume that T_1 and T_2 are equal, but this assumption can lead to large errors in rate prediction. If T_2 is not known or measured, an additional energy balance equation may be used. This equation has been derived, and its use has been discussed by Rocheleau (*6*).

When the rate of effusion from the source is known, the next step is to predict the spatial distribution of species on the substrate.

Distribution of Species on Substrate

The film thickness uniformity and film composition depend, in part, on the molecular or atomic flux variation across a substrate. This flux variation is a function of the directionality of the evaporation source (i.e., the molecular-beam distribution) and the orientation of the substrate relative to the source (*7*). In this section, the fundamental equations that describe the distribution of the incident flux will be introduced, along with the solution of these equations for the evaporation of group II–VI compound-semiconductor materials.

Fundamental Mass Balances. Molecules or atoms evaporated from a source into a vacuum emerge with well-defined spatial distributions. The molecular distributions can be derived from first principles for idealized cases, such as free molecular flow through sharp-edged orifices and tubes. The Clausing (*8*) formulation for free molecular flow through tubes has been verified by Shen (*9*) and Stickney et al. (*10*). This well-defined spatial distribution is formed by atoms or molecules flowing through a nozzle or orifice of the source or evaporating from a small pool into a vacuum and across a hemispherical control volume centered on the source (Figure 5).

All evaporating material is intercepted by the hemisphere, because the molecular beam exists only at angles less than 90° from the center line. Therefore, the mass flow from the source must equal the flux of the molecular beam leaving the control volume across the surface of the hemisphere. Under these conditions, the mass flow from the source is given by

$$\rho_g q = M_w \int_0^{\pi/2} I(\phi) 2\pi R \sin(\phi) R d\phi \tag{12}$$

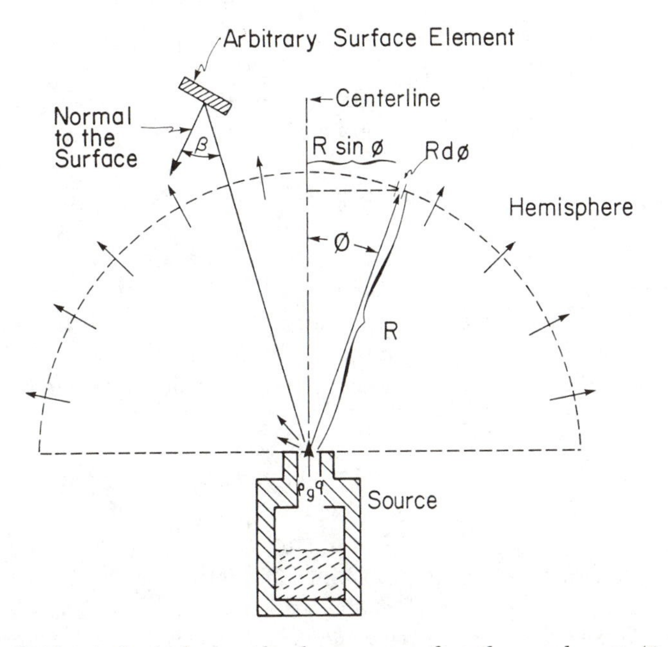

Figure 5. Hemispherical control volume centered at the nozzle exit. (Reproduced with permission from reference 7. Copyright 1985 American Institute of Physics.)

in which $I(\phi)$ is the molecular flux incident on the hemisphere at an angle ϕ, ϕ is the angle from the nozzle center line, and R is the radius of the hemisphere to the point of interest.

To represent and compare data, the beam intensity to the center line of the distribution can be normalized by using equation 13

$$F(\phi) = I(\phi)/I(\phi = 0) \tag{13}$$

in which $F(\phi)$ is a probability distribution function of the molecules leaving the nozzle. This probability distribution function is easily represented by polar plots, which are used by many investigators (*8*, *10*, *11*).

All molecules flowing from the nozzle leave the control volume across the surface of the hemisphere. Consequently, the integrated value of the probability function across the hemisphere is unity, as shown by equation 14

$$1 = C_f \int_0^{2\pi} \int_0^{\pi/2} F(\phi) \sin(\phi) d\phi d\theta \tag{14}$$

in which C_f is the normalization constant and θ is the azimuth angle.

For each probability function, C_f has a unique value that is found by evaluating equation 14.

The substitution of equations 12 and 13 into 14 yields an expression for the beam intensity at any point on the hemisphere:

$$I(\phi) = C_f \rho_g q \, F(\phi)/R^2 M_w \tag{15}$$

Equation 15 can be used only for substrates that have their normal to the surface coincident with the direction of the molecular beam. When the normal to the surface element is at an angle β to the molecular beam, the molecular flux at an angle ϕ from the center line and at an angle of incidence β onto the surface decreases by a factor of cos (β), and the incident molecular flux, r(i), is given by:

$$r(\mathrm{i}) = C_f \rho_g q \, F(\phi) \cos(\beta)/R^2 M_w \tag{16}$$

The incident flux, r(i), can be evaluated from equation 16 if a suitable constitutive equation for the probability distribution F(ϕ) can be obtained. The mass flow rate, $\rho_g q$, can be obtained empirically or calculated by using the procedures outlined in the previous section.

Evaluation of F(ϕ) and r(i). In this section, the distribution of the incident flux is solved for two technologically important processes: electron beam evaporation and evaporation from a high-rate Knudsen cell.

Electron Beam Evaporation. Most electron beam evaporation sources have a probability distribution that can be approximated by the cosine law:

$$F(\phi) = \cos(\phi) \tag{17}$$

Figure 6 shows the molecular beam distribution, F(ϕ), for electron beam evaporation. The cosine law distribution is generally the most uniform distribution achievable from a single source. The beam distribution is relatively flat at $\phi = 0$ but suffers from poor utilization of the evaporating material, because a significant portion of the molecular beam strikes at large angles, $\phi > 30°$. To solve for the incident-flux distribution, r(i), the normalization constant, C_f, as well as the geometrically dependent factors, cos (β) and R, must be substituted into equation 16.

The normalization constant, C_f, is determined by substituting equation 17 into equation 14 to give:

$$C_f = 1/\pi \tag{18}$$

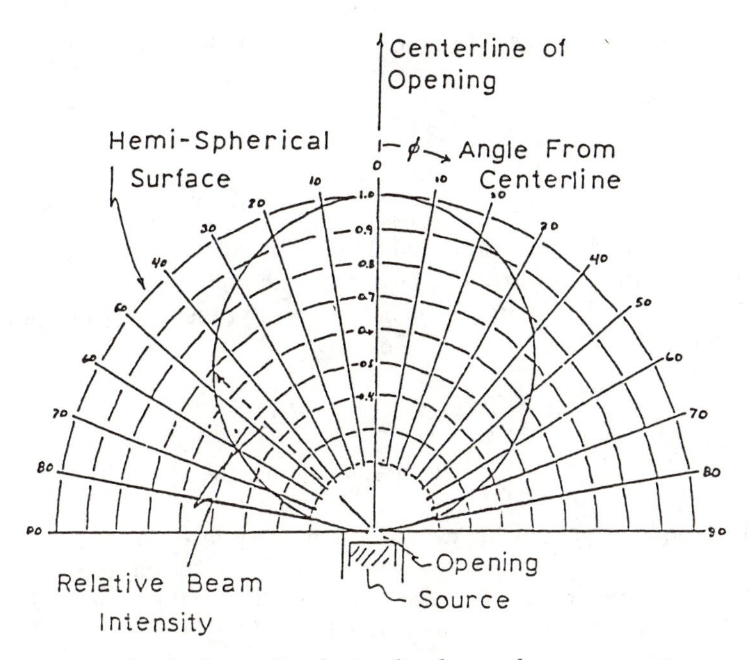

Figure 6. Molecular-beam distribution for electron beam evaporation source obeying the cosine law.

For a planar substrate that is parallel to the source plane at height H above the source, cos (β) and R are defined as follows:

$$\cos(\beta) = \cos(\phi) \tag{19}$$

$$R = H/\cos(\phi) \tag{20}$$

Substitution of equations 18–20 into equation 16 yields the incident-flux distribution:

$$r(\mathrm{i}) = \rho_g q \cos^4(\phi)/\pi H^2 M_w \tag{21}$$

Normalization of equation 21 to the incident flux at $\phi = 0$ [i.e., at maximum r(i)] gives a measure of the relative flux intensity across the substrate:

$$\frac{r(\mathrm{i})}{r(\mathrm{i})\Big|_{\phi=0}} = \cos^4(\phi) \tag{22}$$

By using simple geometric arguments, equation 22 becomes

$$\frac{r(\mathrm{i})}{r(\mathrm{i})\Big|_{\phi=0}} = 1/[1 + (l/H)^2]^2 \tag{23}$$

in which l is the distance on the substrate from the point of the maximum flux. This relative flux distribution can be related directly to the deposited film thickness, as long as the sticking coefficient of the material is uniform across the substrate

$$t(l)/t_{\mathrm{max}} = 1/[1 + (l/H)^2]^2 \tag{24}$$

in which $t(l)$ and t_{max} are the film thickness at distance l and the maximum film thickness, respectively.

Equation 24 typically appears in introductory texts on semiconductor processing (*12*). Despite the fact that the cosine law distribution is relatively flat, equation 24 shows that the relative incident flux or thickness uniformity falls off dramatically. For example, at $l/H = 0.5$, $t(l)/t_{\mathrm{max}} = 0.64$, even though $\cos(\phi) = 0.89$, that is, the incident flux at a planar substrate has fallen to 64% of its maximum value.

A larger substrate-to-source distance, H, can be used if a high degree of uniformity is required. However, increasing H strongly reduces the incident flux, $r(\mathrm{i})$, according to equation 21. Furthermore, for a fixed substrate area, raw material utilization decreases as H increases. Typically, uniformity is improved across large-area substrates by the motion of the substrate through the molecular beam. In this way, directionality in the beam shape can be averaged out.

Co-electron beam evaporation can give alloys of widely varying composition across a single substrate. This variation can be estimated by evaluating equation 16 for each source. One result predicted by this equation is that the composition variation can be minimized by keeping the spacing between the sources small.

High-Rate Sources. Sources such as those shown in Figures 1 and 2 are used in many semiconductor applications and operate at sufficiently high effusion rates to ensure transition or laminar flow through the nozzle or orifice.

Giordmaine and Wang (*13*) investigated molecular beams from long tube arrays operating in the transition flow regime. The shape of the molecular beam, according to Giordmaine and Wang (*13*) and Hanes (*14*), is determined by the depth into the vacuum end of the nozzle at which free molecular flow is encountered.

To illustrate this idea, the Knudsen number was calculated by using the mass flow model described by Rocheleau et al. (5) along the length of an 0.8-cm-diameter nozzle for zinc vapor flowing at 0.01 and 0.1 g/min. The results of this calculation are summarized in Figure 7. Transition or laminar flow occurs through most of the nozzle at $L = 0.4$–3.2 cm. The curves bend sharply upward near the vacuum end of the nozzles at $L < 0.4$ cm as the mean free path increases sharply. Over a short length at this end, free molecular flow conditions are encountered. The probability distribution for the situation illustrated in Figure 7 according to Giordmaine and Wang (*13*) is

$$F(\phi) = 8 \left| \frac{\cos(\phi)}{\pi} \right|^{3/2} \times \int_0^1 (1 - z^2)^{1/2} \int_0^{zk} \exp(-y^2)\, dy\, dz \tag{25}$$

in which z is a dimensionless distance into the nozzle at a source and k is a dimensionless density gradient in a nozzle.

The value of the dimensionless density gradient, k, which appears in

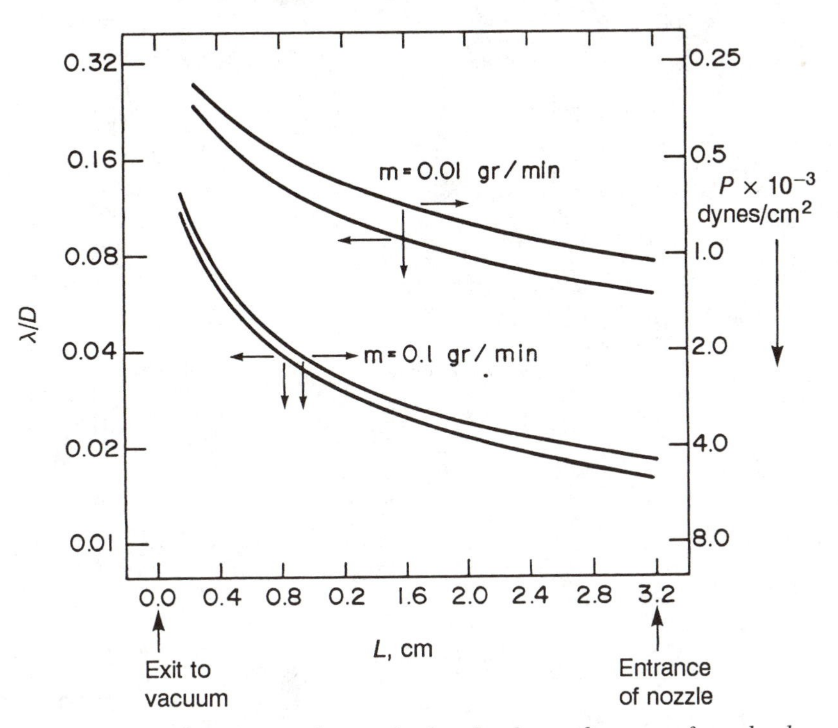

Figure 7. Ratio of mean free path of molecules to diameter of nozzle along length of nozzle for zinc source. (Reproduced with permission from reference 7. Copyright 1985 American Institute of Physics.)

the upper bound of the inner integral of equation 25, is obtained by fitting the following expression to experimental data (7)

$$k = \frac{A}{\tan(\phi)\sin^{1/2}(\phi)\cos^{1/2}(\phi)} \tag{26}$$

in which A is a geometric fitting parameter.

Equation 25 was numerically integrated for a range of values of k. Figure 8 shows the value of the integral $F(\phi)/\cos^{3/2}(\phi)$ function. For $k > 10$, the integral asymptotically approaches unity, and the beam distribution approaches $\cos^{3/2}(\phi)$.

The parameter A is the only material- and process-dependent parameter needed to solve equation 25. The value of A depends on the physical properties of the vapor species (molecular weight and collision diameter) and on

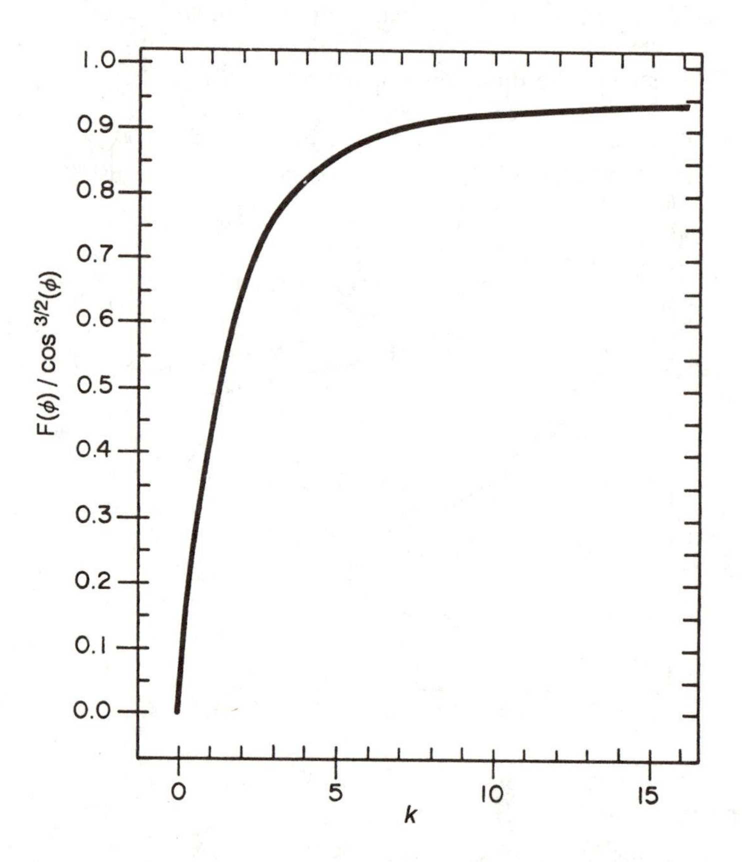

Figure 8. Integral value of equation 25 versus dimensionless density gradient k. *(Reproduced with permission from reference 7. Copyright 1985 American Institute of Physics.)*

the conditions of the evaporation (mass flow through the nozzle and vapor temperature).

Values of *A* can be estimated by using a theoretical expression proposed by Giordmaine and Wang (*13*) or by using an experimentally determined correlation using values of the Knudsen number at a fixed depth from the vacuum end of the nozzle. The procedures for the second method are described by Rocheleau et al. (*5*). This procedure gives the mass flow and pressure drops for evaporation and effusion through a series of flow-restricting orifices. The mean free paths used in the correlation range over two orders of magnitude, from 0.008 to 0.31 cm, and include both viscous and transition flows. Figure 9 shows the parameter *A* plotted against this Knudsen number. The data deviate from the curve with a root-mean-square deviation equal to 8% of the *A* value. Points representing Zn and Cd are very close to each other, a fact indicating that the correlation is material independent.

Figure 9 shows that the maximum or peak value of *A* is greater than

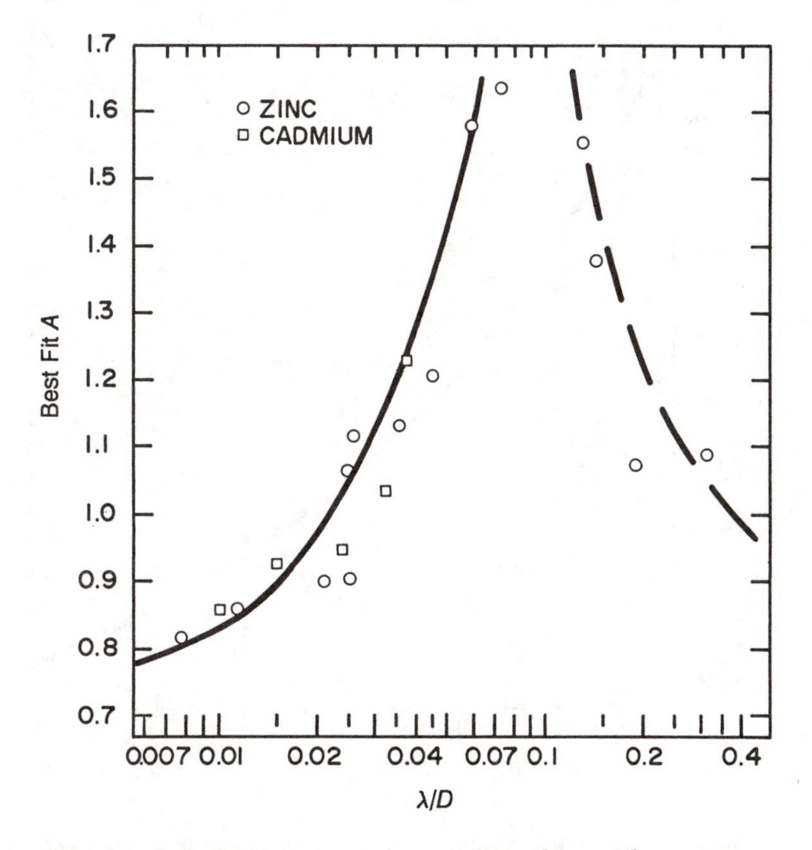

Figure 9. Correlation of fitting parameter A *with Knudsen number at 2.54 cm into the nozzle. (Reproduced with permission from reference 7. Copyright 1985 American Institute of Physics.)*

1.7 at a Knudsen number of 0.10. The correlation falls off abruptly on both sides of this peak. The peak occurs in the transition regime for mass flow. The flow becomes increasingly laminar to the left and free molecular to the right of the peak.

The parameter A determines the probability distribution $F(\phi)$ and, consequently, the normalization constant C_f, according to equation 15. This equation was numerically integrated for a range of A values. The calculated values of C_f are shown as a function of the parameter A in Figure 10. Compared with low values of C_f, high values of C_f mean that the molecular beam is proportionally more intense near the nozzle center line.

The normalization constant, C_f, is not a strong function of the parameter A. For example, a 50% change in A (from 1.0 to 1.5) results in a 17% change in the corresponding value of C_f (from 0.665 to 0.57).

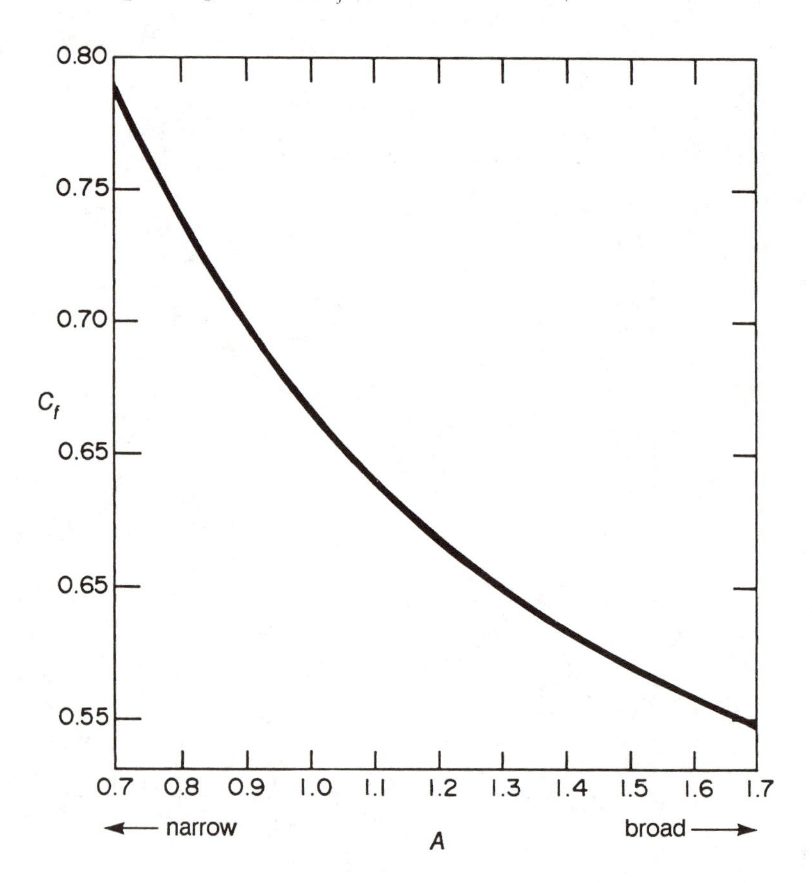

Figure 10. Normalization constant C_f *versus fitting parameter* A. *(Reproduced with permission from reference 7. Copyright 1985 American Institute of Physics.)*

Probability distributions have been measured for Cd and Zn (*1*) and CdTe (*15*) for several nozzle dimensions and temperatures yielding effusion rates of 0.01–0.33 g/min. A representative distribution is shown in Figure 11. The curve drawn through the data points in Figure 11 illustrates agreement between the probability distribution F(ϕ), which was predicted by equations 25 and 26, and the experimental data. The root-mean-square difference between the experimental and predicted probability distribution values is 1–3% of the center line value. This agreement is as good as empirical fits described in the literature that use two or more adjustable parameters (*16*).

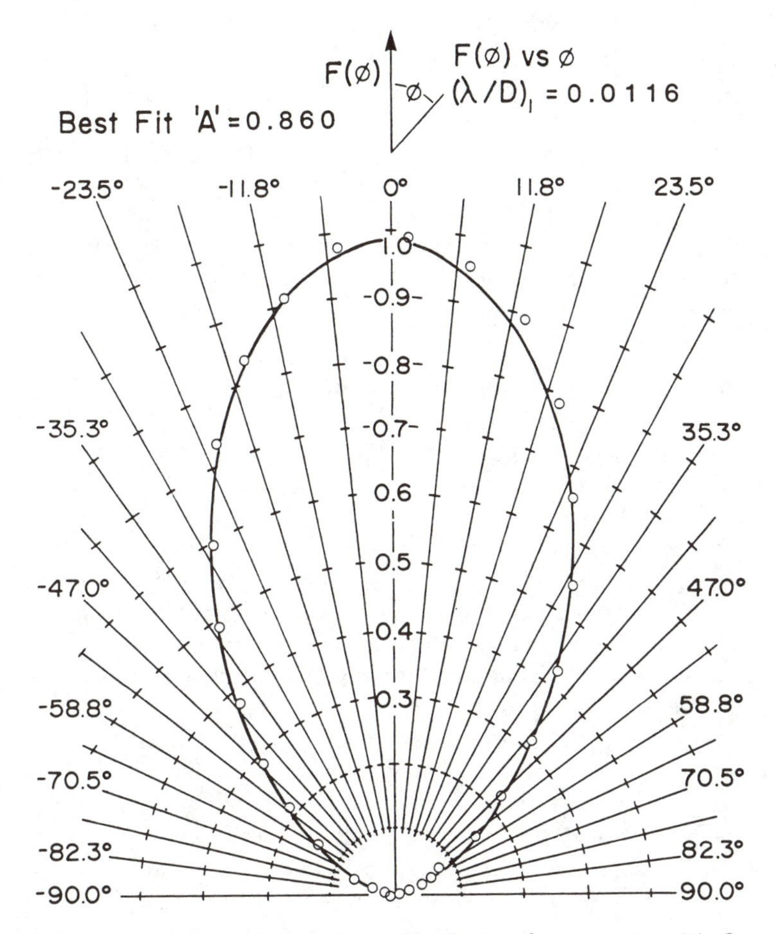

Figure 11. Measured molecular-beam distribution from equation 25. Open circles represent the measured values. The solid curve represents the predicted distribution. (Reproduced with permission from reference 7. Copyright 1985 American Institute of Physics.)

Analysis of Surface Reaction Zone

When an atom or molecule reaches the substrate surface, the formation of product is governed by physical and chemical processes that occur within the surface reaction zone. In this section, the chemical-reaction-engineering framework of Jackson et al. (*17*) is described through an example of the PVD of thin films of the compound-semiconductor ZnS.

According to the generally accepted mechanism for growth from the vapor phase, a fraction of molecules or atoms incident on a substrate are adsorbed. Those molecules or atoms that are not adsorbed are reflected back into the gas phase. The fraction reflected depends on the energy of the incident species and on the thermal-accommodation characteristics of the substrate.

The adsorbed species, which are considered to be adatoms, can diffuse to favorable low-energy sites and react, or they can be emitted into the gas phase. At sufficiently low temperatures, adatoms may have insufficient energy to diffuse and react or to be emitted into the gas phase. These adatoms will be codeposited with the compound film as crystal defects or as a second solid phase. As a result of these competing processes in the surface reaction zone, the growth rate and film composition depend on the flux and energy of the incident species and on the substrate temperature.

Model Equations to Describe Component Balances. The design of PVD reacting systems requires a set of model equations describing the component balances for the reacting species and an overall mass balance within the control volume of the surface reaction zone. Constitutive equations that describe the rate processes can then be used to obtain solutions to the model equations. Material-specific parameters may be estimated or obtained from the literature, collateral experiments, or numerical fits to experimental data. In any event, design-oriented solutions to the model equations can be obtained without recourse to equipment-specific fitting parameters. Thus translation of scale from laboratory apparatus to production-scale equipment is possible.

The control volume of the surface reaction zone, which is at the surface of the growing film (Figure 12), links the physical situation with the mathematical model that follows. Because the control volume is small enough, the incident flux from the sources is uniform within this volume. The net rate of surface diffusion into the control volume is assumed to be negligible compared with the incident flux. An incident component entering the control volume at a rate $r(\mathrm{i}, j)$ is either adsorbed or reflected from the surface, where the rate of reflection is $r(\mathrm{r}, j)$. An adsorbed component may react at a rate $r(\mathrm{rxt}, j)$ to form a compound, be emitted from the surface into the gas phase at a rate $r(\mathrm{e}, j)$, or be codeposited with the compound in an elemental form at a rate $r(\mathrm{d}, j)$.

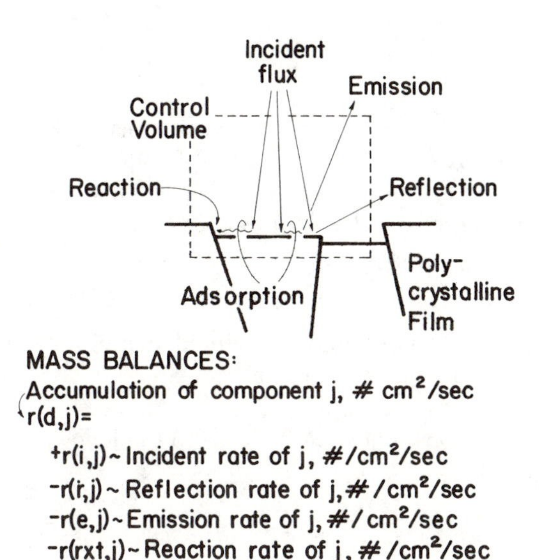

Figure 12. Control volume of surface reaction zone. The component molar balances are defined by equations 27 and 28.

Solutions of the mass balances with appropriate constitutive rate expressions for $r(\mathrm{i}, j)$, $r(\mathrm{r}, j)$, $r(\mathrm{e}, j)$, and $r(\mathrm{rxt}, j)$ for all components within the control volume yield the deposition rate and composition of the film.

The steady-state component molar balance for the control volume of the surface reaction zone shown in Figure 12 is given by

$$\frac{1}{\Phi M_w(j)} \frac{dM(j)}{dt} = r(\mathrm{i}, j) - r(\mathrm{r}, j) - r(\mathrm{e}, j) - r(\mathrm{rxt}, j) \quad (27)$$

in which j represents each of the components in the reacting system (for example, the group II–VI compound-semiconductor ZnS involves three components—Zn, S, and ZnS), Φ is the surface area of the control volume of the surface reaction zone, $M(j)$ and $M_w(j)$ are the mass and molecular weight, respectively, and r is the rate.

The left side of equation 27 represents the molar rate of accumulation (i.e., deposition) of adsorbed component j per unit area (Φ):

$$r(\mathrm{d}, j) = \frac{1}{\Phi M_w(j)} \frac{dM(j)}{dt} \quad (28)$$

For ZnS, several assumptions can be made to simplify the mass balance equation for each component:

1. The basic chemical equation for the formation of ZnS is the

reaction of Zn and S on the surface according to Zn + S $\rightarrow$ ZnS.

2. No reactions occur in the the vapor phase. Therefore the incident flux of ZnS into the control volume, $r(\mathrm{i}, \mathrm{ZnS}) = 0$. Furthermore, the rate of reflection of ZnS, $r(\mathrm{r}, \mathrm{ZnS})$, is 0, because there is no incident beam to reflect.
3. Evaporation of ZnS is negligible at substrate temperatures that are of practical interest, that is, $r(\mathrm{e}, \mathrm{ZnS}) = 0$.
4. Only ZnS is deposited. The accumulation of Zn and S in the control volume is zero, so that $r(\mathrm{d}, \mathrm{Zn}) = r(\mathrm{d}, \mathrm{S}) = 0$.

With these assumptions, equation 27 leads to the following molar balances for Zn, S, and ZnS :

$$r(\mathrm{rxt}, \mathrm{Zn}) = r(\mathrm{i}, \mathrm{Zn}) - r(\mathrm{r}, \mathrm{Zn}) - r(\mathrm{e}, \mathrm{Zn}) \tag{29}$$

$$r(\mathrm{rxt}, \mathrm{S}) = r(\mathrm{i}, \mathrm{S}) - r(\mathrm{r}, \mathrm{S}) - r(\mathrm{e}, \mathrm{S}) \tag{30}$$

$$\frac{1}{\Phi M_w(\mathrm{ZnS})} \frac{dM\,(\mathrm{ZnS})}{dt} = -r(\mathrm{rxt}, \mathrm{ZnS}) \tag{31}$$

From equation 28, the deposition rate of ZnS is given by:

$$r(\mathrm{d}, \mathrm{ZnS}) = \frac{1}{\Phi M_w(\mathrm{ZnS})} \frac{dM(\mathrm{ZnS})}{dt} \tag{32}$$

The reaction rates for Zn, S, and ZnS are related through the chemical equation Zn + S $\rightarrow$ ZnS, so that

$$r(\mathrm{rxt}, \mathrm{Zn}) = r(\mathrm{rxt}, \mathrm{S}) = -r(\mathrm{rxt}, \mathrm{ZnS}) = r(\mathrm{d}, \mathrm{ZnS}) \tag{33}$$

Model Equations to Predict Deposition Rate. Appropriate constitutive expressions are needed to evaluate each of the rate terms in the component molar balances. The final model equations must predict the deposition rate, $r(\mathrm{d}, \mathrm{ZnS})$, as a function of independent control variables—component incident fluxes, $r(\mathrm{i}, \mathrm{Zn})$ and $r(\mathrm{i}, \mathrm{S})$, and the substrate temperature.

In a line-of-sight PVD reactor, the incident flux, $r(\mathrm{i}, j)$, of each component entering the control volume of the surface reaction zone can be independently measured or calculated by using equation 16 and the methods described in the previous sections.

The rate at which the material leaves the control volume is governed by two mechanisms: reflection and emission of the adsorbed species into the gas phase. The incomplete adsorption of an incident component is characterized by the reflection factor δ (*18–20*). Accordingly, the reflection rate, $r(\mathrm{r}, j)$ is assumed to be proportional to the incident rate, as shown by equation 34

$$r(\mathrm{r}, j) = [1 - \delta(j)]r(\mathrm{i}, j) \tag{34}$$

in which $\delta(j)$ is the reflection factor of incident component j. The reflection factor is weakly dependent on the temperatures of the substrate and the incident molecule. It also depends on the composition of the deposited film (*20*). By definition, $\delta(j)$ must be between 0 and 1. In practice, the reflection factor of one component of a compound film can be experimentally determined by using an excess flux of the second component from an elemental source (*21*).

The rate at which the adsorbed components are emitted from the substrate back into the gas phase, $r(\mathrm{e}, j)$, depends on the composition of the surface, (which is expressed as the surface concentration of adatom species j, j^{s}), the binding energy of the adatom to the substrate, and the thermal energy of the adatom:

$$r(\mathrm{e}, j) = E_v(j)(j^{\mathrm{s}}) \tag{35}$$

The emission factor, $E_v(j)$, characterizes the energy of binding to the substrate and the thermal energy of the adatom. The emission factors are component specific and may depend on the structure of the adatom species. As will be shown later, the value of the emission factor does not have to be evaluated, because it is lumped with the forward-reaction rate constant.

In the example of the growth of ZnS, the surface concentrations of adsorbed Zn and S can be obtained by substituting equations 34 and 35 into the respective component balances, equations 29 and 30:

$$[\mathrm{Zn^s}] = [\delta(\mathrm{Zn}) \times r(\mathrm{i, Zn}) - r(\mathrm{rxt, Zn})]/E_v(\mathrm{Zn}) \tag{36}$$

$$[\mathrm{S^s}] = [\delta(\mathrm{S}) \times r(\mathrm{i, S}) - r(\mathrm{rxt, S})]/E_v(\mathrm{S}) \tag{37}$$

Equations 36 and 37 relate the surface concentrations, $[\mathrm{Zn^s}]$ and $[\mathrm{S^s}]$, to the incident fluxes, $r(\mathrm{i})$, and the reaction rates, $r(\mathrm{rxt})$.

Model Equations to Predict Reaction Rate. A final constitutive relation is needed for the reaction rate $r(\mathrm{rxt})$. If the mechanism for the formation of ZnS is assumed to be a bimolecular reaction between adsorbed Zn and S (22), as given by

$$\mathrm{Zn^s} + \mathrm{S^s} \rightarrow \mathrm{ZnS} \tag{38}$$

in which k_f is the rate constant of the forward reaction, then

$$r(\text{rxt, Zn}) = r(\text{rxt, S}) = k_f(\text{ZnS}) \times [\text{Zn}^s][\text{S}^s] \quad (39)$$

The corresponding reverse reaction (ZnS → Zn + S) has been neglected because it is only important at high substrate temperatures, during which dissociative evaporation of the deposited film is significant.

The predictive equation for the rate of deposition of ZnS (equation 40) is obtained by substituting the surface concentrations given by equations 36 and 37 into the reaction rate expression, equation 39:

$$r(\text{d, ZnS}) = K(\text{ZnS})[\delta(\text{Zn})r(\text{i, Zn}) - r(\text{d, ZnS})] \times [\delta(\text{S})r(\text{i, S}) - r(\text{d, ZnS})] \quad (40)$$

An apparent rate constant K(ZnS) is defined by equation 41:

$$K = \frac{k_f(\text{ZnS})}{E_v(\text{Zn})E_v(\text{S})} \quad (41)$$

A convenient way to illustrate the behavior of the model for the example of ZnS deposition is to plot the measured deposition rate, r(d, ZnS), as a function of the incident-flux rate of one element when the incident-flux rate of the second element is fixed. An example is shown in Figure 13 for the deposition of ZnS as a function of the incident-flux rate of sulfur at a substrate temperature of 200 °C. Experimental data points and curves representing the best-fit model predictions are shown for each of four zinc incident-flux rates. A nonlinear least-square procedure was used to obtain the following values for the model parameters that best fit equation 40 to the experimental data: $\delta(\text{Zn}) = 0.6\text{–}0.7$, $\delta(\text{S}) = 0.5\text{–}0.7$, and $K(\text{ZnS}) > 10^{-15}$ cm^2-s/ZnS.

Near the origin of the graph, the film growth rate is proportional to the sulfur incident rate, r(i, S). The slope in this region is approximately equal to the value of the sulfur reflection factor, δ(S), that was obtained from a nonlinear least-square fit of equation 40 to the experimental data. Thus, in the sulfur-limited regime, the deposition rate of ZnS is given by $r(\text{d, ZnS}) \sim \delta(\text{S})\, r(\text{i, S})$, in which $\delta(\text{S}) \sim 0.5$.

Similarly, at a high sulfur incident flux, the deposition rate becomes independent of the sulfur flux. Where the curves are horizontal in Figure 13, the deposition rate is ~70% of the zinc incident rate, a fraction that is very close to the Zn reflection factor, δ(Zn), that was obtained from the least-square fit. In this Zn-limited regime, the deposition rate is given by $r(\text{d, ZnS}) \sim \delta(\text{Zn})\, r(\text{i, Zn})$, in which $\delta(\text{Zn}) \sim 0.7$.

The close agreement between the limiting slopes determined graphically

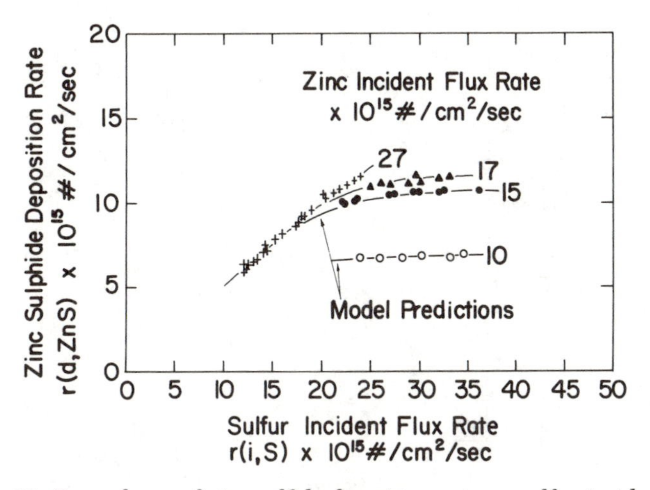

Figure 13. Dependence of zinc sulfide deposition rate on sulfur incident-flux rate at a substrate temperature of 200 °C for four zinc incident fluxes, r(*i, Zn*)*:* 27×10^{15}*,* 17×10^{15}*,* 15×10^{15}*, and* 10×10^{15} *Zn atoms/cm²-s.*

from the sulfur- and zinc-rich regimes and the numerically determined reflection factors indicates that the deposition rate in these regimes is limited by the rate at which the component with lower incident flux is adsorbed onto the substrate.

The model also accounts for the transition or knee in the curves shown in Figure 13. At the knee in the curves, the rate of deposition is controlled by the rates at which the adsorbed zinc and sulfur react. The curvature of the knee depends on the speed of the reaction relative to the emission rate of the elemental components. For a very fast reaction, the curvature is very sharp, as is the case for ZnS deposition at 200 °C (Figure 13).

The analysis of the surface reaction zone has been applied to laboratory-scale PVD of binary and ternary group II–VI compound semiconductors, such as CdS, ZnS, $(Cd_{1-x}Zn_x)S$, CdTe, HgTe, and $(Cd_{1-x}Hg_x)Te$, and the ternary group I–III–VI chalcopyrite $CuInSe_2$ (*17*). For example, Figure 14 shows the comparison between the predicted and measured compositions of ternary $(Zn_xCd_{1-x})S$ alloys. The predicted composition is within 3 atom % of the measured composition across the range of composition from 10 to 90 atom %.

The behavior noted for ZnS and $(Cd_{1-x}Zn_x)S$ has been reported by others for group II–VI (*21*) and III–V (*23*) compound semiconductors. Because the analysis of the surface reaction zone is based on conservation of mass without regard for the mechanism of transporting reactants to the substrate, the framework should be applicable for the engineering analysis of the deposition of a broad group of compound-semiconductor electronic materials by both PVD and CVD.

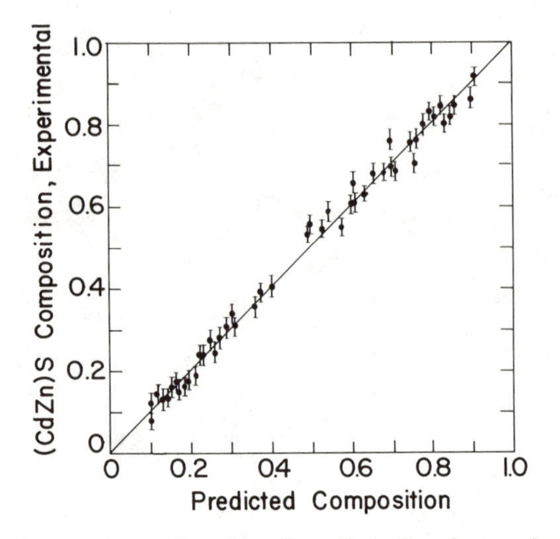

Figure 14. Comparison of measured and predicted zinc fractions, x, *in* $(Cd_{1-x}Zn_x)S$ *alloy films.*

Abbreviations and Symbols

A	geometric fitting parameter (equation 26)
A_s	surface area of source material charge
C	correction factor for short-pipe entrance losses (equation 10a)
C_f	normalization constant (equation 14)
C_o	orifice coefficient (Table I)
C_p	heat capacity of source material (equation 3)
D	nozzle or orifice diameter (Table I)
$E_v(j)$	emission factor of component j (equation 35)
f	fraction of gas molecules diffusely reflected from wall (f = 0.9) (equation 10b)
F_v	view factor (equation 3)
F_e	effective emissivity (equation 3)
$F(\phi)$	probability density distribution function (equation 13)
H	Height of planar substrate above source (equation 20 and Figure 5)
ΔH_R	latent heat of vaporization of source material (equation 3)
$I(\phi)$	molecular flux incident on the hemisphere at an angle ϕ (equation 12)
j	arbitrary elemental or compound species
k	dimensionless density gradient into a nozzle (equation 26)
k_f	forward reaction rate constant (equation 39)
K	correction factor for nonideal orifice (equation 9)
K_p	equilibrium constant for dissociative evaporation (equation 7)

l	distance on substrate from the point of maximum flux (equation 23)
L	pipe or orifice length (equation 9)
$M(j)$	mass of component j (equation 27)
M_w	molecular weight of the material flowing from the source (equation 4)
$M_w(j)$	molecular weight of component j (equations 6 and 27)
n	stoichiometric coefficient (equation 6)
p_1	pressure upstream of orifice or pipe (Table I)
p_2	pressure downstream of orifice or pipe (Table I)
p'	pressure at pipe entrance after contraction (equation 10)
p_m	mean pressure (equation 4)
P_c	vapor pressure in evaporation chamber (Figure 1)
P_e	equilibrium vapor pressure (equation 8)
P_i	vapor pressure in intermediate chamber (Figure 1)
P_o	ambient pressure in vacuum chamber (Figure 1)
q	volumetric flow rate of exit stream (equation 1)
$r(\mathrm{i})$	incident molecular flux at an angle ϕ from the center line, a distance R from the source, and an angle of incidence β onto the surface (equation 21)
$r(\mathrm{d}, j)$	deposition rate or rate of accumulation of component j inside control volume of surface reaction zone (equation 28)
$r(\mathrm{e}, j)$	rate of emission of component j from control volume (equation 27)
$r(\mathrm{i}, j)$	rate of incident flux of component j into control volume of surface reaction zone (equation 27)
$r(\mathrm{r}, j)$	rate of reflection of component j from control volume of surface reaction zone (equation 27)
$r(\mathrm{rxt}, j)$	rate of reaction of component j within control volume of surface reaction zone (equation 27)
R	radius of hemisphere to point of interest (Figure 5) or ideal gas constant
t	time
T_1	temperature of charge (equation 3)
T_2	temperature of bottle wall (equation 3)
V	volume of source material charge (equation 1)
z	dimensionless distance into nozzle of source (equation 25)
β	angle of incidence of molecular beam relative to surface normal (Figure 5)
Γ	pipe or orifice radius (equation 9 and Table I)
$\delta(j)$	reflection factor of component j (equation 34)
θ	azimuth angle (equation 14)
λ/D	Knudsen number
λ_m	mean free path (equation 4)

μ viscosity of gas at mean pressure (equation 4)
ρ bulk density of source material charge (equation 1)
ρ_g mass density of gas phase (equation 1)
σ Stefan–Boltzmann constant (equation 3)
ϕ angle from nozzle center line (Figure 5)
Φ surface area of control volume of surface reaction zone (equation 27)

References

1. Bunshah, R. F. In *Deposition Technologies for Thin Film Coatings*; Bunshah, R. F., Ed.; Noyes: Park Ridge, NJ, 1982; p 4–5.
2. Cho, A. Y.; Arthur, J. R. *Prog. Solid State Chem.* **1975,** *10,* 157–191.
3. Birkmire, R. W.; Hall, R. B.; Phillips, J. E. In *Proceedings, 17th IEEE PV Specialists Conference*; 1984; p 882.
4. Tu, C. W.; Hendel, R. H.; Dingle, R. In *Gallium Arsenide Technology*; Ferry, D. K., Ed.; Sams: Indianapolis, 1985; p 107–153.
5. Rocheleau, R. E.; Baron, B. N.; Russell, T. W. F. *AIChE J.* **1982,** *28,* 656.
6. Rocheleau, R. E., Ph.D. Thesis, University of Delaware, Newark, DE, 1981.
7. Jackson, S. C.; Baron, B. N.; Rocheleau, R. E.; Russell, T. W. F. *J. Vac. Sci. Tech.* **1985,** *A3*(5), 1916–1920.
8. Clausing, P. Z. *Phys.* **1930,** *66,* 471.
9. Shen, L. Y. I. *J. Vac. Sci. Technol.* **1978,** *15,* 10.
10. Stickney, R. E.; Keating, R. F.; Yamamoto, S.; Hastings, W. J. *J. Vac. Sci. Technol.* **1968,** *4,* 10.
11. Dayton, B. B. *Trans. Second American Vacuum Sci. Technol.* **1961,** 5, 5.
12. Glaser, A. B.; Subak-Sharpe, G. E. In *Integrated Circuit Engineering*; Addison–Wesley: Reading, MA, 1979; p 163.
13. Giordmaine, J. A.; Wang, T. C. *J. Appl. Phys.* **1960,** *31,* 463.
14. Hanes, G. R. *J. Appl. Phys.* **1960,** *31,* 2171.
15. Baron, B. N.; Jackson, S. C. Final Report: SERI subcontract XL–4–03146–1, 1986, SERI Publication No. SERI/STR–211–2970.
16. Dobrowlski, J. A.; Waldorf, A.; Wilkinson, R. L. *J. Vac. Sci. Technol.* **1982,** *21,* 881.
17. Jackson, S. C.; Baron, B. N.; Rocheleau, R. E; Russell, T. W. F. *AIChE J.* **1987,** *33*(5), 711–721.
18. Hirth, J. P.; Pound, G. H. *J. Phys. Chem.* **1960,** *64,* 619.
19. Hirth, J. P. In *Vapor Deposition*; Powell, C. F.; Oxles, J. H.; Blocher, J. M., Eds.; Wiley: New York, 1966.
20. Eyring, H.; Wanlass, F. M.; Eyring, E. M. In *Condensation and Evaporation of Solids*; Rutner, E.; Goldfinger, P.; Hirth, J. P., Eds.; Gordon & Breach: New York, 1964.
21. Smith, D. L.; Pickhardt, V. Y. *J. Appl. Phys.* **1975,** *46*(6), 2366.
22. Jackson, S. C., Ph.D. Dissertation, University of Delaware, Newark, DE, 1984.
23. Foxon, C. T. *J. Vac. Sci. Tech. B.* **1983,** *1*(2), 293.

RECEIVED for review December 30, 1987. ACCEPTED revised manuscript January 23, 1989.

5

Chemical Vapor Deposition

Klavs F. Jensen

Department of Chemical Engineering and Materials Science, University of Minnesota, Minneapolis, MN 55455

Chemical vapor deposition (CVD) of thin solid films from gaseous reactants is reviewed. General process considerations such as film thickness, uniformity, and structure are discussed, along with chemical vapor deposition reactor systems. Fundamental issues related to nucleation, thermodynamics, gas-phase chemistry, and surface chemistry are reviewed. Transport phenomena in low-pressure and atmospheric-pressure chemical vapor deposition systems are described and compared with those in other chemically reacting systems. Finally, modeling approaches to the different types of chemical vapor deposition reactors are outlined and illustrated with examples.

CHEMICAL VAPOR DEPOSITION IS A KEY PROCESS in microelectronics fabrication for the deposition of thin films of metals, semiconductors, and insulators on solid substrates. As the name indicates, chemically reacting gases are used to synthesize the thin solid films. The use of gases distinguishes chemical vapor deposition (CVD) from physical deposition processes such as sputtering and evaporation and imparts versatility to the deposition technique.

The reactions underlying CVD typically occur both in the gas phase and on the surface of the substrate. The energy required to drive the reactions is usually supplied thermally by heating the substrate or, in a few instances, by heating the gas. Alternatively, photons from an ultraviolet (UV) light source or from a laser, as well as energetic electrons in plasmas, are used to drive low-temperature deposition processes.

Various reaction schemes, including pyrolysis, reduction, oxidation, disproportionation, and hydrolysis of the reactants, have been used to produce a large variety of thin films relevant to microelectronics processing. Table I

0065–2393/89/0221–0199$15.75/0

Table I. CVD Films for Microelectronic Application

Application	*Type*	*Examples*
Insulator	Oxide	SiO_2, Al_2O_3, TiO_2, Ta_2O_5, B_2O_3, P_2O_5
	Nitride	Si_3N_4, Si_xN_y:H, BN
	Oxynitride	Si_xO_yN, Al_xO_yN
Semiconductor	Element	Si, Ge
	Group II–VI compound	ZnS, ZnSe, CdTe, CdHgTe
	Group III–V compound	GaAs, AlGaAs, AlGaN, InP, GaInAs, GaInAsP, InSb, GaSb
	Oxide	SnO_2, Sn_2O_3, ZnO
Conductor	Metal	W, Mo, Cr, Al
	Silicide	WSi_2, $MoSi_2$, $TiSi_2$
Superconductor		Nb_3Sn, NbN, $YBa_2Cu_3O_x$

lists examples of these films, and Table II gives examples of CVD processes. General reviews of CVD and related thin-film deposition processes are available in a number of books and survey papers (*1–6*). This chapter includes only a short overview of the main processes and concentrates on the fundamental physicochemical phenomena underlying CVD.

CVD Processes

General Process Considerations. To be useful, a CVD process must produce thin films with reproducible and controllable properties including purity, composition, film thickness, adhesion, crystalline structure, and surface morphology. The growth rates must be reasonable, and the deposition must not have significant impact on the microstructures already formed in the substrate. The deposition time must be sufficiently short, and the temperature has to be low enough so that dopant solid-state diffusion does not smear the results of previous processing steps.

The acceptable limits on the properties of the films vary with the application, but stringent demands are characteristic in the processing of materials for electronic applications. The demands increase with the level of integration, the decrease in device size, and the complexity of the device. Semiconductor films for high-performance digital devices (e.g., AlGaAs–GaAs) have to be perfectly single crystalline with impurities in the low part-per-billion range. Film thickness uniformity is generally critical in maintaining the same device characteristics (e.g. threshold voltages) across each substrate and from substrate to substrate. Furthermore, applications for heterojunction digital and optical devices require that the interface concentration between successive layers of semiconductors changes over a few monolayers or is graded in a controlled manner.

The production of films with reproducible and controllable electrical, optical, and mechanical properties requires pure CVD reagents that do not

Table II. CVD Systems

System	Overall Reaction	Deposition: Pressure (Pascal)	Deposition: Temperature (K)	Representative Reference
Si Epitaxy	$SiH_{4-x}Cl_x + H_2 \rightarrow Si + xHCl$, $x = 0, 2, 3$, or 4	101 kPa	1050–1450	7, 8, 16
Low-pressure CVD (LPCVD)	$SiH_2 \rightarrow Si + 2H_2$	50	850–950	20, 21
	$SiH_4 + N_2O \rightarrow SiO_2 + N_2$	50	900–1000	20
	$SiH_2Cl_2 + NH_3 \rightarrow Si_3N_4$	50	1000–1100	167
	$WF_6 + H_2 \rightarrow W + HF$	50	500–700	66, 71
Very Low Pressure CVD	$SiH_4 \rightarrow Si + 2H_2$	1	850–1150	22, 23
Vapor-phase Epitaxy (VPE)	$2HCl + 2Ga \rightarrow 2GaCl + H_2$			
	$12GaCl + 4AsH_3 + 2As + As_4 \leftrightarrows 12GaAs + 12HCl$	101 kPa	800–1100	9
Metallorganic CVD (MOCVD)	$Ga(CH_3)_3 + AsH_3 \rightarrow GaAs + CH_4$	101 kPa	800–900	11–13
	$Ga(CH_3)_3 + Al_2(CH_3)_6 + AsH_3 \rightarrow Al_xGa_{1-x}As + CH_4$	101 kPa	800–1000	11–13
Plasma-enhanced CVD (PECVD)	$SiH_4 \rightarrow a{:}SiH$	50	300–600	26–29
	$SiH_4 + NH_3 \rightarrow Si_xN_y{:}H$	50	400–700	26–29
Photon-assisted CVD	$SiH_4 + N_2O \xrightarrow[Hg]{h\nu} SiO_x$	50–500	300–600	40
	$In(CH_3)_3 + P(CH_3)_3 \xrightarrow{h\nu} InP$	60–500	600–700	35

produce byproducts that incorporate into the growing film and that do not interact with gas-handling and reactor construction materials. The substrate has to be properly cleaned and prepared for the deposition to avoid residual surface impurities that can create defects in the growing film.

The CVD reactor must be designed and operated in such a manner that changes in film thickness, crystal structure, surface morphology, and interface composition can be accurately controlled. The overall process performance depends on the reactor design and process variables such as reactant concentrations, flow rates, energy input, pressure, and substrate conditions.

A model describing the relationship between the performance and the process variables allows the prediction of process results and the optimization of process variables for a specific application. However, the interactions among deposition chemistry, transport processes, and growth modes are complex and, consequently, poorly understood. Therefore, CVD process development has progressed through extensive one-parameter-at-a-time experimentation and empirical design rules.

Several CVD processes have evolved to accommodate the applications of CVD films in microelectronics processing. The various processes are typically characterized in terms of the operating pressure and temperature, as well as the means of energy input. Table II gives examples of typical CVD processes and operating conditions.

CVD Processes at Atmospheric and Reduced Pressures. Atmospheric to slightly reduced pressures (~100–10 kPa) are used primarily to grow epitaxial (i.e., single-crystalline) films of Si and compound semiconductors such as GaAs, InP, and HgCdTe. These processes generally involve high growth temperatures (>850 °C for Si and 400–800 °C for most compound semiconductors), although the reactor walls are cooled to minimize impurity generation.

$SiCl_4$ is the classical reactant for the epitaxial growth of Si, but it has been replaced by $SiHCl_3$, SiH_2Cl_2, and SiH_4 to decrease the deposition temperature and to minimize solid-state diffusion out of the substrate into the growing film. In general, the deposition temperature for epitaxial growth decreases with the Cl content of the reactant from 1150 °C for $SiCl_4$ to 850 °C for SiH_4 (7, 8). Good single-crystalline films are easier to prepare with Cl-containing compounds than with SiH_4, because the reverse etching reaction by HCl preferentially occurs at defect sites.

Vapor-phase epitaxy (VPE) is a well-developed technique for growing group III–V compound semiconductors, specifically GaAs and GaInAsP, from the corresponding hydrides and halides of the individual components (9). The deposition process essentially relies on the temperature dependence of the equilibrium distribution of the desired film material (e.g., GaAs) and the gas-phase species (e.g., GaCl, As_2, and As_4). By imposing a temperature gradient on the reactor, the gas-phase species is formed in a hot region and the equilibrium is shifted towards the desired solid (e.g., GaAs) in a slightly colder region.

This process has been used successfully to grow GaAs and GaInAsP. However, the high reactivity of AlCl and thermodynamic limitations of Al halide compounds make it difficult to deposit Al-containing films by VPE (*1*). This limitation is a serious constraint because many devices use Al-GaAs–GaAs structures.

An alternative CVD process, metallorganic CVD (MOCVD), also called organometallic vapor-phase epitaxy (OMVPE), has attracted considerable attention, because it allows greater flexibility than VPE does in the synthesis of thin, high-purity, epitaxial films of compound semiconductors (*10–13*). This technique entails the transport of at least one of the film constituents as an organometallic compound. For example, AlGaAs can be deposited from organometallic sources of Ga and Al (e.g., $Ga(CH_3)_3$ and $Al_2(CH_3)_6$, respectively) and AsH_3. Alternatively, an organometallic source of As could be used (*14*, *15*). This technique has been used to grow numerous group II–VI and group III–V compound semiconductors, including GaAs, AlGaAs, GaInAsP, GaSb, InSb, ZnSe, and CdHgTe, for optoelectronic and high-speed electronic devices (*10*, *12*).

The horizontal reactor (Figure 1a) is a classical configuration for growth at atmospheric or reduced pressure. This reactor is now primarily used for research and for the epitaxial growth of compound semiconductors, along with the vertical reactor (Figure 1b). The barrel reactor (Figure 1c) is the primary reactor for Si epitaxy (*16*, *17*), and small barrels are beginning to be used in GaAs technology (*18*, *19*). So-called "pancake" reactors also find use in Si technology.

The reactor walls are typically cooled, except for VPE applications, to minimize particulate and impurity problems caused by deposition on the walls. However, this cooling also creates large thermal gradients that induce complex, buoyancy-driven secondary flows. In horizontal and barrel reactors, the susceptor is tilted relative to the main flow direction to improve film uniformity along the length of the susceptor. Uniformity is further controlled in the barrel reactor by adjusting the inlet gas nozzle and spinning the barrel. In vertical reactors, the susceptor is often rotated to reduce film thickness and composition variations, but the rotation speed (5–40 rpm) is generally much lower than that needed to generate an ideal rotating-disk flow (500–2000 rpm).

Low-Pressure CVD Processes. Low-pressure CVD (LPCVD) (~101 Pa) is the main tool for the production of polycrystalline Si dielectric and passivation films used in Si IC (integrated-circuit) manufacture (*1*, *20*, *21*). The main advantage of LPCVD is the large number of wafers that can be coated simultaneously without detrimental effects to film uniformity. This capability is a result of the large diffusion coefficient at low pressures, which allows the growth rate to be limited by the rate of surface reactions rather than by the rate of mass transfer to the substrate. Typically, reactants can be used with no dilution, and therefore growth rates are only an order of magnitude less than those possible at atmospheric conditions, in which high dilution ratios are used to avoid gas-phase nucleation.

Very low pressure processes (~1.3 Pa) have also been used for the growth of single-crystalline Si at relatively low temperatures (*22*, *23*). Low-pressure operation is also advantageous for the growth of compound-semiconductor superlattices by reducing flow recirculations and improving interface abruptness (*24*).

Figures 1e and 1f illustrate two typical LPCVD reactor configurations. These reactors operate at ~50 Pa, and wall temperatures are approximately

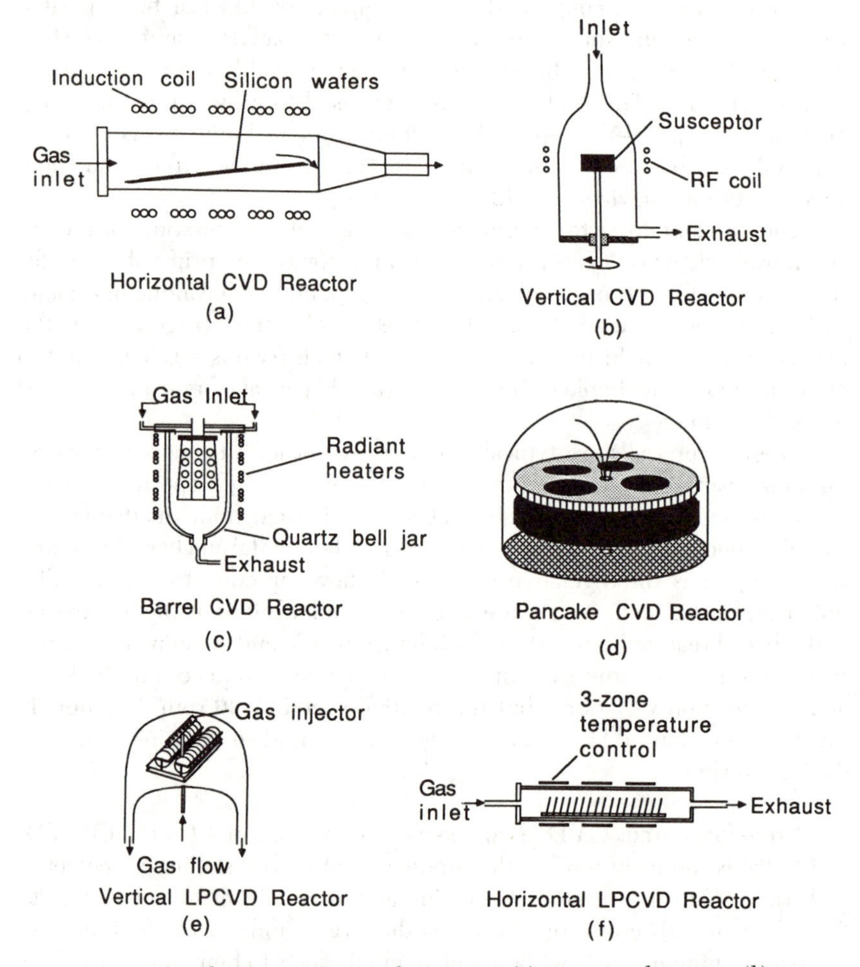

Figure 1. Typical CVD reactor configurations. (a) Horizontal reactor, (b) vertical reactor, (c) barrel reactor, (d) pancake reactor, (e) cross-flow LPCVD reactor, and (f) conventional multiple-wafer-in-tube LPCVD reactor.

equal to those of the deposition surfaces. Thus, deposition also takes place on the walls, a situation that raises potential particulate problems. The horizontal multiple-wafer-in-tube LPCVD reactor (Figure 1e) is the dominant configuration for Si IC manufacture. The vertical-flow reactor (Figure 1f) gives better uniformity and lower particulate counts than the horizontal geometry does but at the expense of low reactant utilization (*25*).

Plasma-Enhanced CVD. Plasma-enhanced CVD (PECVD) has received considerable attention in microelectronics processing because of its ability to grow films at relatively low temperatures and to impart special material properties that cannot be realized with conventional thermal processes (*26–29*). Plasmas used in microelectronics processing are weakly ionized gases composed of electrons, ions, and neutral species. The charged-species concentrations range from 10^9 to $10^{12}/cm^3$, and the ratio of charged species to neutral species ranges from 10^{-6} to 10^{-4}.

These plasmas, also called glow discharges, are generated by applying an external electric field to the process gas at low pressures (0.1–50 Pa). The result is a mixture of high-energy, "hot" electrons (1–10 eV; 10^4–10^5 K) and "cold" ions and neutral species (300 K). This high electron energy relative to the low temperature of neutral species makes discharges useful in driving CVD reactions.

Inelastic collisions between the high-energy electrons and neutral molecules result in, among other processes, electron-impact ionization and molecular dissociation. Electron-impact ionization helps to sustain the discharge, and molecular dissociation creates free radicals that contribute to the deposition processes.

The created ions, electrons, and neutral fragments participate in complex surface reactions that form the basis of the film growth. Positive-ion bombardment of surfaces in contact with the plasma plays a key role by modifying material properties during deposition. A direct-current (dc) bias potential may be applied to the excitation electrode to increase the ion energy and enhance the desired effects of ion bombardment (*30*).

For radiation-sensitive substrates such as compound semiconductors, afterglow deposition systems have been developed (*31*). In these processes, the radicals are formed in the glow discharge and then transported out of the discharge region to a downstream deposition zone. This plasma configuration eliminates ion bombardment and allows the selective activation of reactants by regulating the species that flow through or bypass the discharge.

Additional growth considerations, as well as process modeling and plasma diagnostics underlying plasma-enhanced CVD are further discussed by Hess and Graves in Chapter 8, which is specifically devoted to plasma processing.

Photoassisted CVD. In addition to thermal energy and electron-impact reactions, photons (e.g., UV light) can also drive CVD reactions

(*32–37*). This process has the advantage of low-temperature deposition, which may be needed for the growth of thin films on temperature-sensitive substrates such as compound semiconductors and polymers. If a laser is used as the light source, fine lines of materials can be written directly, and possibly, lithographic steps can be avoided and damaged lines can be repaired.

Photons can promote CVD reactions by different routes (*34*). One or more photons may lead to the gas-phase photolysis of a reactant and the formation of reactive fragments. The light can also be absorbed by adsorbed surface species that tend to undergo reactions leading to thin-film formation. Alternatively, the photons can alter the electronic states of the substrate surface and thereby promote film reactions. Finally, the light can be transformed into heat in the top surface layer and thermally drive the deposition process. This conversion to thermal energy is essentially equivalent to thermal CVD, but if a laser is used, the process has the advantages of increased energy flux, rapid heating, and a spatially well-defined deposition area. Because of their potential direct-line-writing applications, photolytic- and pyrolytic-laser-assisted CVD processes are areas of active research (*30–39*), and most reactor systems are small special-purpose laboratory reactors.

Photosensitization is used for large-area photochemically stimulated CVD, because the generation of a sufficient photon flux over a large area to drive the chemistry directly is difficult. Usually, Hg excited by an external Hg lamp is used as a sensitizer. The energy in the excited Hg is then transferred to other gas-phase species that decompose and react to form a thin film. The process is used in horizontal reactors for the deposition of SiO_x and SiN_yH_z from SiH_4, N_2O, and NH_3 (*40–42*) and to assist the deposition of CdHgTe, in which Hg is a natural gas-phase constituent (*43*).

Other CVD Processes. CVD also finds extensive use in the production of protective coatings (*44, 45*) and in the manufacture of optical fibers (*46–48*). Whereas the important question in the deposition of protective coatings is analogous to that in microelectronics (i.e., the deposition of a coherent, uniform film), the fabrication of optical fibers by CVD is fundamentally different. This process involves gas-phase nucleation and transport of the aerosol particles to the fiber surface by thermophoresis (*49, 50*). Heating the deposited particle layer consolidates it into the fiber structure. Often, a thermal plasma is used to enhance the thermophoretic transport of the particles to the fiber walls (*48, 51*). The gas-phase nucleation is detrimental to other CVD processes in which thin, uniform solid films are desired.

CVD Fundamentals

Chemical vapor deposition of thin films involves gas-phase and surface reactions combined with transport processes. Figure 2 gives a schematic rep-

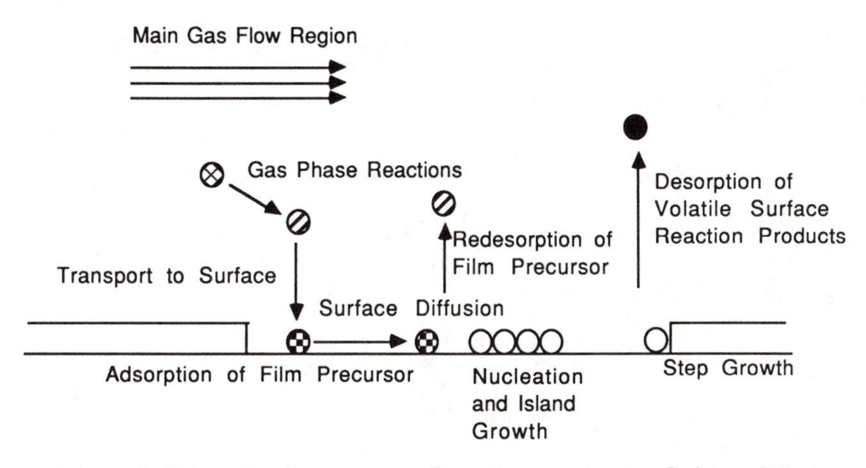

Figure 2. Schematic of transport and reaction processes underlying CVD.

resentation of the different elements of the growth process, which can be summarized in the following steps:

1. mass transport in the bulk gas flow region from the reactor inlet to the deposition zone,
2. gas-phase reactions leading to the formation of film precursors and byproducts,
3. mass transport of film precursors to the growth surface,
4. adsorption of film precursors on the growth surface,
5. surface diffusion of film precursors to the growth sites,
6. incorporation of film constituents into the growing film (island),
7. desorption of byproducts of the surface reactions, and
8. mass transport of byproducts in the bulk gas flow region away from the deposition zone towards the reactor exit.

Similar reaction sequences have been identified in other chemically reacting systems, specifically catalytic combustion (*52*, *53*), solid-fuel combustion (*54*), transport and reaction in high-temperature incandescent lamps (*55*), and heterogeneous catalysis (*56* and references within). The elementary reactions in hydrocarbon combustion are better understood than most CVD gas-phase reactions are. Similarly, the surface reaction mechanisms underlying hydrocarbon catalysis are better known than CVD surface reactions.

The deposition of Si by the reduction of SiH_4 is a convenient example of a CVD process that clearly displays the reaction steps just listed. SiH_4

diluted in H_2 is the starting material. As the SiH_4 is transported into the hot gas phase adjacent to the substrate, it pyrolyzes to form SiH_2 and H_2 according to the following reaction:

$$SiH_4 \leftrightarrows SiH_2 + H_2 \quad (1)$$

The silylene molecule, SiH_2, is very reactive and rapidly inserts itself into H_2, silane, and higher silanes with almost no activation energy (*57–59*), as shown by the following reactions:

$$SiH_4 + SiH_2 \leftrightarrows Si_2H_6 \quad (2)$$

$$Si_2H_6 + SiH_2 \leftrightarrows Si_3H_8 \quad (3)$$

The higher silanes may lose hydrogen by a reaction similar to that given in equation 1:

$$Si_2H_6 \leftrightarrows Si_2H_4 + H_2 \quad (4)$$

All the silicon hydrides are adsorbed on the Si surface. The unsaturated species and higher silanes are adsorbed more readily than silane (*60*). The adsorbed silicon species diffuse on the surface to growth sites, where the Si is incorporated in the growing film and the byproduct, hydrogen, is released and eventually desorbs as hydrogen molecules. The relative rates of surface diffusion, nucleation, and adsorption govern the crystalline morphology of the growing film. This and other fundamental issues are discussed in subsequent sections, along with further information on specific CVD systems, including systems for Si and GaAs deposition.

Nucleation and Growth Modes. The three primary growth modes for thin films are illustrated in Figure 3 (*61*). In three-dimensional island growth, referred to as Volmer–Weber growth, small clusters are nucleated directly on the substrate surface. The clusters grow into islands of the film material that eventually coalesce to form a continuous film (Figure 3a). This growth mode takes place when the film atoms are more strongly bound to each other than to the substrate. This growth mode applies to silicon growth on insulators (e.g., Si on SiO_2, Si_3N_4, or Al_2O_3) (*8*, *62–64*) and is also a common growth mode for metals on insulators. Aluminum CVD is an extreme example of a CVD process in which a catalyst such as $TiCl_4$ is needed to nucleate the clusters (*65*, *66*).

Two-dimensional layer-by-layer growth (Figure 3c), also called Franck–van der Merwe growth, occurs when the film atoms are equally or less strongly bonded to each other than to the substrate. This growth mode applies to homoepitaxy on clean substrates (e.g., Si on Si). The presence of

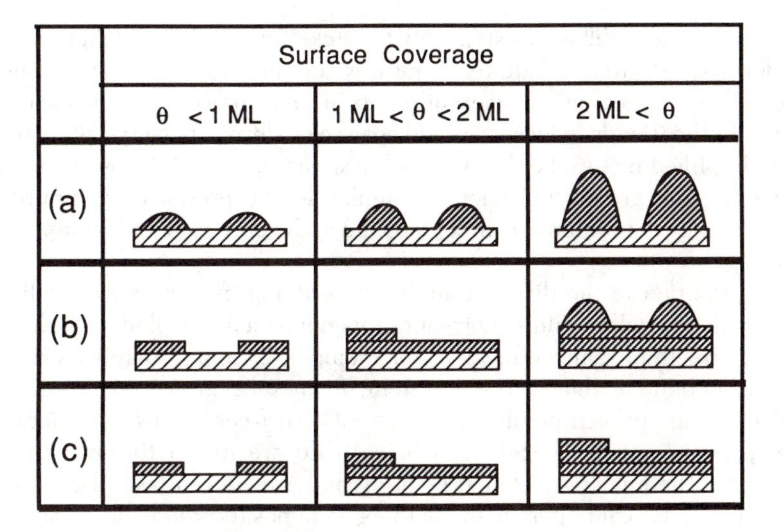

Figure 3. Crystal growth modes: (a) three-dimensional island (Volmer–Weber) growth, (b) layer-plus-island (Stranski–Kastanov) growth, and (c) layer-by-layer (Franck–van der Merwe) growth.

impurities, specifically carbon on the surface, gives rise to three-dimensional island growth (*61, 67*). Often, slightly off-axis (1°–3°) oriented substrates are used to suppress island nucleation and to promote layer growth. Heteroepitaxy of lattice-matched systems, such as $Al_xGa_{1-x}As$ on GaAs, also follow the layer-by-layer growth mode.

The layer-plus-island growth mode (Figure 3b), also called Stranski–Krastanov growth, is a combination of the other growth modes. After the growth of one or a few monolayers, subsequent layer growth becomes unfavorable, and islands form on top of the initial layers. This transition from two-dimensional to three-dimensional growth is not well understood, but for some systems, the transition may be due to an increase in elastic energy that prevents the lattice constant or crystalline structure of the film from being continued into the bulk crystal. This growth mode applies to lattice-mismatched compound-semiconductor systems in which thin (<100 Å) strained layers may be deposited and defects emerge in thicker layers. This issue is important for strained-layer superlattices (*68, 69*). This growth mode occurs more frequently than was originally expected (*61*).

The growth temperature has a strong influence on the structure of the deposited film. At low temperatures (and high growth rates), the surface diffusion is slow relative to the arrival rate of film precursors. In this situation, the adsorbed precursor molecule is likely to interact with an impinging precursor molecule before it has a chance to diffuse away on the surface, and an amorphous film is formed. At high temperatures (and low growth rates), the surface diffusion is fast relative to the incoming flux. The adsorbed species can then diffuse to step growth sites, and single crystalline layers are formed. At intermediate temperatures (and intermediate growth rates), nucleation occurs at many different points on the surface. Adsorbed species diffuse to the islands, which grow and coalesce to form a polycrystalline film. This simplified picture holds for simple systems such as Si deposition (*7, 8*). However, the growth behavior is complex in the presence of impurities (intentionally or unintentionally added) and in the growth of multicomponent films.

In addition to the different applications of amorphous, polycrystalline, and single crystalline films, nucleation phenomena are exploited in the selective growth of semiconductors. For example, by appropriately balancing the Cl amount in the Si–H–Cl system, Si may be grown on exposed Si surfaces while preventing deposition on SiO_2 surfaces (*62–64, 70*). Because the nucleation rate on the SiO_2 is sufficiently slow relative to the layer growth on the Si, the nuclei are etched before island growth can take place. However, if the Cl content is high, nothing is deposited, and the Si layer is etched. On the other hand, if the Cl content is too low, growth occurs nonselectively on all surfaces.

A similar selective growth process is observed for the CVD of W on Si–SiO_2 substrates (*66, 71, 72*) and for MOCVD of GaAs on GaAs–SiO_2 (or W) substrates (*73, 74*). The GaAs growth system also uses the difference in nucleation rates on the GaAs and SiO_2 surfaces, but the retarded growth of nuclei on the SiO_2 may be caused by the thermal decomposition of GaAs rather than an etching reaction.

In classical nucleation theory, nucleation is described in terms of the Gibbs free energy change involved in making the nucleus and the energy change necessary to increase the surface area of the cluster. The balance between these terms leads to a critical cluster size (r^*) beyond which the cluster will continue to grow. The critical value can be related to the surface tension (γ), atomic volume ($\overline{V}$), and supersaturation (p/p_{eq}) in the following manner (*8*):

$$r^* = \frac{2\gamma\overline{V}}{kT \ln (p/p_{eq})} \tag{5}$$

For most CVD reactions, the supersaturation is so high that calculated values of r^* are of atomic dimensions. For such reactions, the classical theory is not appropriate, and detailed atomic treatments must be considered. Because of the interesting fundamental questions underlying nucleation and the important applications of thin films, interest in modeling adsorption, surface diffusion, and nucleation has been considerable. These efforts are described in several, well-documented reviews (*61, 75–78*).

Chemical Equilibrium. Although CVD is a nonequilibrium process controlled by chemical kinetics and transport phenomena, equilibrium analysis is useful in understanding the CVD process. The chemical reactions and phase equilibria determine the feasibility of a particular process and the final state attainable. Equilibrium computations with intentionally limited reactants can provide insights into reaction mechanisms, and equilibrium analysis can be used also to estimate the defect concentrations in the solid phase and the composition of multicomponent films.

Source Considerations. Many CVD sources, especially sources for organometallic CVD, such as $Ga(CH_3)_3$ and $Ga(C_2H_5)_3$, are liquids at near room temperatures, and they can be introduced readily into the reactor by bubbling a carrier gas through the liquid. In the absence of mass-transfer limitations, the partial pressure of the reactant in the gas stream leaving the bubbler is equal to the vapor pressure of the liquid source. Thus, liquid–vapor equilibrium calculations become necessary in estimating the inlet concentrations. For the MOCVD of compound-semiconductor alloys, the computations have also been used to establish limits on the control of bubbler temperature to maintain a constant inlet composition and, implicitly, a constant film composition (*79*). Similar gas–solid equilibrium considerations govern the use of solid sources such as $In(CH_3)_3$.

Gas-Phase Chemical Equilibrium. Because CVD operating temperatures are high and molar flow rates and deposition rates are generally low, near-equilibrium conditions are often approached in various reactor regions. The equilibrium composition at constant temperature and pressure is generally computed by one of two ways: (1) direct minimization of the Gibbs free energy of the system subject to elemental abundance and mole number nonnegativity constraints or (2) transformation of the species mole number variables into a new set of extent-of-reaction variables and then minimization of the Gibbs free energy in terms of these new variables. Several reviews are available on the computation of complex reaction equilibria by these techniques (*80–82*).

Because of the relative ease of determining the system state by equilibrium calculations relative to experiments or detailed kinetics models, chemical equilibrium analysis has been the traditional approach to CVD process modeling. An extensive literature exists for the Si–Cl–H CVD system (*7*) and the As–Ga–Cl–H VPE process (*1*). The analysis of MOCVD systems has been limited by the lack of thermodynamic data. A recent equilibrium analysis of the MOCVD of GaAs (*83, 84*) is a good source of data for the GaAs system.

As an example, Figure 4 shows the equilibrium prediction of the Si yield from a $SiCl_4$–H_2 reactant mixture as a function of the $SiCl_4$ partial

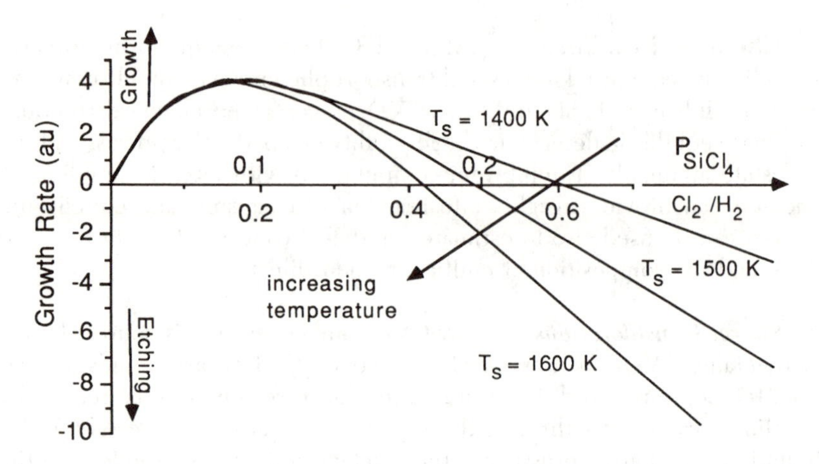

Figure 4. Curves of growth rate for Si as function of the Cl_2/H_2 ratio predicted from equilibrium calculations (Reproduced with permission from reference 7. Copyright 1978 Elsevier.)

pressure and the Cl_2/H_2 ratio for three deposition temperatures (7). This interesting system involves both deposition and etching according to the following overall reactions:

$$SiCl_4(g) + 2H_2(g) \leftrightarrows Si(s) + 4HCl(g) \text{ (deposition)} \quad (6)$$

$$SiCl_4(g) + Si(s) \leftrightarrows 2SiCl_2(g) \text{ (etching)} \quad (7)$$

$$SiCl_4(g) + Si(s) + 2H_2(g) \leftrightarrows 2SiH_2Cl_2(g) \text{ (etching)} \quad (8)$$

$$3SiCl_4(g) + Si(s) + 2H_2(g) \leftrightarrows 4SiHCl_3(g) \text{ (etching)} \quad (9)$$

Both deposition and etching may occur, and the window in $SiCl_4$ partial pressures and temperatures for which deposition is possible must be established. Figure 4 shows that at low partial pressures of $SiCl_4$, the growth rate increases with $SiCl_4$. At higher partial pressures of $SiCl_4$, the etching reaction leading to $SiCl_4$ dominates. Additional analysis shows that the etching by formation of chlorosilanes takes place at low temperatures (<1200 K) (7). Typically, silicon CVD from chlorosilanes is operated close to the point where deposition and etching balance. CVD under these conditions leads to excellent single-crystalline material, because the growth rate is low and the etching reaction preferentially attacks defect sites. Furthermore, because these growth conditions are near equilibrium, equilibrium analysis has been instrumental in understanding the Si CVD process and in identifying operating windows.

The growth of Si from silane has no equivalent etching process; therefore, this process is more difficult to control and is not amenable to the same equilibrium treatment as the Si–Cl–H system. Nevertheless, computations with detailed silane kinetics show that the gas phase is in a quasi-thermodynamic equilibrium in the hot zone above the susceptor. In this equilibrium situation, the species concentrations are determined by the initial silane decomposition step (equation 1) and the chemical equilibria between the higher saturated and unsaturated silanes (*85–87*). Besides greatly simplifying the analysis of CVD reactors, this treatment has the advantage that only the rate constant for the initial silane decomposition has to be known, and that rate constant has been measured experimentally.

Solid-Phase Chemical Equilibrium. For the growth of multicomponent films, the solid film composition must be predicted from the gas-phase composition. In general, this prediction requires detailed information about transport rates and surface incorporation rates of individual species, but the necessary kinetics data are rarely available. On the other hand, the equilibrium analysis only requires thermodynamic data (e.g., phase equilibrium data), which often are available from liquid-phase-epitaxy studies, as discussed by Anderson in Chapter 3.

Thermodynamic predictions of the solid-phase composition have been very successful for the growth by MOCVD of group III–V compound semiconductors (e.g., $InAs_{1-x}Sb_x$ and $GaAs_{1-x}Sb_x$) even though the gas-phase reactions are far from equilibrium (*88–91*). The procedure is also useful for estimating solid–vapor distribution coefficients of group II–VI compound semiconductors (e.g., $Cd_{1-x}Hg_xTe$ and $ZnSe_{1-x}S_x$) grown by MOCVD (*92*). In the analysis, the gas phase is considered to be an ideal mixture, that is

$$\mu_i^{(g)} = \mu_i^{(g)o} + RT \ln (p_i/p_i^o) \tag{10}$$

where μ_i and p_i are the chemical potential and partial pressure, respectively, of component i, R is the gas constant, and T is temperature (in Kelvin units). The superscripts g and o refer to the gaseous state and the standard state, respectively.

Two models are frequently used to predict the activity coefficient of the solid: the regular solution model (*93*) and the DLP (delta–lattice–parameter) model (*94*). With both models, the activity coefficient of component i, γ_i, is calculated in terms of the interaction parameter, Ω, by the expression

$$\ln \gamma_i = (1 - x_i)^2\Omega/RT \tag{11}$$

where x_i is the mole fraction of component i in the solid phase. The chemical potential of component i in the solid phase is then

$$\mu_i^{(s)} = \mu_i^{(s)o} + RT \ln (x_i\gamma_i) \tag{12}$$

The interaction parameter, Ω, is a fitting parameter in the regular solution model that can be found from liquid–solid equilibrium data (*93*). With the DLP model, the interaction parameter is calculated from the lattice parameters of the binary compounds. For a compound semiconductor $A_{1-x}B_xC$, Ω is computed from the lattice constants a_{AC} and a_{BC} of the binary compounds from the following expression

$$\Omega_{AC-BC} = 5 \times 10^7 \frac{(a_{AC} - a_{BC})^2}{[1/2(a_{AC} + a_{BC})]^{4.5}} \tag{13}$$

By equating the chemical potentials in the two phases and imposing conservation and stoichiometry constraints, the solid-phase composition may be predicted from equations 10–13 (*see* references 88 and 91 for examples).

On the basis of the DLP model, the solid is expected to be an ideal solution for systems in which both components have similar lattice constants. This case is true for the AlAs–GaAs system (Figure 5), in which the solid-phase composition equals the gas-phase composition. Compounds in which the two components have very different behavior will show highly nonlinear composition variations, and miscibility gaps may occur (*90, 91*). As an example, Figure 5 also shows that the solid-phase composition of $InAs_{1-x}Sb_x$ is a nonlinear function of the gas-phase composition.

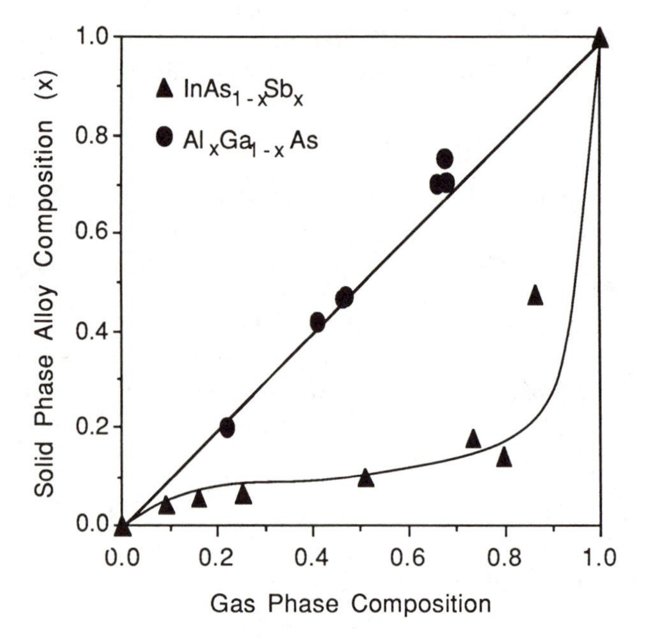

Figure 5. Solid composition versus vapor composition for the group III–V alloys $Al_xGa_{1-x}As$ and $InAs_{1-x}Sb_x$ (89).

Thermodynamic analysis can be useful also in predicting the effect of gas-phase composition on defect concentrations in the solid and, implicitly, on the electrical properties of the deposited film (*1*, *95*, *96*). This technique has been used to predict the concentration and change in electrical carriers from electrons to holes in PbS with increasing sulfur pressure over the PbS crystal (*96*).

Thermodynamic analysis is a useful tool in understanding CVD processes but should be used with caution and careful attention to the assumptions underlying the application. Because CVD is a nonequilibrium process, the thermodynamic predictions are often only semiquantitative and mainly serve to provide insights into the process. Accurate process prediction must include chemical kinetics and transport rate considerations.

Chemical Reaction Mechanisms and Kinetics. CVD chemistry is complex, involving both gas-phase and surface reactions. The role of gas-phase reactions expands with increasing temperature and partial pressure of the reactants. At high reactant concentrations, gas-phase reactions may eventually lead to gas-phase nucleation that is detrimental to thin-film growth. The initial steps of gas-phase nucleation are not understood for CVD systems, not even for the nucleation of Si from silane, which has a potential application in bulk Si production (*97*). In addition to producing film precursors, gas-phase reactions can have adverse effects by forming species that are potential impurity sources.

Traditionally, CVD reaction data have been reported in terms of growth rates and their dependence on temperature. The data are often confounded by mass-transfer effects and are not suitable for reactor analysis and design. Moreover, CVD reaction data provide little insight, if any, into impurity incorporation pathways. Therefore, the replacement of traditional macroscopic deposition studies with detailed mechanistic investigations of CVD reactions is an area of considerable interest. A recent, excellent review of CVD mechanistic studies, particularly of Si CVD, is available (*98*), and the present discussion will be limited to highlighting mechanisms of Si CVD and of GaAs deposition by MOVCD as characteristic examples of the combined gas-phase and surface reaction mechanisms underlying CVD.

Gas-Phase Reaction Mechanisms and Kinetics. *Decomposition of SiH_4.* The gas-phase decomposition mechanism for SiH_4 likely involves a large number of species and reactions as complex as those of hydrocarbon combustion systems. A mechanism of 27 reactions has been proposed on the basis of a sensitivity analysis of a detailed pyrolysis scheme involving 120 elementary reactions (*85*, *86*). This mechanism predicts the existence of Si and Si_2 in the gas phase, a prediction that is supported by laser-induced-fluorescence measurements (*99*, *100*). However, Si and Si_2 are not major contributors to the overall film growth. SiH_4, SiH_2, and Si_2H_6 may be the key species for the prediction of Si growth rates from SiH_4–H_2 mixtures (*85–87*). Nevertheless, the presence of Si_2 at concentrations that are orders of magnitude higher than the equilibrium concentration indicates the complexity of the SiH_4 pyrolysis mechanism.

The initial pyrolysis reaction of SiH_4 has been explored extensively (*98*), and the growing consensus is that the thermal decomposition of SiH_4 involves the elimination of H_2 to form silylene, as shown in equation 1 (*98, 101*). Possibly, thermal pyrolysis is controlled by heterogeneous reactions on hot surfaces (*102, 103*), but this hypothesis is controversial, and considerable experimental evidence for a gas decomposition mechanism exists (*98*). However, at low pressures and high temperatures, heterogeneous decomposition will likely be important in the overall mechanism (*104*).

The decomposition of SiH_4 could also occur by loss of a hydrogen radical.

$$SiH_4 \rightarrow SiH_3 + H \tag{14}$$

However, the large bond dissociation energy makes this reaction an unlikely thermal pathway (*105*). In general, the Si_nH_m species in which m is an odd number is less thermodynamically stable than the species in which m is an even number (*106, 107*).

Silylene can insert itself into H_2, SiH_4, and higher silanes with almost no activation energy (*59, 98, 108*). The higher silanes also decompose and undergo rearrangement reactions. The main decomposition path for disilane is to silane plus silylene, with H_2 elimination to silylsilene competing at high temperatures (*109*). In a hydrogen ambient, a mechanism based on SiH_4, SiH_2, and Si_2H_6 may be sufficient to model Si CVD (*87*), but in a nonhydrogen ambient, hydrogen-deficient higher silanes must be included (*99, 100*).

Decomposition of Chlorosilanes. Although the growth of Si from chlorosilanes has been widely studied because of its importance in the microelectronics industry, the detailed gas-phase decomposition mechanism is not known. The mechanism is believed to include at least the following reactions (*110–113*):

$$SiCl_4 + H_2 \leftrightarrows SiHCl_3 + HCl \tag{15a}$$

$$SiHCl_3 + H_2 \leftrightarrows SiH_2Cl_2 + HCl \tag{15b}$$

$$SiH_2Cl_2 \leftrightarrows SiHCl + HCl \tag{15c}$$

$$SiH_2Cl_2 \leftrightarrows SiCl_2 + H_2 \tag{15d}$$

Decomposition of Ga and As Compounds. The gas-phase reactions underlying the MOCVD of GaAs are complex and only partially understood.

The initial decomposition of $Ga(CH_3)_3$ involves the loss of methyl radicals (*114–121*). These methyl radicals can subsequently react with H_2 or the arsenic source, such as AsH_3, abstract H from an organometallic or hydrocarbon species, or recombine. The reaction with H_2 leads to H radicals that can react with the parent organometallic compound to accelerate its decomposition. The following are some of the reactions:

$$Ga(CH_3)_3 \rightarrow Ga(CH_3)_2\cdot + CH_3\cdot \tag{16a}$$

$$Ga(CH_3)_2\cdot \rightarrow GaCH_3 + CH_3\cdot \tag{16b}$$

$$H_2 + CH_3\cdot \rightarrow CH_4 + H\cdot \tag{16c}$$

$$Ga(CH_3)_3 + H\cdot \rightarrow Ga(CH_3)_2\cdot + CH_4 \tag{16d}$$

$$CH_3\cdot + CH_3\cdot \rightarrow C_2H_6 \tag{16e}$$

$$CH_3\cdot + OM \rightarrow CH_4 + OM' \tag{16f}$$

In reactions 16a–f, OM and OM′ designate organometallic or hydrocarbon species or fragments. The presence of $CH_3\cdot$ radicals has been verified by infrared (IR) diode laser spectroscopy (*116*), and the reaction mechanisms have been investigated by replacing the usual H_2 ambient by D_2 (*118–120*). Reactions 16a–f form a typical free-radical mechanism with initiation (16a and b), chain-transfer (16c), propagation (16d), and termination (16e and f) reactions.

The H· radical reaction with the parent compound (16c) is a likely explanation for the lower decomposition temperatures in H_2 relative to N_2 (*119–121*). The exact mechanism for the reaction of H· with $Ga(CH_3)_3$ is not known, but the experimentally observed acceleration is also predicted by simulations of $Ga(CH_3)_3$ decomposition based on a detailed kinetics mechanism for GaAs growth (*122*). The effect of H· will change with pressure and residence time, and this variation may explain the observation of only minor differences in decomposition temperatures at low pressures and short residence times (*118*).

Total decomposition to Ga atoms has been inferred from IR measurements (*117, 120, 121*). but no direct observations have been made of Ga atoms during the thermal gas-phase pyrolysis of $Ga(CH_3)_3$. $GaCH_3$ is considered to be the most stable species, (*114, 123*), whereas In atoms may be formed during the decomposition of $In(CH_3)_3$ (*116*). The difference between the decompositions of $In(CH_3)_3$ and $Ga(CH_3)_3$ correlates well with the stronger methyl–metal bond in $Ga(CH_3)_n$ (*116*).

Arsine decomposes heterogeneously at MOCVD conditions (*117, 120, 121, 123*), but the mechanism is not completely understood. Early kinetics studies (*125, 126*) indicate that arsine decomposes through the adsorption of arsine and subsequent loss of H to the surface. However, recent coherent anti-Stokes Raman-scattering investigations point toward a more complex mechanism (*127, 128*). Arsine interacts strongly with the $Ga(CH_3)_3$ in the gas phase through gas-phase methyl radical attack (*116, 129*) and adduct formation (*120, 121*).

$$CH_3\cdot + AsH_3 \rightarrow CH_4 + AsH_2\cdot \qquad (17a)$$

$$Ga(CH_3)_3 + AsH_3 \rightarrow H_3As:Ga(CH_3)_3 \qquad (17b)$$

The adduct can either decompose to the original constituents or rearrange with the loss of CH_4.

$$H_3As:Ga(CH_3)_3 \rightarrow H_2As:Ga(CH_3)_2 + CH_4 \qquad (18)$$

Eventually, a polymeric substance of the form $-(As{:}GaCH_2)_n-$ may be formed. Adduct formation is thought to be responsible for the observed lower decomposition temperature of $Ga(CH_3)_3$ in the presence of AsH_3 relative to $Ga(CH_3)_3$ in a carrier gas (*120, 121*). Furthermore, with AsH_3 and D_2, the primary reaction product appears to be CH_4 rather than CH_3D, as expected on the basis of the free-radical mechanism (equations 16a–f). However, the formation of CH_4 may be due to reactions of $Ga(CH_3)_x$ and $CH_3\cdot$ with adsorbed AsH_x species (*122*). Estimates indicate that the adduct is too unstable to play a major role in the growth chemistry (*129*), but this conclusion is subject to uncertainties in the thermochemical data base.

In addition to decomposition by loss of free radicals, $Ga(C_2H_5)_3$ and Ga compounds with larger alkyl groups can undergo decomposition through a β-elimination reaction (*118, 130*).

$$Ga(C_2H_5)_3 \rightarrow Ga(C_2H_5)_2 + C_2H_5\cdot \qquad (19a)$$

$$Ga(C_2H_5)_3 \rightarrow GaH(C_2H_5)_2 + C_2H_4 \qquad (19b)$$

The β-elimination reaction (equation 19b) has the advantage of bringing H to the surface and forming stable hydrocarbon products; the formation of hydrocarbon products reduces carbon incorporation into the growing film (*131*).

In addition to understanding the reactions of commonly used precursors, researchers have considerable interest in developing new precursors, particularly for As and P. This interest in new precursors has been driven primarily by the desire to replace the highly toxic AsH_3 and PH_3 gases with more easily handled liquid sources. Examples of compounds investigated as replacements for arsine include $As(CH_3)_3$ (*132*), $AsH(C_2H_5)_2$ (*133*), $As(C_2H_5)_3$ (*134*), and $AsH_2(C_4H_9)$ (*135–138*). The compounds with H bonded to As give promising results but do not possess electrical properties comparable with those routinely obtained with AsH_3.

Rates of Gas-Phase Reactions. Reaction rates have been reported for only a few CVD gas-phase reactions, and most reports are primarily for the silane system. Because of the high temperatures and low pressures used in CVD, the direct use of reported gas-phase rate constants must be done with care. In addition to mass-transfer and wall effects, process pressure may be another factor affecting reaction rates. Process pressure affects major CVD processes, such as the deposition of Si from SiH_4 and GaAs from $Ga(CH_3)_3$, reactions that involve unimolecular decomposition. The collisional activation, deactivation, and decomposition underlying these reactions can be summarized qualitatively by the following reactions (*139, 140*):

$$M + A \underset{k_{-1}}{\overset{k_1}{\rightleftarrows}} M + A^* \tag{20a}$$

$$A^* \xrightarrow{k_2} \text{products} \tag{20b}$$

where A represents the reactant and M is a collision partner (including A). A pseudo-steady-state assumption for the energized molecule A* leads to the Lindemann–Hinshelwood rate form

$$-\frac{dp_A}{dt} = \left(\frac{k_1 k_2 p_M}{k_1 + k_{-1} p_M}\right) p_A \tag{21}$$

where p_A is the pressure of the reactant A, p_M is the total pressure, k is rate constant, and t is time.

Thus, the reaction is first order in the high-pressure limit, with the rate constant $k_\infty = k_1 k_2 / k_{-1}$. At low pressures, the reaction becomes limited by collision, and the rate constant becomes proportional to the total pressure, p_M.

Several theories exist for predicting the intermediate fall-off region (*139, 141*), including the widely used RRKM (Rice–Ramsberger–Kasel–Marcus) theory (*139, 140, 142*). Figure 6 illustrates RRKM predictions of the pseudo-first-order reaction rate constant for SiH_4 pyrolysis to SiH_2 and H_2 for various pressures (*142*) The figure shows that errors of several orders of magnitude may result if k_∞ is used at low pressures (e.g., in LPCVD reactor simulations). Similar errors arise in connection with collisionally stabilized recombination reactions that also occur frequently in CVD systems, such as

$$A + B \leftrightarrows AB^* + M \rightarrow AB \tag{22}$$

Surface Reaction Mechanisms and Kinetics. *Si Growth.* Like the situation for gas-phase reactions, the most studied and best understood surface reaction system is the growth of Si on Si substrates through the use of silane. The main results of these studies are summarized in the following sections;

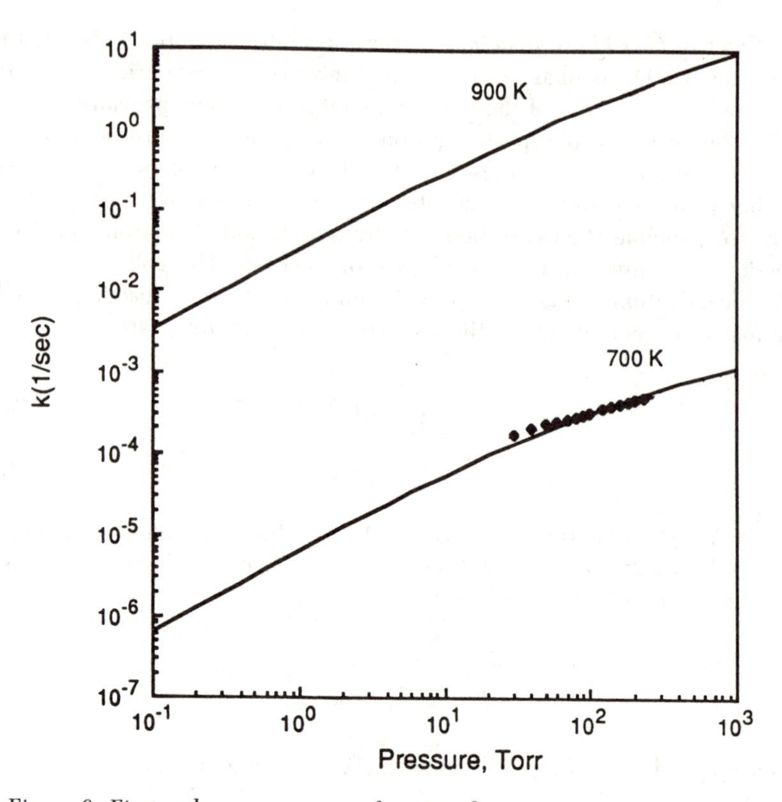

Figure 6. First-order rate constant for SiH_4 decomposition as a function of pressure and temperature. The solid points represent rate measurements by Purnell and Walsh (218). *1 torr = 133.322 Pa.*

however, a more extensive discussion is given in the review paper by Jasinski et al. (*98*).

Ultrahigh-vacuum (UHV) surface spectroscopy has been used with molecular beams of SiH_4 and mass spectroscopy to elucidate the Si growth mechanism (*67, 143*). Joyce et al. (*67*) found that Si growth is preceded by an induction period when surface oxide was removed as SiO. The subsequent film growth proceeds by growth and coalescence of adjacent nuclei with no apparent formation of defects. Henderson and Helm (*144*) proposed a step-flow model in which adatoms from SiH_4 surface reactions diffuse to kink sites.

Farrow (*145*) determined the reaction probability of SiH_4 over a wide range of temperatures. The data have been used subsequently in the modeling of Si CVD reactions (*85, 86*). On the basis of modulated-molecular-beam experiments, Farnaam and Orlander (*146*) proposed a SiH_4 surface reaction mechanism involving SiH_4 reacting with surface Si to form two adsorbed

SiH_2 molecules and the subsequent decomposition of these molecules with the evolution of molecular hydrogen.

$$SiH_4(g) + Si(s) \rightarrow 2SiH_2(adsorbed) \rightarrow 2Si(s) + 2H_2(g) \qquad (23)$$

However, recent molecular-beam experiments by Buss et al. (*147*) indicate that the surface reactions are more complex and highly nonlinear. Recent thermal desorption studies of SiH_4, Si_2H_6, and Si_3H_8 (*60, 148*) show that SiH_4 has a low sticking coefficient, σ, on <111> Si ($\sigma = 5 \times 10^{-5}$ at room temperature), whereas the sticking coefficients of Si_2H_6 and Si_3H_8 are approximately 0.1. Because the higher silanes are products of the gas-phase pyrolysis of SiH_4, their high sticking coefficients relative to that of SiH_4 have considerable implications for the overall growth behavior during Si CVD (*87, 149*).

The presence of intentionally or unintentionally added impurities on the Si surface strongly affects the nucleation and growth of Si. Trace amounts of carbon change the growth mode from step growth to three-dimensional growth leading to defects and, in the most severe cases, polycrystalline growth (*67*). As previously mentioned, surface oxides lead to growth induction periods and may also cause three-dimensional growth. By paying special attention to Si surface cleaning, epitaxial growth can be achieved at temperatures as low as 550 °C in very low pressure CVD (*22, 23, 150*).

Dopants, specifically AsH_3, PH_3, and B_2H_6, strongly influence the kinetics of Si growth. AsH_3 and PH_3 reduce the growth rate, whereas B_2H_6 enhances the SiH_4 surface reaction (*145*). Several mechanisms have been proposed for these effects, on the basis of either surface poisoning by site blockage (*145*) or electronic effects (*151*). Competitive chemisorption experiments show that PH_3 adsorption occurs with a sticking coefficient of near unity and that surface sites are consequently blocked for SiH_4 chemisorption (*152, 153*). By chemical analogy, a similar mechanism is expected for the AsH_3–SiH_4 system, in agreement with the observed reduction in growth rates (*145*). In contrast, B_2H_6 chemisorbs (dissociatively) only at high temperatures (600 °C), a fact (*153*) indicating that an electronic mechanism rather than a surface site coverage mechanism is active.

GaAs Growth. Relatively little is known about the surface reactions underlying GaAs growth. Langmuir–Hinshelwood mechanisms have been proposed (*154, 155*), but no quantitative comparisons between actual growth data and model predictions have been made. Further insights into surface reactions underlying the growth by MOCVD of compound semiconductors are likely to result from two recent growth techniques: metallorganic molecular beam epitaxy (MOMBE) (also called chemical beam epitaxy [CBE]) (*156, 157*) and atomic layer epitaxy (ALE) (*158, 159*).

MOMBE involves the use of organometallic source compounds in a MBE chamber; thus, UHV surface analysis can be used to explore the surface reactions of organometallic compounds, as illustrated by Robertson et al. (*160*). ALE entails saturating the surface alternatingly with Ga and As compounds. This technique has the practical advantage of allowing layer-by-layer growth and the potential for studying the adsorption processes of organometallic species. UHV spectroscopic studies reveal that methyl radicals are formed during the surface reactions of $Al_2(CH_3)_6$ and $Ga(CH_3)_3$ (*161–163*). However, the importance of methyl radicals during MOCVD growth conditions in which the mean free paths are small is not clear.

A knowledge of the surface reaction mechanism is essential to the control of carbon incorporation into the growing film. The most widely accepted mechanism is based on the work of Kuech and Veuhoff (*164*). According to this mechanism H must be present on the surface to eliminate the adsorbed CH_3 as CH_4. The methyl groups are primarily adsorbed on Ga sites, with many of them arriving to the surface as $GaCH_3$. During the growth of GaAs from AsH_3 and $Ga(CH_3)_3$, the H comes from adsorbed AsH_x and not from the H_2 carrier gas. This mechanism is supported by the following experimental observations: (1) Carbon is preferentially incorporated on the As sites, to make the material p-type; (2) the H–H bond is much stronger than the As–H bond; (3) similar film properties are obtained with He and H_2 as carrier gases; and (4) carbon incorporation is reduced when the $AsH_3/Ga(CH_3)_3$ ratio is increased.

This summary of the present understanding of CVD chemistry shows that CVD kinetics awaits the same kind of concerted experimental effort that has enhanced the general understanding of combustion chemistry and heterogeneous catalysis. Currently, the effort to explain common CVD systems, such as Si and GaAs deposition, is growing. Meanwhile, with the present scarcity of rate parameters, two avenues for predicting the performance of CVD processes exist. The first approach involves the complementation of existing data with estimated rate coefficients, as done by Coltrin et al. (*85*, *86*) for Si and by Tirtowidjojo and Pollard (*165*, *166*) and Mountziaris and Jensen (*121*) for the MOCVD of GaAs. The second approach involves the formulation of an overall kinetics model and the extraction of parameters from growth rate data, as illustrated by Roenigk and Jensen (*167*).

CVD Transport Phenomena. Like other chemically reacting systems such as combustion and heterogeneous catalysis, the transport processes taking place during CVD can be characterized by dimensionless parameter groups that arise from the scaling of the governing transport equations. These groups are summarized in Table III, along with their physical interpretation and typical order of magnitude in CVD processes. Their temperature dependence and importance in CVD transport phenomena have been discussed in general terms by Rosenberger (*168*).

Table III. Dimensionless Groups in CVD Modeling Equations

Name	*Definition*	*Physical Interpretation*	*Typical Order of Magnitude*
Knudsen	$\mathrm{Kn} = \frac{\lambda}{L}$	mean free path/characteristic length	$\lesssim 1$
Mach	$\mathrm{Ma} = \frac{<v>}{u_\infty}$	linear velocity/speed of sound	$<10^{-2}$
Prandtl	$\mathrm{Pr} = \frac{\nu}{\alpha}$	momentum diffusivity/thermal diffusivity	0.7
Schmidt	$\mathrm{Sc} = \frac{\nu}{D}$	momentum diffusivity/mass diffusivity	1–10
Reynolds	$\mathrm{Re} = \frac{<v>L}{\nu}$	momentum flux by convection/momentum flux by diffusion	10^{-1}–10^{2}
Peclet (thermal)	$\mathrm{Pe}_h = \mathrm{RePr}$	thermal flux by convection/thermal flux by diffusion	10^{-1}–10^{2}
Peclet (mass)	$\mathrm{Pe}_m = \mathrm{ReSc}$	mass flux by convection/mass flux by diffusion	10^{-1}–10^{3}
Grashof (thermal)	$\mathrm{Gr}_t = \frac{g\beta_t L^3 \Delta T}{\nu^2}$	buoyancy force/viscous force	1–10^{5}
Grashof (solutal)	$\mathrm{Gr}_s = \frac{g\beta_c L^3 \Delta c}{\nu^2}$	buoyancy force/viscous force	1–10^{3}
Rayleigh (thermal)	$\mathrm{Ra}_t = \mathrm{Gr}_t\mathrm{Pr}$	buoyancy force/viscous force	1–10^{5}
Rayleigh (solutal)	$\mathrm{Ra}_s = \mathrm{Gr}_s\mathrm{Sc}$	buoyancy force/viscous force	1–10^{3}
Dämkohler (gas phase)	$\mathrm{Da}_g = \frac{R(C,T)_{ref}L}{C_{ref}<v>}$	characteristic time for flow/characteristic time for gas-phase reaction	10^{-3}–10^{3}
Dämkohler (surface)	$\mathrm{Da}_s = \frac{R(C,T)_{ref}L}{C_{ref}D}$	characteristic time for diffusion to surface/characteristic time for surface reaction	10^{-3}–10^{3}

CVD reactors operate at sufficiently high pressures and large characteristic dimensions (e.g., wafer spacing) such that Kn (Knudsen number) $<< 1$, and a continuum description is appropriate. Exceptions are the recent vacuum CVD processes for Si (*22, 23*) and compound semiconductors (*156, 157, 169*) that work in the transition to the free molecular flow regime, that is, Kn > 1. Figure 7 gives an example of SiH_4 trajectories in nearly free molecular flow (Kn $\sim$ 10) in a very low pressure CVD system for silicon epitaxy that is similar to that described by Meyerson et al. (*22, 23;* Meyerson and Jensen, manuscript in preparation). Wall collisions dominate, and be-

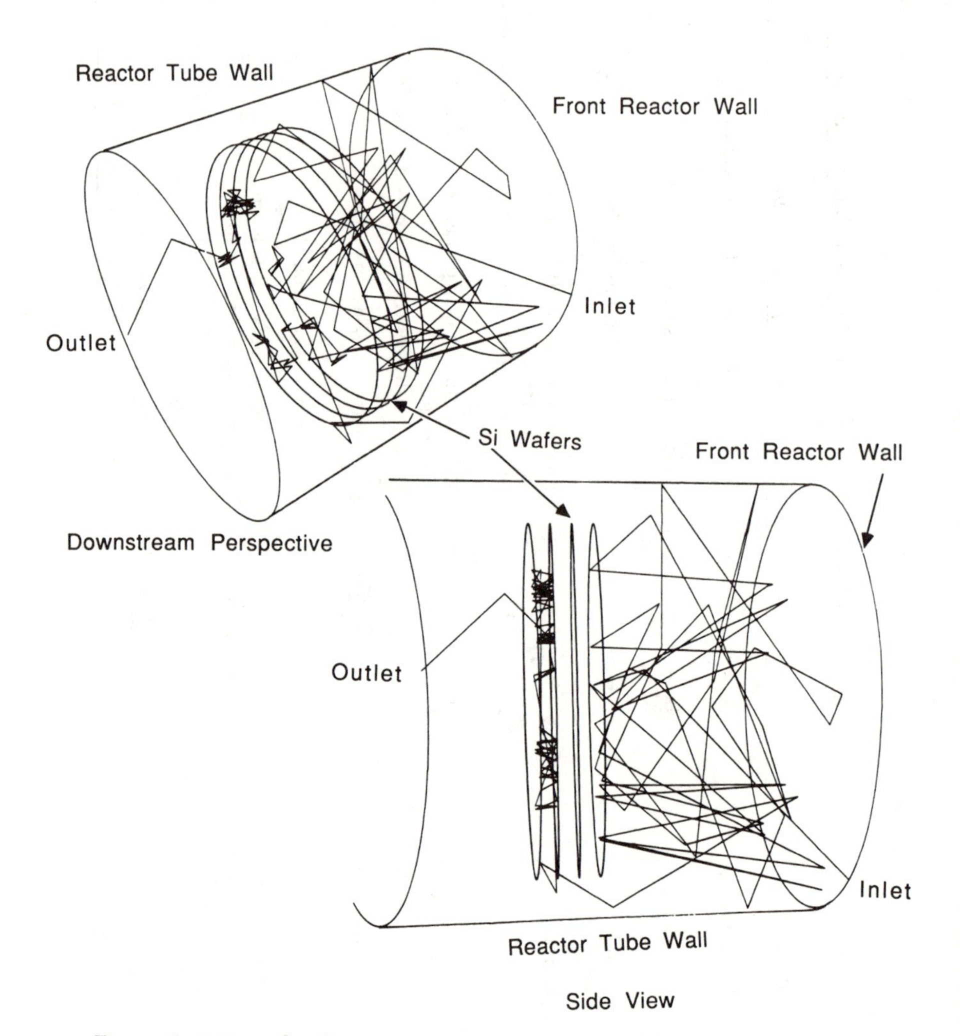

Figure 7. SiH_4 molecular trajectories in free molecular flow in a very low pressure (1 Pa) hot-wall CVD reactor for Si epitaxy.

cause of the low reaction probability of SiH_4, the molecule undergoes a large number of collisions before it finally reacts with Si and H_2 on a Si wafer.

The free molecular flow also implies that downstream impurities are distributed rapidly throughout the system. Because the gas velocities are low in conventional CVD reactors, Re (Reynolds number) < 100, and the flows are laminar. Furthermore, the Mach number is very small (Ma $<<$ 1), which implies that, except for the gas expansion with increasing temperature, the gas may be modeled as an incompressible fluid. The Prandtl number (Pr) is of order unity (typically, Pr $= 0.7$), and the Schmidt number (Sc) is at most an order of magnitude larger than the Prandtl number. Therefore, the difficulties associated with low-Prandtl-number fluids and large differences in the Prandtl and Schmidt numbers found during crystal growth from the melt do not arise in CVD modeling (*see* Chapter 2 by Brown).

The Peclet numbers are useful for estimating the relative contributions of convection and diffusion to mass and heat transfer. If Pe is large (>10), convection dominates, and a plug-flow model may be appropriate for simple reactor computations. When Pe is small ($<<1$), diffusion dominates, and the system behaves like a well-stirred reactor. Thus, Pe may be used to estimate whether downstream impurities can diffuse into the deposition zone.

The Grashof and Rayleigh numbers measure the strength of natural convection in the system. Buoyancy-driven flows due to thermal or concentration gradients are likely to occur if the thermal or solutal Rayleigh (or Grashof) numbers are large, respectively. The parameters vary linearly with temperature and concentration gradients and with the cube of the reactor height. Thus, reactors with large heights are prone to natural-convection effects. More exact criteria are discussed in this section for specific reactor configurations.

The Damköhler numbers are useful measures of the characteristic transport time relative to the reaction time. If the surface Damköhler number (sometimes referred to as the CVD number; *see* reference 7) is large, mass transfer to the surface controls the growth. For small Damköhler numbers, surface kinetics governs the deposition. Similarly, if the gas-phase Damköhler number is large, the reactor residence time is an important factor, whereas if it is small, gas-phase reactions control the deposition.

Dimensionless Quantities and Reactor Types. Transport phenomena in CVD reactors can be described in terms of two broad groups: (1) hot-wall, low-pressure reactors and (2) cold-wall, reduced- and atmospheric-pressure reactors.

Hot-Wall Reactors. Because of the large mass diffusivities and nearly isothermal conditions (except for the entrance zone) in hot-wall, low-pressure reactors (50 Pa), multicomponent diffusion and chemical reactions are critical factors in the overall reactor behavior, whereas details of the flow field are less important (*167, 170*). In the conventional multiple-wafer-in-tube horizontal low-pressure reactor (Figure 1e), the main transport mechanism in the space between the wafers is multicomponent diffusion, and Stefan flow due to volume expansion or contraction during reactions, such as when one molecule of SiH_4 reacts to form solid Si and two molecules of H_2. The mass Peclet number (Pe_m) for the annular-flow region of this reactor configuration is of order unity, and diffusion and convection of mass play equal roles. In the vertical LPCVD geometry (Figure 1f), the gas passes through the wafers in cross flow to minimize depletion effects. Typical Pe_m values are 1–10; therefore, a full description of convection, multicomponent diffusion, and chemical reaction is necessary to model the system behavior.

Cold-Wall Reactors. The large thermal gradients encountered in cold-wall, reduced- and atmospheric-pressure reactors lead to buoyancy-driven secondary flows superimposed on the forced flow entering the reactor. Complex flow structures, including return-flow and longitudinal-roll cells in horizontal reactors and recirculation cells in vertical reactors, have been visualized by TiO_2 smoke tests (*24, 171–174*) and have been inferred from laser holographic observations of perturbations in averaged density gradients (*175, 176*). Flow visualization techniques provide considerable insight into the flow phenomena, but smoke tests at growth temperatures are limited by the thermophoretic transport of seed particles away from hot regions (i.e., the susceptor) towards cold regions (*177*). In fact, thermophoretic migration was behind the original observation of a particle-free layer next to the susceptor in a horizontal reactor that led to the so-called stagnant-layer model (*178*). Laser holography is limited by giving spatially averaged density gradients that may not reflect the actual flow of reactants, because the reactants are often present in low concentrations only and have significantly different transport coefficients compared with that of the carrier gas (e.g., H_2). Therefore, to gain further insights into the complex flow structures in cold-wall reactors, flow visualizations must be augmented by fluid-flow computations (*179*).

Horizontal Reactors. Horizontal reactor flow may involve both transverse and longitudinal rolls, as well as time-periodic flows. Insights into these phenomena may be gained from previous analysis of idealized, analogous systems, as well as from recent experiments and computations. Analytical studies of flow between two plates of infinite size differentially heated from below (*180*) and horizontal channel flow (*181*) indicate that the development of transverse and longitudinal rolls depends on the relative and absolute magnitudes of the dimensionless Rayleigh and Reynolds numbers, Ra and Re, as well as the aspect ratio.

For values of Ra greater than the critical Rayleigh number (Ra_c = 1708 for plates of infinite size), transverse rolls exist in a finite-width channel for low values of Re. At sufficiently high values of Re, the transverse rolls are replaced by longitudinal rolls. Traveling waves of transverse rolls may appear before the transition, and such periodic flows have been observed experimentally (*182*) and have been predicted by two-dimensional transport models (*183, 184*). Evidence for combined longitudinal and transverse rolls has been reported (*185*). For large values of Re and small values of Ra, the flow will be dominated by forced convection. Gas expansion effects not included in the classical analyses can further complicate the flow by creating single transverse rolls, so-called return flows (*175, 186, 187*). These phenomena occur within the range of values of Re and Ra typical for CVD reactor operation and have considerable implications for film thickness uniformity and junction abruptness.

To demonstrate the main features of the flow in horizontal CVD reactors, the deposition of silicon from silane is used as an example (*87*). The conditions are as follows: an 8-cm-wide reactor with either adiabatic side walls or side walls cooled to the top wall temperature of 300 K, a 1323 K hot susceptor (bottom wall), a total pressure of 101 kPa, and an initial partial pressure of silane in H_2 of 101 Pa. The growth rate of silicon is strongly influenced by mass transfer under these conditions. Figure 8 shows fluid-particle trajectories and spatially varied growth rates for three characteristic cases.

In the first case (Figure 8a), the side walls are adiabatic, and the reactor height (2 cm) is low enough to make natural convection unimportant. The fluid-particle trajectories are not perturbed, except for the gas expansion at the beginning of the reactor that is caused by the thermal expansion of the cold gas upon approaching the hot susceptor. On the basis of the mean temperature, the effective Rayleigh number, Ra_t, is 596, which is less than the Rayleigh number of 1844 necessary for the existence of a two-dimensional, stable, steady-state solution with flow in the transverse direction that was computed for equivalent Boussinesq conditions (*188*).

Previous computations (*189*) show that the critical value of Ra_t for non-Boussinesq conditions is approximately the same as that for a Boussinesq fluid in a box heated from below, at least when H_2 is the carrier gas. Thus, results from the stability analysis of the classical Rayleigh–Benard problem of a two-dimensional fluid layer heated from below (*see* reference 190 for a review) may be used to indicate the type of behavior to be expected in a horizontal reactor with insulated side walls. As anticipated from this analysis, an increase in the reactor height from 2 to 4 cm raises the value of Ra_t to 4768, which is beyond the stability limit, $Ra_{t,critical}$ = 2056, for a box of aspect ratio 2 (*188*). The trajectories show the development of buoyancy-driven axial rolls that are symmetric about the midplane and rotating inward. For larger values of Ra_t (>6000), transitions to three-dimensional or time-de-

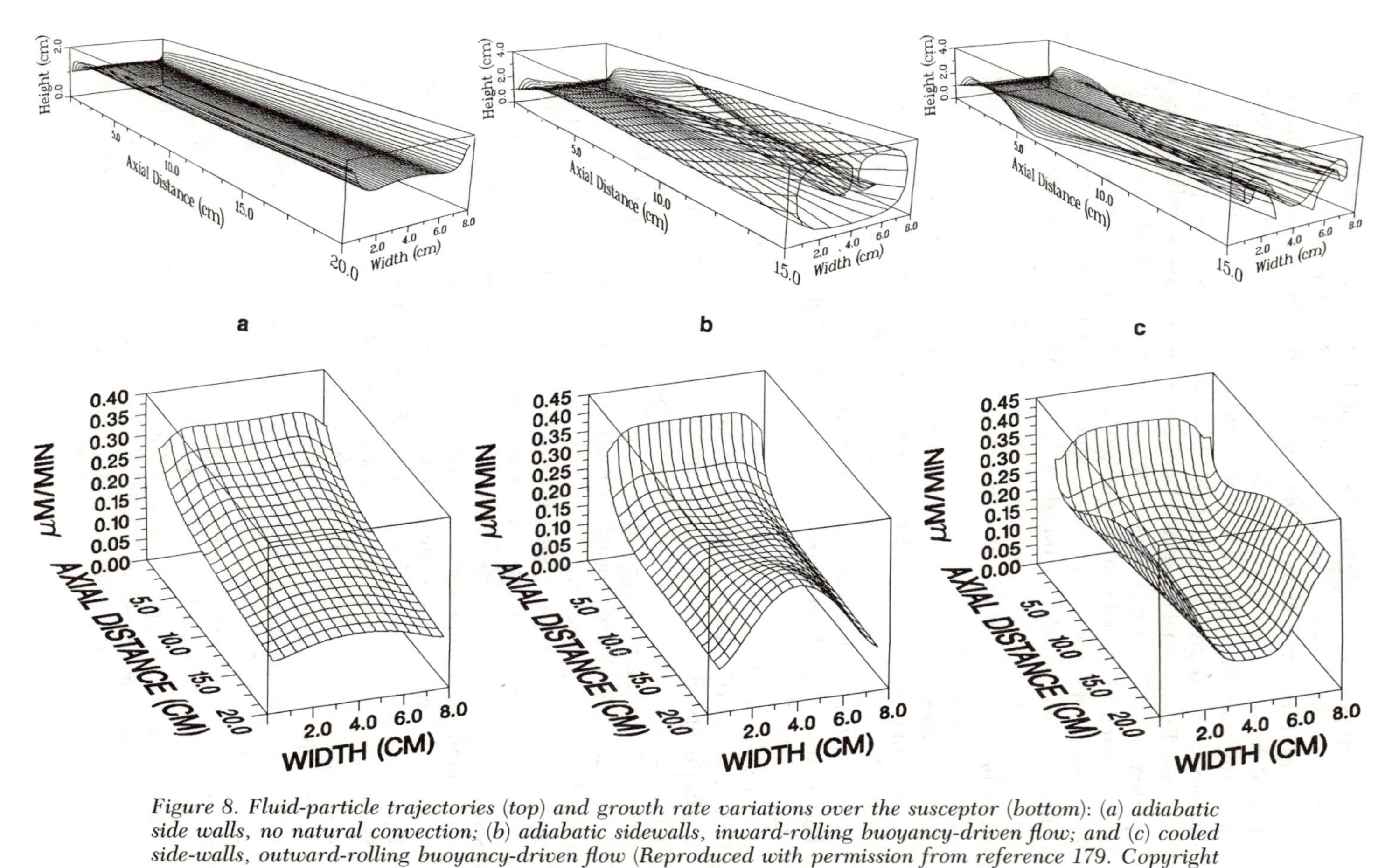

Figure 8. Fluid-particle trajectories (top) and growth rate variations over the susceptor (bottom): (a) adiabatic side walls, no natural convection; (b) adiabatic sidewalls, inward-rolling buoyancy-driven flow; and (c) cooled side-walls, outward-rolling buoyancy-driven flow (Reproduced with permission from reference 179. Copyright 1987 American Society of Mechanical Engineers.)

pendent flows are expected on the basis of the linear-stability analysis (*190*). Holographic observations by Giling (*175*) for N_2 indicate the existence of such complex flow structures.

Effect of Thermal Boundary Conditions. When the side walls are cooled instead of being insulated, there is no critical Ra_t number, and any transverse temperature gradient will lead to a buoyancy-driven secondary flow. Compared with the previous example (Figure 8b), the rolls are reversed and now rotate outward. These examples demonstrate the strong influence of the thermal boundary conditions on CVD reactor flows.

The plot of growth rate in Figure 8a shows that even without buoyancy-driven secondary flows, a considerable variation in the growth rate in the transverse direction exists. The decrease in the axial velocity near the side walls leads to both a shorter thermal entrance length and a greater depletion near the walls compared with the behavior in the middle of the reactor. These perturbations from two-dimensional behavior induced by the side walls extend away from the side walls to a distance about equal to the reactor height. Thus, two-dimensional models may not be sufficient to predict CVD reactor performance even in the absence of buoyancy-driven rolls.

The presence of rolls creates further nonuniformity problems. In the case of insulated side walls (Figure 8b), the inward-rotating rolls lead to increased deposition in the region around the midplane, whereas the opposite effect is observed for the cold-side-wall case (Figure 8c) in which the rolls rotate outward. For that situation, the deposition rate decreases around the midplane of the reactor and increases near the walls. This dramatic effect of flow structures on deposition rate uniformity has been shown experimentally for the MOCVD of GaAs by van de Ven et al. (*191*).

At values of Re lower than those used in the previously discussed examples, return flows may occur (*186, 187, 191*). Figure 9 illustrates a fully three-dimensional flow with both a return flow and a longitudinal roll (*192*). The side wall is adiabatic, but the flow rolls outward, in contrast to the case study just described (Figure 8b). This reversal is brought about by the presence of the return cell, which is created by gas expansion at the leading edge of the susceptor. In this case, the return cell will broaden concentration gradients between adjacent layers by increasing the reactant residence time. This case is one example of the complex three-dimensional flow structures that may exist in CVD reactors.

In addition to influencing convection rolls, the thermal boundary conditions are critical in predicting cold-finger-entrance effects (*193*). Figure 10 illustrates the predicted and measured isotherms for a low (2-slm [standard liters per minute]) and a high (8-slm) inlet H_2 flow. The temperatures were obtained by Raman spectroscopy (*193, 194*). At high flow rates (Figure 10b), a cold finger protrudes into the deposition zone because of the heating of the upper-reactor quartz wall by radiation from the susceptor. Thus, an

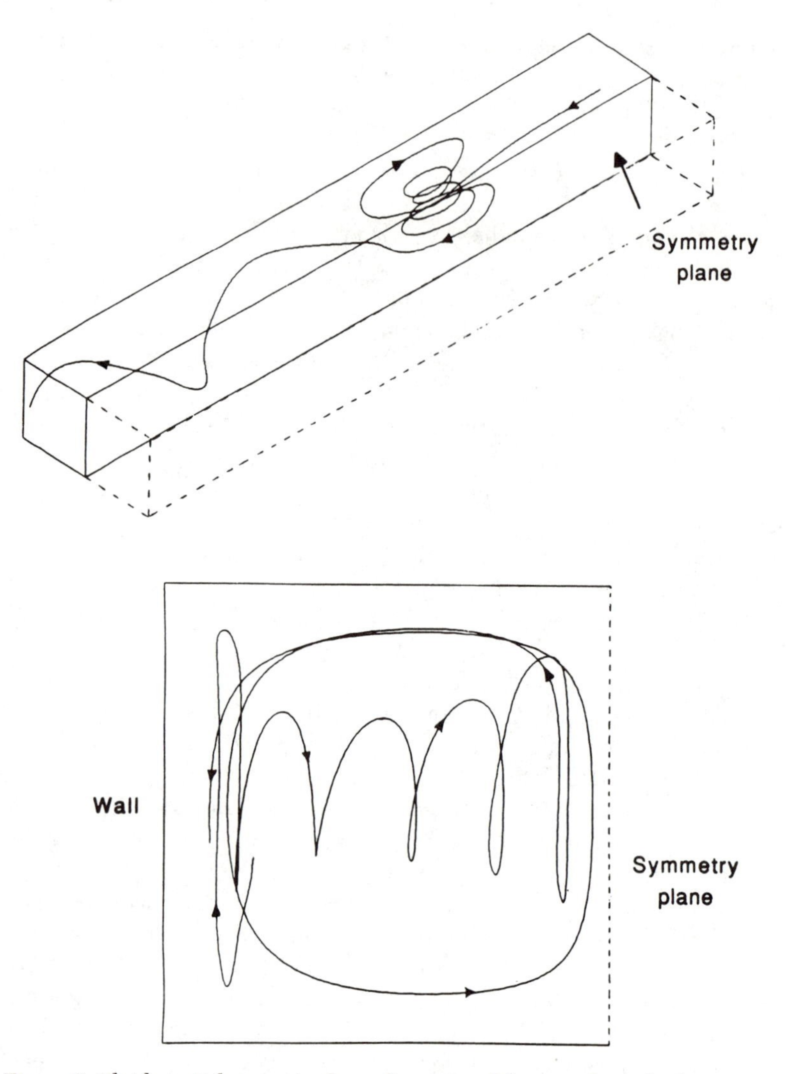

Figure 9. Fluid-particle trace in three-dimensional flow involving both a transverse roll (return flow) and longitudinal rolls. The reactor is symmetric with respect to the midplane (192).

accurate heat-transfer analysis that includes radiation is needed to explain the data. This cold finger may play a major role in achieving uniformity in horizontal reactors by preserving reactants for downstream deposition regions. A similar effect can be produced by replacing the H_2 carrier gas by N_2, which has a lower thermal conductivity (*193*).

Flow Patterns and Film Properties. Uniform film thickness and sharp interfaces can be realized in vertical reactors by creating a boundary layer

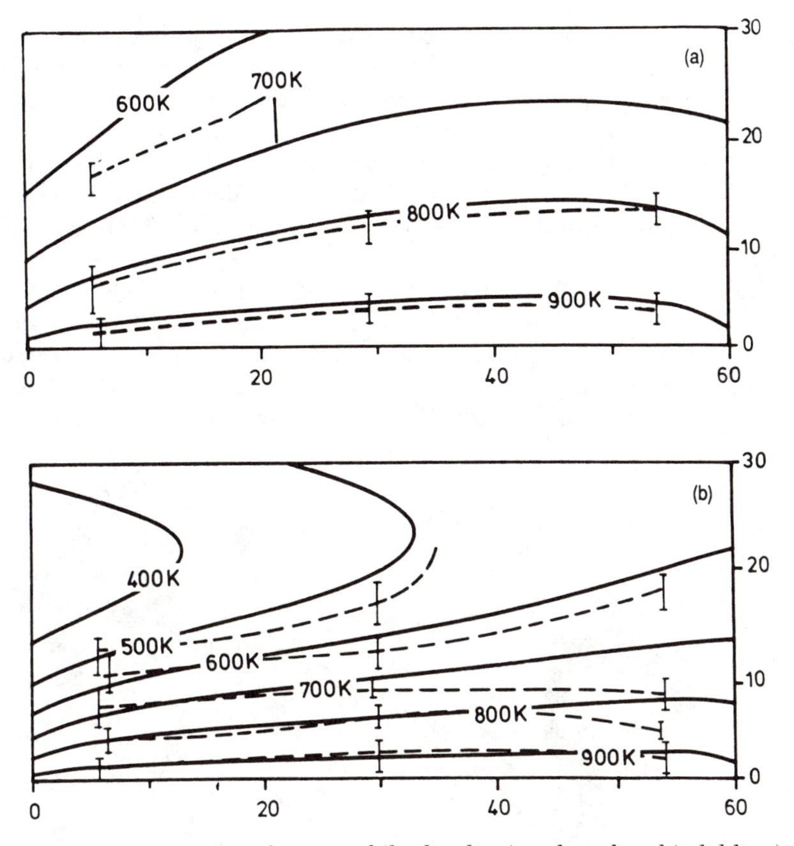

Figure 10. Comparison of measured (broken lines) and predicted (solid lines) temperatures in a horizontal reactor for two inlet flow velocities of H_2: (a) 2 slm (standard liters per minute) and (b) 8 slm (193).

of uniform thickness and by avoiding flow eddies. Ideally, the stagnation point, impinging-jet, and rotating-disk flows have uniform vertical gas velocities that produce a boundary layer of constant thickness in which the gas moves outward radially. However, the actual vertical reactor flow is strongly influenced by buoyancy effects caused by a destabilizing density gradient and by the presence of reactor walls (*24, 171, 173, 195*). Heating the susceptor produces an unstable density gradient such that the least dense gas is next to the susceptor surface. In the presence of an unstable density gradient, the reactor flow patterns may become dominated by buoyancy-driven recirculations.

In practice, the recirculation cells are often eliminated by increasing the inlet flow rate, a procedure that also improves film thickness uniformity. However, because of nonlinear interactions among buoyancy, viscosity, and inertia terms, the transitions between the flow patterns may be abrupt, and multiple stable-flow fields may exist for the same parameter values (*24, 195,*

196). This phenomenon is illustrated in Figure 11, which shows the average rate of mass transfer to the substrate in terms of the Nusselt number for various susceptor temperature and inlet gas flow rates.

At high inlet flow rates, the flow is dominated by forced convection with high mass-transfer rates, whereas at low inlet flow rates, buoyancy-driven recirculations prevail, with resulting low Nusselt numbers. For conditions in the cross-hatched area of Figure 11, which correspond to typical operating conditions for metallorganic CVD reactors, multiple flow patterns are possible, depending on the start up of the reactor. Heating of the susceptor followed by turning on the inlet flow is likely to produce a flow field dominated by buoyancy-driven recirculations, whereas the opposite procedure may lead to a situation dominated by forced convection. The actual reactor behavior is probably further complicated by transitions to time-periodic and

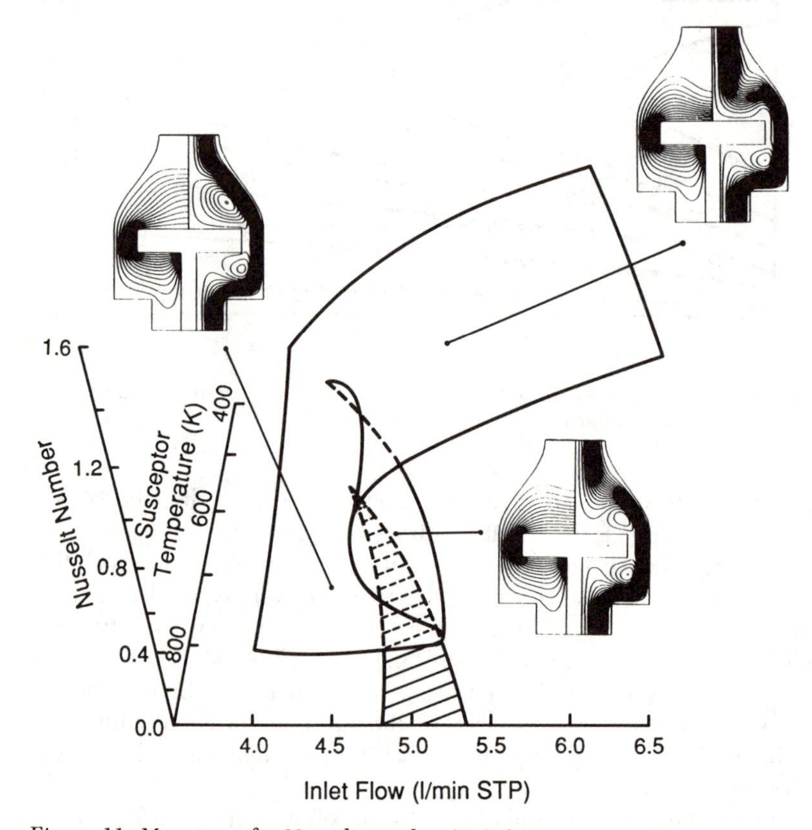

Figure 11. Mass-transfer Nusselt number (Nu) for various susceptor temperatures and inlet flow rates for a vertical CVD reactor with a 900 K susceptor and 300 K reactor walls and operating at 10.1 kPa (Reproduced with permission from reference 24. Copyright 1987 Elsevier).

fully three-dimensional flows. Nevertheless, the key point is that the reactor behavior will depend on its start up history.

The susceptor is often rotated slowly (2–40 rpm) in vertical CVD reactors to eliminate heating nonuniformities. However, rotation of the susceptor at higher speeds (>500 rpm) to emulate a rotating-disk flow may be advantageous; rotation at higher speeds creates a uniform mass-transfer layer in the absence of wall effects. Moreover, the pumping action of the rotating susceptor creates a forced-convection-dominated flow without the disadvantages of increasing the inlet flow rate as previously discussed. At the same time, film uniformity and growth rate are increased. This behavior is shown in Figure 12. The susceptor rotation also stabilizes the flow. However, very rapid spinning of the susceptor generates flow recirculations next to the reactor walls (*197*, *198*).

Factors Affecting Junctions between Successive Films. The growth of sharp or accurately graded compositional variations between successive films is one of the key issues in CVD reactor operation, in addition to uniform film thickness and composition. To obtain sharp junctions by switching among the reactant streams, mixing and interdiffusion in the gas-handling system and the reactor must be minimized. The intermixing in the gas-handling system between two streams of different composition may be estimated from the classical Taylor–Aris dispersion analysis (*199*, *200*). The approximate mixing length is defined by the following equation

$$z = [4Ld(1/\mathrm{Pe} + \mathrm{Pe}/192)]^{1/2} \tag{24}$$

where $\mathrm{Pe} = vd/D$, D is the molecular diffusion coefficient, v is the linear velocity, d is the diameter of the tube, and L is the length of the tube after the mixing point. If t_o is the time to grow a monolayer, then the maximum tube length allowed to realize a junction that is N layers wide is given by

$$L_{\max} = \frac{(Nt_o v)^2}{4d(1/\mathrm{Pe} + \mathrm{Pe}/192)} \tag{25}$$

If the reactor flow is dominated by forced convection and free of laminar eddies, the same expression (equation 25) can be used to estimate the maximum separation between the mixing point and the growth interface. For a given mass flow, the Peclet number, Pe, is independent of pressure, whereas the linear velocity increases with decreasing pressure; therefore, the maximum allowable length increases with the square of the decreasing pressure, and operation at reduced reactor pressures is advantageous when sharp interfaces are desired.

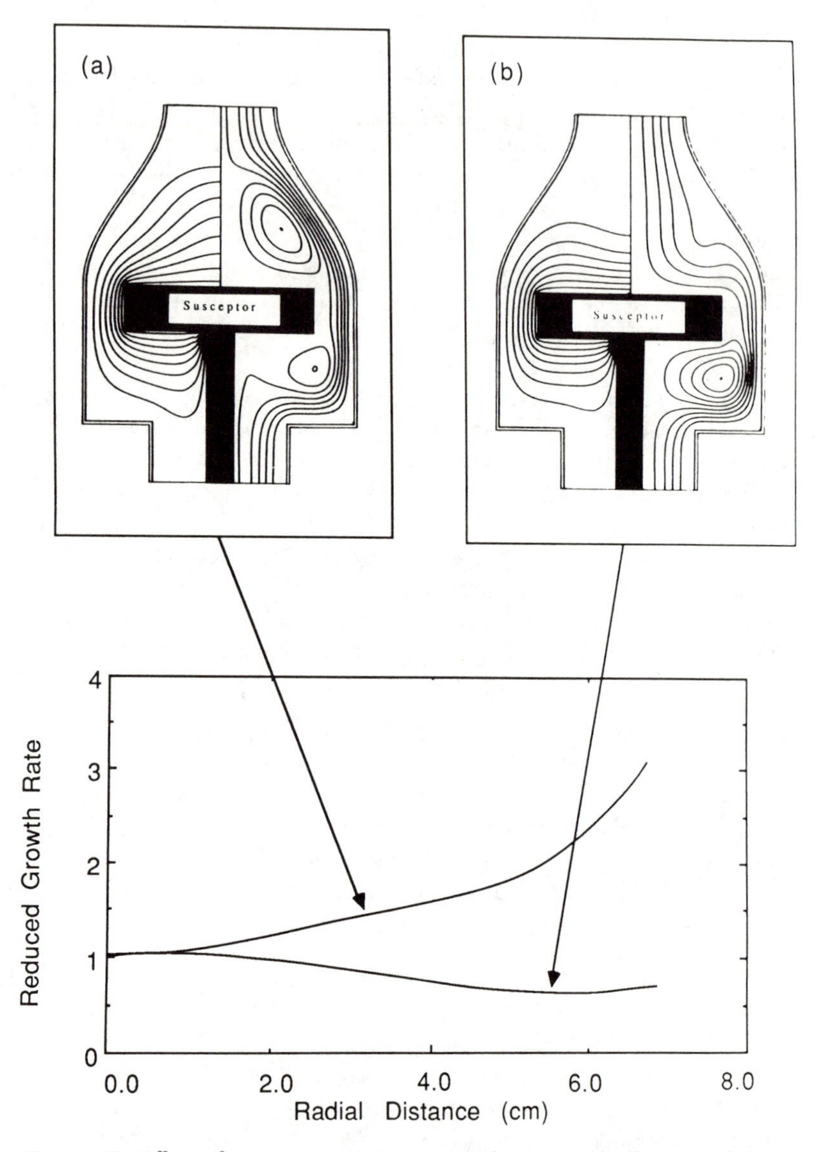

Figure 12. Effect of susceptor rotation on isotherms, streamlines, and deposition rates on GaAs MOCVD relative to the center line value: (a) no rotation, center line growth rate of 2.0 μm/h and (b) 900 rpm, center line growth rate at 8.0 μm/h (same conditions as for Figure 11).

The switching time between different gas-phase compositions is adversely affected by pressure spikes (*201*) and the presence of recirculations and return flows. Recirculations lead to increased reactor residence time for reactants, because gas trapped in the recirculations is isolated from the main flow and the composition adjusts only by the slow diffusion across the separating streamlines. For example, in a simulation of the growth of an AlAs–GaAs heterojunction, recirculations led to transition times in the order of 10 s, corresponding to an approximate interface width of 100 Å, whereas 5-Å junctions were realized under forced-convection conditions (*24*, *195*). The dispersion problem also occurs in stop-growth procedures, in which the flow of the source species is stopped, the overall flow is balanced, and the species is flushed out before a new species is introduced. For this procedure, the reactor residence time must be known.

Operation at Reduced Pressure. To obtain uniform deposition rates and abrupt interfaces, CVD systems are often operated at reduced pressures (10.1 kPa) but with similar mass flow rates as systems operated at atmospheric pressure. As mentioned earlier, operation at reduced pressure increases the linear flow rate but keeps the Peclet number constant. This situation leads to shorter residence times and sharper junctions.

For constant mass flow, the value of of the thermal Rayleigh number (Ra_t) decreases with the square of pressure, and thus longitudinal buoyancy-driven rolls are less likely to occur at reduced-pressure conditions than at atmospheric conditions. Similarly, scaling analysis of simple axisymmetric flows relevant to vertical-reactor operation suggests that the ratio of forced convection to thermal convection varies as $Re/Gr_t^{1/2}$ (*171*). (Gr_t is the thermal Grashof number.) Because the value of Re is independent of pressure for constant mass flow to the reactor, recirculations driven by thermal convection are expected to disappear with decreasing pressure. This effect is illustrated in Figure 13 (*202*), which shows the disappearance of the dominant recirculation as the pressure is reduced from 101 to 10.1 kPa.

Other Factors That Destabilize CVD Systems. The preceding discussion has focused on the effects of buoyancy-driven flows caused by thermal gradients. However, because of the often large mass differences between reactants and the carrier gas (e.g., Hg and H_2 for the growth of CdHgTe), buoyancy-driven flows can also occur because of concentration gradients. In addition, thermal and solutal (concentration) effects can interact to further destabilize the system. This thermosolutal convection phenomenon is well known in bulk crystal growth (*203*, *204*) but has not been recognized generally in CVD. The additional characteristic dimensionless group for the analysis of this problem is the solutal Rayleigh number given in Table III.

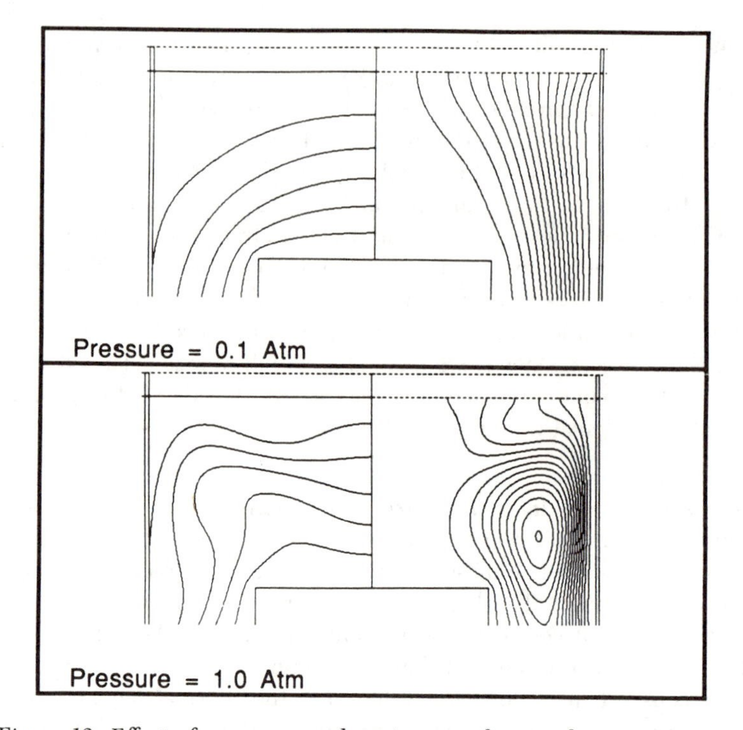

Figure 13. Effect of pressure on the existence of recirculations in a vertical CVD reactor. The right side of the figure shows the streamlines, and the left side shows the isotherms (1 atm = 101.325 kPa).

A case study based on the growth of a Hg-containing compound by MOCVD in an isothermal vertical reactor at 250 °C is as a convenient example to illustrate solutal convection. Because of the depletion of Hg at the growth interface, the reaction has a destabilizing density gradient, which, for high Hg inlet concentrations (10.1 kPa), leads to strong solutal convection that results in nonuniform and reduced growth rates (Figure 14). When the inlet Hg concentration is low (10.1 Pa), recirculation does not occur, and the central region of the substrate exhibits uniform growth rate (*205*). Although the Hg depletion at the growth interface rarely will be sufficient to create a concentration gradient that is large enough to drive a solutal convection cell, Hg condensation can generate cells that severely affect the growth of Hg-containing semiconductor alloys, such as CdHgTe (*206*). Palmateer et al. (*207*) showed that solutal effects also may be important in the growth of heterojunctions when the switching of composition involves variations in gas density, as in the growth of InP–InGaAsP structures.

The considerable mass difference between reactants and carrier gas combined with often large thermal gradients means that thermal diffusion may contribute to the overall mass transfer. Thermal diffusion drives reactants

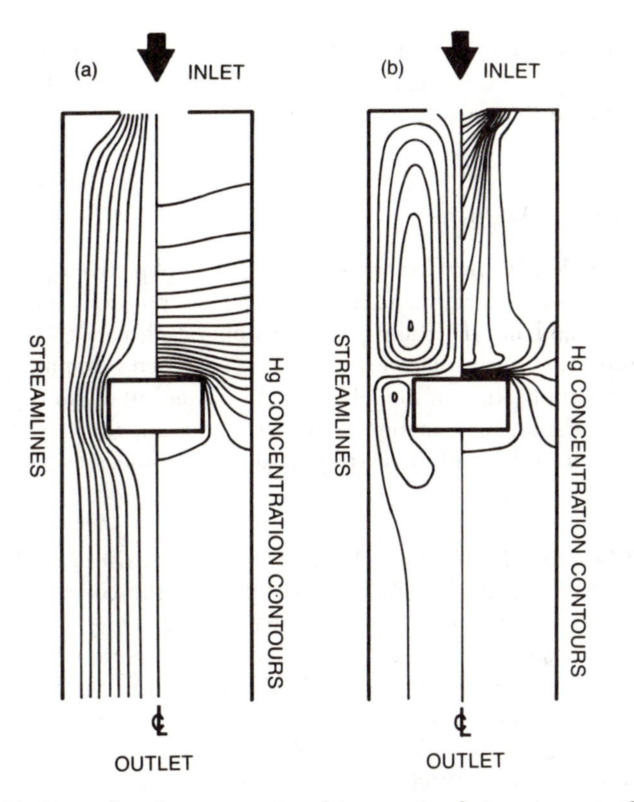

Figure 14. Examples of concentration-driven recirculations in an isothermal (550 K) vertical CVD reactor for the growth of a Hg-containing compound at 101 kPa showing streamlines (left side) and isoconcentration curves (right side). The inlet Hg partial pressures were (a) 10.1 kPa and (b) 10.1 Pa (Reproduced with permission from reference 205. Copyright 1988 Elsevier.)

away from the hot substrate towards cold reactor regions and thereby reduce the growth rate by 10–50 typically, depending on growth conditions (*86, 87, 168, 208*).

Heat transfer is an extremely important factor in CVD reactor operation, particularly for LPCVD reactors. These reactors are operated in a regime in which the deposition is primarily controlled by surface reaction processes. Because of the exponential dependence of reaction rates on temperature, even a few degrees of variation in surface temperature can produce unacceptable variations in deposition rates. On the other hand, with atmospheric CVD processes, which are often limited by mass transfer, small susceptor temperature variations have little effect on the growth rate because of the slow variation of the diffusion with temperature. Heat transfer is also a factor in controlling the gas-phase temperature to avoid homogeneous nucleation through premature reactions. At the high temperatures (700–1400 K) of most

CVD processes, radiation is a major component of heat transfer. The cold-finger effect illustrated in Figure 10 is caused by radiative heat transfer from the susceptor to the top reactor wall (*193*).

CVD Reactor Analysis

General Modeling Considerations. The objective of CVD reactor modeling is to relate performance measures (film deposition rate, uniformity, composition, and interface abruptness) to operating conditions (pressure, temperature, and reactant concentrations) and reactor geometry. Besides the practical application of models in performance prediction, process optimization, parameter estimation, and reactor design, the models also provide insights into the underlying physicochemical processes.

CVD Reactor Models. CVD reactor models consist of nonlinear, coupled, partial differential equations that represent the conservation of momentum, energy, total mass, and individual species. The general derivation and form of these equations are given in standard references on transport phenomena (e.g., reference 209). Table IV summarizes the balanced equations for most CVD processes. The solution to the momentum balance, equation IV.1, gives the velocity components (v_x, v_y, and v_z), whereas the continuity equation (IV.2) gives the local pressure, p. The energy balance (equation IV.3) defines the temperature. In the formulation of the energy balance, contributions from viscous energy dissipation and the Dufour flux have been omitted, because they are generally negligible for CVD conditions (*208, 209*).

An equation of state is needed to relate the temperature, pressure, and density. Because CVD processes are operated usually at atmospheric or

Table IV. General Balanced Equations for CVD Processes

Name	*Definition*	*Number*
Momentum balance	$\rho v \cdot \nabla v = -\nabla p + \rho g + \nabla \cdot \left[\mu \nabla v + \mu (\nabla v)^T - \frac{2}{3} \mu I \nabla \cdot v\right]$	IV.1
Continuity equation	$\nabla \cdot (\rho v) = 0$	IV.2
Energy balance	$\rho C_p v \cdot \nabla T = \nabla \cdot (k \nabla T) - \sum \left[J_i \cdot \nabla \overline{H}_i + \overline{H}_i \sum_{j=1}^{n^g} \nu_{ij}{}^g R_j{}^g\right]$	IV.3
Equation of state	$\rho = \frac{\overline{M} p}{RT}$	IV.4
Species balances	$0 = \nabla \cdot J_i + cv \cdot \nabla x_i - cv \cdot \nabla \ln \overline{M} - \sum_{j=1}^{n^g} \nu_{ij}{}^g R_j{}^g$	IV.5
	$x_i - \alpha_i{}^T \nabla \ln T = \sum_{\substack{k=1 \\ k \neq i}} \frac{(x_i J_k - x_k J_i)}{c D_{ik}}$	IV.6

NOTE: The boundary conditions depend on the specific system and are discussed in the text (*166, 167*).

reduced pressure, the ideal gas law (equation IV.4) is a reasonable choice. Gas expansion effects due to density changes with heating of the gas phase play a major role in the flow behavior and must be included. Thus, replacement of the ideal gas law by a simple linearization of the temperature dependence of the density, as done in the Boussinesq approximation, is rarely appropriate. Also, the temperature variations of the physical parameters must be included. However, because of the low Mach numbers, compressibility effects that otherwise would unnecessarily complicate the solutions of the reactor models may be neglected. In addition, for typical CVD conditions, the flows are laminar and in quasi steady state relative to the film growth dynamics.

The balance over the *i*th species (equation IV.5) consists of contributions from diffusion, convection, and loss or production of the species in n^g gas-phase reactions. The diffusion flux combines ordinary (concentration) and thermal diffusions according to the multicomponent diffusion equation (IV.6) for an isobaric, ideal gas. Variations in the pressure induced by fluid mechanical forces are negligible in most CVD reactors; therefore, pressure diffusion effects need not be considered. Forced diffusion of ions in an electrical field is important in plasma-enhanced CVD, as discussed by Hess and Graves (Chapter 8).

In equation IV.6, D_{ik} represents the ordinary diffusion coefficient for binary interactions, and α_i is the thermal diffusion ratio. The reactants are often present in small amounts (<1%) relative to the carrier gas; thus, the multicomponent diffusion expression (equation IV.6) may be replaced by a simple Fickian diffusion expression that includes thermodiffusion

$$J_i = -cD_{im}(\nabla x_i + \alpha_i^{\mathrm{T}} \nabla \ln T) \tag{26}$$

where J_i is the flux of compound i relative to mass average velocity, c is concentration, D_{im} is the multicomponent diffusivity of trace i in mixture m, x_i is the mole fraction of component i, and T is temperature.

For cases in which the dilute approximation cannot be used, alternative expressions to equation IV.6 have been investigated for combustion systems. Such alternative expressions may also have computational advantages for CVD systems (*86*, *210*, *211*). The solution of equations IV.5 and IV.6, along with the following constraint

$$\sum_{i=1}^{S} x_i = 1 \tag{27}$$

gives the mole fractions of each species in the gas phase.

Boundary Conditions. The boundary conditions are specific for the reactor configuration. No slip is an appropriate velocity boundary condition at nonreacting reactor walls. Because of the incorporation of atoms into the film and the release of gaseous byproducts at the growth interface, a finite normal velocity exists at the substrate surface. For dilute reactants, this velocity may be neglected, and the no-slip boundary may be adopted. Boundary conditions for the temperature are critical in determining the behavior of the CVD system, but they have not been given the importance they deserve. In the absence of detailed heat-transfer modeling, the following are common boundary conditions: a measured wall temperature profile, a fixed Biot number, and insulated side walls corresponding to no cooling or a constant wall temperature corresponding to infinite cooling. A complete treatment of thermal boundary conditions including conduction in the reactor wall and radiative heat transfer is given by Fotiadis et al. (*193*).

The boundary conditions on the species concentrations for nonreacting and reacting surfaces are given by equations 28 and 29, respectively.

$$N_i \cdot \mathbf{n} = 0 \tag{28}$$

$$N_i \cdot \mathbf{n} = -\sum_{j=1}^{n^s} \nu_{ij}{}^s \mathrm{R}_j{}^s \tag{29}$$

In the preceding equations, N_i is the flux of component i, $\mathbf{n}$ is the outward normal to the surface, $\nu_{ij}{}^s$ is the stoichiometric coefficient, and $R_j{}^s$ is the jth surface reaction.

The growth rate can then be estimated in terms of the net incorporation of the film species in the various surface reactions, that is

$$\text{growth rate (length/time)} = \overline{V}_{\text{film}} \sum_{i=1}^{n^s} \sum_{j=1}^{S} n_i{}^{\text{film}} \nu_{ij}{}^s R_j{}^s \tag{30}$$

where $\overline{V}_{\text{film}}$ is the molar volume of the film and $n_i{}^{\text{film}}$ is the number of atoms in species i.

Rate Expressions. A major difficulty in CVD reactor modeling is the choice of appropriate rate expressions, R_j, for the gas-phase-species balance and the surface boundary conditions. As described previously (*see* Nucleation and Growth Modes), most of CVD chemical kinetics is unknown. Therefore, rate parameters may have to be estimated from experimental growth data as part of the reactor-modeling effort.

For CVD processes at atmospheric or reduced pressure, the reactants are usually used in low concentration in H_2 or some inert carrier gas. Therefore, volume changes due to the change in the number of moles between reactants and products are negligible. In addition, in CVD processes, unlike in combustion systems, energy contributions caused by heats of reaction are insignificant. Therefore, the flow and energy solutions may be separated from the mass-transfer analysis, a simplification that reduces storage and computational requirements and allows different chemical mechanisms to be be considered for the same flow situation.

This simplification is not possible for some CVD systems in which large density changes are associated with the deposition process. The growth of CdHgTe is a typical example that shows how the depletion of Hg next to the substrate creates an unstable density gradient that drives recirculations (*205*), as discussed earlier and illustrated in Figure 14. LPCVD processes use little or no diluent and often involve several species, and multicomponent diffusion may be an important factor (*21*). Fortunately, these reactors are isothermal, and the relative insensitivity of reactor performance to details of the fluid flow greatly simplifies the analysis.

Numerous modeling studies of CVD reactors have been made and are summarized in recent review papers (*1*, *212*). Table 3 in reference 212 lists major examples of CVD models up to mid-1986. Therefore, rather than giving an exhaustive list of previous work, Table V presents a summary of the major modeling approaches and forms the basis for the ensuing discussion, which is most appropriately handled in terms of two groups: (1) hot-wall LPCVD systems and (2) cold-wall, near-atmospheric-pressure reactors. In LPCVD reactors, diffusion and surface reaction effects dominate, whereas in cold-wall reactors operated at near-atmospheric pressures, fluid flow and gas-phase reactions are important in predicting performance, as discussed earlier in relation to transport phenomena.

LPCVD Reactor Models. ***First-Order Surface Reaction.*** The traditional horizontal-wafer-in-tube LPCVD reactor resembles a fixed-bed reactor, and recent models are very similar to heterogeneous-dispersion models for fixed-bed reactors (*21*, *167*, *213*). To illustrate CVD reactor modeling, this correspondence can be exploited by first considering a simple first-order surface reaction in the LPCVD reactor and then discussing complications such as complex reaction schemes, multicomponent diffusion effects, and entrance phenomena.

The model is based on the schematic representation of the commercial reactor shown in Figure 1e. The wafers are supported concentrically and perpendicular to the flow direction within the tube. The heats of reaction associated with the deposition reactions are small because of the low growth rates obtained with LPCVD (~2 Å/s). Furthermore, at high temperatures (1000 K) and low pressures (100 Pa), radiation is the dominant heat-transfer mechanism. Therefore, temperature differences between wafers and the furnace wall will be small. This small temperature difference eliminates the need for an energy balance. Moreover, buoyancy-driven secondary flows are unlikely. In fact, because of the rapid diffusion, the details of the flow field

Table V. Representative CVD Reactor Modeling Studies

Reactor System	*Deposition System*	*Representative Reference*	*Modeling Approach*
Low-pressure, horizontal multiple wafer-in-tube	$SiH_4 \rightarrow$ poly-Si	174	Two-dimensional model of binary diffusion and reaction problem between wafers with analytical and numerical solutions.
	$SiH_2Cl_2 + NH_3 \rightarrow Si_3N_4$	167	One-dimensional model of wafer space, one-dimensional model of flow region, multiple component diffusion, gas phase and surface reactions, entrance effects, comparison with data, and numerical solution.
Impinging jet at atmospheric conditions	$BCl_3 \rightarrow B$ $Ga(CH_3)_3 + AsH_3 \rightarrow GaAs$	165, 208, 224	Similarity solution of ideal axisymmetric flow, multicomponent diffusion, evaluation of thermodiffusion, multiple gas-phase and surface reactions, and thermodynamic analysis of reaction pathways and species.
Rotating disk at atmospheric and reduced pressure conditions	$SiH_4 \rightarrow$ epitaxial Si	225	Two-dimensional, axisymmetric flow and heat transfer analysis of detailed chemistry model with 17 species and 27 elementary reactions combined with similarity solution for flow problem.
	$Ga(CH_3)_3 + AsH_3 \rightarrow GaAs$	195, 198	Two-dimensional, axisymmetric flow, mass- and heat-transfer analysis, reactor geometry effects, flow transitions, and mass-transport-limited growth.

Vertical reactor or stagnation point flow reactor at atmospheric and reduced pressure conditions	Arbitrary gas and surface chemical mechanisms	227	Two-dimensional, axisymmetric mass- and heat-transfer, arbitrary gas-phase and surface reaction kinetics, effect on growth uniformity, and thermodiffusion considerations.
	$Ga(CH_3)_3 + AsH_3 \rightarrow GaAs$	24	Two-dimensional, axisymmetric flow, mass- and heat-transfer analysis, reactor geometry effects, buoyancy driven flows, flow transitions, mass-transport-limited growth, and transients in the growth of heterojunctions.
Horizontal reactor at atmospheric and reduced pressure conditions	$SiH_4 \rightarrow$ epitaxial Si	85, 86	Two-dimensional boundary layer treatment of flow, mass- and heat-transfer, detailed chemistry model with 17 species and 27 elementary reactions, comparison with data, and thermodiffusion effects.
	$Ga(CH_3)_3 + AsH_3 \rightarrow GaAs$ and $SiH_4 \rightarrow$ epitaxial Si	87, 189	Three-dimensional parabolic flow, mass- and heat-transfer treatment, secondary buoyancy-driven flows, thermodiffusion effects, sloped susceptor, effect of thermal boundary conditions, three species and four elementary reactions for Si deposition, and mass-transfer-limited growth for GaAs deposition.

need not be accounted for, and simple plug flow may be assumed in the annular region. Because the wafer spacing is small (5 mm) relative to the wafer diameter (100–150 mm), diffusion may be assumed to be the main mode of transport in the space between the wafers.

With the assumptions just given, the reactant concentration is described by the following balanced equation in the wafer region

$$\frac{1}{r}\frac{\partial c}{\partial r}\left(r\frac{\partial c}{\partial r}\right) + \frac{\partial^2 c}{\partial z^2} = 0 \tag{31}$$

where r is the radial coordinate, c is concentration, and z is the axial coordinate and with the following boundary conditions

$$\left.\frac{\partial c}{\partial r}\right|_0 = c_b \qquad \left.c\right|_{r=R_w} = c_b \tag{32a}$$

and

$$\pm D\frac{\partial c}{\partial z} = k_s c \text{ at } z = z_k \text{ and } z = z_k + \Delta \tag{32b}$$

where c_b is the bulk concentration, R_w is the Si wafer radius, D is the diffusion coefficient, k_s is the surface reaction rate constant, and z_k is the axial position of the kth wafer.

This linear system of equations can be solved readily by separation of variables (*170*). However, for the small ratios of spacing to wafer diameter (0.05) usually used in LPCVD reactors, the axial variation will be significant only for extremely fast reactions for which the reactor geometry is inappropriate (*170*). Therefore, the equations may be averaged over the axial direction. After making the equation dimensionless, one obtains the following expression

$$\frac{1}{\xi}\frac{d}{d\xi}\left(\xi\frac{dy}{d\xi}\right) = \phi^2 y \tag{33}$$

with

$$\left.\frac{dy}{d\xi}\right|_0 = 0 \qquad \left.y\right|_{\xi=1} = 1 \tag{34a}$$

and

$$y = c/c_o \quad \xi = r/R_w \quad \phi^2 = 2R_w{}^2 k_s/\Delta D \tag{34b}$$

In the preceding equations, ξ is a dimensionless radial position, ϕ is the Thiele modulus, y is dimensional concentration, c_o is input concentration, and R_w is Si wafer radius.

This model is equivalent to the model for a first-order reaction in an infinite cylinder of catalysts (*214*). Analogous to the solution of the catalyst particle problem, the notation of an effectiveness factor can be introduced as the ratio of reaction on each pair of wafers (back and front) to the deposition rate at the wafer edge, that is,

$$\eta = 2\int_0^1 y\xi d\xi = 2I_1(\phi)/[\phi I_o(\phi)] \tag{35}$$

where η is the effectiveness factor and I is the modified Bessel function of the first kind.

Thus, if $\eta < 1$, the deposition proceeds more rapidly at the wafer edge than at the interior of the wafer, and a nonuniform film thickness results. This result is the so-called bull's-eye effect. Figure 15a illustrates the variation in η with ϕ for a first-order reaction and a large aspect ratio, $A = R_w/\Delta$. For the first-order reaction, the film thickness has the well-known form

$$\delta(\xi)/\delta(\xi = 1) = I_o(\phi\xi)/I_o(\phi) \tag{36}$$

which is graphed in Figure 15b for different values of ϕ. The diffusion limitation is apparent at high values of ϕ.

Modeling of Annular Flow Region. Continuing with this simple example, we next consider the annular flow region. The balanced equation for this region has the form

$$\frac{1}{r}\frac{\partial}{\partial r}\left(r\frac{\partial c}{\partial r}\right) + \frac{\partial^2 c}{\partial z^2} - v\frac{dc}{dz} = 0 \tag{37}$$

with the following boundary conditions:

$$-D\frac{\partial c}{\partial r}\bigg|_{R_t} = kc(r = R_t) \tag{38a}$$

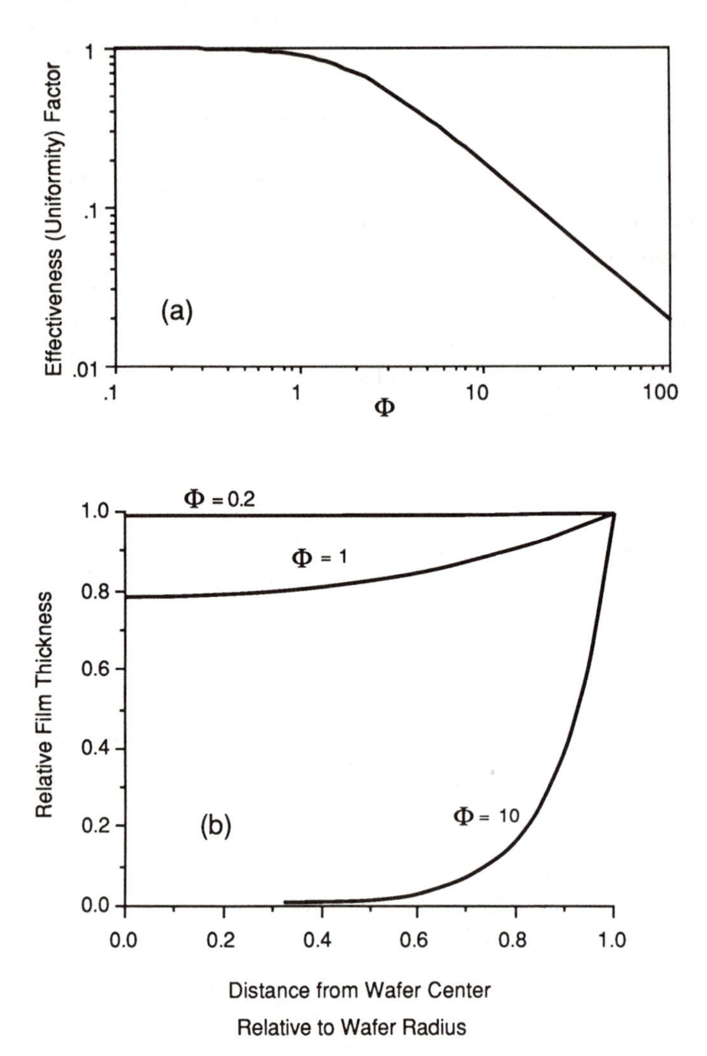

Figure 15. (*a*) *Effectiveness factor, η, as a function of φ for various aspect ratios.* (*b*) *Examples of film thickness variations for different values of φ.*

$$D \left. \frac{\partial c}{\partial r} \right|_{R_w} = [(R_w/\Delta)\eta + \alpha(R_t/R_w)]k_s c(r = R_w) \tag{38b}$$

$$-D \left. \frac{\partial c}{\partial z} \right|_0 = v[c_o - c(z = 0)] \quad \left. \frac{\partial c}{\partial z} \right|_L = 0 \tag{38c}$$

In the preceding equations, v is linear velocity and R_t is the LPCVD tube radius.

Besides the deposition on the wafers, the radial boundary conditions (equations 38a and b) account for deposition on the reactor wall and the support boat. The first term in equation 38b represents the deposition on the wafers, and the second term gives the deposition on the boat. α is the boat area relative to the tube area. The axial boundary conditions (equation 38c) are the usual Danckwerts boundary conditions.

Again, the full set of equations can be solved, but appropriate simplifications are more instructive. For common LPCVD reactions and conditions the Damköhler and Peclet numbers are small ($<<1$ and ~ 1, respectively). Moreover, the annulus is narrow (10–20 mm) compared with the deposition zone (500 mm). Therefore, radial variation may be neglected, and the following dimensionless equation results

$$\frac{1}{\text{Pe}}\frac{dx}{d\zeta^2} - \frac{dx}{d\zeta} - (\text{Da}_1 + \eta\text{Da}_2)x = 0 \qquad (39)$$

where $\text{Pe} = vL/D$, $\text{Da}_1 = R_t\alpha k_s/R_w v$, and $\text{Da}_2 = R_w k_s/\Delta v$. Equation 39 is the axial dispersion model, which can readily be solved to predict the depletion of reactant and the corresponding decrease of thickness along the length of the tube. Thus, this simple example demonstrates the close analogy between the classical fixed-bed-reactor models and simple LPCVD reactor descriptions. This analogy can be exploited in terms of temperature profiling, recycling, and moving-bed strategies for improving the film thickness variation along the tube (*21, 167, 213*). However, the recycling and moving-bed approaches are not practical because of particulate problems. Moreover, a number of compounding factors limits the use of the simple effectiveness factor analogy.

Nonuniformity in LPCVD. LPCVD processes typically involve several reactants with little or no dilution, and the deposition is associated with significant volume changes. Thus, multicomponent diffusion effects may be important (*21, 167*). Thermal entrance effects and reactions occurring in the tube before the deposition can also influence the overall reactor performance (*215*). Flow simulations indicate that eddies created near wafer edges have only a minor effect on the deposition rate for normal operating conditions (*170*). However, with large molecules or small particles, these edges may lead to increased film thickness near wafer edges.

Another issue in LPCVD reactor modeling is the transition to molecular flow for which the continuum formulation breaks down. This transition may be important in the modeling of very low pressure CVD Si epitaxy (*22, 23*). Monte Carlo simulations of free molecular flow in a very low pressure CVD reactor for Si epitaxy were illustrated in Figure 7 and discussed in an earlier section (*see* Transport Phenomena). The continuum approach also breaks down for the modeling of step coverage of micrometer-sized features. In that case, Monte Carlo simulations may be used to study changes in step coverage with process conditions (*216*).

LPCVD chemistry is complex, with reactions involving both gas-phase and surface chemistry. Because the process ideally operates in a regime controlled by chemical reactions, it is very sensitive to chemical pathways and rates. Furthermore, as discussed earlier (*see* Chemical Reaction Mechanisms and Kinetics), the gas-phase kinetics is influenced by pressure effects. The deposition of pure and in situ-doped polycrystalline Si is an illustrative example of how variations in chemical pathways and rates can dramatically effect a CVD process. Polycrystalline Si can be grown uniformly at approximately 100 Å/min with relatively few difficulties (*21, 217*). The growth occurs primarily by heterogeneous decomposition of SiH_4. If the classical rate data from Purnell and Walsh (*218*) are extrapolated to low pressures, in which case SiH_4 decomposition is a first-order reaction, the influence of gas-phase reactions is overpredicted by two orders of magnitude (Figure 6). This overprediction clearly demonstrates the importance of including pressure effects in the kinetics.

From a device-manufacturing view point, the ability to dope polycrystalline Si during deposition is advantageous. However, the addition of small amounts of PH_3 ($<1\%$) to the SiH_4 feed gas greatly affects the process, with the rate of deposition changed from 100 Å/min with high uniformity to a much reduced rate of 5 Å/min with severe radial variation (*217*). This behavior cannot be explained by the simple effectiveness factor analysis previously outlined. When the growth rate decreases, the uniformity is expected to improve, but it is in fact degraded. The explanation for the phenomenon resides in the combined SiH_4 gas-phase and surface chemistry.

SiH_4 Surface Reactions. The initial reactions involved in the gas-phase decomposition of SiH_4 have been described (*see* Gas-Phase Reaction Mechanisms and Kinetics). The decomposition involves the highly reactive species SiH_2, which readily inserts itself into SiH_4 to form higher order silanes, such as disilane and trisilane. SiH_4 is adsorbed and reacts on solid surfaces with a low reaction probability (*144–148*), whereas SiH_2, Si_2H_4, and Si_2H_6 react readily. PH_3 poisons the Si surface and blocks the SiH_4 surface reaction (*152*). However, surface poisoning by PH_3 need not affect the more reactive SiH_2 and higher silanes. Thus, the deposition scheme for in situ-doped polycrystalline Si involves a slow gas-phase reaction followed by a very rapid surface reaction (i.e., with a small effectiveness factor), which explains the low but highly nonuniform growth rates.

With the original silane chemistry data, models predict that the reactive species leading to nonuniform growth is SiH_2 (*219, 220*). However, new laser spectroscopy data by Inoue et al. (*59*) and by Jasinski et al. (*108*) make the SiH_2 insertion reaction three orders of magnitude faster than predicted by the classical analysis. These new data indicate that SiH_2 is present only in small amounts and that the nonuniformity is therefore caused by the higher silanes, particularly disilane and silylsilene (*215*). This example demonstrates the importance of detailed information on reaction paths and rates. Analogous mechanisms involving gas-phase species are behind the nonuniformities in other LPCVD systems,such as Si_3N_4 (*167*).

Modeling of Cold-Wall, Atmospheric-Pressure CVD Reactors. ***Horizontal CVD Reactors.*** CVD at reduced and atmospheric pressures is strongly affected by fluid-flow phenomena and gas-phase reactions. As described in a previous section (*see* Transport Phenomena), entrance effects and buoyancy-driven flows caused by large thermal differences between wall and susceptor temperatures can generate high growth rates and composition nonuniformities. These effects are particularly pronounced during epitaxial growth when the substrate temperature is high and mass transfer often controls the deposition rate.

Simple Analytical Models. To derive simple analytical models for horizontal reactors, two flow simplifications have been used: boundary layer similarity models and film theory (*see* Table 3 in reference 212). In these treatments, a constant concentration shape is assumed from the start of the deposition zone or from an axial position after the initial concentration profile development zone. Thereafter, the shape stays constant, with only the absolute magnitude of the concentration changing with axial position.

For mass-transfer-controlled growth, this assumption necessarily leads to an exponential form for the axial variation in film thickness, δ, relative to the film thickness of the leading edge of the susceptor, δ_0. The result is

$$\delta/\delta_0 = \exp[-(Dz/dh\langle v\rangle)] \tag{40}$$

where h is the height of the reactor, $\langle v \rangle$ is the average linear velocity, and d depends on the concentration profile assumed. As for other film theory models, D/d may be thought of as the mass-transfer coefficient for the system (*221*). For the stagnant-layer model, d corresponds to the stagnant-layer thickness, that is, the characteristic distance over which diffusion takes place (*178*).

The treatment can be modified to include effects of the temperature development and tilting of the susceptor by using the temperature dependence of the diffusion coefficient and adjusting d and $\langle v\rangle$ (*191*). In this manner, the experimental data can be correlated, but the model has limited capability for predicting behavior beyond the particular set of experiments used to fit the model. In fact, because of the low values of the Reynolds number (<50) in typical horizontal CVD reactors, film theory and simple boundary layer models are inappropriate. The boundary layer grows quickly and exceeds the height of the reactor before the reactor exit. Moreover, thermal expansion and buoyancy-driven entrance flow phenomena (*see* Transport Phenomena) make the use of classical boundary layer theory meaningless. Thus, the good agreement between average growth rates and predictions from models of the form in equation 40 may be due to fortuitous off-setting effects of simplifying assumptions rather than to a correct physical picture (*168*).

Two-Dimensional Models. Two-dimensional fully parabolized transport models have also been used to predict velocity, concentration, and temperature profiles along the length of horizontal reactors. The work by Coltrin et al. (*85, 86*) is notable, being one of the first efforts to include detailed kinetics models in CVD reactor simulation, analogous to what has been done for combustion modeling. By using sensitivity analysis, a mechanism was derived consisting of 17 species and 27 elementary reactions from a detailed SiH_4 pyrolysis scheme involving 120 elementary reactions. Experimentally determined kinetics parameters were available for a few steps, and the remaining rate parameters were estimated. In addition, a full multicomponent transport formalism useful for general CVD reactor modeling was introduced, along with thermodiffusion. The model simulations demonstrated the importance of gas-phase reactions and underscored the necessity of including detailed homogeneous and heterogeneous kinetics in CVD reactor models. Furthermore, comparisons of model predictions with diagnostic experiments using laser showed good agreement, even for minor species such as Si atoms (*99, 100*).

Three-Dimensional Models. As previously discussed (*see* Transport Phenomena) in connection with Figure 8, transverse variations in film thickness exist in addition to axial variations even in the absence of buoyancy-driven secondary flows because of the influence of the side walls. Moreover, the presence of buoyancy-driven rolls superimposed on the main flow leads to considerable additional spatial variations in the growth rate. For example, for cooled side walls, the growth is suppressed in the region around the midplane of the reactor and increases near the side walls (*see* Figure 8c), whereas for insulated side walls, the opposite growth rate behavior is observed for values of the Rayleigh number exceeding the critical value (*see* Figure 8b).

The three-dimensional, fully parabolic flow approximation for momentum and heat- and mass-transfer equations has been used to demonstrate the occurrence of these longitudinal roll cells and their effect on growth rate uniformity in Si CVD from SiH_4 (*87*) and GaAs MOCVD from $Ga(CH_3)_3$ and AsH_3 (*189*). However, gas expansion in the entrance zone combined with flow obstructions, such as a steeply sloped susceptor, can also produce fully three-dimensional return flows, as illustrated in Figure 9. Therefore, to obtain a realistic picture of horizontal reactor performance, three-dimensional transport descriptions must be included, in addition to detailed reaction schemes.

Vertical CVD Reactors. Models of vertical reactors fall into two broad groups. In the first group, the flow field is assumed to be described by the one-dimensional similarity solution to one of the classical axisymmetric flows: rotating-disk flow, impinging-jet flow, or stagnation point flow (*222*). A detailed chemical mechanism is included in the model. In the second category, the finite dimension of the susceptor and the presence of the reactor walls are included in a detailed treatment of axisymmetric flow phenomena, including inertia- and buoyancy-driven recirculations, whereas the chemical mechanism is simplified to a few surface and gas-phase reactions.

Pollard and co-workers are the primary exponents of the first approach (*165, 166, 208, 223, 224*). By taking advantage of the reduced one-dimensional similarity solution, they investigated a number of modeling issues, including the sensitivity of the model predictions to inaccuracies in thermodynamic data (*224*) and to thermodiffusion effects (*208*). The mechanism underlying MOCVD of GaAs has also been explored by formulating a detailed gas-phase and surface reaction scheme (166). Recently, Coltrin et al. used the same approach to extend the detailed Si deposition mechanism to a rotating-disk reactor (*225*).

The second approach to the modeling of vertical CVD reactors aims at understanding the limits of the classical analysis and predicting the effects of buoyancy, susceptor rotation, susceptor edge, and reactor geometry on film thickness uniformity and interface composition abruptness between successive layers (*24, 171, 173, 179, 196, 198, 205, 226, 227*). The transport phenomena involved and the effects of deposition uniformity have already been discussed (*see* Transport Phenomena). The reactor models have been limited to a few, in some cases generic, gas-phase and surface reactions. As in the case of horizontal reactors, the detailed transport and chemistry descriptions must be merged to obtain realistic descriptions of vertical CVD reactors that can be used for design and optimization computations.

Modeling of Miscellaneous CVD Reactors. In addition to the classical CVD reactor configurations discussed in the preceding sections, a wide variety of CVD reactor configurations have been used, including barrel and pancake-type reactors for epitaxy and vertical cross-flow LPCVD reactors. Barrel reactors have often been modeled as horizontal reactors, because the flow geometry of one barrel side is similar to that of a horizontal reactor (Table 3 in reference 212). However, the similarity disappears if buoyancy effects and barrel rotation are included in the analysis.

Relatively few analyses have been done of other reactor geometries. Transport and reaction processes in the classical CVD systems must be understood in order to resolve misconceptions in the literature and optimize existing processes. However, new reactor designs must be considered in the development of reactors capable of yielding uniform film thickness and composition over a large area, as well as providing sharp interfaces between layers. In addition, microscopic factors, such as morphological stability and step coverage, must be included in the modeling effort.

Modeling of Nonconventional CVD Processes. Modeling of nonconventional CVD processes, such as plasma-enhanced CVD and photon-stimulated CVD, follows the same framework as the modeling of standard CVD processes, with modifications to account for the different energy input. For plasma-enhanced CVD, the model must include the transport of charged species (electrons and ions), the energy distribution of electrons, ion–molecule interactions, and electron-impact reactions, in addition to the usual transport and reactions of neutral species. This complex modeling problem is discussed in detail by Hess and Graves in Chapter 8.

Photons can promote CVD reaction by different routes (*34*, *35*). One or more photons may lead to the photolysis of a reactant and the formation of highly reactive photofragments in the gas phase. The light can also be absorbed by adsorbed species, which then undergo reactions, or it can alter the surface states of the solid and promote reaction. Finally, the light can be transformed into heat in the top surface layer, and the resulting heat drives the deposition process directly. These processes are being studied primarily for laser direct-writing and etching applications, and they would benefit from combined reaction-and-transport modeling. However, except for thermal processes, relatively few analyses have been made of these photon-induced processes.

Pyrolytic-laser-assisted CVD is analogous to thermally driven CVD, but instead of a diffuse heating source, a focused laser beam is used to define deposition areas spatially (*32*, *38*, *39*) or to heat the gas phase selectively (*228*). The use of laser has the added advantages of increased energy flux and rapid heating. To avoid photochemistry, the gas phase must be transparent to the radiation.

The analysis of the thermal process is interesting for several reasons. First, measurement of the temperature in the irradiated spot is difficult; therefore, the temperature must be estimated by modeling. Analytical expressions for the temperature rise have been developed (*229*, *230*), but extension of these approaches to cases with irregularly shaped deposits on a substrate with time- and temperature-dependent properties is difficult.

Second, the cross-sectional profile of the growing film can show severe nonlinear behavior when the laser power is varied. At low power, the profile follows the Gaussian shape of the laser beam intensity, but at high power, volcano-shaped deposits form (*38*). By using a detailed finite-element model of the transient heat transfer in the substrate coupled with gas-phase heat and mass transfers, the substrate temperature can be accurately predicted, and volcano effect can be shown to be caused by depletion and nonlinear surface kinetics (*231, 232*). Examples of the different shapes of deposits are illustrated in Figure 16.

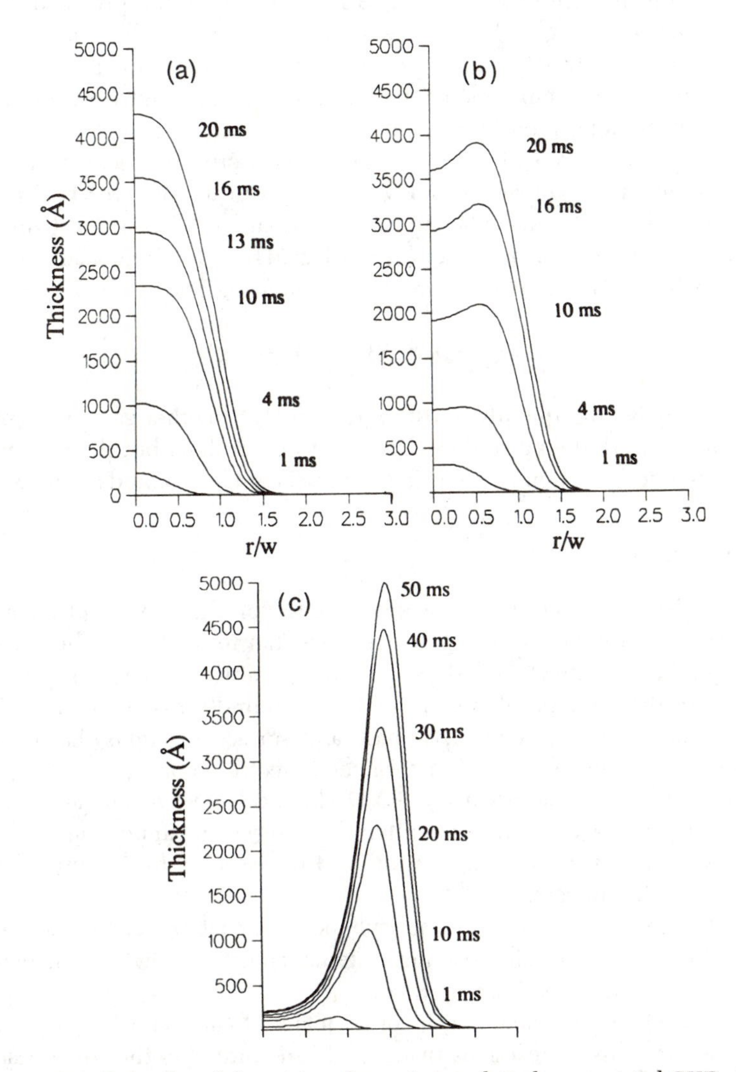

Figure 16. Examples of deposition shapes in pyrolytic-laser-assisted CVD: (a) first-order surface reaction, no mass-transfer effects; (b) first-order surface reaction, depletion effects; and (c) Langmuir–Hinshelwood surface kinetics, no mass transfer effects. The ratio r/w *is the radial position relative to the beam width* (231, 232).

Third, the line can become unstable during laser writing, and instead of a single line, a periodic pattern of discrete deposits is obtained (*233–235*). This pattern is analogous to bifurcations in other spatially distributed systems, such as catalytic fixed-bed reactors, and can be analyzed in the same manner (*235*).

In laser-assisted thermal CVD by gas-phase heating, the laser is used to vibrationally excite the gas (e.g., SiH_4) and active film precursors (e.g., SiH_2). The modeling of these processes revolves around the transport phenomena that control the access of the film precursors to the surface, as exemplified by the finite-element analysis by Patnaik and Brown of amorphous silicon deposition (*228*).

Finally, photosensitization is used in wide-area photochemical CVD, because generation of sufficient UV photons over a large area to drive the chemistry directly is difficult. Hg vapor is usually used as a sensitizer, or it is naturally present, as in the growth of HgCdTe (*43*). Hg is readily excited by an external Hg lamp according to the following reaction:

$$Hg + h\nu(253.7\ \text{nm}) \rightarrow Hg^* \tag{41}$$

The energy is subsequently transferred from Hg* to other gas-phase species by collisions. Modeling studies of this process are few, but the concepts of photochemical reaction engineering (*236*, *237*) can be adapted to this system.

Conclusion

Chemical vapor deposition is a key process for the growth of electronic materials for a large variety of devices essential to modern technology. Its flexibility and relatively low deposition temperatures make CVD attractive for future device applications in Si and compound-semiconductor technologies. The process involves gas-phase and surface reactions that must be controlled to achieve desired material and electronic properties.

Except for silane chemistry, CVD chemical mechanisms and kinetics are poorly characterized. Currently, the interest in understanding CVD chemistry is growing, and the results will be essential to the future development of the process.

CVD reactors involve transport phenomena that are analogous to those found in other chemically reacting systems, specifically heterogeneous catalytic reactors and combustion systems.

LPCVD reactor modeling involves many of the same issues of multicomponent diffusion reactions that have been studied in the past decade in connection with heterogeneous catalysis. Complex fluid-flow phenomena strongly affect the performance of atmospheric-pressure CVD reactors. Two-dimensional and some three-dimensional flow structures in the classical horizontal and vertical CVD reactors have been explored through flow visualization and detailed transport computations, and an increased understanding of the role of gas expansions, buoyancy effects, and reactor enclosure design in the overall transport phenomena is emerging. However, fully three-dimensional structures remain to be completely understood.

The applications of nonconventional CVD processes, such as laser and plasma processing, are likely to expand, and these applications require an increased fundamental understanding of the underlying chemistry and transport processes.

Abbreviations and Symbols

a_{AC}	lattice constant of compound AC
c	concentration
c_b	bulk concentration
c_o	input concentration
C_p	heat capacity
D	molecular diffusivity
D_{ik}	binary molecular diffusivity
D_{im}	multicomponent diffusivity of trace *i* in mixture *m*
g	refers to gaseous state when used as a superscript
H_i	molar enthalpy of component *i*
I_o	modified Bessel function of the first kind, of order zero
I_1	modified Bessel function of the second kind, of order one
J_i	flux of compound *i* relative to mass average velocity
k	thermal conductivity
k_s	surface reaction rate constant
k, k_1, k_2	rate constants
k_∞	rate constant at high-pressure limit
L	length of deposition zone in LPCVD reactor
$\mathbf{n}$	outward normal to the deposition surface
n^g	number of gas-phase reactions
n^s	number of surface reactions
n_i^{film}	number of film atoms in species *i*
N_i	flux of component *i*
M	average molecular weight
o	refers to the standard state when used as a superscript
p	pressure
p_A, p_M	partial pressures of A and M, respectively
p_i	partial pressure of component *i*
p_{eq}	equilibrium pressure
r	radial coordinate
r^*	critical nucleation radius
R	gas constant
R_j^g	*j*th gas-phase reaction

R_j^s	*j*th surface reaction
R_t	LPCVD tube radius
R_w	Si wafer radius
s	refers to the surface when used as a superscript
S	number of species
T	temperature
t	time
v	linear velocity
$\langle v \rangle$	average linear velocity
$\overline{V}$	atomic volume
$\overline{V}_{\mathrm{film}}$	molar volume of film
x_i	mole fraction component *i*
y	dimensional concentration
z	axial coordinate
z_k	axial position of *k*th wafer
α	boat area relative to tube area
α_i^T	thermal diffusivity ratio
γ	surface tension
γ_i	activity coefficient of component *i*
δ	film thickness
δ_0	film thickness at leading edge of substrate
Δ	wafer spacing
ζ	dimensionless radial position
η	effectiveness factor
μ	viscosity
μ_i	chemical potential of component *i*
ν_i	stoichiometric coefficient
ν_{ij}^s	stoichiometric coefficient for surface reactions
ξ	dimensionless radial position
ρ	density
ϕ	Thiele modulus
Ω	interaction parameter (*see* equation 11)

Acknowledgments

I am grateful for support from the National Science Foundation, the Camille and Henry Dreyfus Foundation, the Guggenheim Foundation, and the Minnesota Supercomputer Institute. I am also grateful to the following colleagues who have contributed to this paper through research and discussion with me: L. Da-Cheng, E. O. Einset, D. I. Fotiadis, K. Giapis, S. Kieda, D. W. Kisker, T. F. Kuech, D. R. McKenna, B. S. Meyerson, H. K. Moffat, T. J. Mountziaris, P. E. Price, Jr., K. F. Roenigk, D. Skouby, and W. Richter.

References

1. Hess, D. W.; Jensen, K. F.; Anderson, T. J. *Rev. Chem. Eng.* **1985,** *3*, 97–186.
2. *Thin Film Processes*; Vossen, J. L.; Kern, W., Eds.; Academic: New York, 1978.
3. Kern, W.; Ban, V. S. In *Thin Film Processes*; Vossen, J. L.; Kern, W., Eds.; Academic: New York, 1978; p. 257–331.
4. Bunshah, R. F. *Deposition Technologies for Films and Coatings*; Noyes: Park Ridge, NJ, 1982.
5. *Handbook of Thin Film Technology*; Maissel, K. I.; Glang, R. W., Eds.; McGraw–Hill: New York, 1970.
6. Sherman, A. *Chemical Vapor Deposition for Microelectronics*; Noyes: Park Ridge, NJ, 1987.
7. Bloem, J.; Giling, L. J. *Current Topics in Materials Science* **1978,** *1*, 147–342.
8. Bloem, J. *J. Cryst. Growth* **1980,** *50*, 581–604.
9. Olsen, G. H. In *Integrated Circuits: Chemical and Physical Processing*; Stroeve, P., Ed.; ACS Symposium Series 290; American Chemical Society: Washington, DC, 1985; pp 221–240.
10. Kuech, T. F. *Mater. Sci. Rep.* **1987,** *2*, 1–49.
11. Dapkus, P. D. *Annu. Rev. Mater. Sci.* **1982,** *12*, 243–269.
12. Dupuis, R. D. *Science (Washington, DC)* **1984,** *226*, 623–629.
13. Ludowise, M. J. *J. Appl. Phys.* **1985,** *58*, 31–R55.
14. Speckman, D. M.; Wendt, J. P. *Appl. Phys. Lett.* **1987,** *50*, 676–678.
15. Lum, R. M.; Klingert, J. K.; Lamont, M. G. *Appl. Phys. Lett.* **1987,** *50*, 284–286.
16. Cullen, G. W.; Corboy, J. F.; Metzl, R. *RCA Rev.* **1983,** *44*, 187–216.
17. Corboy, J. F.; Pagliaro, R., Jr. *RCA Rev.* **1983,** *44*, 231–249.
18. Tandon, J. L.; Yeh, Y. C. M.; *J. Electrochem. Soc.* **1985,** *132*, 662–668.
19. Ikeada, M.; Kojima, S.; Kashiwayanagi, Y. *J. Cryst. Growth* **1986,** *77*, 157–162.
20. Rosler, R. S. *Solid State Technol.* **1977,** *20(4)*, 63–69.
21. Roenigk, K. F.; Jensen, K. F. *J. Electrochem. Soc.* **1985,** *132*, 448–454.
22. Meyerson, B. S. *Appl. Phys. Lett.* **1986,** *48*, 797–799.
23. Meyerson, B. S.; Gamin, E.; Smith, D. A.; Nguyen, T. N. *J. Electrochem. Soc.* **1986,** *133*, 132–135.
24. Fotiadis, D. I.; Kremer, A. M.; McKenna, D. R.; Jensen, K. F. *J. Cryst. Growth* **1987,** *85*, 154–164.
25. Foster, D.; Learn A.; Kamins, T. *Solid State Technol.* **1986,** *29*(5), 227–232.
26. Hess, D. W. *J. Vac. Sci. Technol., A* **1984,** *2*(2), 244–252.
27. Reif, R. *J. Vac. Sci. Technol., A* **1984,** *2*(2), 429–435.
28. Sherman, A. *Thin Solid Films* **1984,** *113*, 135–149.
29. Veprek, S. In *Current Topics in Materials Science*; Kaldis, E., Ed.; North-Holland: Amsterdam, 1980; Vol. 4, pp 151–236.
30. Greene, J. E.; Barnett, S. A. *J. Vac. Sci. Technol.* **1982,** *21*(2), 285–302.
31. Lucovsky, G.; Richard, P. D.; Tsu, D. V.; Lin, S. Y.; Plarkumas, R. J. *J. Vac. Sci. Technol., A* **1986,** *4*, 681–688.
32. Brauerle, D. *Chemical Processing with Lasers*; Springer–Verlag: Heidelberg, 1986.
33. Bauser, E. In *Crystal Growth of Electronic Materials*; E. Kaldis, Ed; Elsevier: Amsterdam, 1984; pp 41–55.
34. Chuang, T. J. *J. Vac. Sci. Technol.* **1982,** *21*(3), 798–806.
35. Ehrlich, D. J.; Tsao, J. Y. *J. Vac. Sci. Technol., B* **1983,** *1*(4), 969–85.
36. Osgood, R. M., Jr. *Annu. Rev. Phys. Chem.* **1983,** *34*, 77–101.
37. Houle, F. A. *Appl. Phys., A* **1986,** *41*, 315–330.

38. Allen, S. D.; Tringubo, A. B. *J. Appl. Phys.* **1982,** *54*(3), 1641–643.
39. Allen, S. D.; Jan, R. Y.; Edwards, R. H.; Mazuk, S. M.; Vernon, S. D. *SPIE* **1984,** *459*, 42–48.
40. Chen, J. Y.; Henderson, R. C.; Hall, J. T.; Peters, J. W. *J. Electrochem. Soc.* **1984,** *131*(9), 2146–151.
41. Hamano, K.; Numazawa, Y.; Yamazaki, K. *Jpn. J. Appl. Phys.* **1984,** *23*(9), 1209–215.
42. Tarui, Y.; Hidaka, J.; Aota, K. *Jpn. J. Appl. Phys.* **1984,** *23*(*11*), L–827–829.
43. Irvine, S. J. C.; Giess, J.; Gough, J. S.; Blackmore, G. W.; Royle, A.; Mullin, J. B.; Chew, N. G.: Cullis, A. G. *J. Cryst. Growth* **1986,** *77*, 437–451.
44. Yee, K. K. *Int. Metals Rev.* **1978,** *23*(*1*), 19–42.
45. Quinto, D. T. *J. Vac. Sci. Technol., A* **1988,** *6*, 2149–157.
46. Kruppers, D.; Lydtin, H. In *Chemical Vapor Deposition - Sixth International Conference*; Donaghey, L. F.; Rai-Choudhury, P.; Tauber, R. N., Eds.; Electrochemical Society: Pennington, NJ, 1977; pp 461–476.
47. Midwinter, J. E. *Optical Fibers for Transmission*; Wiley: New York, 1979.
48. *Adventures in Optical Fiber Communication*; Nagel, S. R.; MacChesney, J. B.; Walker, K. L., Eds.; Academic: Orlando, 1985.
49. Walker, K. L.; Geyling, F. T.; Nagel, S. R. *J. Am. Chem. Soc.* **1980,** *63*, 552–558.
50. Walker, K. L.; Harvey, J. W.: Geyling, F. T.; Nagel, S. R. *J. Am. Ceram. Soc.* **1980,** *63*, 96–102.
51. Kruppers, D. *Proceedings, Seventh International Conference on Chemical Vapor Deposition*; Electrochemical Society: Pennington, NJ, 1979; pp 159–175.
52. Ablow, C. M.; Schechter, S.; Wise, H. *Combust. Sci. Technol.* **1980,** *22*, 107–117.
53. Pfefferle, L. D.; Pfefferle, W. C.; *Catal. Rev. Sci. Eng.* **1987,** *29*, 219–264.
54. Sotirchos, S. V.; Srimivas, B.; Amundson, N. R. *Rev. Chem. Eng.* **1984,** *2*, 175–237.
55. Rosner, D. E. In *High Temperature Lamp Chemistry*; Electrochemical Society: Pennington, NJ, 1988; pp 88–84, 111–138.
56. *Chemical Reaction and Reactor Engineering*; Carberry, J. J.; Varma, A., Eds.; Marcel Dekker: New York, 1986.
57. Inoue, G.; Suzuki, M. *Chem. Phys. Lett.* **1985,** *122*, 361–364.
58. Walsh, R. *Acc. Chem. Res.* **1987,** *14*, 246–252.
59. Jasinski, J. M.; Chu, J. O. *J. Chem. Phys.* **1988,** *88*, 1678–687.
60. Gates, S. M. *Surf. Sci.* **1988,** *195*, 307–309.
61. Venables, J. A.: Spiller, G. D. T.: Hanbrucka, M. *Rep. Prog. Phys.* **1984,** *47*, 399–459.
62. Classen, W A. P.; Bloem, J. *J. Electrochem. Soc.* **1980,** *127*, 194–202.
63. Classen, W. A. P.; Bloem, J. *J. Electrochem. Soc.* **1980,** *127*, 1836–1843.
64. Classen, W. A. P.; Bloem, J. *J. Electrochem. Soc.* **1981,** *128*, 1353–1359.
65. Cooke, M. J.; Heinecke, R. A.; Stern, R. C. *Solid State Technol.* **1982,** *25*(*12*), 62–65.
66. Levy, R. A.; Green, M. L. *J. Electrochem. Soc.* **1987,** *134*, 37C–49C.
67. Joyce, B. A. *Rep. Prog. Phys.* **1974,** *37*, 363–420.
68. Biefeld, R. M. *J. Cryst. Growth* **1986,** *77*, 392–399.
69. Matthews, J. W.; Blakeslee, A. E. *J. Vac. Sci. Technol.* **1977,** *14*, 989–994.
70. Jastrzebski, L. *J. Cryst. Growth* **1983,** *63*, 493–516.
71. Broadbent, E. K.; Stacy, W. T. *Solid State Technol.* **1985,** *28*(*12*), 51–59.
72. McConica, C. M.; Krishnamani, K. *J. Electrochem. Soc.* **1986,** *133*, 2542–2548.
73. Gale, R. P.; McClelland, R. W.; Fan, J. C. C.; Bozler, C. O. *Appl. Phys. Lett.* **1982,** *41*, 545–547.

74. Asai, H.; Ando, S. *J. Electrochem. Soc.* **1985,** *132*, 2445–2453.
75. Gilmer, G. H.; Broughton, J. Q. *Annu. Rev. Mater. Sci.* **1986,** *16*, 487–516.
76. Kern, R.; Lelay, G.; Msetois, J. *J. Curr. Topics Mater. Sci.* **1979,** *3*, 139–419.
77. van der Eerden, J. P. *Crystals*; Springer–Verlag: Heidelberg, 1983, 9, 115–144.
78. Madhukar, A.; Chaisas, S. V. In *CRC Crit. Rev. Solid State Mater. Sci.* **1988,** *14*, 1–130.
79. Betsch, R. J. *J. Cryst. Growth* **1986,** *77*, 210–218.
80. Smith, W. R. *Ind. Eng. Chem. Fund.* **1980,** *19*, 1–10.
81. Van Zeggeren, F.; Storey, S. H. *The Computation of Chemical Equilibria*; Cambridge University Press: Cambridge, 1970.
82. White, C. W.; Seidler, W. D. *AIChE J.* **1981,** *27*, 466–471.
83. Tirtowidjojo, M.; Pollard, R. *J. Cryst. Growth* **1986,** *77*, 200–209.
84. Tirtowidjojo, M.; Pollard, R. In *Processing of Electronic Materials*; Law, C. G.; Pollard, R., Eds.; Engineering Foundation: New York, 1987; pp 89–107.
85. Coltrin, M. E.; Kee, R. J.; Miller, J. A. *J. Electrochem. Soc.* **1984,** *131*, 425–434.
86. Coltrin, M. E.; Kee, R. J.; Miller, J. A. *J. Electrochem. Soc.* **1986,** *133*, 1206–1214.
87. Moffat, H. K.; Jensen, K. F. *J. Electrochem. Soc.* **1988,** *135*, 459–471.
88. Stringfellow, G. B. *J. Cryst. Growth* **1983,** *62*, 225–229.
89. Stringfellow, G. B.; Cherng, M. J. *J. Cryst. Growth* **1983,** *64*, 413–415.
90. Stringfellow, G. B. *J. Cryst. Growth* **1984,** *68*, 111–122.
91. Stringfellow, G. B. In *Processing of Electronic Materials*; Law, C. G.; Pollard, R., Eds.; Engineering Foundation: New York, 1987; pp 114–133.
92. Kisker, D. W.; Zawadzki, A. G. *J. Cryst. Growth* **1988,** *89*, 379–390.
93. Panish, M. B.; Ilegems, M. *Prog. Solid State Chem.* **1972,** *7*, 39–83.
94. Stringfellow, G. B. *J. Cryst. Growth* **1974,** *27*, 21–34.
95. Kroger, F. A. *The Chemistry of Imperfect Crystals*; 2nd rev. ed.; North-Holland: Amsterdam, Netherlands, 1974.
96. Swalin, R. A. *Thermodynamics of Solids*; 2nd ed.; Wiley: New York, 1972.
97. Alam, M. K.; Flagan, R. C. *Aerosol Sci. Tech.* **1986,** *5*(2), 237–248.
98. Jasinski, J. M.; Meyerson, B. S.; Scott, B. A. *Annu. Rev. Phys. Chem.* **1987,** *38*, 109–140.
99. Breiland, W. G.; Ho, P.; Coltrin, M. E. *J. Appl. Phys.* **1986,** *60*, 1505–1513.
100. Breiland, W. G.; Coltrin, M. E.; Ho, P. *J. Appl. Phys.* **1986,** *59*, 3267–3273.
101. White, R. T.; Espino-Rios, R. L.; Rogers, D. S.; Ring, M. A.; O'Neal, H. E. *Int. J. Chem. Kin.* **1985,** *17*, 1029–1065.
102. Robertson, R.; Hills, D.; Gallagher, A. *Chem. Phys. Lett.* **1984,** *103*, 397–404.
103. Robertson, R.; Gallagher, A. *J. Chem. Phys.* **1986,** *85*, 3623–3630.
104. Scott, B. A.; Estes, R. D.; Jasinski, J. M. *J. Chem. Phys.* **1988,** *89*, 2544–2549.
105. Doncaster, A. M.; Walsh, R. *Int. J. Chem. Kin.* **1981,** *13*, 503–514.
106. Ho, P.; Coltrin, M. E.; Binkley, J. S.; Melius, C. F. *J. Phys. Chem.* **1985,** *89*, 4647–4657.
107. Ho, P.; Coltrin, M. E.; Binkley, J. S.; Melius, C. F. *J. Phys. Chem.* **1986,** *90*, 3399–3406.
108. Inoue, G.; Suzuki, M. *Chem. Phys. Lett.* **1985,** *122*, 361–364.
109. Dzarnoski, J.; Rickborn, S. F.; O'Neal, H. E.; Ring, M. A. *Organometallics* **1982,** *1*, 1217–1220.
110. Ban, V. S. *J. Electrochem. Soc.* **1978,** *125*, 317–320.
111. Nishizawa, J.; Saito, M. *J. Cryst. Growth* **1981,** *52*, 213–218.
112. Sedgewick, T. O. In *Proceedings, 6th International Conference on CVD*, Electrochemical Soceity: Pennington, NJ, 1977; pp 79–89.
113. Ho, P.; Breiland, W. G. *Appl. Phys. Lett.* **1983,** *44*, 125–126.
114. Jacko, M. G.; Price, S. J. W. *Can. J. Chem.* **1963,** *41*, 1560–1567.

115. Jacko, M. G.; Price, S. J. W. *Can. J. Chem.* **1964,** *42,* 1198–1205.
116. Butler, J. E.; Bottka, N.; Silman, R. S.; Gaskill, D. K. *J. Cryst. Growth* **1986,** *77,* 73–78.
117. Denbaars, S. P.; Maa, B. Y.; Dapkus, P. D.; Danner, A. D.; Lee, H. C. *J. Cryst. Growth* **1986,** *77,* 188–194.
118. Lee, P.; Omstead, T. R.; McKenna, D. R.; Jensen, K. F. *J. Cryst. Growth 1987 85,* 165–174.
119. Larsen, C. A.; Buchan, N. I.; Stringfellow, G. B. *Appl. Phys. Lett.* **1988,** *52,* 480–482.
120. Stringfellow, G. B. In *Mechanisms of Reactions of Organometallic Compounds with Surface*; Cole-Hamilton, D., Ed.; NATO Adv. Study Institute: in press.
121. Dapkus, P. D.; DenBaars, S. P.; Chen, Q.; Maa, B. Y. In *Mechanisms of Reactions of Organometallic Compounds with Surface,* Cole-Hamilton, D., Ed.; NATO Adv. Study Institute: in press.
122. Mountziaris, T. J.; Jensen, K. F. In *Chemical Perspectives of Microelectronics Processing;* Gross, M. E.; Jasinski, J.; Yates, J. T., Eds.; Proc. Mat. Res. Soc. 131, paper E6–6, 1988.
123. Mitchell, S. A.; Hacket, P. A.; Rayner, D. M.; Humphries, M. R. *J. Chem. Phys.* **1985,** *83,* 5028–5038.
124. Nishizawa, J.; Kurabayashi, T. *J. Electrochem. Soc.* **1983,** *130,* 413–417.
125. Tamaru, K. *J. Phys. Chem.* **1955,** *59,* 777–780.
126. Frolov, I. A.; Kitaev, E. M.; Druz', B. L. *Zhur. Fiz. Khim.* **1977,** *51,* 1106–1108.
127. Luckerath, R.; Tommack, P.; Hertling, A.; Koss, H. J.; Balk, P.; Jensen, K. F.; Richter, W. *J. Cryst. Growth* **1988,** *93* 151–158.
128. Luckerath, R.; Richter, W.; Jensen, K. F. In *Mechanisms of Reactions of Organometallic Compounds with Surface*; Cole-Hamilton, D., Ed.; NATO Adv. Study Institute: in press.
129. Tirtowidjojo, M.; Pollard, R. *J. Cryst. Growth* **1986,** *77,* 200–206.
130. Coates, C. E.; Green, M. L. H.; Wabb, K. *Organometallic Compounds*; Methuen: London, 1967; p 319.
131. Kuech, T. F.; Veuhoff, E.; Kuan, T. S.; Deline, V.; Potemski, R. *J. Cryst. Growth* **1986,** *77,* 257.
132. Lum, R. M.; Klingert, K. K.; Lamount, M. G. *J. Cryst. Growth* **1988,** *89,* 137.
133. Bhat, R.; Koza, M.A.; Skromme, B. *J. Appl. Phys. Lett.* **1987,** *50,* 1194–1196.
134. Speckman, D. M.; Wendt, J. P. *Appl. Phys. Lett.* **1987,** *50,* 676–678.
135. Chen, C. H.; Larsen, C. A.; Stringfellow, G. B. *Appl. Phys. Lett.* **1987,** *50,* 218–220.
136. Lum, R. M.; Klingert, K. K. *Appl. Phys. Lett.* **1987,** *50,* 284–286.
137. Lee, P. W.; Omstead, T. R.; McKenna, D. W.; Jensen, K. F. *J. Cryst. Growth* **1988, 134–142.**
138. Omstead, T. R.; Vand Sickle, P.; Lee, P. W.; Jensen, K. F. *J. Cryst. Growth* **1988, 20–28.**
139. Robinson, P. J.; Holbrook, K. A. *Unimolecular Reactions*; Wiley: London, 1972.
140. Meyerson, B. S.; Jasinski, J. M. *J. Appl. Phys.* **1987,** *61,* 785–787.
141. Troe, J. *J. Phys. Chem.* **1979,** *83,* 114–126.
142. Roenigk, K. F.; Jensen, K. F.; Carr, R. W. *J. Phys. Chem.* **1987,** *91,* 5732–5739; **1988,** 92, 4254.
143. Joyce, B. A.; Bradley, R. R. *Philos. Mag.* **1966,** *15,* 1167–1187.
144. Henderson, R. C.; Helm, R. F. *Surf. Sci.* **1972,** *30,* 310–334.
145. Farrow, R. F. C. *J. Electrochem. Soc.* **1974,** *121,* 899–907.
146. Farnaam, M. J.; Orlander, D. R. *Surf. Sci.* **1984,** *145,* 390–406.

147. Buss, R. F.; Ho, P.; Breiland, W. G.; Coltrin, M. E. *J. Appl. Phys.* **1988,** *63,* 2808–2819.
148. Gates, S. M.; Scott, B. A.; Beach, D. B.; Imbitil, R.; Demuth, J. C. *J. Vac. Sci. Technol., A* **1987,** *5,* 628–630.
149. Meyerson, B. S., Scott, B. A.; Tsui, R. *Chemtronics* **1987,** *1,* 155–161.
150. Donahue, T. J.; Reif, R. *J. Electrochem. Soc.* **1986,** *133,* 1691–1697.
151. Chang, C. A. *J. Electrochem. Soc.* **1976,** *123,* 1245–1247.
152. Meyerson, B. S.; Yu, M. L. *J. Electrochem. Soc.* **1984,** *131,* 2366–2368.
153. Yu, M. L.; Vitkavage, D. J.; Meyerson, B. S. *J. Appl. Phys.* **1986,** *59,* 4032–4037.
154. Schyler, D. J.: Ring, M. A. *J. Organomet. Chem.* **1976,** *114,* 9–19.
155. Reep, D. H.; Ghandi, S. K. *J. Electrochem. Soc.* **1983,** *130,* 675–680.
156. Prutz, N.; Veuhoff, E.; Heinecke, H., Heyen, M.; Lruth, H.; Balk, P. *J. Vac Sci. Technol., B* **1985,** *3,* 671–673.
157. Tsang, W. T. *J. Electron. Mater.* **1986,** *15,* 235–247.
158. Tischler, M. A.; Bedair, S. M. *J. Cryst. Growth* **1986,** *77,* 89–94.
159. Nishizawa, J.; Abe, H.; Kurabayashi, T. *J. Electrochem. Soc.* **1985,** *132,* 1197–1200.
160. Robertson, A., Jr.; Chiu, T. H.; Tsang, W. T.; Cunningham, J. E. *J. Appl. Phys.* **1988,** *64,* 877–887.
161. Squire, D. W.; Dulcey, C. S.; Lin, M. C. *Chem. Phys. Lett.* **1985,** *116,* 525–528.
162. Squire, D. W.; Dulcey, C. S.; Lin, M. C. *J. Vac. Sci. Technol., B* **1985,** *3,* 1513–1519.
163. Squire, D. W.; Dulcey, C. S.; Lin, M. C. *Chem. Phys. Lett.* **1986,** *131,* 112–117.
164. Kuech, T. F.; Veuhoff, E. *J. Cryst. Growth* **1984,** *68,* 148–154.
165. Tirtowidjojo, M.; Pollard, R. In *Proceedings, First Foundation Conference on Processing of Electronic Materials*; Engineering Foundation: New York, 1987; pp 89–107.
166. Tirtowidjojo, M.; Pollard, R. *J. Cryst. Growth* **1988,** *93,* 108–115.
167. Roenigk, K. F.; Jensen, K. F. *J. Electrochem. Soc.* **1987,** *134,* 1777–785.
168. Rosenberger, F. In *Proceedings, Tenth International Conference on CVD*; Cullen, G. W., Ed.; Electrochemical Society: Pennington, NJ, 1987; pp 11–22.
169. Fraas, L. M.; McLeod, P. S.; Partain, L. D.; Cape, J. A. *J. Vac. Sci. Technol., B* **1986,** *4,* 22–28.
170. Middleman, S.; Yeckel, A. *J. Electrochem. Soc.* **1986,** *133,* 1951–1956.
171. Wang, C. A.; Groves, S. H.; Palmateer, S. C.; Weybourne, D. W.; Brown, R. A. *J. Cryst. Growth* **1986,** *77,* 139–143.
172. Takahashi, R.; Koza, Y.; Sugawara, K. *J. Electrochem. Soc.* **1972,** *119,* 1406–1412.
173. Wahl, G. *Thin Solid Films* **1977,** *40,* 13–26.
174. Stock, L.; Richter, W. *J. Cryst. Growth* **1986,** *77,* 144–150.
175. Giling, L. J. *J. Electrochem. Soc.* **1982,** *129,* 634–643.
176. Williams, J. E.; Peterson, R. W. *J. Cryst. Growth* **1986,** *77,* 128–135.
177. Talbot, L.; Cheng, R. K.; Schefer, R. W.; Willis, D. R. *J. Fluid Mech.* **1980,** *101,* 737–758.
178. Eversteyn, F. C.; Severin, P. J. W.; Van den Brekel, C. H. J.; Peek, H. L. *J. Electrochem. Soc.* **1970,** *117,* 925–931.
179. Jensen, K. F.; Fotiadis, D. I.; Moffat, H. K.; Einset, E. O.; Kremer, A. M.; McKenna, D. R. In *Interdisciplinary Issues in Materials and Manufacturing*; Samanta, S. K., Komanduri, K., McMeeking, R., Chen, M. M., Tseng, A., Eds.; American Society of Mechanical Engineers: New York, 1987; pp 565–586.
180. Gage, K. S.; Reid, W. H. *J. Fluid Mech.* **1968,** *33,* 21–32.

181. Platten, J. K.; Legros, J. C. *Convection in Liquids*, Springer–Verlag: New York, 1984; Chapter 8.
182. Luijkx, J.-M.; Platten, J. K.; Legros, J. C. *Int. J. Heat Mass Transfer* **1981,** *24,* 1287–1291.
183. Evans, G.; Greif, R., submitted for publication in *Int. J. Heat Mass Transfer.*
184. Westphal, G. H.; Shaw, D. H.; Hartzell, R. A. *J. Cryst. Growth* **1982,** *56,* 324–331.
185. Chiu, K. C.; Ouazzani, J.; Rosenberger, F. *Int. J. Heat Mass Transfer* **1987,** *30,* 1655–1662.
186. Visser, E. P.; Klein, C. R.; Govers, C. A. M.; Hoogendorn, C. J.; Giling, L. J., submitted for publication in *J. Cryst. Growth.*
187. van Opdorp, C.; Leys, M. R. *J. Cryst. Growth* **1987,** *84,* 271–288.
188. Jackson, C. P.; Winters, K. H. *Int. J. Num. Methods Fluids* **1984,** *4,* 127–145.
189. Moffat, H. K.; Jensen, K. F. *J. Cryst. Growth* **1986,** *77,* 108–119.
190. Busse, F. S. *Rep. Prog. Phys.* **1978,** *41,* 1929–1967.
191. van de Ven, J.; Rutten, G. J. M.; Raaymakers, M. J.; Giling, L. J. *J. Cryst. Growth* **1986,** *76,* 352–372.
192. Einset, E. O.; Fotiadis, D. I.; Jensen, K. F., submitted for publication in *J. Cryst. Growth.*
193. Fotiadis, D. I.; Boekholt, M.; Jensen, K. F.; Richter, W., submitted for publication in *J. Cryst. Growth.*
194. Richter, W.; Hünermann, L. *Chemtronics* **1987,** *2,* 175–182.
195. Jensen, K. F.; Fotiadis, D. I.; McKenna, D. R.; Moffat, H. K. In *Supercomputer Research in Chemistry and Chemical Engineering*; Jensen, K. F.; Truhlar, D. G., Eds.; ACS Symposium Series 353; American Chemical Society: Washington, DC, 1987, pp 353–376.
196. Kusumoto, Y.; Hayashi, T.; Komiya, S. *Jpn. J. Appl. Phys.* **1985,** *24,* 620–625.
197. Jensen, K. F.; Fotiadis, D. I.; Lee, P. W.; McKenna, D. R.; Moffat, H. K. *Soc. Photo-Opt. Instrum. Eng. Proc.* **1987,** *796,* 178–190.
198. Patnaik, S.; Brown, R. A.; Wang, C., submitted for publication in *J. Cryst. Growth.*
199. Taylor, G. I. *Proc. Roy. Soc. (London),* A **1953,** *219,* 186–203.
200. Aris, R. *Proc. Roy. Soc. (London),* A **1956,** *235,* 67–77.
201. Thrush, E. J.; Whiteaway, G.; Wale-Evans, G.; Wright, D. R.; Cullis, A. G. *J. Cryst. Growth* **1984,** *68,* 412–421.
202. Fotiadis, D. I.; Kieda, S.; Jensen, K. F., submitted for publication in *J. Cryst. Growth.*
203. Arnado, P. M.; Brown, R. A. *J. Cryst. Growth* **1987,** *80,* 155–190.
204. McFadden, G. B.; Rehm, R. G.; Coriell, S. R.; Chuck, W.; Morrish, K. A. *Mettal. Trans., A* **1984,** *15,* 2125–2137.
205. Kisker, D. W.; McKenna, D. R.; Jensen, K. F. *Mater. Lett.* **1988,** *6,* 123–128.
206. McKenna, D. R.; Jensen, K. F.; Kisker, D. W., submitted for publication in *J. Cryst. Growth.*
207. Palmateer, S. C.; Groves, S. H.; Wang, C. A.; Weybourne, D. W.; Brown, R. A. *J. Cryst. Growth* **1987,** *83,* 202–210.
208. Jenkinson, J. P.; Pollard, R. *J. Electrochem. Soc.* **1984,** *131,* 2911–2917.
209. Bird, R. B.; Stewart, W. E.; Lightfoot, E. N. *Transport Phenomena*; Wiley: New York, 1960.
210. Coffee, T. P.; Meimerl, J. M. *Combust. Flame* **1981,** *43,* 273–289.
211. Dixon-Lewis, G. In *Combustion Chemistry*; Gardiner, W. C., Jr., Ed.; Springer–Verlag: New York, 1984, pp 21–126.
212. Jensen, K. F. *Chem. Eng. Sci.* **1987,** *42,* 923–958.
213. Jensen, K. F.; Graves, D. B. *J. Electrochem. Soc.* **1983,** *130,* 1950–1957.

214. Aris, R. *The Mathematical Theory of Diffusion and Reaction in Permeable Catalysts*; Clarendon: Oxford, Vol. 1.
215. Roenigk, K. F. Ph.D. Thesis, University of Minnesota, 1987.
216. Ikegawa, M.; Kobayashi, J. In *Proceedings, 7th Symposium on Plasma Processing*; Electrochemical Society: Pennington, NJ, 1988.
217. Meyerson, B. S.; Olbricht, B. S. *J. Electrochem. Soc.* **1984,** *131,* 2361–365.
218. Purnell, J. H.; Walsh, R. *Proc. Roy. Soc. (London)*, A **1966,** *293,* 543–561.
219. Jensen, K. F.; Hitchman, M. L.; Ahmed A., In *Proceedings, Fifth European Conference on CVD*; University of Uppsala: Uppsala, Sweden, 1985, pp 144–151.
220. Yeckle, A., Middleman, S. *J. Electrochem. Soc.* **1987,** *134,* 1275–1281.
221. Cussler, E. L. *Diffusion*; Cambridge University Press: Cambridge, 1984.
222. Schlichting, H. *Boundary Layer Theory*; McGraw–Hill: New York, 1979.
223. Pollard, R.; Newman, J. *J. Electrochem. Soc.* **1980,** *127,* 744–752.
224. Michaelidis, M.; Pollard, R. *J. Electrochem. Soc.* **1984,** *131,* 861–868.
225. Kee, R. J.; Evans, G. H.; Coltrin, M. E. In *Supercomputer Research in Chemistry and Chemical Engineering*; Jensen, K. F.; Truhlar, D. G., Eds.; ACS Symposium Series 353; American Chemical Society: Washington, DC, 1987; pp 334–352.
226. Houtman, C.; Graves, D. B; Jensen, K. F. *J. Electrochem. Soc.* **1986,** *133,* 961–970.
227. Wahl, G.; Schmaderer, F.; Huber, R.; Weber, R. In *Proceedings, Tenth International Conference on CVD*; Cullen, G. W., Ed.; Electrochemical Society: Pennington, NJ, 1987; pp 42–52.
228. Patnaik, S.; Brown, R. A. *J. Electrochem. Soc.* **1987,** *135,* 697–706.
229. Lax, M. *J. Appl. Phys.* **1977,** *48,* 3919–3924.
230. Lax, M. *Appl. Phys. Lett.* **1978,** *33,* 786–788.
231. Skouby, D.; Jensen, K. F. *J. Appl. Phys.* **1988,** *63,* 154–164.
232. Skouby, D.; Jensen, K. F. *Proc. Mat. Res. Soc.* **1988,** *101,* 107–112.
233. Gross, M. E.; Fisanick, G. J.; Gallagher, P. K.; Schnoes, K. J.; Katzir, A. *J. Appl. Phys.* **1985,** *57,* 1139–1142.
234. Gross, M. E.; Applebaum, A.; Schnoes, K. J. *J. Appl. Phys.* **1986,** *60,* 529–533.
235. Price, P. E., Jr.; Jensen, K. F. In *Laser and Particle Beam Chemical Processes on Surfaces*, Johnson, A. W.; Loper, G. L.; Sigmon, T. W., Eds.; Proc. Mat. Res. Soc. 129, paper B14.1.
236. Spadoni, G.; Stramigioli, C.; Santarelli, P. *Chem. Eng. Sci.* **1980,** *35,* 925–931.
237. Alfano, R. L. ; Romero, L.; Cassano, A. E. *Chem. Eng. Sci.* **1986,** *41,* 421–44, 1137–1154.

RECEIVED for review December 30, 1987. ACCEPTED revised manuscript November 16, 1988.

6

Diffusion and Oxidation of Silicon

Richard B. Fair

Microelectronics Center of North Carolina, Research Triangle Park, NC 27709, and Department of Electrical Engineering, Duke University, Durham, NC 27706

Oxidation and diffusion in silicon are processes that significantly affect the fabrication of microelectronic devices. However, our knowledge of the fundamental principles governing these processes is inadequate, and this inadequacy affects our ability to understand and model submicrometer ultralarge-scale-integration technologies. These advanced processes require p–n junctions of 1000-Å depth and oxides of 100-Å thickness. The existing theories and models do not adequately describe the physical mechanisms that dominate diffusion and oxidation in these regimes. The theories, new ideas, issues, and unknowns about these processes are reviewed in this chapter.

SILICON-PROCESSING TECHNOLOGY has depended heavily on thermal oxidation and the diffusion of impurities since the 1950s. The use of diffusion techniques to form p–n junctions was disclosed in a 1952 patent by Pfann (*1*). Since then, numerous approaches have been studied on how to introduce dopants into silicon with the goal of controlling the electrical properties of the junction, concentrations of dopants, uniformity and reproducibility, and cost of manufacture. Thermal oxides were used initially to selectively mask dopants during diffusion steps. Additional research indicated that the passivation properties of thermal oxides may be used to advantage in devices. Two very important developments in semiconductor technology grew out of research in oxides: the planar process invented by Hoerni (*2*) and the MOS (metal–oxide–semiconductor) transistor, which was first disclosed by Kahng and Atalla (*3*).

Oxidation and diffusion continue to be important in submicrometer VLSI (very-large-scale integration) technology. Modern integrated devices require

0065–2393/89/0221–0265$14.85/0

ultrashallow junctions of 1000-Å depth and thin oxides of 100-A thickness. These requirements make it necessary for semiconductor scientists and engineers to refine further the knowledge base governing these processes. Short-time or low-temperature steps are needed to produce shallow junctions and thin oxides. These processing regimes have never been explored before, and recent data have shown that unusual transient effects occur during dopant diffusion. The models based on existing theories of oxidation do not apply well to oxides whose thicknesses are less than 350 Å.

Although oxidation and diffusion are closely interrelated processes, they will be discussed separately in this chapter. However, the effects of oxidation on diffusion and those of doping on oxidation will be covered to emphasize the interrelationships that exist. The treatment of the subject is not exhaustive; my intention is to give the reader a background of the subject and to inform the reader about the issues and the gaps in our knowledge of these processes.

Diffusion

This section focuses on the diffusion of impurities in silicon. Because of the large solid solubility in silicon of group III and group V doping impurities, diffusion proceeds by interactions with point defects: silicon vacancies and silicon self-interstitials. Each high-temperature processing step can change the levels of vacancies and self-interstitials and, therefore, the diffusion of impurities. These effects can be understood at two levels—the atomic level and the continuum level. A discussion of diffusion from both points of reference is provided in the following sections.

The process of introducing impurities into silicon is called *predeposition*. Chemical predeposition is described in terms of a solution to the diffusion equation. Predeposition by ion implantation is described in terms of ion penetration into silicon, distributions of implanted impurities, lattice damage, etc.

Continuum Theory. Solid-state diffusion is described in terms of a continuity equation known as Fick's second law:

$$\frac{\partial}{\partial x}\left(D\frac{\partial C}{\partial x}\right) = \frac{\partial C}{\partial t} \tag{1}$$

Equation 1 describes the rate of change of the concentration of impurity with time. The diffusion coefficient D is expressed in square centimeters per second, and the concentration C is expressed usually in number of atoms per cubic centimeter.

When D is constant, the surface concentration of the diffusing impurity is fixed, the concentration of the impurity at $x = \infty$, $C(\infty, t)$, is 0 for all time, and the concentration at any point in the crystal at $t = 0$, $C(x, 0)$, is 0. Under these conditions, the solution to equation 1 is given by (*4*)

$$C(x, t) = C_o \operatorname{erfc}\left(\frac{x}{2\sqrt{Dt}}\right) \tag{2}$$

where erfc is the complementary error function. If $C(\infty, t) = C_B$ and $C(x, 0) = C_B$, where C_B is the background doping concentration in the semiconductor, then

$$C(x, t) = C_o \operatorname{erfc}\left(\frac{x}{2\sqrt{Dt}}\right) + C_B \tag{3}$$

Thus, for the boundary conditions just described, the concentration of impurity as a function of space and time is given by a complementary error function (erfc) whose argument is $x/\sqrt{Dt}$. The complementary error function is a tabulated function.

Diffusion processes that are performed with a constant surface concentration are normally referred to as predeposition steps. Predepositions are usually done in N_2 furnace ambients with a small percentage of O_2, and the doping species is introduced into the furnace in gaseous form. The dopant concentration in the N_2 gas stream is varied to change the surface concentration in the silicon. Typical predeposition temperatures are 900–1000 °C, and typical predeposition periods are 30–60 min. The sources of dopants include liquids, solids, and gases.

Boron. For the predeposition of boron, the most prevalent species in the gas phase in the furnace is B_2O_3. Once B_2O_3 is deposited on the silicon surface, the oxide reacts with the silicon to bring about doping, as shown in equation 4.

$$B_2O_3 + \frac{3}{2}\,Si \leftrightarrows 2B + \frac{3}{2}\,SiO_2 \tag{4}$$

B_2O_3 may come from either one of the reactions described in equations 5 and 6.

$$2H_3BO_3 \xrightarrow[185\ °C]{\Delta} B_2O_3 + 3H_2O \tag{5}$$

$$4BN + 3O_2 \rightarrow 2B_2O_3 + 2N_2 \tag{6}$$

The source of boron nitride in equation 6 may be the disks about the size of a silicon wafer that are placed next to the wafers in the diffusion furnace.

By varying the partial pressure of the gas phase of the dopant in the furnace, the concentration of impurities in the silicon can be changed. Henry's law relates the concentration of dopants that are introduced in the furnace to the surface concentration by the relation $C_o = Hp_s$, where C_o is the surface concentration of dopant, H is Henry's constant, and p_s is the partial pressure of the dopant gas.

Figure 1 is a plot of boron concentration versus p_s that illustrates Henry's law. When the solid solubility of boron in silicon is reached, Henry's law no longer applies. Thus, most predeposition steps are operated at a sufficiently high partial pressure in the dopant gas phase to achieve solid solubility of the dopant in the silicon. This requirement provides a natural control for reproducible diffusion results.

Phosphorus. For the predeposition of phosphorus, the predominant species in the gas phase is P_2O_5. The doping reaction with P_2O_5 is shown in equation 7.

$$P_2O_5 + \frac{5}{2}\,Si \leftrightarrows \frac{5}{2}\,SiO_2 + 2P \qquad (7)$$

Sources of P_2O_5 vapor are solid P_2O_5, red phosphorus, $POCl_3$, PBr_3, $NH_4H_2PO_4$, or PN.

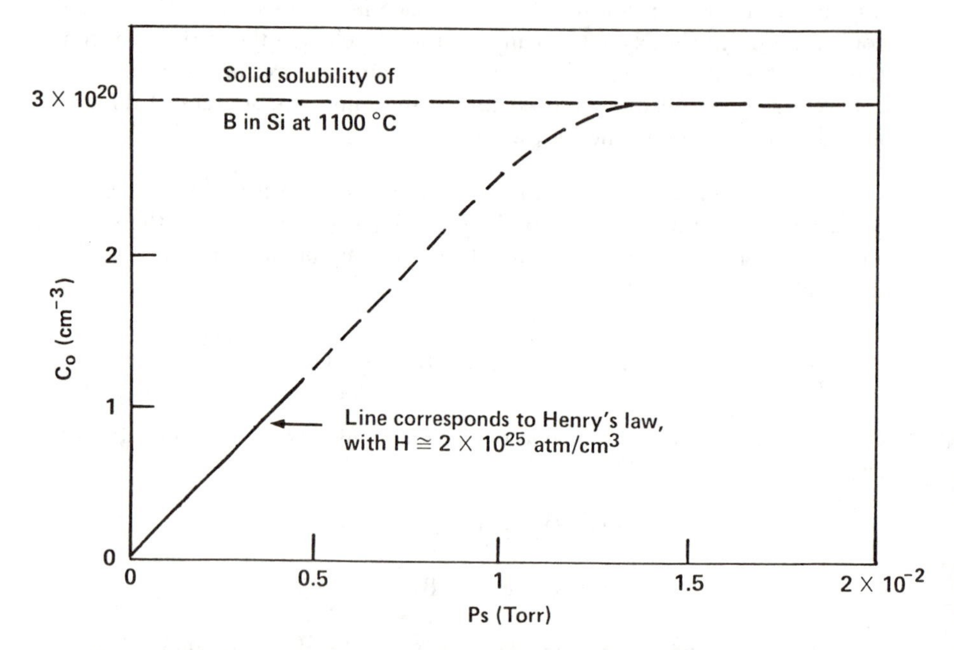

Figure 1. Surface concentration of boron as a function of the partial pressure of B_2O_3 in the ambient at 1100 °C. (Reproduced with permisssion from reference 118. Copyright 1988 Noyes Publications.)

The goal of the predeposition step is to deposit some number of atoms per square centimeter, $Q(t)$, in the silicon substrate. That number is calculated by integrating the total concentration per cubic centimeter from 0 to ∞ as shown in equation 8.

$$\begin{aligned} Q(t) &= \int_o^\infty C(x, t)dx \\ &= C_o \int_o^\infty \text{erfc}\,(x/2\sqrt{Dt})dx \\ &= C_o \left(\frac{2}{\sqrt{\pi}} \sqrt{Dt}\right) \end{aligned} \tag{8}$$

Once the predeposition is completed with Q atoms per square centimeter, the next step is to redistribute the atoms to give the desired junction depth.

If Q atoms per square centimeter are deposited on the semiconductor surface with the boundary conditions $C(x, t = 0) = Q\delta(x)$ and $C(\infty, t) = 0$ (*4*), then the distribution of impurities after diffusion for a time t is given by a Gaussian function solution to equation 1 (equation 9).

$$C(x, t) = \frac{Q}{\sqrt{\pi Dt}} \exp\left(\frac{-x^2}{4Dt}\right) \tag{9}$$

The Gaussian distribution can be used to describe the impurity profile that results from a drive-in step with no dopant gas in the furnace. The drive-in step itself is performed in several types of ambients: dry oxygen, steam, nitrogen, or argon. The drive-in temperatures range from 900 to 1200 °C.

The analytical solutions to Fick's continuity equation represent special cases for which the diffusion coefficient, D, is constant. In practice, this condition is met only when the concentration of diffusing dopants is below a certain level ($\sim 1 \times 10^{19}$ atoms/cm^3). Above this doping density, D may depend on local dopant concentration levels through electric field effects, Fermi-level effects, strain, or the presence of other dopants. For these cases, equation 1 must be integrated with a computer. The form of equation 1 is essentially the same for a wide range of nonlinear diffusion effects. Thus, the research emphasis has been on understanding the complex behavior of the diffusion coefficient, D, which can be accomplished by studying diffusion at the atomic level.

Atomic Theory of Diffusion. ***Diffusion Mechanisms.*** The atomic theory of diffusion describes how an atom moves from one part of a crystal to another. The lattice sites in a crystal are assumed to be fixed locations of the atoms making up the crystal. The atoms oscillate around these lattice sites, which are their equilibrium positions. These oscillations lead to finite chances that an atom will move from its lattice site to another position in the crystal. Atoms can move from one site in the crystal to another through the vacancy mechanism, the interstitial mechanism, and the interstitialcy mechanism (Figure 2).

On the basis of thermodynamic considerations, some of the lattice sites in the crystal are vacant, and the number of vacant lattice sites generally is a function of temperature. The movement of a lattice atom into an adjacent vacant site is called vacancy diffusion. In addition to occupying lattice sites, atoms can reside in interstitial sites, the spaces between the lattice sites. These interstitial atoms can readily move to adjacent interstitial sites without displacing the lattice atoms. This process is called interstitial diffusion. The interstitial atoms may be impurity atoms or atoms of the host lattice, but in either case, interstitial atoms are generally present only in very dilute amounts. However, these atoms can be highly mobile, and in certain cases, interstitial diffusion is the dominant diffusion mechanism.

A mechanism related to interstitial diffusion is the interstitialcy mechanism. In this process, an interstitial atom moves into a lattice site by dis-

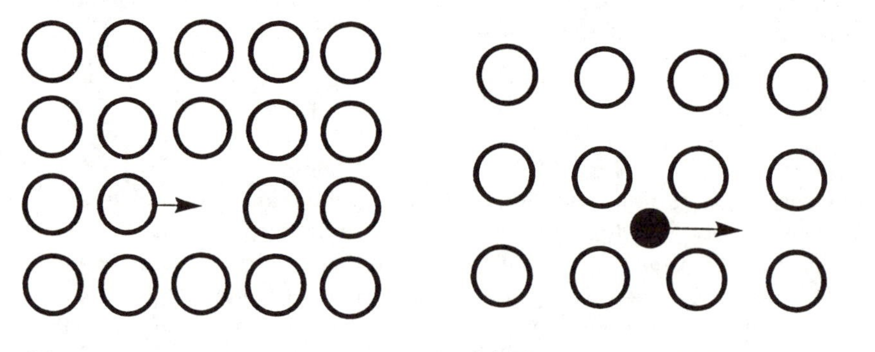

(a) The vacancy diffusion mechanism. (b) The interstitial diffusion mechanism.

(c) The interstitialcy mechanism.

Figure 2. Dominant diffusion mechanisms in silicon.

placing the atom on that site onto an adjacent interstitial site. Although several other diffusion mechanisms may exist in semiconductors, the three mechanisms just described are dominant in silicon.

The Flux Equation in Diffusivity. The number of atoms that cross a unit area in unit time is known as the flux. In one dimension, the atoms only move to the right or to the left when they change position along the x axis (Figure 3). In this simple case, the atoms are assumed to be located in planes at x_o and $x_o + a_o$, as shown in the figure. The flux J is simply the product of the concentration C and the velocity v.

$$J_x = Cv \tag{10}$$

The net flux is the difference between the flux to the right and and that to the left.

$$J_x = \frac{1}{2} v(C_{x_o} - C_{x_o+a_o}) \tag{11}$$

where C_{xo} and C_{xo+a_o} are the concentrations at x_o and $x_o + a_o$, respectively. The factor ½ in equation 11 arises from the fact that at any one plane, half

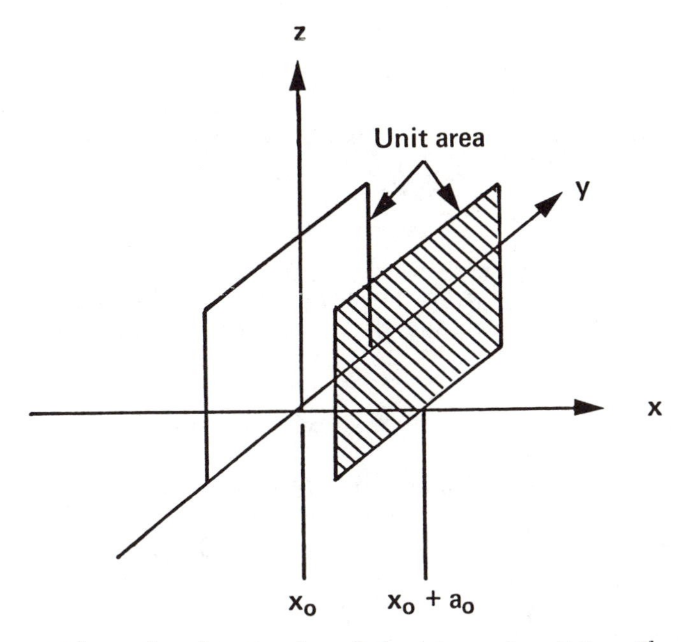

Figure 3. Flux in the x *direction through the unit area in unit time. The planes of unit area are located at* $x = x_o$ *and* $x = x_o + a_o$. *(Reproduced with permission from reference 118. Copyright 1988 Noyes Publications.)*

of the atoms move in the $+x$ direction and other other half move in the $-x$ direction. When a_o approaches 0,

$$a_o\left(\frac{C_{x_o} - C_{x_o+a_o}}{a_o}\right) = -a_o\frac{dC}{dx} \tag{12}$$

and equation 11 becomes

$$J_x = -\frac{1}{2}va_o\frac{dC}{dx} \tag{13}$$

For motion by discrete jumps between planes a_o apart, the velocity is the product of the number of jumps per second, Γ, and the distance a_o of each jump. Equation 13 may now be written as

$$J_x = -\frac{1}{2}a_o^2\Gamma\frac{dC}{dx} \tag{14}$$

and the quantity $\frac{1}{2}a_o^2\Gamma$ is called the diffusivity or diffusion coefficient D.

$$D = \frac{1}{2}a_o^2\Gamma \tag{15}$$

Equation 15 shows that for diffusion by a particular mechanism, the calculation of the diffusivity is reduced to the calculation of the jump frequency Γ. The jump frequency by the vacancy mechanism is given by

$$\Gamma = X_v w \tag{16}$$

where w is the frequency at which an atom and an adjacent neighboring vacancy exchange and X_v is the probability that the adjacent site is vacant. From statistical thermodynamics, the vacancy atom fraction is given by equation 17

$$X_v = \exp(\Delta S_f/k)\exp(-\Delta H_f/kT) \tag{17}$$

where ΔS_f is the entropy of formation of the vacancy and ΔH_f is the enthalpy of vacancy formation. These terms are related to the Gibbs free energy change for vacancy formation (ΔG_f) through equation 18.

$$\Delta G_f = \Delta H_f - T\Delta S_f = -kT\ln X_v \tag{18}$$

The frequency w is more difficult to derive from fundamental principles, but a discussion of the physics involved in evaluating w provides useful insights into the quantities that affect diffusion.

In self-diffusion by the vacancy mechanism, a lattice atom moves from a normal lattice site to a vacancy. As shown in Figure 4, the atom must move from the normal lattice site in a to the saddle point position in b to reach the vacancy at c. The energy at the saddle point is greater than that at the equilibrium lattice sites, and the atoms must be sufficiently activated in order to move to b and then to c. The fraction of the lattice atoms activated to the saddle point is related to the Gibbs free energy change between positions a and b. The atom fraction of activated atoms, X_m, is expressed by

$$X_m = \exp(\Delta S_m/k) \exp(-\Delta H_m/kT) \tag{19}$$

where ΔS_m and ΔH_m are the entropy and enthalpy of motion, respectively.

The frequency w at which an atom and an adjacent neighboring vacancy exchange can be written as

$$w = X_m \gamma \tag{20}$$

where the frequency γ is generally not known and is usually taken as the lattice vibration frequency of an atom about its equilibrium site, which is in the order of 10^{13}/s. Combining equations 16, 17, 19, and 20 yields the expression for the jump frequency for vacancy self-diffusion.

$$\Gamma = \gamma \exp[(\Delta S_f + \Delta S_m)/k] \exp[-(\Delta H_f + \Delta H_m)/kT] \tag{21}$$

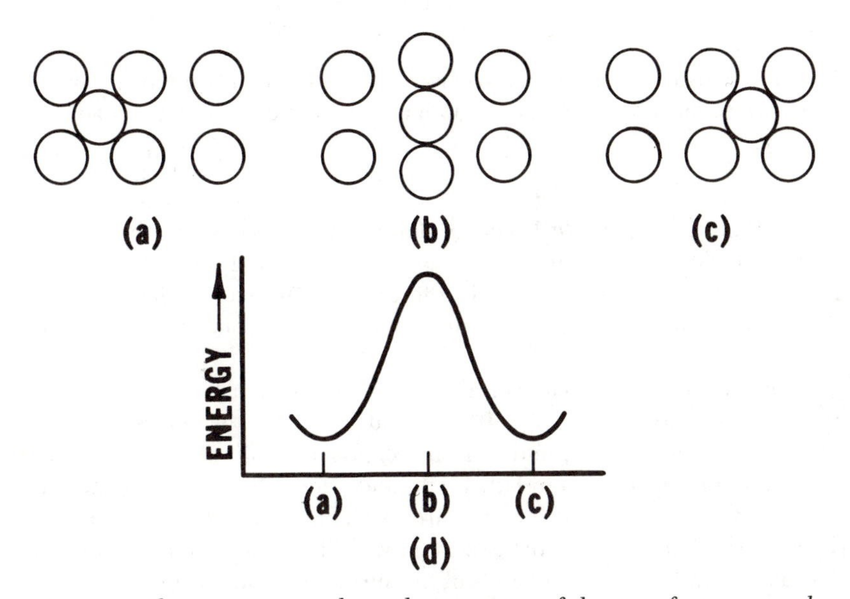

Figure 4. The sequence a–c shows the movement of the atom from a normal lattice site to an adjacent vacancy. Part d shows the variation of free energy as the atom moves from a to c. (Reproduced with permission from reference 119. Copyright 1981 Academic Press.)

Experimentally, diffusivity is given by the Arrhenius expression

$$D = D_o \exp(-E_A/kT) \tag{22}$$

where E_A is the activation energy. Thus, the diffusivity is given by

$$D = \frac{1}{2} a_o^2 \gamma \exp[(\Delta S_f + \Delta S_m)/k] \exp[-(\Delta H_f + \Delta H_m)/kT] \tag{23}$$

A comparison of equations 22 and 23 gives the prefactor D_o as

$$D_o = \frac{1}{2} a_o^2 \gamma \exp[(\Delta S_f + \Delta S_m)/k] \tag{24}$$

and

$$E_A = \Delta H_f + \Delta H_m \tag{25}$$

Therefore, diffusivity is basically the product of the lattice vibration frequency, vacancy concentration, and activated-lattice concentration (equation 26).

$$D = \frac{1}{2} a_o^2 X_v X_m \gamma \tag{26}$$

Also, the activation energy for vacancy diffusion depends upon the energy necessary to form the vacancy and to move the lattice atom into an adjacent vacancy.

Multiple-Charge-State Vacancy Model. On the basis of the previous discussion, diffusion depends upon the concentration of point defects, such as vacancies or self-interstitials, in the crystal. Therefore, diffusion coefficients can be manipulated by raising or lowering the concentration of point defects.

For the vacancy mechanism, the single vacancy in silicon is believed to exist in four charge states: V^+, V^x, V^-, and $V^=$, where + refers to a donor level, x is a neutral level, and – is an acceptor level (5, 6). The creation of a vacancy introduces a new lattice site and, thus, four new valence band states in the crystal. These states are available as acceptors but are not shallow. The lattice distortion associated with the vacancy splits states from the valence and conduction bands of the surrounding atoms by a few tenths of an electronvolt into the forbidden cap. States split from the valence band become donors, and those split from the conduction band become acceptors.

At low temperatures, one deep donor level, V^+, a few tenths of an electronvolt above the valence band edge, a single acceptor level, V^-, near midgap, and a double donor level, $V^=$, very near the condition band edge must be present (Figure 5) (*7*). The levels depicted in Figure 5 represent a best guess based on experiment (*8–10*).

Experiments show that both silicon self-diffusion and the diffusion of group III and group V impurities in silicon depend upon the position of the Fermi level. The initial assumption in the vacancy diffusion model of self-diffusion is that an observed diffusivity arises from the simultaneous movement of neutral and ionized vacancies. Each charge type vacancy has a diffusivity whose value depends upon the charge state, and the relative concentrations of vacancies depend upon the Fermi level (*11*). Calculated changes in relative concentrations of charge species versus the Fermi level, E_f, at 300 and 1400 K are shown in Figure 6 (*7*). Whereas at low temperature V^x is the dominant species in intrinsic silicon, at high temperatures, both V^+ and V^- are more numerous, and there is no value of E_f for which V^x dominates.

Another important concept is that every time an ionized vacancy is formed the crystal must return the neutral vacancy population back to equilibrium by generating an additional vacancy. This reequilibration is necessary to keep the concentration of uncharged vacancies, an intrinsic property of the crystal, unchanged. Hence, the concentration will be given only as a function of temperature through equation 17. However, the population of ionized vacancies can be controlled through the law of mass action or by

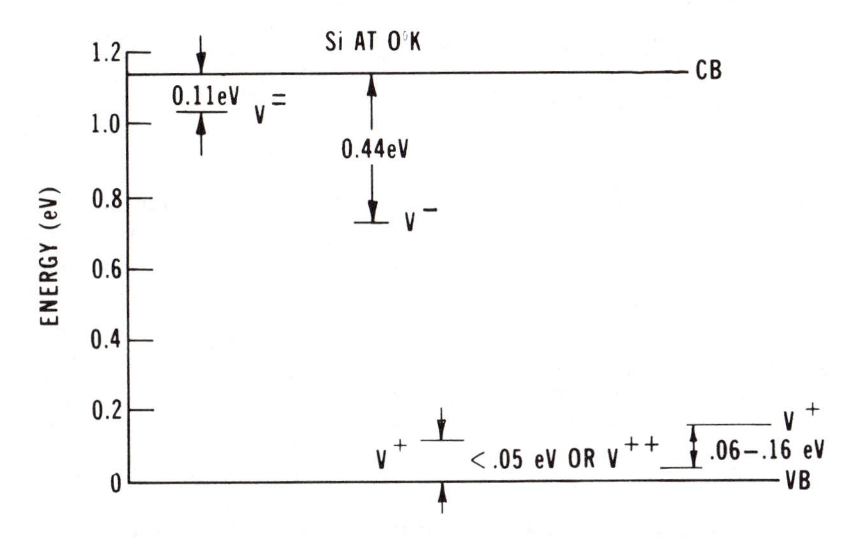

Figure 5. Estimated vacancy energy levels in the silicon band gap at 0 K. Abbreviations: CB, conduction band; VB, valence band. (Reproduced with permission from reference 119. Copyright 1981 Academic Press.)

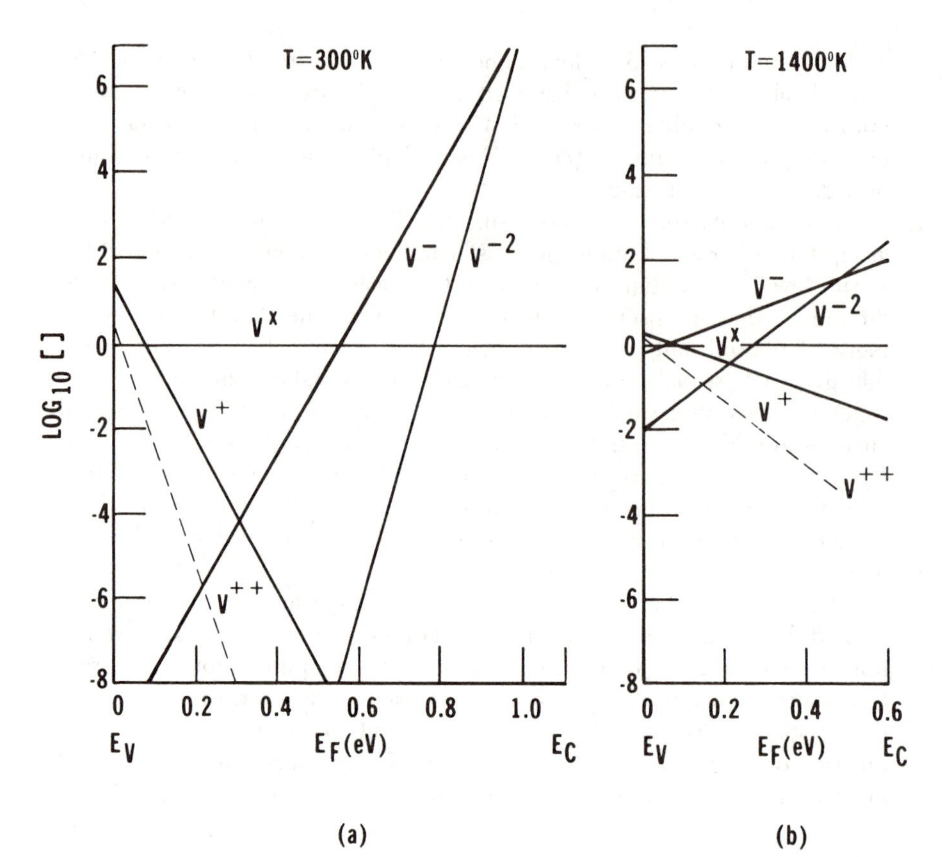

Figure 6. Calculated changes in the ratios of ionized vacancies to neutral vacancies at 300 and 1400 K. Abbreviations: E_V, *valence band energy;* E_F, *Fermi energy;* E_C, *conduction band energy. (Reproduced with permission from reference 119. Copyright 1981 Academic Press.)*

control of the Fermi level. With either method of control, as the doping becomes more n-type or more p-type, the total vacancy concentration increases as the population of ionized vacancies increases. Because impurity and self-diffusion coefficients depend upon the concentration of vacancies, the diffusion coefficients will also increase with doping. Such concentration-dependent diffusion can occur when the doping level exceeds the intrinsic electron concentration, n_i at the diffusion temperature. An illustration of concentration-dependent diffusion is shown in Figure 7.

The Role of Point Defects in Silicon Processing. ***The Balancing Act in Silicon Processing.*** Both silicon oxidation and the diffusion of impurities occur at high temperatures and involve point defects such as va-

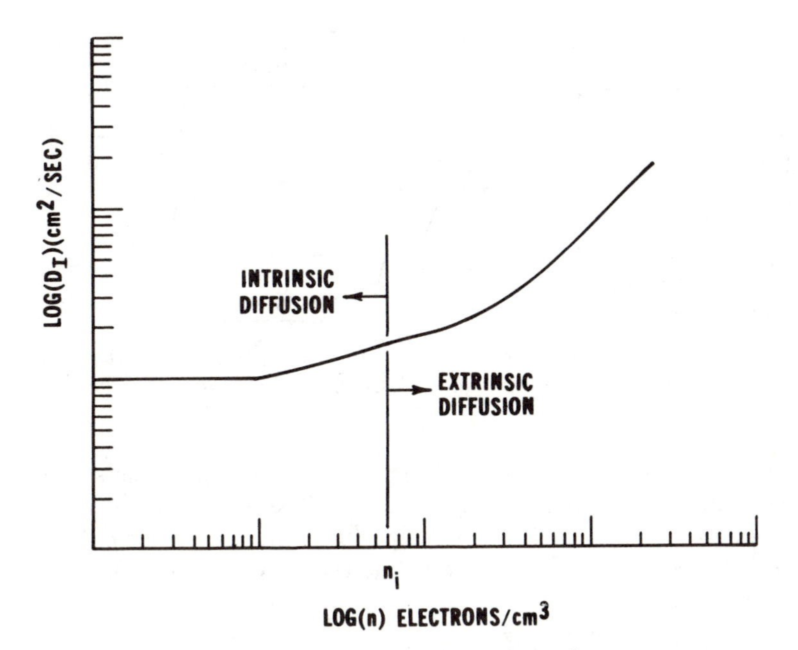

Figure 7. Donor impurity diffusion coefficient (D_I) *vs. electron concentration (electrons per cm*^{3}*) showing regions of intrinsic and extrinsic diffusion.* (*Reproduced with permisssion from reference 119. Copyright 1981 Academic Press*).

cancies or self-interstitials. The first level of process design involves the concept of doping and junction formation, threshold voltage control, or the gain control of a transistor. Another goal of doping is low sheet resistance.

The primary goal of oxidation is the controlled growth of SiO_2 layers. A critical concern in oxidation is the growth of stable oxides with electrical integrity. To create a nonplanar structure, it is necessary to consider the viscous flow characteristics of the oxide and whether the viscosity is low enough to release stress. In general, the process engineer spends a lot of time dealing with these first-order requirements, but the rest of the time is spent in trying to balance factors that are generally not well understood.

The diagram of the point defect balancing act is shown in Figure 8. The arrows in the figure indicate the directions of interactions. For example, diffusion may change the concentration of point defects, and point defects themselves can affect diffusion. Oxidation produces defects, and point defects can affect oxidation. The balancing act involves the generation of point defects and the effect of this generation on these major processes. Diffusion may introduce strain into the lattice that can affect surface quality. As these processes produce point defects, extended structural defects may grow in the silicon.

Point defects can also influence the precipitation of oxygen. Oxygen is

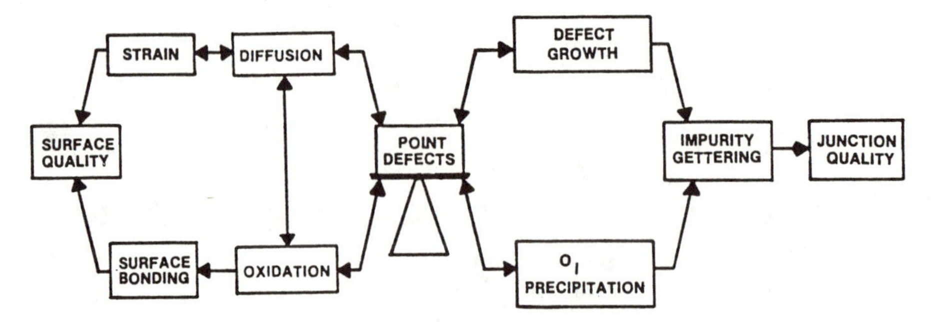

Figure 8. The point defect balancing act in silicon processing. (Reproduced with permission from reference 118. Copyright 1988 Noyes Publications.)

incorporated into the crystal during crystal growth, and during subsequent heat treatments, this oxygen may precipitate. These precipitates create good gettering sites for attracting metal impurities and, thus, remove them from active device regions. Such internal gettering would have an impact on junction quality.

Point Defects. Point defects are defined as atomic defects. Atomic defects such as metal ions can diffuse through the lattice without involving themselves with lattice atoms or vacancies (Figure 9), in contrast to atomic defects such as self-interstitials. The silicon self-interstitial is a silicon atom that is bonded in a tetrahedral interstitial site. Examples of point defects are shown in Figure 9.

One of the major controversies in solid-state science is the nature of the dominant native point defect in silicon. Is the dominant native point defect in silicon the monovacancy or the silicon self-interstitial? Well-developed arguments have been proposed for each type, but the current consensus is that both types are present and important.

Monovacancy. Statistical thermodynamics requires that if a vacancy is formed by removing an atom from the crystal and depositing it on the surface, then the free energy of the crystal must decrease as the number of created vacancies increases until a minimum in this free energy is reached. Because a minimum in the free energy exists for a certain vacancy concentration in the crystal, the vacancy is a stable point defect. The following facts about vacancies have been obtained experimentally: (*12*).

1. Electron paramagnetic resonance measurements only identify vacancies or vacancy complexes in Si irradiated by electrons. The absence of Si self-interstitials has been ascribed to rapid athermal migration even at 2 K (*13*).

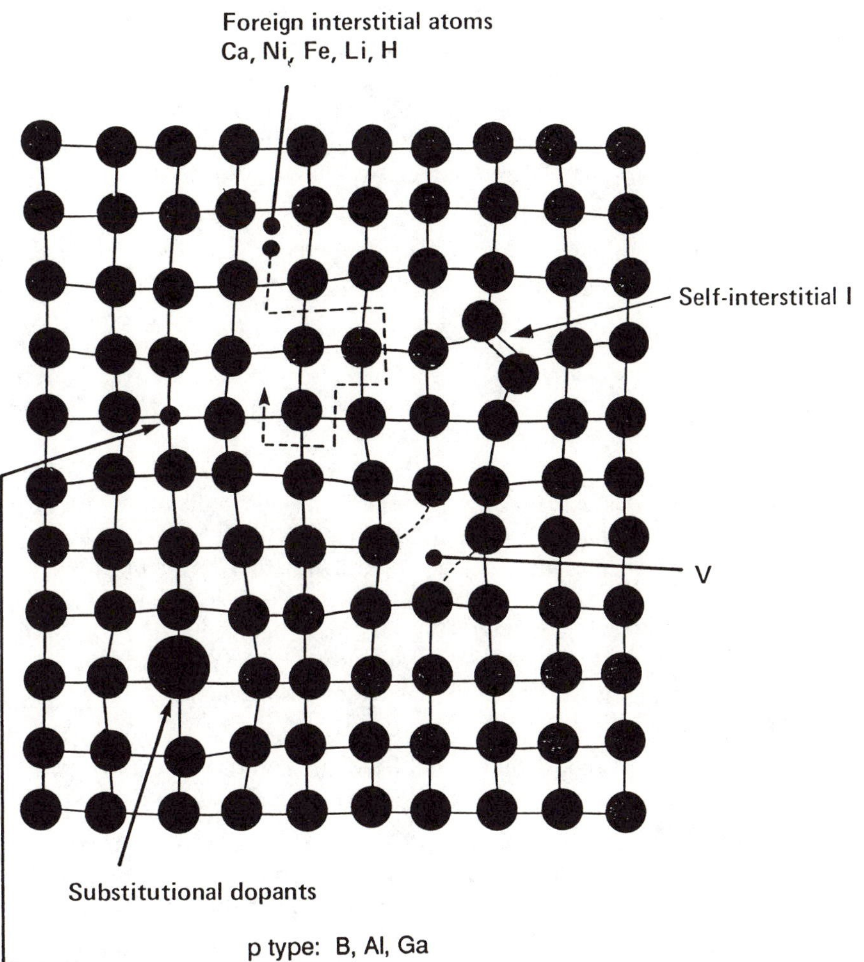

Figure 9. Examples of point defects in the silicon lattice. (Reproduced with permission from reference 118. Copyright 1988 Noyes Publications.)

2. Diffusion phenomena, as well as calculations of diffusion entropy and enthalpy, have been successfully explained by ascribing multiple ionization levels to vacancies that are the same as those observed for the vacancy in low-temperature irradiation experiments (*14–16*). Vacancies and self-interstitials are interchangeable as far as diffusion is concerned, provided that both have charge states.

3. Theoretical estimates of the heats and entropies of formation of vacancies correspond well with those of the native defects observed in diffusion and quenching experiments (*15, 17–19*).

4. Channeling studies of impurity–defect interactions in Si show that under helium ion bombardment the trapping efficiency of impurities for radiation-produced defects is very low, near 30 K (*19*). Vacancies are not mobile in Si below this temperature, whereas interstitials still are. This observation implies that the impurity–defect interactions involve vacancies.
5. Positron annihilation lifetime measurements have been performed on float-zone Si at high temperatures and show that vacancy-like defects are formed (*20*).

The Silicon Self-Interstitial Atom. A similar consistent statistical thermodynamic analysis of the existence of self-interstitials shows that silicon self-interstitials are stable point defects. The following arguments further support the silicon self-interstitial.

1. The majority of dislocation loops and stacking faults observed by transmission electron microscopy of Si are judged to be of extrinsic or interstitial character. Although there are four proposed mechanisms by which extrinsic-type dislocations may be formed without any self-interstitials being present (*12*), most workers believe that self-interstitial precipitation is the dominant mechanism in extrinsic-type dislocations.
2. The picture of self-interstitials in Si developed by Seeger and Frank (*21*) is consistent with observations indicating self-interstitial migration at low and high temperatures.
3. Evidence for the liquid-drop character of B-swirl defects in Si comes from the observation that upon melting, droplets of liquid Si are formed in the interior of the solid phase (*22*).
4. In n–p structures formed by sequential diffusions of B and P, dislocation climb occurred at the same time that the emitter-push effect was seen in the B layer (*23*). This result implies that the same point defect is responsible for both phenomena.
5. Stacking-fault growth below P diffusion and the enhanced diffusion of the buried layer occur simultaneously (*24*).
6. Calculations of total energy show that self-interstitials form and migrate in Si with a total activation energy roughly the same as that of self-diffusion (*25*).

After the balance sheet of pros and cons surrounding the question of the native defect in Si is reviewed, the question remains. What is the native defect responsible for the diffusion of impurity and growth of defects in Si? So far we only have clues.

The current majority opinion is that both types of point defects are important. Thermal equilibrium concentrations of point defects at the melting point are orders of magnitude lower in Si than in metals. Therefore, a direct determination of their nature by Simmons–Balluffi-type experiments (*26*) has not been possible. The accuracy of calculated enthalpies of formation and migration is within ±1 eV, and the calculations do not help in distinguishing between the dominance of vacancies or interstitials in diffusion. The interpretation of low-temperature experiments on the migration of irradiation-induced point defects is complicated by the occurrence of radiation-induced migration of self-interstitials (*27, 28*).

In addition, the structure and properties of point defects at low temperatures and at high temperatures may be different (*29*). The observation of extrinsic-type dislocation loops in dislocation-free, float-zone Si indicate that self-interstitials must have been present in appreciable concentrations at high temperature during or after crystal growth (*30, 31*). However, it is unclear whether these self-interstitials were present at thermal equilibrium or were introduced during crystal growth by nonequilibrium processes.

In view of the uncertainties regarding the native point defect in Si, it is necessary in discussions of self-diffusion and dopant diffusion to take account of both types of defects.

Point Defect Models of Diffusion in Silicon. Under conditions of thermal equilibrium, a Si crystal contains a certain equilibrium concentration of vacancies, C_V^o, and a certain equilibrium concentration of Si self-interstitials, C_I^o. For diffusion models based on the vacancy, $C_V^o >> C_I^o$ and the coefficients of dopant diffusion and self-diffusion can be described by equation 27 (*15*)

$$D_i = D_i^x + D_i^- + D_i^= + D_i^+ \tag{27}$$

where D_i is the measured diffusivity and D_i^x, D_i^-, $D_i^=$, and D_i^+ are the intrinsic diffusivities of the species through interactions with vacancies in the neutral, single-acceptor, double-acceptor, and donor charge states, respectively. These individual contributions to the total measured diffusivity were described in a previous section.

For diffusion models based on self-interstitials, $C_I^o >> C_V^o$. Dopant diffusion and self-diffusion are assumed to occur via an interstitialcy mechanism (*32*). Mobile complexes consisting of self-interstitials in various charge states and impurities are assumed to exist.

In principle, both vacancies and self-interstitials may occur simultaneously and somewhat independently. Indeed, any relationship between C_V^o and C_I^o may be dominated by the Si surface, which can act as a source or sink for either species. If a local dynamical equilibrium exists between recombination and spontaneous bulk generation, vacancies (V) and self-interstitials (I) would react as follows

$$V + I \leftrightarrows O \tag{28}$$

where O denotes the undisturbed lattice. The law of mass action at equilibrium for this reaction is given by

$$C_I C_V = C_I^o C_V^o \tag{29}$$

At sufficiently long times and high temperatures, equation 29 is fulfilled (*33*, *34*). However, a substantial amount of time may be required to reach dynamical equilibrium (*34*, *35*). This observation suggests that vacancy–self-interstitial recombination is an activated process. In addition, under conditions in which point defects are injected, equation 29 may not be valid.

If both types of point defects are important, then diffusion processes may involve both types, and the following relation applies

$$D_i = D_i^I + D_i^V \tag{30}$$

where D_i^I is the interstitialcy contribution and D_i^V is the vacancy contribution to the total measured diffusivity, D_i. Vacancies and self-interstitials can cooperate in effecting the diffusion of impurity by the Watkins replacement mechanism (*36*) shown in Figure 10. Interstitial dopant impurities can be created by the exchange between a self-interstitial and a substitutional dopant atom. The newly created interstitial impurity will migrate until it finds a vacancy, and then it is free to diffuse again as a substitutional impurity.

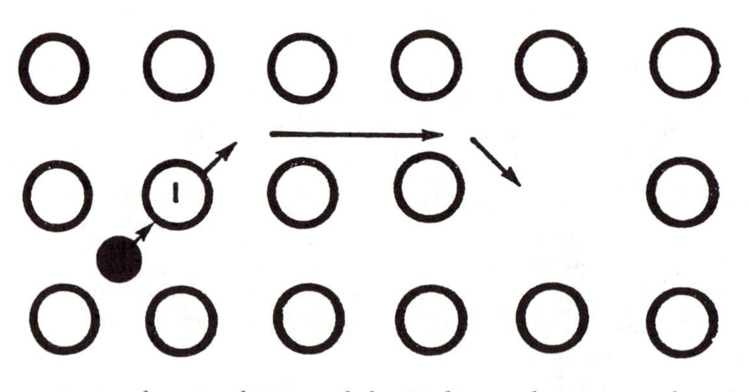

Figure 10. A schematic diagram of the Watkins replacement mechanism. (Reproduced with permission from reference 118. Copyright 1988 Noyes Publications.)

Vacancies and self-interstitials can exist in equilibrium with each other in the silicon lattice. The concentration of each species can be described by equilibrium equations of the following type.

$$C_V{}^o = \exp\,(S_f{}^V/k)\,\exp\,(-\Delta H_f{}^V/kT) \tag{31a}$$

$$C_I{}^o = \exp\,(S_f{}^I/k)\,\exp\,(-\Delta H_f{}^I/kT) \tag{31b}$$

For silicon self-diffusion, the total diffusion coefficient could be expressed as

$$D_{\mathrm{Si}} = f_V D_V C_V{}^o + f_I D_I C_I{}^o \tag{32}$$

where f_V and f_I are the fractional contributions of vacancies and self-interstitials, respectively, to self-diffusion. A substantial debate exists as to the values of these fractional coefficients (Figure 11). The concept that impurity diffusion is dominated by vacancies only was held until 1968, when Seeger and Chick (37) proposed that both self-interstitials and vacancies can contribute to diffusion in silicon. However, the concept of vacancies and interstitials coexisting in silicon leads to several unresolved questions. Is there

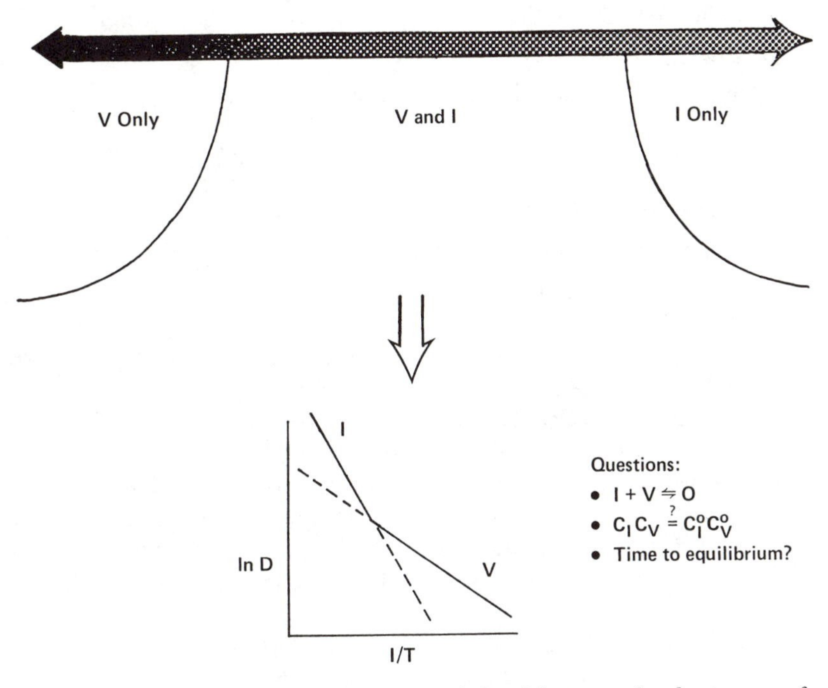

Figure 11. A diagram of the spectrum of the debate on the dominance of vacancies vs. self-interstitials.

dynamic equilibrium between self-interstitials and vacancies? What is the time to establish this dynamic equilibrium?

Experimental Observations. Various experiments yield numerous indirect results that can help determine whether vacancies or self-interstitials are involved in diffusion. The following is a partial list of such experiments.

- Oxidation studies
  - —enhanced and retarded diffusion
  - —backside oxidation
  - —role of Si_3N_4 surface films and nitridation
  - —dependence of oxidation-enhanced and oxidation-retarded diffusion on doping
  - —effect of chlorine in the oxidizing ambient
  - —effects of doping on oxidation
- Determination of the effect of doping on oxidation stacking fault shrinkage or growth
- Determination of the effect of diffusion on O precipitation
- Codiffusion studies
- Transmission electron microscopic studies of precipitates and defects
- Determination of the relation between total doping and electrically active doping
- Dopant profile measurements
- Determination of the role of stress on diffusion

For example, during oxidation, enhanced diffusion of phosphorus, boron, and arsenic are observed, as well as retarded diffusion of antimony. However, if direct nitridization of the silicon surface occurs, then the inverse effects are observed, that is, enhanced antimony diffusion and retarded phosphorus diffusion. Also, oxidation-enhanced diffusion is significantly affected by doping. As either p- or n-type doping concentration increases above n_i, oxidation-enhanced diffusion diminishes. If chlorine is introduced into the oxidizing ambient, oxidation-enhanced diffusion is likewise diminished.

Not only is enhanced diffusion of impurities observed during oxidation, but in addition, stacking faults can grow. A stacking fault is a plane of dislocated material that may intersect the silicon surface but that also has a bounding partial dislocation. These faults grow when sufficient numbers of self-interstitials are generated such that the concentration of self-interstitials in the lattice is higher than that in the bounding partial-dislocation core

(Figure 12). Because oxidation is a process that generates excess self-interstitials, stacking faults will grow during oxidation.

Other experiments that have been performed include irradiating uniformly doped silicon wafers with protons and observing the diffusion of the dopant after irradiation. Additional discussion of these effects will follow.

Diffusion in the Presence of Excess Point Defects. ***Oxidation-Enhanced Diffusion.*** Oxidation generally enhances the diffusion of group III and group V elements except for antimony (Figure 13). Oxidation-enhanced diffusion is generally observed by depositing a silicon nitride mask on the silicon surface that will prohibit oxidation in the regions that it covers.

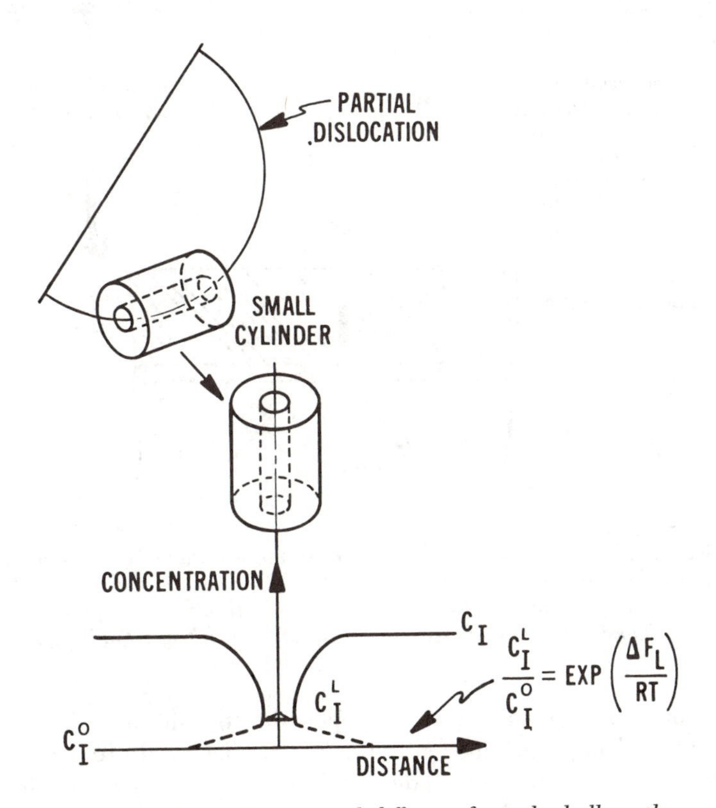

Figure 12. A model of self-interstitial diffusion from the bulk to the partial dislocation bounding a stacking fault. Under nonoxidizing conditions, the concentration of self-interstitials at the fault line, C_I^L, is greater than the equilibrium bulk interstitial concentration, C_I^o. Under oxidizing conditions, C_I is greater than C_I^L until the retrogrowth temperature is reached. (Reproduced with permission from reference 45. Copyright 1981 The Electrochemical Society, Inc.)

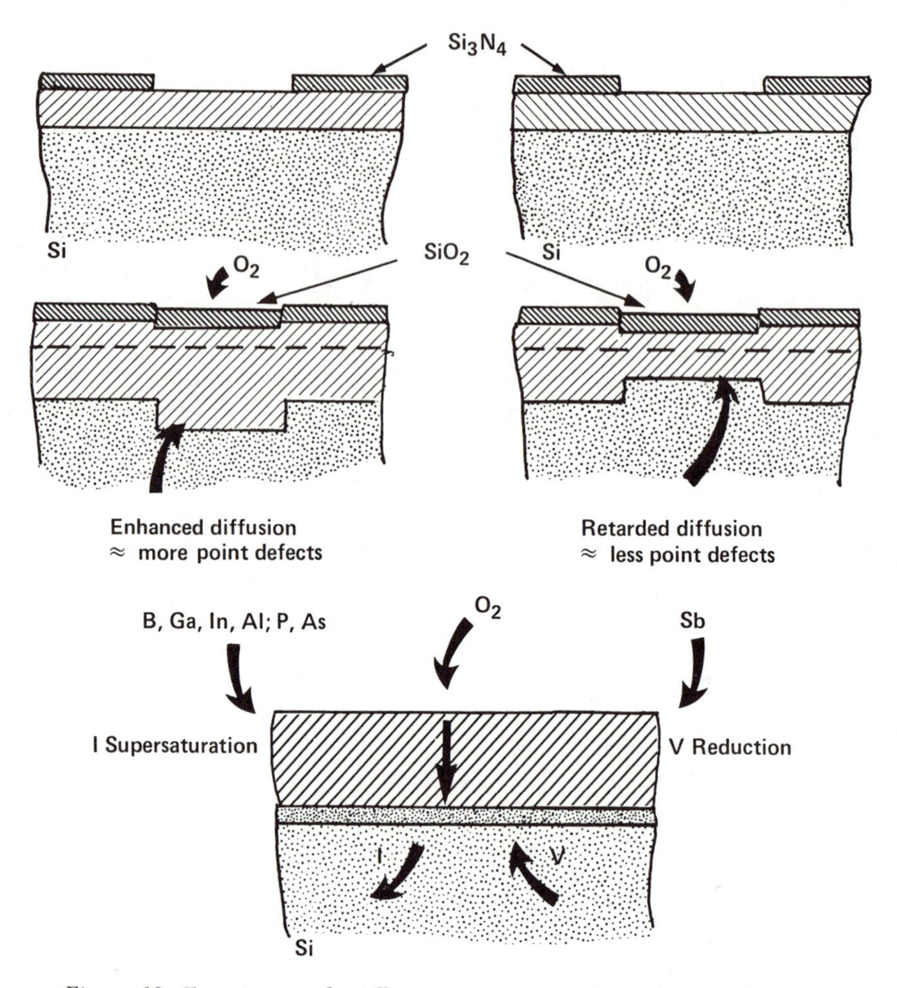

Figure 13. Experiments that illustrate oxidation-enhanced or oxidation-retarded diffusion of dopants in silicon. The supersaturation of self-interstitials associated with the oxidation process drives both effects. (Reproduced with permission from reference 118. Copyright 1984 Noyes Publications.)

Oxidation is performed in a window opened to the silicon surface, and the differential changes in junction depth can be observed. To explain these results, Hu (38) proposed a model with the following essential points:

1. Oxidation of Si at the Si–SiO_2 interface is usually incomplete to the extent that approximately 1 Si atom in 1000 is unreacted.
2. The unreacted Si, severed from the lattice by the advancing Si–SiO_2 interface, becomes mobile. These atoms can enter

the Si lattice interstices and cause a flux of self-interstitials away from the interface.

3. Growth of oxidation-induced stacking faults proceeds by the absorption of the generated self-interstitials. Oxidation-enhanced diffusion can occur as a result of the presence of the excess interstitials via the Watkins (*36*) replacement mechanism or by an interstitialcy process.

If the Watkins replacement mechanism is ignored, the diffusivity (D) of an impurity atom under conditions of nonequilibrium point defect concentrations is given by

$$D = D_i^I(C_I/C_I^o) + D_i^V(C_V/C_V^o) \tag{33}$$

where C_I and C_V are the excess self-interstitial and vacancy concentrations, respectively. The fractional interstitialcy factor is defined as (*34*)

$$f_i = D_i^I/D_i^V \tag{34}$$

and we can write

$$D_i = f_i(C_I/C_I^o) + (1 - f_i)(C_V/C_V^o) \tag{35}$$

Calculations of the fractional interstitialcy components for B, P, As, and Sb are shown in Table I (*33, 39–42*). A significant spread in the values of f_i is obtained. The value of f_i has been correlated with the amount of energy required to make a substitutional dopant atom become interstitial. Energies of interstitial formation in Si are shown in Table II. The larger the energy

Table I. Fractional Interstitialcy Components of Diffusion via Self-Interstitials in Silicon at 1000–1100 °C

Element	*Fair* (39)	*Antoniadis* (34)	*Matsumoto* (41)	*Gosele* (40)	*Mathiot* (42)
B	0.17	0.32	0.41	0.8–1.0	0.18
Al	0.2			0.6–0.7	
P	0.12	0.40	0.35–0.5	0.5–1.0	0.19
As	0.09	0.43	0.45–0.75	0.2–0.5	0.16
Sb	0.13	0.015		0.02	

Table II. Estimated Interstitial Formation Energies in Silicon

Element	*Interstitial Formation Energy (eV)*
Si	2.2
Al^{2+}	2.21
B	2.26
P	2.4
As	2.5

of interstitial formation, the smaller is the fractional interstitialcy component of diffusion.

The diffusion of Sb is retarded during oxidation of the Si surface (*33*). This retardation can be explained by assuming that Sb diffuses predominantly by a vacancy mechanism and that the self-interstitials generated at the oxidizing surface combine with vacancies to reduce their concentration.

Two recent experiments have yielded considerable support for an interstitialcy mechanism for P diffusion. Fahey et al. (*43*) observed that direct nitridation of Si produces a supersaturation of vacancies such that P diffusion is substantially reduced in the underlying Si. On the other hand, when an oxide is grown over the doped Si and then oxynitridation is performed at the SiO_2 surface, excess self-interstitials are generated and P diffusion is greatly enhanced. These results are shown in Figure 14 and are compared with those for P diffusion in a neutral ambient. Similar results were obtained with B. The average enhanced or retarded diffusivities of P, As, and Sb after direct nitridation are shown in Figure 15. From these data the authors concluded that the fractional interstitialcy component of P diffusion is 70–100%.

Dependence of Oxidation-Enhanced Diffusion on Doping. Taniguchi (*44*) found that oxidation-enhanced diffusion decreases as the concentration

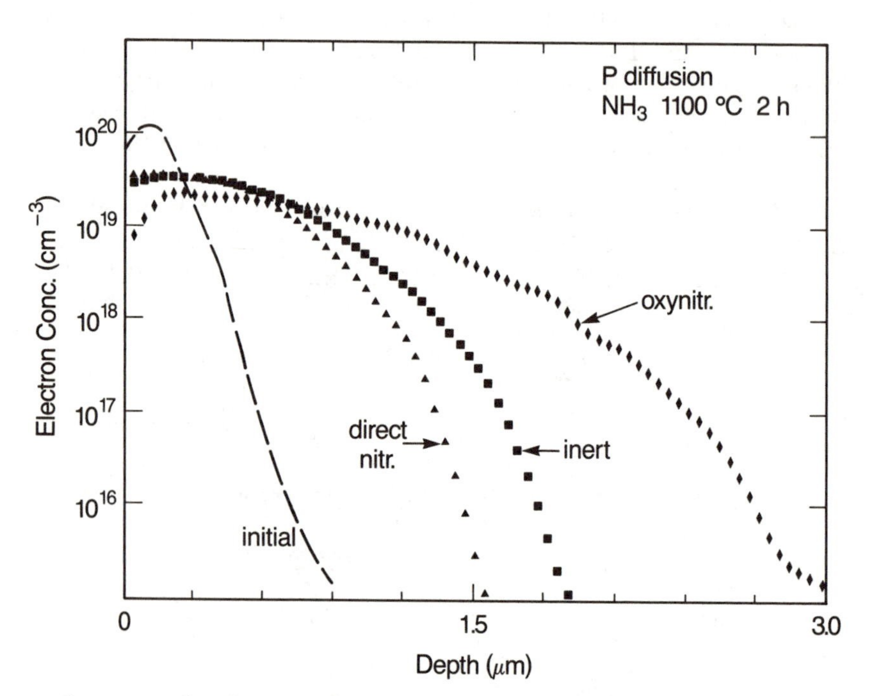

Figure 14. Phosphorus profiles after direct nitridation, oxynitride formation, and inert ambient diffusion at 1100 °C.

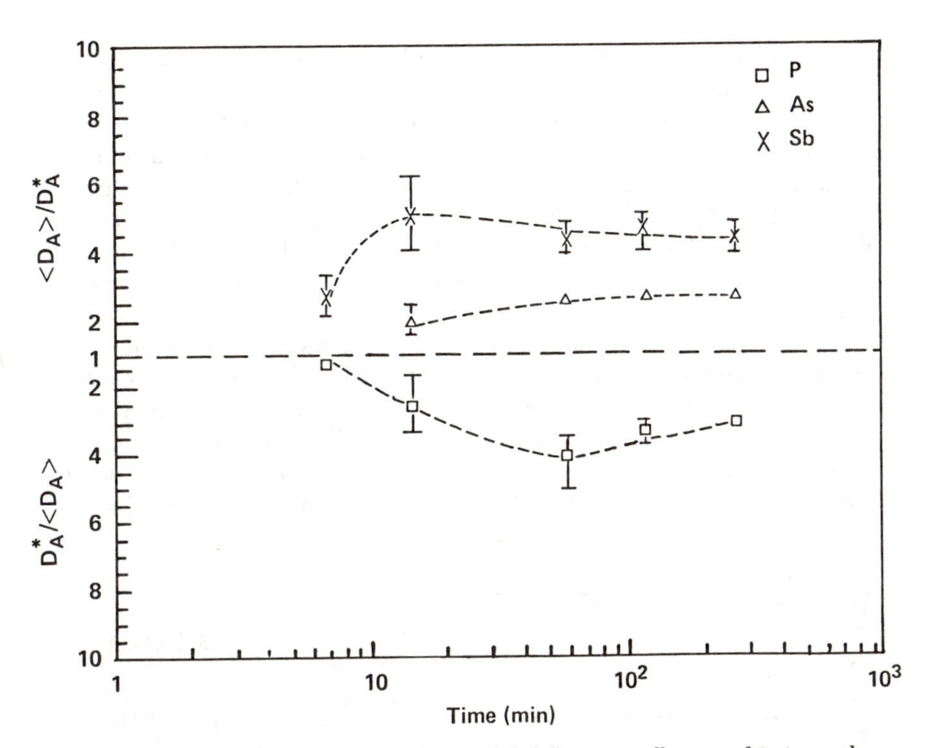

Figure 15. Average enhanced- and retarded-diffusion coefficients of P, As, and Sb following direct nitridation experiments. Data are from Fahey (43). Abbreviations: <D_A>*, time-averaged diffusion coefficient;* D_A**, intrinsic diffusion coefficient. (Reproduced with permission from reference 118. Copyright 1984 Noyes Publications.)*

of the diffusing impurity increases beyond the point where concentration-dependent diffusion occurs. This effect was explained in terms of the reduction of oxidation-produced self-interstitials by recombination with the increasing supply of vacancies. Fair (45) assumed that the equilibrium vacancy concentration is unaffected initially by the self-interstitials generated at the oxidizing surface. However, the quasi steady-state value of interstitial supersaturation is inversely proportional to the vacancy concentration, which increases with doping above n_i. The oxidation-enhanced dopant diffusivity, D_e, is then

$$\begin{aligned} D_e &= D_{SI} + \Delta D_o \\ &= D_i(C_V/C_V^o) + D_i f_i (C_I/C_I^o)_i (C_V^o/C_V) \end{aligned} \quad (36)$$

where $(C_I/C_I^o)_i$ is the self-interstitial supersaturation under intrinsic doping conditions and C_V/C_V^o is the vacancy enhancement when doping exceeds n_i.

Equation 36 is divided into the contributions to the diffusion of substitutional impurity under nonoxidizing conditions, D_{SI}, and the enhanced contribution due to oxidation, ΔD_o. Figure 16 shows the data of Taniguchi et al. (*44*) for oxidation-enhanced diffusion of P and B versus the total number of dopant impurities per square centimeter, Q_T. The calculated values of D_{SI} and ΔD_o are shown in comparison with the experimental data. Reasonable agreement is obtained. Thus, Taniguchi's model of self-interstitial recombination with vacancies is consistent with the models of high-concentration diffusion of B and P used by Fair in his calculations.

Additional work is needed to refine these models. For example, rapid vacancy generation and vacancy–interstitial recombination are assumed. These effects combine to modulate the self-interstitial supersaturation. Thus, the supply of self-interstitials at the oxidizing interface cannot keep up with recombination effects. Rapid recombination may be justified at high doping levels, because species such as V^+ and I^- or V^- and I^+ may be plentiful.

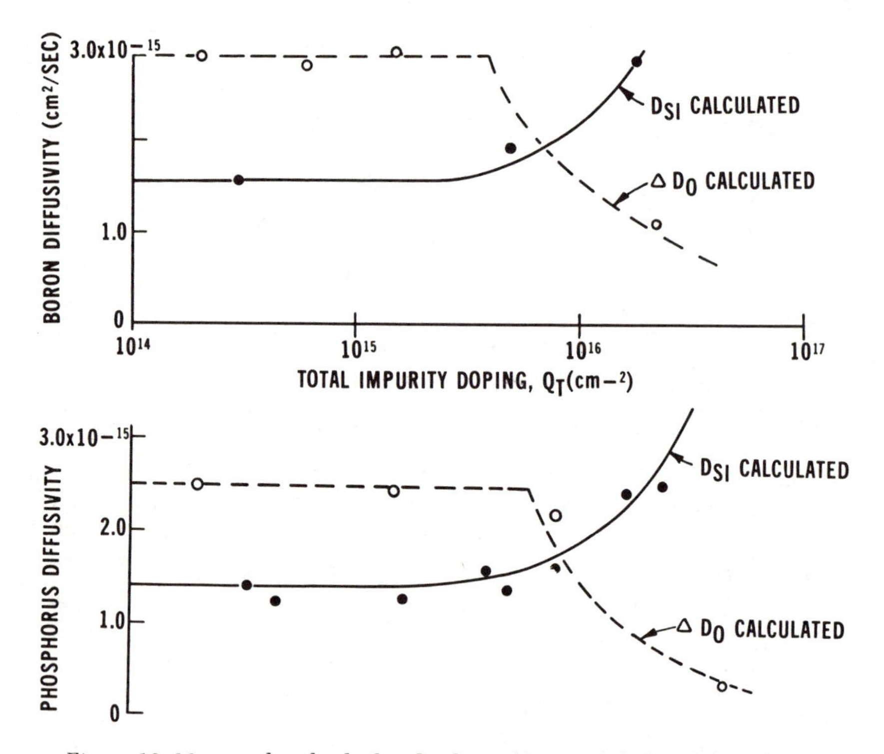

Figure 16. Measured and calculated values of boron and phosphorus diffusivities as a function of total impurity doping. Data are divided into contributions to substitutional impurity diffusion under nonoxidizing conditions, D_{SI} *, and the enhanced contribution due to oxidation,* ΔD_o*. Data are from Taniguchi et al.* (44)*. (Reproduced with permission from reference 45. Copyright 1981 The Electrochemical Society, Inc.)*

The cross-section for charged-defect recombination will likely be larger than that for neutral-defect recombination. In addition, dopant ions will likely provide additional recombination sites when the dopant is paired with a point defect.

Effect of Chlorine on Oxidation-Enhanced Diffusion. If chlorine is added to oxygen in the furnace in sufficient concentrations such that stacking-fault retrogrowth occurs (*46*), then oxidation-enhanced diffusion becomes negligible (*47*). This result is believed to be due to the generation of vacancies at this Si–SiO_2 interface when Cl reacts with Si atoms on lattice sites to produce SiCl by the following reaction

$$\mathrm{Si} + \frac{1}{2}\,\mathrm{Cl_2} \rightarrow \mathrm{SiCl} + V \tag{37}$$

The vacancy generated is then available to recombine with a Si self-interstitial produced by oxidation:

$$I + V \rightarrow O \tag{38}$$

As a result, the supersaturation of self-interstitials in the silicon surface and the bulk is reduced or eliminated, thereby inhibiting stacking-fault growth and enhanced diffusion (Figure 17).

Point Defect Generation During Phosphorus Diffusion. *At Concentrations above the Solid Solubility Limit.* The mechanism for the diffusion of phosphorus in silicon is still a subject of interest. Hu et al. (*46*) reviewed the models of phosphorus diffusion in silicon and proposed a dual vacancy–interstitialcy mechanism. This mechanism was previously applied by Hu (*38*) to explain oxidation-enhanced diffusion. Harris and Antoniadis (*47*) studied silicon self-interstitial supersaturation during phosphorus diffusion and observed an enhanced diffusion of the arsenic buried layer under the phosphorus diffusion layer and a retarded diffusion of the antimony buried layer. From these results they concluded that during the diffusion of predeposited phosphorus, the concentration of silicon self-interstitials was enhanced and the vacancy concentration was reduced. They ruled out the possibility that the increase in the concentration of silicon self-interstitials was due to the oxidation of silicon, which was concurrent with the phosphorus predeposition process.

More recently, Tsai et al. (*24*) studied the diffusion of buried As and Sb layers in response to surface diffusion of P. Float-zone Si wafers (<100> orientation, p-type, and 100 Ω-cm) were used in this study.

As^+ ions were implanted at 100 keV to a dose of $1 \times 10^{14}/\mathrm{cm}^2$ to half of the wafers, and Sb^+ ions were implanted at 150 keV to a dose of 5 ×

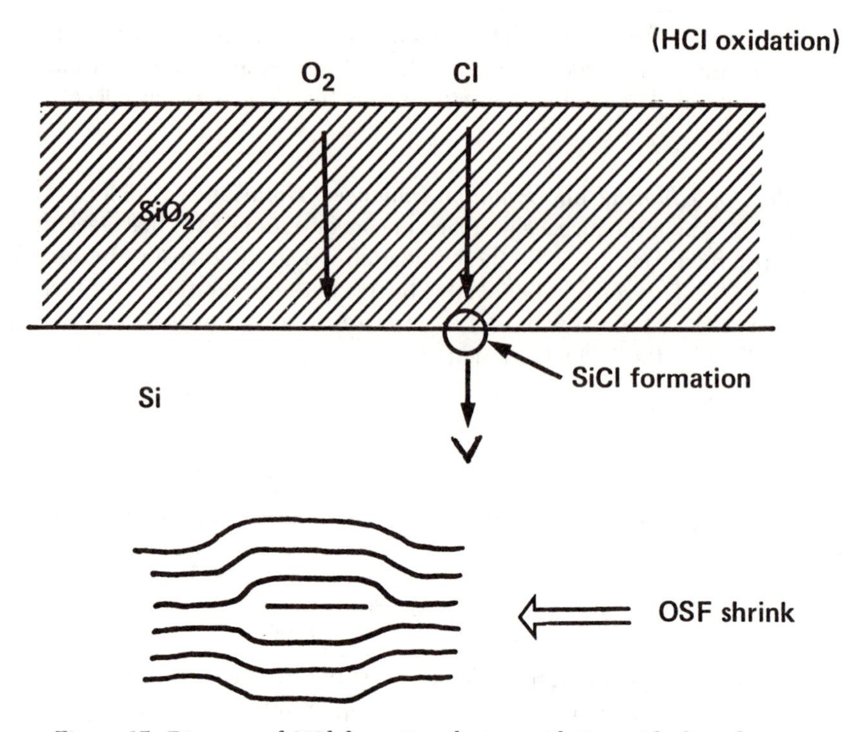

Figure 17. Diagram of SiCl formation during oxidation with the subsequent injection of vacancies. The injection of vacancies reduces the concentration of self-interstitials in the bulk and causes oxidation stacking faults (OSF) to shrink. (Reproduced with permission from reference 118. Copyright 1988 Noyes Publications.)

$10^{13}/cm^2$ to the other half of the wafers. After being cleaned, the wafers were annealed at 900 °C for 30 min in a nitrogen ambient to activate the implanted ions and to anneal out the major implant damage. This annealing condition was similar to that used by Harris et al. (*47*). P-type epitaxial films (100 Ω-cm) were grown on these wafers to a thickness close to 10 μm by using a standard atmospheric reactor. About 9000 to 10,000 Å of oxide was deposited on most of the wafers to form the diffusion mask against phosphorus. The oxide films were deposited in a LPCVD (low-pressure chemical vapor deposition) system by using $SiCl_2H_2$ and oxygen at 900 °C. These wafers were then made dense in oxygen at 800 °C for 30 min. Oxide windows were formed, and P was diffused from a $POCl_3$ source. All of these chemical source diffusions were performed at surface concentrations above the solid solubility limit. Ion implantation created doped layers with P concentrations below solid solubility. Samples were analyzed by using TEM (XTEM) micrographs, spreading resistance, defect etching, and SIMS.

A typical result is shown in Figure 18, with micrographs showing variations of stacking faults inside a narrow P-diffused region and the enhance-

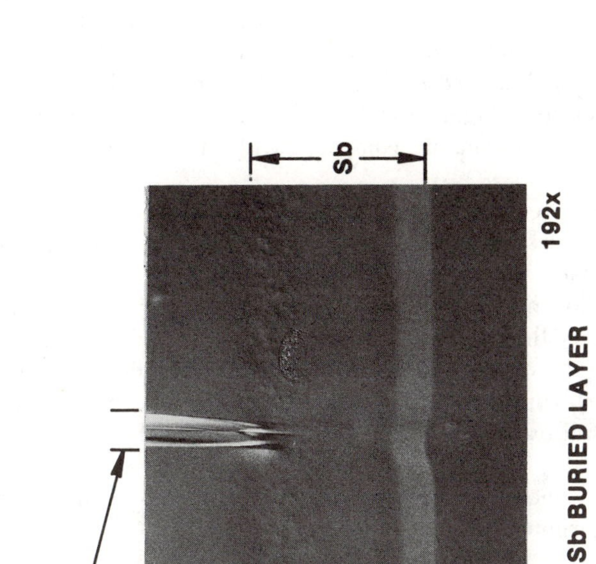

Figure 18. Micrographs showing variation of stacking faults inside a narrow phosphorus diffusion region and the slight enhancement of arsenic diffusion and retardation of antimony diffusion. P diffusion was at 1150 °C for 60 min. (Reproduced with permission from reference 24. Copyright 1987 The Electrochemical Society, Inc.)

ment of As diffusion and retardation of Sb diffusion. Measurements of enhanced As diffusion and retarded Sb diffusion were made over a wide temperature range with a fixed surface concentration of P. These diffusion coefficients are plotted in Figure 19 where D_e is the As diffusivity and D_r is the Sb diffusivity. These data are compared with intrinsic diffusion coefficients.

The data in Figure 19 suggest that high-concentration P diffusion with surface doping in excess of solid solubility produces silicon self-interstitials that diffuse to the silicon surface and into the bulk crystal. This conclusion is based on the following observations: (1) Stacking-fault growth were found in the regions under the window in which P diffuses. (Stacking faults grow by absorbing self-interstitials). (2) The diffusion of As was enhanced, whereas the diffusion of the Sb buried layers was retarded, results which are consistent with observations of oxidation-enhanced diffusion in which self-interstitials are produced.

The activation energy for enhanced diffusion due to an interstitialcy mechanism is determined from equation 39

$$D_e(I) = D_i(I)G_I \tag{39}$$

where G_I is a self-interstitial supersaturation factor. G_I is believed to be related to self-interstitial formation due to P precipitation and diffusion. Thus from Figure 19, G_I has an activation energy of −2.59 eV. This result implies that self-interstitial generation decreases with increasing temperature, which is why no enhanced diffusion was observed in this study at 1200 °C.

The activation energy for vacancy-dominated diffusion is given by equation 40

$$D(V) = D_i(V)G_V \tag{40}$$

where

$$G_V = C_V/C_V{}^o \tag{41}$$

and $C_V/C_V{}^o$ is the vacancy supersaturation ratio. At thermal equilibrium,

$$C_I C_V = C_I{}^o C_V{}^o \tag{42}$$

where C_I is the self-interstitial concentration above or below the intrinsic concentration $C_I{}^i$. Because

$$G_I = C_I/C_I{}^o \tag{43}$$

then

$$G_V = 1/G_I \tag{44}$$

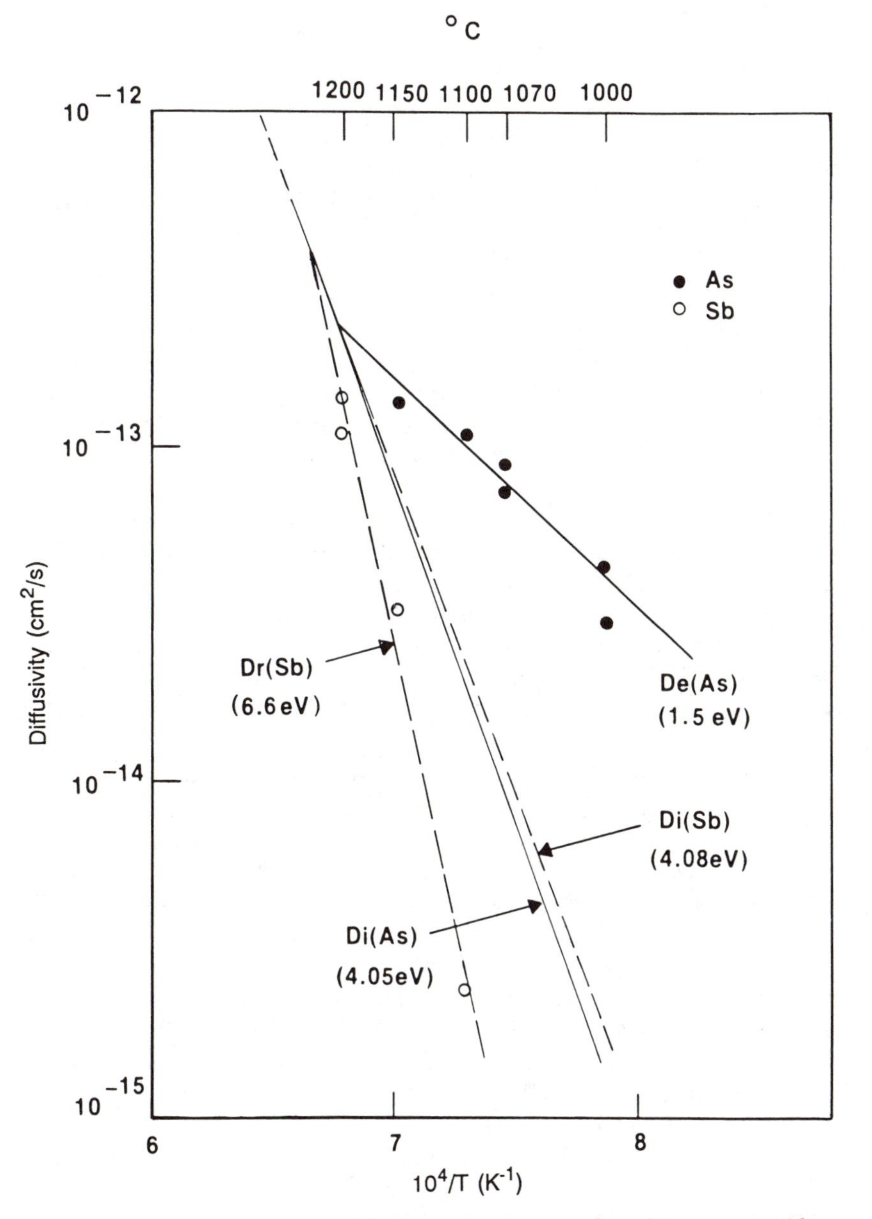

Figure 19. Plots of intrinsic diffusion coefficients of Sb and As vs. 1/T. Also shown are diffusivity data for Sb and As buried layers from Table I in the presence of a phosphorus surface diffusion. Abbreviations: D_r, *reduced Sb diffusivity;* D_e, *enhanced As diffusivity;* D_i, *intrinsic diffusivity. (Reproduced with permission from reference 24. Copyright 1987 The Electrochemical Society, Inc.)*

and equation (40) can be written as

$$D(V) = D_i(V)/G_I \tag{45}$$

Applying equation 45 to the Tsai study is risky, because externally injected self-interstitials render equation 42 invalid. However, as a general guide to understanding the origin of reduced Sb diffusion in the presence of excess self-interstitials, the activation energy of Sb diffusion from equation 45 is given by

$$D_r(V) \propto \exp(-4.08 \text{ eV}/kT)/\exp(2.59 \text{ eV}/kT) \propto \exp(-6.67 \text{ eV}/kT) \tag{46}$$

This result agrees well with the observed value of 6.62 eV and supports the notion that the reduced diffusion of Sb is a result of a reduced vacancy population caused by self-interstitial generation from the P-diffused layer.

At Concentrations below the Solid Solubility Limit. Defect generation during phosphorus diffusion at concentration levels below solid solubility was also investigated by Tsai et al. (*48*). Diffusions of ion-implanted P were performed in low-oxygen ambient and in nitrogen with a silicon nitride cap. Markers of buried layers of arsenic and antimony were used to study the effect of point defect generation on the diffusions. Defects were revealed by using the Schimmel etch. The following results were obtained: (1) Stacking faults were formed below the phosphorus-diffused region in a low-oxygen ambient and at 1050–1150 °C. (2) With long diffusion times, the diffusion of buried layers of As was enhanced and that of buried layers of Sb was retarded. (3) Diffusion in nitrogen with a silicon nitride cap yielded few stacking faults, reduced P diffusion, and neither enhancement nor retardation of the As or Sb buried layers, respectively, for the same processing conditions that produced the results in (2). (4) At phosphorus concentrations above solid solubility, diffusion in N_2 with a nitride film cap enhanced the diffusion of the As buried layer, and stacking faults were also observed. Thus, the intrinsic diffusion of phosphorus generates excess silicon self-interstitials. The silicon nitride cap apparently ties up the silicon surface with Si–N bonds, making it more difficult to generate the self-interstitials required for P diffusion.

In addition to generating excess self-interstitials, high-concentration P diffusion also causes the local equilibrium vacancy concentration to increase. Nishi et al. (*49*) showed that significant enhancement of Sb diffusion occurs inside the highly doped P profile, a result suggesting a strong enhancement of the vacancy concentration there.

Excess Point Defects and Low-Thermal-Budget Annealing. Submicrometer VLSI (very-large-scale integration) technologies require low thermal budgets (the product of dopant diffusivity and diffusion time) to limit the diffusional motion of dopants. Two options exist to reduce the thermal budget: (1) reduced process temperatures with extended furnace-annealing times or (2) high process temperatures with rapid thermal-annealing (RTA) times. However, for both options, enhanced dopant diffusion is sometimes observed (*50–54*). For example, low-temperature furnace anneals of ion-implanted B layers in Si can exhibit substantial diffusion depending upon the completeness of activation of the implant (*51*). Models that have been proposed to explain these results (*55, 56*) lack generality and do not predict the correct time dependence of annealing, nor do they deal with electrical activation and damage removal (*51*).

Recently, Michel (*51*) proposed that the enhanced diffusion of B during low-temperature furnace annealing results from the presence of an active species that is mobile. As annealing proceeds, the inactive B species, which may be defect clusters, converts to the active species exponentially with time. The time constant has an activation energy of ~5 eV, as determined by Seidel and MacRae (*57*). As a result of the annealing of defect clusters, the enhanced diffusion of B is assumed to occur uniformly throughout the implanted region but at a limited period that decays exponentially. Interactions that may result from damage produced by high-dose implants as proposed by Fair et al. (*53*) are ignored. Thus, Michel's model applies primarily to low-dose B implants where the peak concentration of B is below the solid solubility at the annealing temperature.

The effects of damage by ion implantation on the low-temperature diffusion of dopant can also be studied by implanting Si^+ or Ge^+ ions into predeposited layers in Si. Recently, Servidori et al. (*58*) studied the influence of lattice defects induced by Si^+ implantation. Using triple crystal X-ray diffraction and TEM, they confirmed (1) that below the original amorphous surface–crystal interface, interstitial dislocation loops and interstitial clusters exist and (2) that epitaxial regrowth leaves a vacancy-rich region in the surface.

Therefore, dopants exhibit different amounts of enhanced or retarded diffusion during annealing according to which region the dopant is in and whether it diffuses via a vacancy or self-interstitialcy mechanism. For the case of B predeposited from a BBr_3 source to an initial junction depth of 1200 Å, subsequent Si^+ implantation and annealing at 750–900 °C caused retarded diffusion. However, for deep predepositions (3400 Å), this processing produced substantial enhanced diffusion. Thus, any mathematical model must include the spatial dependence of implant damage, the nature of the damage (whether the damage is rich in vacancies or self-interstitials), a calculation of the damage annihilation during annealing, and estimates of point defect production during annealing. The different kinds of defects produced by both low- and high-dose B implants are shown schematically in Figure 20.

Point defect production as a result of implantation damage of the kind shown in Figure 20 gives rise to anomalous B diffusion during subsequent annealing. A summary of recent models that explain these effects follows

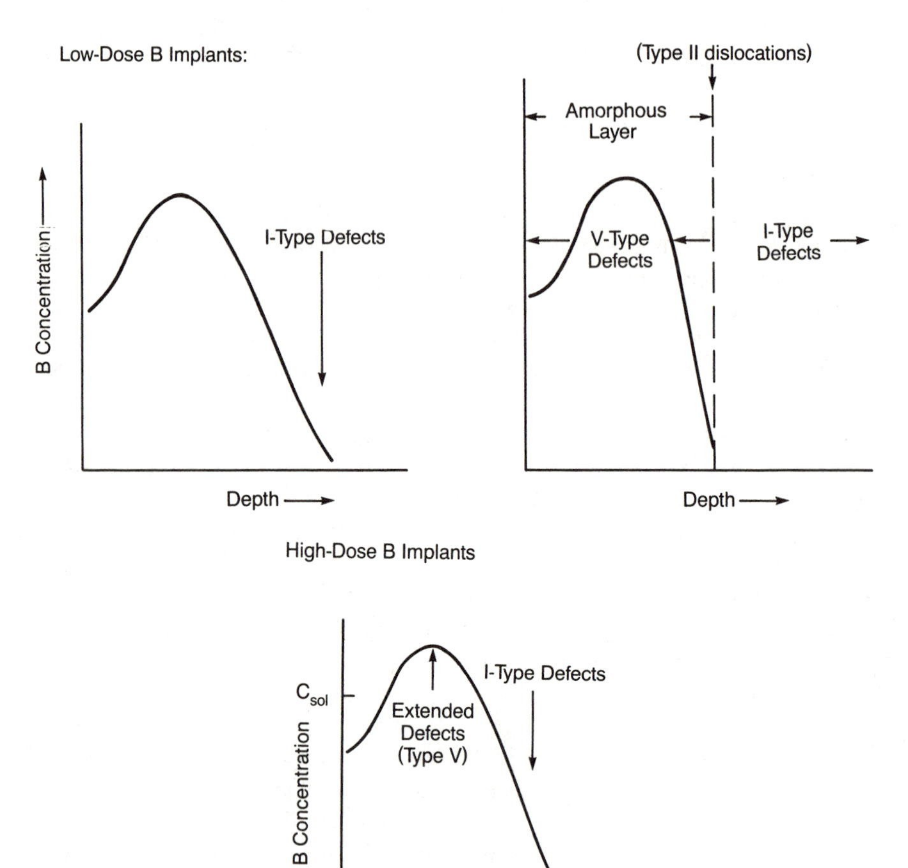

Figure 20. Ion implantation defect production models for low- and high-dose B implants into crystalline and preamorphized Si. C_{sol} is the solid solubility of B.

(59). These models have been incorporated in the PREDICT program for computer simulation (*60*).

- B implants and low-temperature furnace annealing with transient diffusion that is associated with the activated removal of implant damage in the tail region of the implant. The magnitude of the enhanced, transient diffusivity increases with implant dose and energy but reaches saturation at 2×10^{-13} cm^2/s.
- Preamorphization and postamorphization with Si^+ or Ge^+ implants. The solid solubility of B within the regrown surface

layer and the diffusion of B beyond the depth of the original amorphous layer–crystal interface are enhanced. Also, reduced diffusion may occur in the vacancy-rich regrown surface layer.

- High-dose B implants with all-temperature furnace annealing and rapid thermal annealing (RTA). Local diffusivity depends on extended defect formation and annihilation. Also, B clustering occurs above solid solubility.
- Low-dose B and P implants with RTA. Time-dependent diffusion is associated with annealing of implant damage.

Each of these effects are described in the following sections, with the discussion limited to the case where the B concentration does not exceed solid solubility at the annealing temperature.

Low-Temperature Furnace Annealing. The enhanced diffusion of low-dose B implants in Si during low-temperature annealing can be dramatic. Examples are given in Figures 21–23. For Figure 21, two channel B implants, one at 35 keV (6.6×10^{12} cm^{-2}) and the other at 75 keV (1.4×10^{11}/cm^2), were performed through a poly/gate oxide structure, and a two-step anneal (600 °C, 30 min; 700 °C, 30 min) was performed (*61*). After the anneal, the surface concentration decreased by a factor of 2, and the tail portion of the B profile moved by approximately 400 Å. For Figure 22, a 2×10^{14}/cm^2, 5-keV implant was performed through 100 Å of SiO_2 and furnace annealed for 35 min. Very little diffusion occurred for B concentrations above 3×10^{18}/cm^3 (approximate n_i at 800 °C), whereas a movement of 150–250 Å occurred in the B tail (A. E. Michel, unpublished).

Figure 23 shows profiles of as-implanted B (1×10^{14}/cm^2 at 800 keV and 1.2 and 2.0 MeV) (*62*). After a two-step anneal (500 °C, 1 h; 850 °C, 0.5 h), substantial diffusion in the order of 900–1600 Å occurred. Under normal conditions, the diffusion length for this case should be less than 150 Å.

The diffusional displacement of B is a function of implant dose and energy. The energy dependence is illustrated in Figure 24, which shows the diffusion of B at a concentration of 1×10^{17}/cm^3 versus R_p, the projected range of B implantation. The implants were 1×10^{14}–2×10^{14} B atoms per cm^2 annealed at 800–850 °C for approximately 0.5 h. The displacement increases with implant depth and then reaches saturation. The calculated curve in Figure 24 is based on the concentration of excess self-interstitials in the tail of the implant that increases directly with range, up to a maximum value.

The low-temperature enhanced diffusion of B can be modeled by calculating an effective diffusivity that is then applied to the calculation of the B profile by using the PREDICT program (59). The duration of enhanced diffusion is related to the damage annealing time. Empirically, the removal

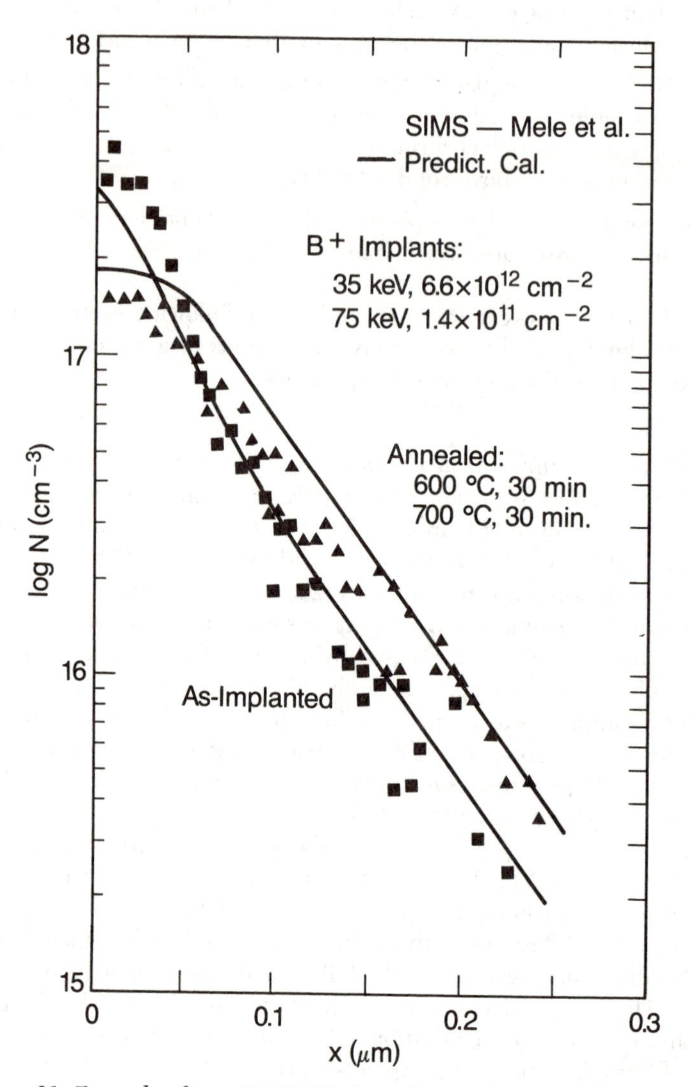

Figure 21. Example of two MOSFET channel B implants performed through a poly gate/oxide structure and annealed at 600 °C for 30 min and at 700 °C for 30 min. The substantial enhanced diffusion is shown modeled with calculations from the Predict program. Data are from Mele et al. (61). *Abbreviation and symbols: SIMS, secondary ion mass spectrometry;* ■, *measured after implant;* ▲, *measured after anneals.* (*Reproduced with permission from reference 59. Copyright 1988 Institute of Electrical and Electronics Engineers, Inc.*)

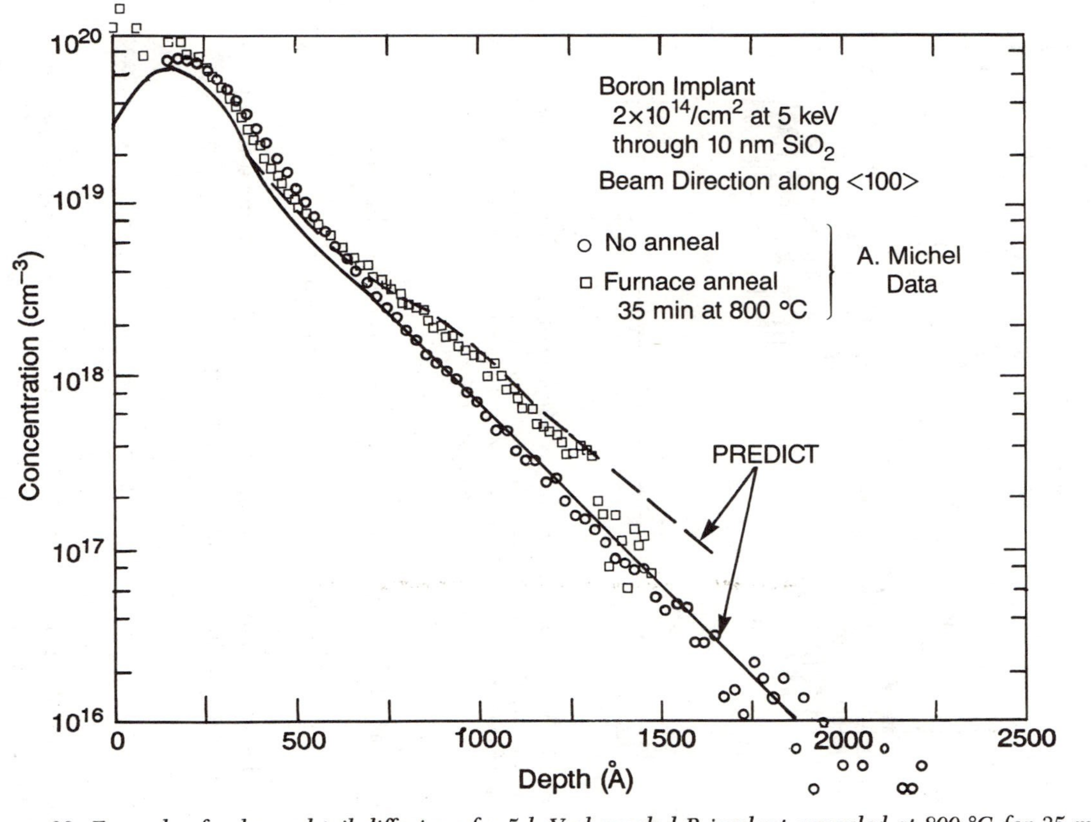

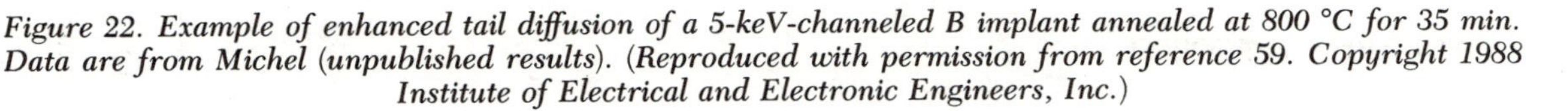
Figure 22. Example of enhanced tail diffusion of a 5-keV-channeled B implant annealed at 800 °C for 35 min. Data are from Michel (unpublished results). (Reproduced with permission from reference 59. Copyright 1988 Institute of Electrical and Electronic Engineers, Inc.)

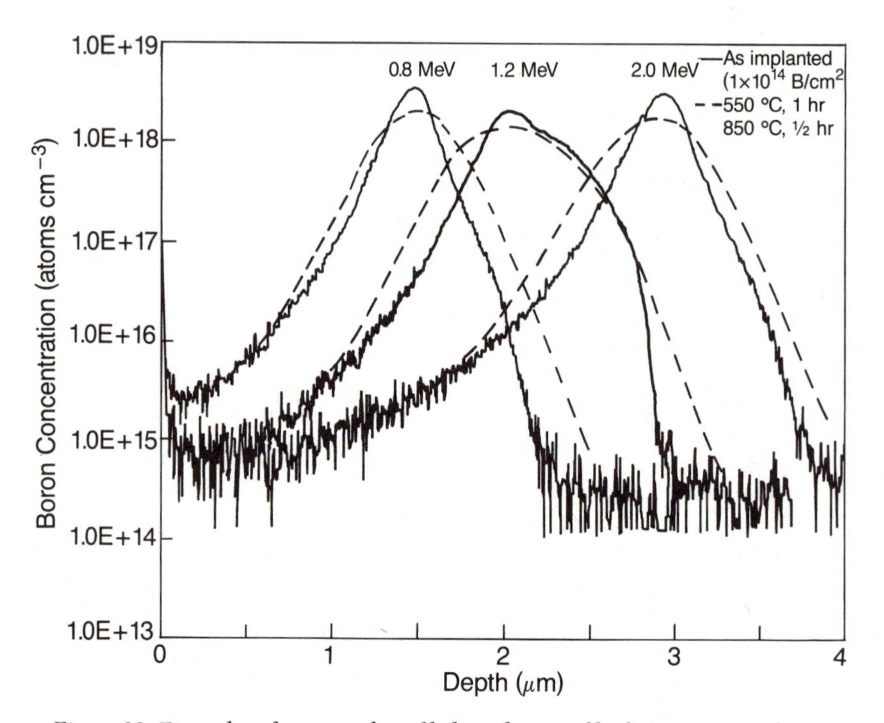

Figure 23. Examples of measured profile broadening of high-energy B implants at low annealing temperatures. SIMS data are from Ingram et al. (62). (Reproduced with permission from reference 59. Copyright 1988 Institute of Electrical and Electronics Engineers, Inc.)

of excess point defects occurs at time t_a, which is defined by equation 47

$$t_a = 4 \times 10^{-22} \exp(5 \text{ eV}/kT)(\text{second}) \tag{47}$$

The fraction of the implant damage that anneals, f_a, is assumed to depend directly on time at the annealing temperature:

$$f_a = \frac{t}{t_a} \tag{48}$$

The annealing of damage occurs until $f_a = 1$. If more than one temperature is used in a process, then

$$f_a = \frac{t_1}{t_{a1}} + \frac{t_2}{t_{a2}} + \ldots \tag{49}$$

Thus, the annealing is cumulative.

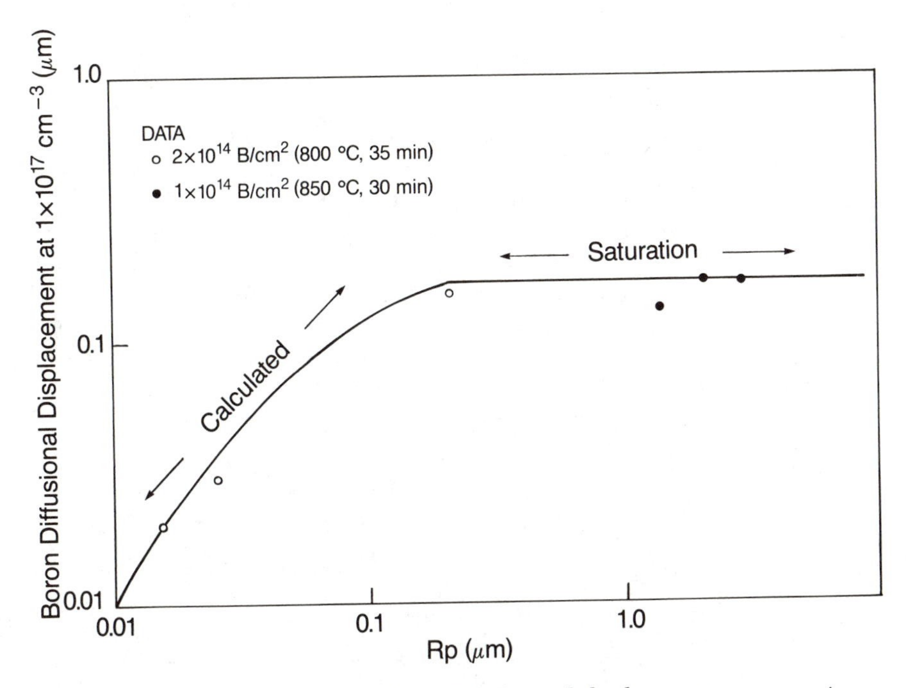

Figure 24. Measured and calculated B diffusional displacement versus projected ion implantation range (R_p) *after annealing 1 × 10¹⁴–2 × 10¹⁴/cm² implants at 800–850 °C. (Reproduced with permission from reference 59. Copyright 1988 Institute of Electrical and Electronics Engineers, Inc.)*

For <100> and <111> Si, an additive diffusivity describes B diffusion for implant doses $\leq 2 \times 10^{14}/\text{cm}^2$

$$D_{\text{enh}} = 9 \times 10^{-12}(\text{dose}/10^{13})^{1/4}\,(R_p/7 \times 10^{-6})\,\exp\,(-0.6\ \text{eV}/kT) \quad (50)$$

where R_p is in centimeters. If the B dose is greater than $2 \times 10^{14}/\text{cm}^2$, then

$$D_{\text{enh}} = 1.4 \times 10^{-11}(R_p/7 \times 10^{-6})\,\exp\,(-0.6\ \text{eV}/kT) \quad (51)$$

D_{enh} is expressed in square centimeters per second. The dose dependence of D_{enh} has been verified for implants in the range from 5×10^{12} to $1 \times 10^{16}/\text{cm}^2$. The calculated profiles in Figures 21 and 22 are based on this model, with D_{enh} used as an additive term to the normal temperature-dependent diffusion.

Reverse Time Effect of Diffusion. Michel (*51*) observed two new effects associated with the low-temperature annealing of B-implanted layers: (1) an apparent "reverse" time effect for furnace diffusion when preceded by rapid thermal annealing and (2) low-temperature enhanced diffusion only for B at concentrations below n_i (Figure 25).

The concentration dependence of enhanced B diffusion has been observed previously but only when the peak B concentration exceeded solid solubility (53, 63). The reduced B diffusion in the profile peak has been ascribed to B clustering and extended defects introduced by high-dose implants. However, the implant used for the data in Figure 25 does not exceed solid solubility. Two other possibilities exist: (1) the B peak is in a vacancy-rich region produced by implantation and cannot diffuse because of an insufficient supply of self-interstitials or (2) the generated self-interstitials during damage annealing change charge state when $C_B > n_i$, and this change affects B diffusion. No data exist to resolve this question.

The reverse time effect of diffusion was modeled by using equations 47–51. The 900 °C RTA step causes the damage anneal time, t_a, to decrease relative to the 800 °C anneal. Thus, the duration of transient enhanced diffusion decreases as the RTA anneal time increases. Calculations using this model are shown in Figure 25.

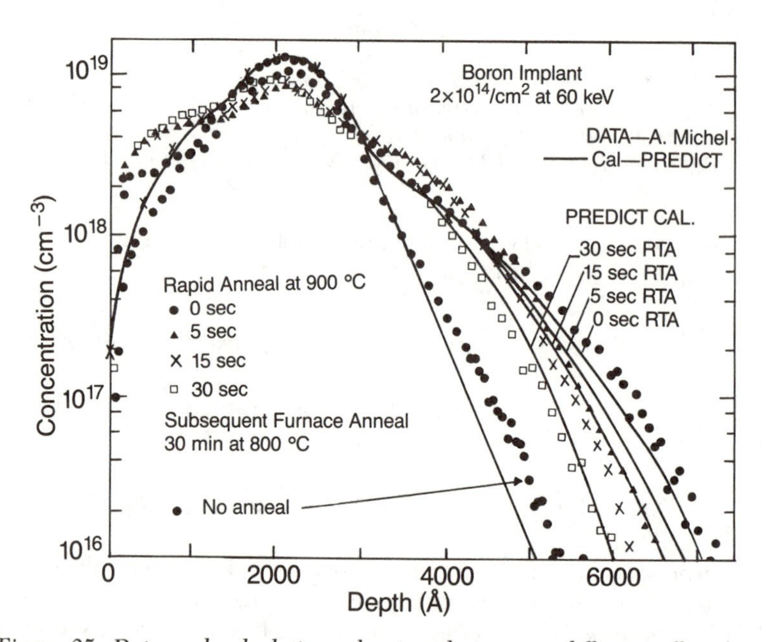

Figure 25. Data and calculations showing the reverse diffusion effect from implantation damage annealing. As the RTA time increases, the amount of enhanced diffusion during the subsequent 800 °C furnace anneal decreases. Data are from Michel (51). (Reproduced with permission from reference 59. Copyright 1988 Institute of Electrical and Electronics Engineers, Inc.)

Pre- and Postanneal Implantations of Si. Implantation of Si^+ ions into predeposited B layers in Si produce either enhanced or retarded diffusion of B during subsequent annealing (*58, 63–66*). Servidori et al. (*58*) showed that reduced diffusion of B predeposits occurs at 750 and 900 °C if the initial B junction depth is 1200 Å, and that enhanced diffusion occurs if the initial junction depth is 3400 Å. The Si^+ implants produced amorphous layers with depths of about 2000 Å. These results and other data indicate that enhanced B diffusion occurs in the end-of-range region of the Si implant where excess self-interstitials and interstitial defects exist. If the Si^+ dose is sufficient to create an amorphous Si layer that is shallower than the implant depth, then the enhancement of B diffusion during subsequent annealing is relatively independent of Si^+ dose (A. E. Michel, unpublished findings). Thus, for B concentrations $< n_i$,

$$D_{enh} = 5 \times 10^{-13} \text{ cm}^2/\text{s} \tag{52}$$

for the first 100 s of the anneal for temperatures >725 °C. This athermal diffusivity is added to the normal temperature-dependent terms to get the total diffusion coefficient.

A sample calculation using these conditions is shown in Figure 26. B^+ was implanted ($1 \times 10^{15}/\text{cm}^2$ and 50 keV) and annealed by RTA at 1150 °C for 10 s. Then Si^+ was implanted to create an amorphous layer, and the RTA cycle was repeated. These data were compared with a B profile obtained with no Si^+ implant but with the same heat treatment. The enhanced diffusion of B was calculated by using equation 52, and good agreement was achieved. This model has been verified for RTA and furnace anneals at temperatures as low as 750 °C.

When the B-doped layer is completely contained within the amorphous layer caused by the Si^+ implant, retarded diffusion is observed. Godfrey et al. (*63*) showed that at 950 °C, 1-h furnace anneals of 25-keV B implants (10^{12}–10^{16} B atoms per cm^2) produced shallower results when the B implants were done in preamorphized Si. The B diffusion coefficient for annealing of preamorphized samples that are B implanted is given by

$$D_B = 0.56 \exp(-3.42 \text{ eV}/kT) \tag{53}$$

On the other hand, without preamorphization, the intrinsic diffusivity of B implants is

$$D_B = 0.0019 \exp(-2.7 \text{ eV}/kT) \tag{54}$$

For example, at 950 °C, the ratio D_B (preamorphized)/D_B (crystalline) is 0.32.

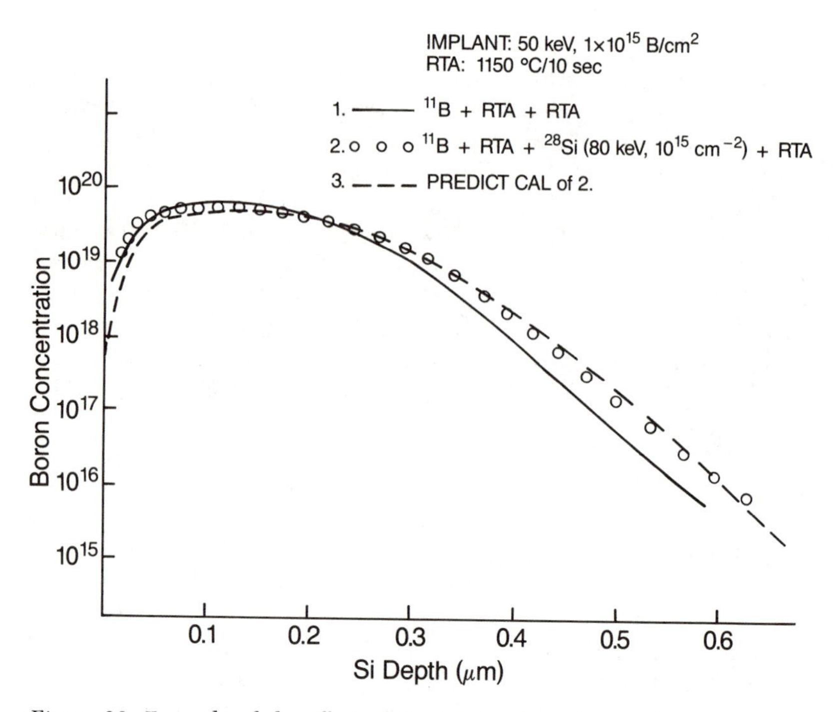

Figure 26. Example of the effect of Si^+ ion implantation on the subsequent diffusion of RTA-annealed B. Data are from Cho et al. (65). *(Reproduced with permission from reference 59. Copyright 1988 Institute of Electrical and Electronics Engineers, Inc.)*

Equation 54 for D_B applies only for B implant doses of $>7 \times 10^{13}/cm^2$, <100> Si substrates, and temperatures between 900 and 1180 °C. The lower activation energy in equation 54 probably reflects the self-interstitial dominance of B diffusion in the interstitial-rich implanted region. However, a high-dose Si^+ implant can produce a vacancy-rich amorphous region that may account for the reduced B diffusion within the regrown region and D_B according to equation 53.

Transient Diffusion During Rapid Thermal Annealing. When the B implant dose is less than $3 \times 10^{14}/cm^2$ and an RTA is performed, transient diffusion is observed (66). Examples are shown in Figure 27 for implants of 1×10^{14}–2×10^{14} B atoms per cm^2 performed at energies from 1 to 60 keV. The previously published data of Sedgwick (66) are included. This case is modeled similarly as the low-dose B implant–furnace anneal case, except for the magnitude of D_{enh}. Thus, if neutral and donor point defects contribute to B diffusivities D_i^x and D_i^+, respectively, then the total B diffusion coefficient (D_B) is given by (59)

$$D_B = (D_i^x + D_{enh}) + D_i^+ \tag{55}$$

where

$$D_{enh} = 3.6 \times 10^{-13}(\text{dose}/10^{13})^{1/4}(R_p/7 \times 10^{-6}) \qquad (56)$$

The implant range dependence has been verified for B implants from 1 to 60 keV (Figure 27). Equation 56 is applied to the calculations until the implant damage has been annealed, and then $D_{enh} = 0$. The maximum allowed value of D_{enh} is 1.5×10^{-12} cm^2/s. The damage anneal time criterion was derived from data that show initial rapid B diffusion during RTA and then a marked slow-down (*51*).

Low-dose P implants exhibit similar enhanced diffusion during RTA (*67*) in the 800–1150 °C range. The magnitude of this effect is a diffusivity of 6×10^{-13} cm^2/s for 10 s (*68*). However, Morehead and Hodgson (*56*) modeled low P diffusion during RTA with an effective temperature-dependent diffusivity given by equation 57.

$$D_{enh} = 0.007 \exp(-2.2 \text{ eV}/kT) \qquad (57)$$

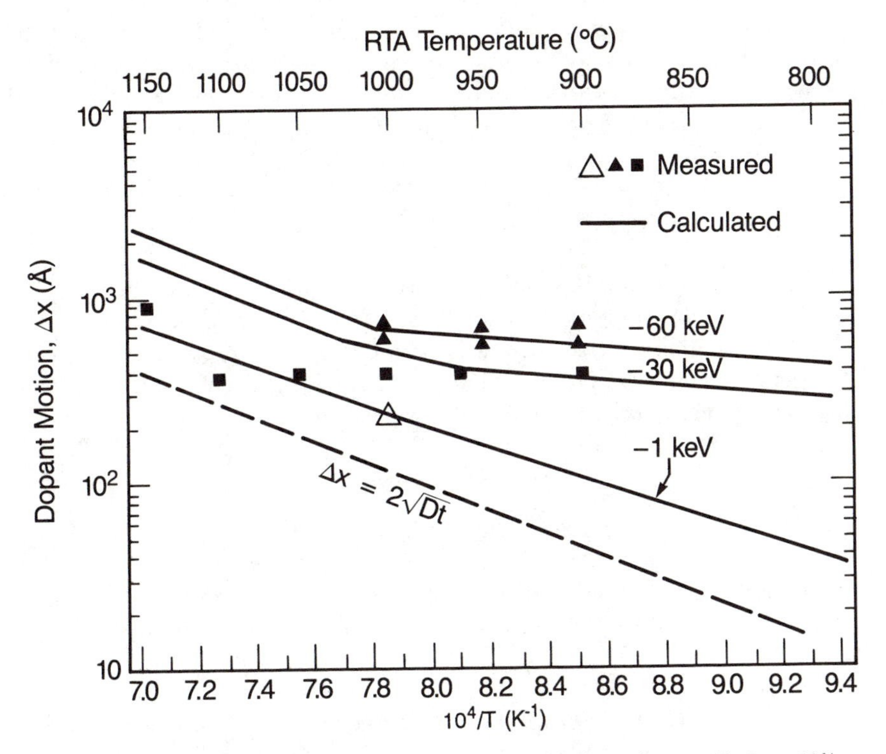

Figure 27. Examples of diffusion of low-dose B implants (1×10^{14}–2×10^{14} B atoms/cm^2) during RTA for 10 s. Calculated curves are based on a time-dependent model described in the text.

This diffusivity was applied to the mobile fraction of P that is capable of diffusing. In the PREDICT program, equation 57 is used for the first 2 s at peak RTA temperature over the entire P profile with good results (68), provided that the anneal temperature is below 1020 °C.

The 2.2 eV in equation 57 may represent the migration enthalpy of the phosphorus–self-interstitial pair (56) with the excess self-interstitials produced by damage annealing.

Conclusion. Low-thermal-budget process modeling is far more complex than previously believed, because ion-implantation-related point defects and extended defects decay over significant portions of the annealing time. The enhanced or retarded diffusion of dopants is, therefore, dependent upon (1) the amounts and types of defects produced by ion implantation, (2) whether the Si wafer has been preamorphized, and (3) the annealing procedure. Because no first-principle models exist to account for all of the variables involved, empirical models have been developed for each dopant, substrate, implantation condition, and annealing combination. The basic models used in the PREDICT program have been presented in their simplest form. Additional considerations are required for a successful implementation of these models, such as the dependence on the annealing ambient. For example, a slightly oxidizing ambient may slow down the annihilation of implantation-produced defects and, thus, reduce the effects described in this section.

Oxidation of Silicon

The thermal oxidation process is an essential feature of planar-device fabrication and plays an important role in the diffusion of dopants in Si. In the thermal oxidation process, Si reacts with either oxygen or water vapor at temperatures between 600 and 1250 °C to form SiO_2. The oxidation reaction may be represented by the following two reactions:

$$Si + O_2 \rightarrow SiO_2 \tag{58}$$

$$Si + 2H_2O \rightarrow SiO_2 + 2H_2 \tag{59}$$

In the following sections, oxidation models for calculating oxide thickness and process variables that influence oxidation, as well as oxide structure, are discussed.

Oxide Thickness Versus Time. Silicon oxidation has been modeled by using the linear–parabolic macroscopic formulation of Deal and Grove (69). As a starting point for the study of this model, the kinetics of oxidation will be briefly reviewed, and then the current theories on the mechanism of oxide growth will be discussed.

Several definitive experiments have shown that during the thermal oxidation of silicon the oxidizing species (some form of oxygen) diffuses through the oxide layer and reacts with the silicon to produce more oxide at the $Si–SiO_2$ interface (*70, 71*). For oxidation to occur, three consecutive oxidant fluxes must exist: (1) transport from the furnace ambient to the outer oxide surface, (2) diffusion through the oxide layer of thickness x_o, and (3) reaction with silicon at the interface. At steady state, all three fluxes are equal.

Details of the mathematical treatment of the fluxes can be found in reference 69. After the oxidant flux is related to oxide growth rate, equation 60 is derived

$$\frac{dx_o}{dt} = \frac{k_s(C^*/N_1)}{1 + (k_s/h) + (k_s x_o/D_{\text{eff}})} \tag{60}$$

where k_s is the chemical surface reaction rate constant for oxidation, C^* is the equilibrium oxidant concentration in the gas ambient, N_1 is the number of oxidant molecules incorporated into a unit volume of growing oxide, h is the gas-phase transport coefficient of oxidant, and D_{eff} is the effective diffusion coefficient of oxidant in the oxide.

To solve the differential equation for x_o, a set of initial conditions is required. If the initial oxide thickness is x_i at $t = 0$, then the solution to equation 60 is (*69*)

$$x_o^2 + Ax_o = B\,(t + \tau) \tag{61}$$

where

$$A = 2D_{\text{eff}}\,(k_s^{-1} + h^{-1}) \tag{62}$$

$$B = 2D_{\text{eff}}C^*/N_1 \tag{63}$$

$$\tau = (x_i^2 + Ax_i)/B \tag{64}$$

The parameter τ represents a shift in the time coordinate that corrects for the presence of the initial oxide layer x_i.

By solving equation 61, the time dependence of x_o is obtained. To study this result, two limiting cases can be examined that are separated by the characteristic time, $A^2/4B$, or the corresponding oxide thickness, $x_o \simeq D_{\text{eff}}/k_s$. Thus, for $t >> A^2/4B$, the solution to equation 61 becomes

$$x_o^2 = Bt \tag{65}$$

Because of the appearance of equation 65, B is called the parabolic rate constant. This limiting case is the diffusion-controlled oxidation regime that occurs when oxidant availability at the Si–SiO_2 interface is limited by transport through the oxide (thick-oxide case).

For short times and thin oxides, if $(t + \tau) << A^2/4B$, the solution to equation 61 becomes

$$x_o = (B/A)(t + \tau) \tag{66}$$

This solution describes the linear growth regime, and B/A is the linear rate constant where

$$B/A = k_s h C^*/N_1(k_s + h) \tag{67}$$

The linear growth rate depends on the interface reaction and is limited by the value of k_s and the availability of oxidant for the interface reaction.

The rate constants that are derived from the Deal–Grove model allow a reasonable characterization of oxidation as a function of temperature, furnace ambient, silicon doping concentration, and silicon crystal orientation. But the model does not give a detailed understanding of the mechanisms that produce these dependencies, nor is it relevant to oxides with thicknesses less than 350 Å. This deficiency of the model is particularly troublesome, because VLSI devices are currently being produced with gate oxide thicknesses in the order of 100 Å. This anomalous oxidation region has been attributed to nonconforming kinetics in the early stages of oxidation, when growth is assumed to proceed via a linear (surface-controlled) process. However, recent analysis of the Deal–Grove model by Reisman et al. (72) has shown that, except at zero oxide thickness, no linear regime is possible.

Analysis of a vast amount of data by Reisman et al. (72) has shown that silicon oxide thickness versus time can be modeled by a general power law of the form

$$x_o = a(t_g + \tau)^b \tag{68}$$

where x_o is the final SiO_2 thickness, a and b are constants, t_g is the time for growth measured in a given experiment, and τ is the time to grow an oxide of thickness x_i, already present on the Si surface. The result of this model contrasts with that of the Deal–Grove model for which oxide thickness is the solution to the quadratic equation, equation 61.

A comparison of the power law description of oxide thickness (equation 68) and the Deal–Grove model is shown in Figure 28. The data of Deal and Grove taken at 700 °C are plotted on linear scales, and equation 66 is plotted with measured values of B/A and τ. Shown for comparison is equation 68 with the appropriate constants obtained by a fractional weighted least-square

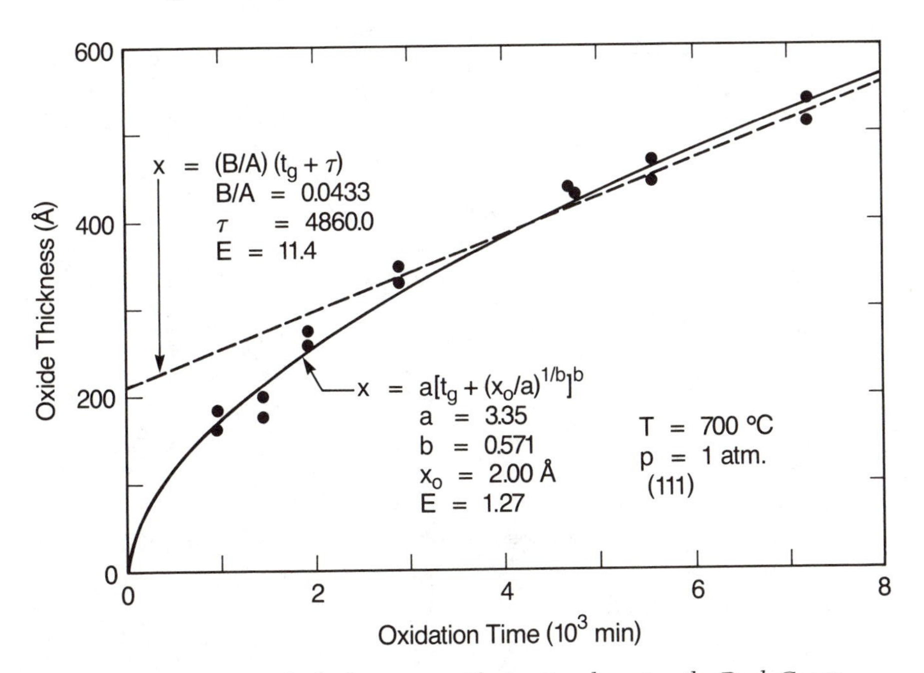

Figure 28. Fits to oxide thickness vs. oxidation time by using the Deal–Grove model and a power law expression (72).

fit to the data. Because no anomalous thin-oxide regime is observed, the power law better describes the data.

A universal fit of the power law given by equation 68 to all the data for dry oxidation accumulated for the past 20 years was found (72). Values for the time-independent exponent b varied from 0.24 to 0.90, depending on temperature and oxygen pressure. The fit of these data to a power law suggests that oxidation is limited by either surface reaction or mass transport. If the limiting value of b is 1, then oxidation is limited by surface reaction. If the limiting value of b is 0.5 (parabolic growth regime), then mass transport of oxygen limits oxidation. Nicollian and Reisman (73) argue that b exceeds 0.5 in many cases. Thus, oxidation is limited by surface reaction, and mass transport of oxidant must be very rapid through amorphous SiO_2.

The key feature of the model of Nicollian and Reisman (73) is that the surface-reaction-limited oxidation process has viscous oxide flow as the rate-limiting mechanism. During oxidation, volume expansion occurs when silicon is oxidized, and the viscous flow of oxide accompanies this expansion. The rate of growth at the interface is given by

$$\frac{dx_0}{dt} = k_f\, p^n - k_r\, p_r^{\,m} \tag{69}$$

where k_f and k_r are the forward and reverse specific reaction rate coefficients, respectively, p^n is instantaneous partial pressure of SiO for the forward reaction of order n, and p^m is the instantaneous partial pressure of SiO for the reverse reaction of order m:

$$SiO_2\ (s) + Si\ (s) \rightarrow 2SiO\ (g) \tag{70}$$

The rate of the reverse reaction is assumed to be negligible compared with that of the forward reaction (73).

In the viscous-flow model, oxidation rate and viscosity are related through k_f:

$$\frac{dx_o}{dt} = k_f\,(t, p, T)\ p^n \tag{71}$$

A time-dependent k_f is required to explain the physical meanings of the coefficients a and b in equation 68. Thus,

$$k_f\,(t, p, T) = k_A \exp\left[-\Delta E_A\,(t, p, T)/kT\right] \tag{72}$$

where k_A is the specific rate coefficient of the forward reaction at infinite temperature and ΔE_A is the activation energy of the reaction

$$\Delta E_A\,(t, p, T) = \Delta E_v\,(t, p, T) + \Delta E_F \tag{73}$$

Therefore, the activation energy is not constant and consists of ΔE_F for the reaction energy and ΔE_v, which is the energy required to produce free volume by viscous flow.

The viscosity of fused silica decreases with increasing water vapor pressure (74), but no oxygen pressure dependence has been reported. However, Nicollian and Reisman (73) reason that the increase in oxidation rate with increasing oxygen pressure implies a decrease in average viscosity, an inference that explains the pressure dependence of ΔE_v. The time dependence of ΔE_v is believed to result from the structural reconfiguration of newly formed oxide from an initially ordered structure to an amorphous structure. This reordering is accompanied by a continuing increase in average oxide viscosity with time. However, no direct experimental data exist at present to support these fundamental concepts.

Numerous other models have been proposed to explain the deviation of dry oxidation from linear–parabolic kinetics. For example, field-assisted oxidant diffusion during the oxidation of metals was proposed by Cabrera and Mott (75) and used by Deal and Grove (69) to explain the results for thin oxides. Ghez and van der Meulen (76) proposed the dissociation of molecular oxygen into atomic oxygen at the Si–SiO_2 interface and the reaction of both species to form SiO_2. Other suggestions of reaction mechanisms for thin oxides include reactions involving ionic and molecular oxygen (*77*); diffusion of two oxidant species, with one species dominant in the thin-oxidation regime (*78,79*); and oxide-thickness-dependent diffusion of the oxidant related to the existence of structural channels through SiO_2 (*80*). Another parallel diffusion model based on $^{18}O_2$ tracer experiments has been reported by Han and Helms (*81*). In this model, the interface reaction rate constants of the oxidant species were quite different. Determination of the rate constants led to parabolic–linear–parabolic kinetics.

The models described so far can explain most of the available data, but no model by itself is completely consistent with all data. Helms et al. (*82*) correctly pointed out that considerable experimental data are available to test models. The assessments of models should consider the following points:

- Accurate fits to available data
- Appropriate parameters should follow linear Arrhenius behavior as a function of temperature (*69*). The model must demonstrate oxidant pressure dependence (*78*), "memory" effects as a function of growth temperature (*83*), and substrate doping effects on oxidation (*84*).
- Consistency of the model with optical data showing index of refraction as a function of growth temperature (*85*), data on O_2 solubility and diffusion coefficient in SiO_2, and isotopic tracer data (*86*).

Process Variables in Oxidation and Oxide Structure. ***Oxide Structure.*** The basic structural unit of thermally grown SiO_2 is a silicon atom surrounded tetrahedrally by four oxygen atoms. These tetrahedra are joined together at their corners by oxygen bridges to form the quartz network. In the amorphous structure, a tendency to form the characteristic rings with six silicon atoms exists.

Amorphous silica is a more open structure than crystalline quartz, because only 43% of the space is occupied. Consequently, a wide variety of impurities can readily enter amorphous SiO_2 and diffuse through the layer. Additional loosening of the SiO_2 network has been observed when H_2O is present in the oxide (*87, 88*). Related to the loosening effect of H_2O on SiO_2 is the report that the diffusion coefficient of oxygen in H_2O-treated bulk SiO_2 is greater than that in dry O_2 (*89*). This effect has been used to explain the increase in the parabolic rate constant for dry O_2 oxidation when trace amounts of H_2O are introduced into the furnace ambient (*90*).

Additional clues about the physics and chemistry of silicon oxidation can be obtained by studying the Si–SiO_2 interface. Helms (*91*) reviewed the current understanding of the morphology and electronic structure of the interface. One important result of Helm's discussion is that it is now believed that a nonstoichiometric region of a few angstroms thick exists at the interface (*92–94*). Thus, the transition from crystalline silicon to the amorphous oxide phase is quite abrupt (~5 Å) at any point. Although the chemical composition of this transition layer has not been determined, it is estimated that excess silicon atoms ($\sim 10^{15}/cm^2$) reside in that layer (*91*). Consequently, the influence of electric fields on oxidation may occur because the transition region acts as an electrochemical cell in which the oxygen-bearing species becomes ionized and moves according to the normal field and concentration gradients in that region.

Process Variables. The thermal oxidation process is a direct function of process variables, including the condition of the silicon surface. The following are important factors that affect thermal oxidation:

- silicon surface cleaning,
- trace amounts of water,
- concentration of Cl-bearing species,
- temperature control and ramp up/ramp down,
- silicon orientation, and
- silicon doping.

In addition, these process variables influence the electrical properties of the $Si–SiO_2$ interface, that is, they influence the quantities of fixed charges, interface states, and charge traps.

Silicon Surface Cleaning. The commonly used procedures for cleaning Si wafers involve the use of acidic and basic hydrogen peroxide solutions. Schwettmann et al. (*95*) investigated three cleaning solutions: $2H_2SO_4–H_2O_2$, $NH_4OH–H_2O_2–H_2O$, and $HCl–H_2O_2–H_2O$. After 10 min in the prescribed solution, the wafers were rinsed and dried. Oxidation was carried out at 1000 °C in dry O_2. Measurements of oxide thickness were made versus oxidation time, and these results are shown in Figure 29. The linear rate constant of the Deal–Grove model (*69*) varied by a factor of 2, depending on the cleaning solution. The parabolic rate constant was not affected.

Auger analysis and ESCA (electron spectroscopy for chemical analysis) of the cleaned wafers showed no contaminants. However, the ammonium hydroxide solution produced a high mobile-ion content in the grown films.

More recently, Gould and Irene (*96*) studied surface-cleaning effects on the oxidation of silicon wafers for oxide thicknesses up to 4300 Å. They found

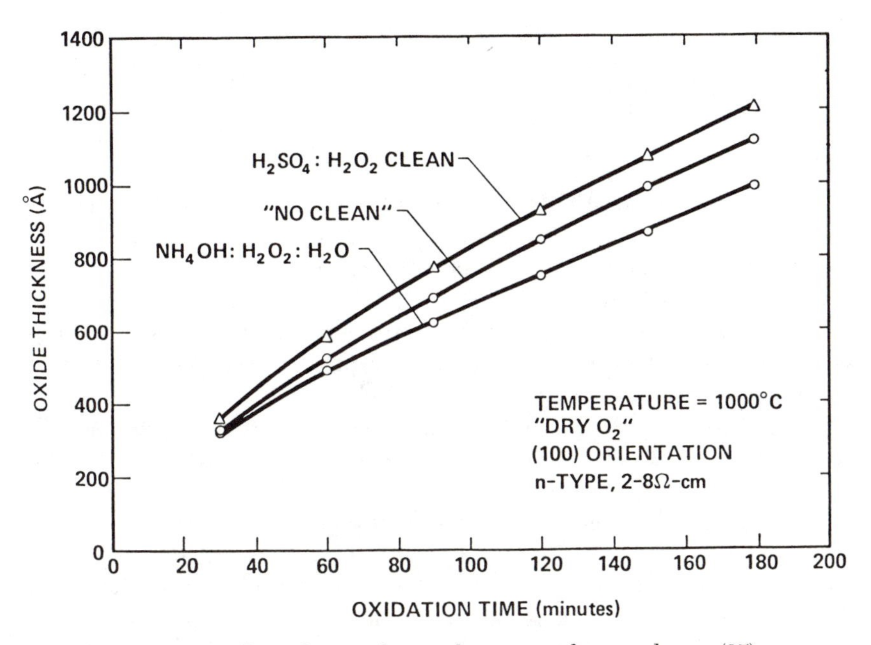

Figure 29. Effect of preoxidation clean on oxide growth rate (95).

results similar to those of Schwettmann et al. (95) with regard to thicker oxides being grown with base–acid–HF cleaning solutions compared with cleaning solutions containing base only. However, the cleaning procedures affected oxidation rates well beyond the linear growth region as defined by the Deal–Grove model (69). On the basis of measurements of refractive indices and oxide density calculations, they concluded that the samples with the highest oxidation rates had the lowest oxide densities and vice versa. The oxides with lower density should enable faster diffusion of oxygen through the oxide and, therefore, faster oxidation.

Gould and Irene (96) state that the oxide structural changes they measured do not provide conclusive results, but the density differences may be significant. These results may also provide support for the model of Nicollian and Reisman (73), in which oxidation is presumed to be limited by interface reaction for all oxide thicknesses.

Effects of Water. The diffusion of water in SiO_2 may be complicated by dissociation that produces hydroxyl groups as shown by equation 74.

$$H_2O + Si^{4+} - O - Si^{4+} \rightarrow 2Si^{3+} - HO \qquad (74)$$

Evidence for this reaction has been reviewed recently by Revesz (97). Thus, the oxidation of silicon in water vapor involves the diffusion of OH and removal of H_2 according to the following reaction (*98*):

$$Si^{3+} - HO - Si^{4+} \rightarrow Si^{4+} - O - Si^{4+} + \frac{1}{2} H_2 \qquad (75)$$

Si^{3+} represents the localized reduction of the network SiO_{2-x}. The value of x was measured from UV absorption to be $\sim 3 \times 10^{-5}$ (*98*).

The most predominant effect of H_2O in the oxidizing ambient is to increase the parabolic rate constant (*90*). As a result, the effect of the interface reaction as the rate-controlling process increases with increasing H_2O content. A relatively small H_2O concentration (25 ppm) in O_2 is already sufficient to increase the parabolic rate constant by factors of 1.3 and 1.6 for <111>- and <100>-oriented silicon wafers, respectively (*99*). The linear rate constant increases more gradually over the range of added H_2O (0–2000 ppm) (*90*).

As pointed out earlier, the effect of trace amounts of water on silicon oxidation has been ascribed to the loosening of the SiO_2 structure. Possibly, hydrogen plays a catalytic role in the interface reaction between silicon and H_2O (*100*). Hydrogen is almost always present in SiO_2 films, coming from numerous possible sources (*100*). However, because the thermal oxidation of silicon even in oxygen is not completely understood (*101*), the effects of H_2O and other hydrogen-bearing species cannot be explained in detail at present.

Perhaps hydroxyl groups or H_2O molecules, or both, diffuse through the SiO_2 film and react with silicon, and hydrogen is formed at the Si–SiO_2 interface, resulting in the generation of extrinsic defects at the interface. For example, hydrogen may react with SiO_2 at high temperatures to yield trivalent Si defects or SiH groups (*99*). Thus, for $T \sim 1000$ °C,

$$Si - O - Si + 2H \rightarrow Si - OH \;\; H - Si \qquad (76)$$

Another example is the reaction of hydrogen with surface silicon atoms, Si_s, that have a dangling bond (unpaired electron). Thus, for $T \leq 500$ °C,

$$Si_s\cdot + H \rightarrow Si_s - H \qquad (77)$$

Therefore, SiOH behaves as an electron acceptor, whereas SiH behaves as an electron donor. These defects may contribute to the electrical properties of the Si–SiO_2 interface. Consequently, proper control of oxidation and annealing conditions with respect to hydrogen (water) is a crucial step in technology.

Effects of Chlorine. The addition of chlorine-bearing species (HCl, Cl_2, and C_2HCl_3 [trichloroethylene or TCE]) to the oxidation ambient results in improved properties of the Si–SiO_2 system. Chlorine lowers the interface state density, stabilizes the surface potential, and enhances the dielectric strength of the oxide (*102*). Also, such additives increase the oxidation rate compared with that in pure dry O_2 at all temperatures (*102–104*). Examples of the increases in the parabolic and the linear rate constants with HCl concentration are shown in Figure 30 for O_2–HCl mixtures (*103*). Data for both <111> and <100> silicon are included. Essentially no orientation effect is observed in the parabolic rate constant *B* (Figure 30a), and increasing the HCl concentration above 1% results in a linear increase in *B*. The large initial increase in *B* and the linear increase with subsequent additions are believed to be partly due to water generation by the following reaction (*103*):

$$4HCl + O_2 \rightarrow 2H_2O + 2Cl_2 \tag{78}$$

However, oxidation in O_2–Cl_2 mixtures also results in an enhanced oxidation rate, a fact indicating that the chlorine species itself is mainly responsible for the enhanced oxidation (*102*).

Silicon Orientation Effects. Thermal oxidation rate is influenced by the orientation of the Si substrate. The effect involves the linear rate constant used in the Deal–Grove model (*69*). The ratio of this constant, *B*/*A*, for

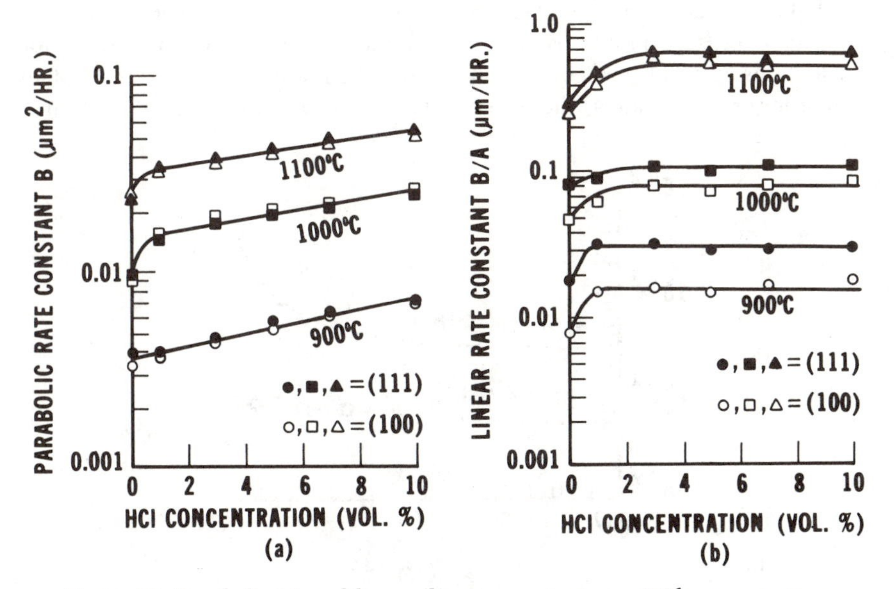

Figure 30. Parabolic (a) and linear (b) rate constants vs. HCl concentration. (Reproduced with permission from reference 103. Copyright 1977 The Electrochemical Society, Inc.)

<111> Si to that for <100> Si is given by

$$\frac{B/A\text{<111>}}{B/A\text{<100>}} = \frac{C_1\text{<111>} \exp(-2.0 \text{ eV}/kT)}{C_1\text{<100>} \exp(-2.0 \text{ eV}/kT)} \approx 1.7 \tag{79}$$

Thus, <100> surfaces oxidize more slowly than <111> surfaces do. The lower oxidation rate of <100> surfaces is due probably to the fewer sites with which oxygen can react. B/A depends on the number of silicon bonds per cubic centimeter available. The linear oxidation rate for various forms of Si follows the sequence <110> > <111> > <311> > <511> > <100> (*105*). A crossover in rate such that the rate for <111> Si is greater than that for <110> Si occurs at 700 and 1000 °C but not at 1100 °C.

The orientation dependence of the interface reaction has been attributed to the number of Si–Si bonds available for reaction (*76, 106, 107*), the orientation of the bonds (*76, 106*), the presence of surface steps (*108, 109*), stress in the oxide film (*110, 111*), and the attainment of maximum coherence across the Si–SiO_2 interface (*76, 111*). However, no strong correlations have been established between these properties and oxidation rate, although Lewis and Irene (*105*) developed a qualitative correlation between the order of the initial rates and the density of atoms on planes parallel to the surface.

Effects of Silicon Doping. Silicon heavily doped with donor or acceptor impurities can exhibit oxidation rates that are considerably enhanced relative to lightly doped silicon (*84, 112*). For example, the dependencies of the rate constants on substrate phosphorus doping level are shown in Figure 31 for oxidations of <111> silicon at 900 °C. B/A increases sharply by more than an order of magnitude as the phosphorus level increases beyond $\sim 10^{20}/\text{cm}^3$.

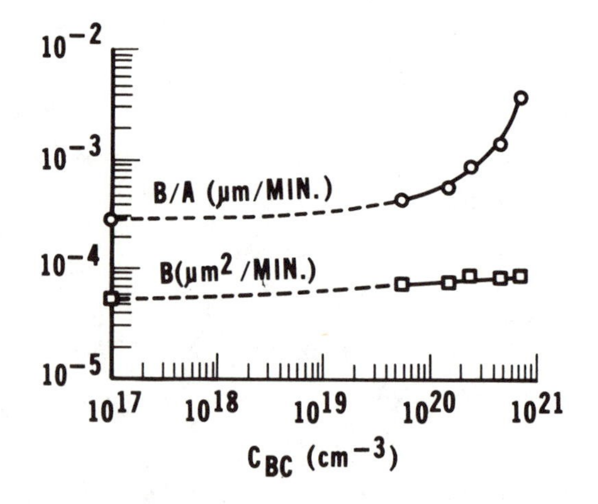

Figure 31. Rate constants vs. silicon substrate. (Reproduced with permission from reference 116. Copyright 1979 The Electrochemical Society, Inc.)

The parabolic rate constant, B, is affected only slightly. These results show that the doping effect only influences the interface reaction.

The temperature dependence of the linear rate constants for various electron concentrations in phosphorus-doped silicon indicate that the doping has only a slight effect on the associated activation energy. Thus, because B/A is proportional to the rate of the interface reaction, the doping effect is buried in the chemical, electrical, or, possibly, mechanical dependence of surface rate on doping.

The oxidation reactions occurring at the Si–SiO_2 interface may be chemical reactions involving holes, silicon, and ionized oxygen. However, for these reactions to occur, space or sites must be available to incorporate oxygen into the growing SiO_2 film. Dobson (*113, 114*) proposed that the oxidation reaction may include the filling of a vacancy at the interface of the silicon lattice by an oxygen atom to annihilate a silicon vacancy, or the oxygen atom may occupy a silicon site to create a silicon interstitial. The net result would be a flow of silicon atoms away from the interface to incorporate the growing oxide by making room for the oxygen atoms. Alternatively, a flux of silicon vacancies flowing to the interface from the bulk may provide reaction sites for the oxygen species. The oxidizing interface then becomes an interstitial source or a vacancy sink. This process is illustrated in Figure 32 (*115*). Silicon interstitials may flow into the SiO_2 and meet and react with the oxidation species.

If the kinds of point-defect-related events just discussed do occur, we expect a relationship between oxidation rate and the dependence of these defects on doping level. Ho and Plummer (*116, 117*) proposed a model that accounts for the increase of B/A on this basis. The model assumes that the oxidation of lightly doped silicon proceeds via nonvacancy-dominated interface reactions such as the incorporation of an oxygen atom onto a silicon site. However, because the total vacancy concentration increases with increasing doping concentration above n_i, it is proposed that above a certain doping level, vacancy-assisted oxidation becomes dominant at the interface (Figure 32). These results are consistent with the earlier discussion on the doping dependence of oxidation-enhanced diffusion, where it was assumed that the equilibrium concentration of vacancies increased with doping and acted to reduce the injected self-interstitial concentration through recombination events. Alternatively, the increased vacancy may provide a competing oxidation mechanism that reduces the oxidation reaction that produces self-interstitials.

Conclusion. Existing empirical models such as the Deal–Grove model have been used for many years as a means of estimating oxide thickness as a function of extracted rate constants. The effects of processing on oxidation are included in these rate constants, but generally only by means of measurement. No fundamental basis exists yet for the direct calculation of these

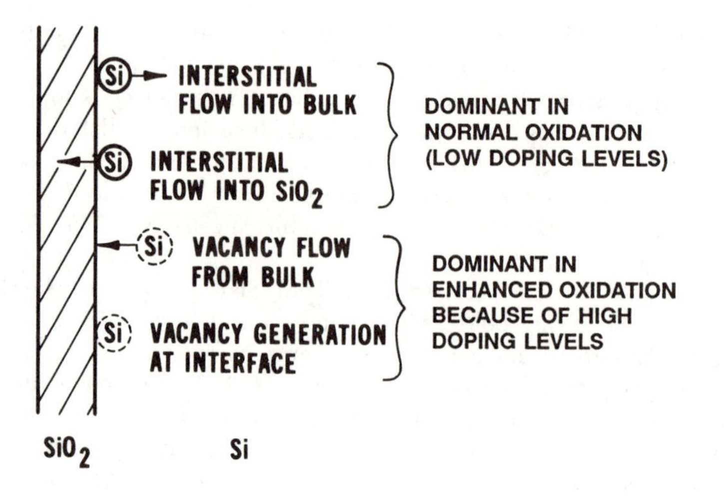

Figure 32. Role of silicon point defects in the oxidation reaction (115).

rate constants as a function of any variables. This problem results from the fact that no exact detailed model of oxidation exists on which to base such calculations. However, recent proposals seem promising as a basis for such calculations.

References

1. Pfaan, W. G. Semiconductor Signal Translating Device, U.S. Patent No. 2,597,028 (1952).
2. Hoemi, J. A., IRE Electron Devices Meeting, Washington, DC, 1960; U.S. Patents 3,025,589 (1962) and 3,064,167 (1962).
3. Kahng, D.; Atalla, M. M. IRE–IEEE Solid-State Device Research Conference, Carnegie Institute of Technology, Pittsburgh, Pa., 1960.
4. Crank, J., *Mathematics of Diffusion*; Oxford University Press: London, 1956.
5. Watkins, G. D. In *Radiation Damage in Semiconductors*; Dunod: Paris, 1964; p 97.
6. Watkins, G. D. In *Radiation Damage in Semiconductors*; Institute of Physics: Bristol, 1973; p 228.
7. Van Vechten, J. A. In *Lattice Defects in Semiconductors 1974*; Institute of Physics: Bristol, 1975; p 212.
8. Watkins, G. D. In *Lattice Defects in Semiconductors 1974*; Institute of Physics: Bristol, 1975; p 1.
9. Naber, J. A.; Mall, C. E.; Leadon, R. E. In *Radiation Damage and Defects in Semiconductors*; Institute of Physics: Bristol, 1973; p 26.
10. Elstner, L.; Kamprath, W. *Phys. Status Solidi* **1967,** *22*, 541.
11. Shaw, D. *Phys. Status Solidi B* **1975,** *72*, 11.
12. Van Vechten, J. A. *Phys. Rev. B* **1978,** *17*, 3197.
13. Weigel, C.; Peak, D.; Corbett, J. W.; Watkins, G. D.; Messmer, R. P.; *Phys. Rev. B.* **1973,** *8*, 2906.
14. Fair, R. B.; Tsai, J. C. C. *J. Electrochem. Soc.* **1977,** *124*, 1107.

15. Fair, R. B. In *Impurity Doping Processes in Silicon*; Wang F. F. Y., Ed.; North Holland: Amsterdam, 1981; pp 315–442.
16. Van Vechten, J. A.; Thurmond, C. D. *Phys. Rev. B.* **1976,** *14*, 3551.
17. Van Vechten, J. A. In *Lattce Defects in Semiconductors, 1974*; Institute of Physics: Bristol, 1973; p 1.
18. Phillips, J. C.; Van Vechten, J. A. *Phys. Rev. Lett.* **1973,** *30*, 220.
19. Swanson, M. L.; Davies, J. A.; Quenneville, A. F.; Saris, F. W.; Wiggers, L. W. *Radiat. Eff.* **1978,** *35*, 51.
20. Dannefaer, S.; Hogg, B.; Kerr, D. In *Proceedings of the 13th International Conference on Defects in Semiconductors*; Kimerling, I. C.; Parsey, J. M., Eds.; Metallurgical Society of AIME: Warrendale, 1985; p 225.
21. Seeger, A.; Frank, W. *Conf. Ser. Inst. Phys.* **1973,** *16*, 262.
22. Chicawa, J.; Shirai, S. *J. Cryst. Growth* **1977,** *39*, 328.
23. Strunk, H.; Gosele, U.; Kolbesen, B. O. *Appl. Phys. Lett.* **1979,** *34*, 530.
24. Tsai, J. C. C.; Schimmel, D. G.; Fair, R. B.; Maszara, W. *J. Electrochem. Soc.* **1987,** *134*, 1508.
25. Pantelides, S. T.; Car, R.; Kelly, P. J.; Oshryama, A. Amer. Phys. Soc. Meeting, Detroit, March 1984.
26. Simmons, R. O.; Ballufi, R. W. *Phys. Rev.* **1962,** *125*, 862.
27. Kimerling, L. C.; Larg, D. V. *Conf. Ser. Inst. Phys.* **1975,** *23*, 589.
28. Watkins, G. D.; Troxell, J. R.; Chattterjee, A. P. *Conf. Ser. Inst. Phys.* **1979,** *46*, 16.
29. Seeger, A.; Frank, W.; Gosele, U. *Conf. Ser. Inst. Phys.* **1979,** *46*, 1948.
30. Foll, H.; Gosele, U.; Kolbesen, B. O. *J. Cryst. Growth* **1977,** *50*, 90.
31. Petroff P. M.; deKock, A. J. R. *J. Cryst. Growth* **1975,** *30*, 117.
32. Gosele, U; Strunk, H. *Appl. Phys.* **1979,** *20*, 265.
33. Mizuo, S.; Higuchi, H. *Jpn. J. Appl. Phys.* **1981,** *20*, 739.
34. Antoniadis, D. A.; Moskowitz, I. *J. Appl. Phys.* **1982,** *53*, 9214.
35. Tan, T. Y.; Gosele, U. *Appl. Phys. Lett.* **1982,** *40*, 616.
36. Watkins, G. D. In *Effects Des Rayonnements Sur Les Semiconductors*; Dunod: Paris, 1965.
37. Seeger. A.; Chik, K. P. *Phys. Status Solidi* **1968,** *29*, 455.
38. Hu, S. M. *J. Appl. Phys.* **1974,** *45*, 1567.
39. Fair, R. B. *J. Appl. Phys.* **1980,** *51*, 5828.
40. Gosele, U; Tan T. Y. In *Defects in Semiconductors II*; Mahajan, S.; Corbett, J. W., Eds.; North Holland: Amsterdam, 1983.
41. Matsumato, S.; Ishikawa, Y.; Niimi, T. *J. Appl. Phys.* **1983,** *54*, 5049.
42. Mathiot, D.; Pfister, J. C. *J. Appl. Phys.* **1984,** *55*, 3518.
43. Fahey, P.; Barbuscia, G.; Mosleki, M.; Dutton, R. W. *Appl. Phys. Lett.* **1985,** *46*, 784.
44. Taniguchi, K.; Kurosawa, K.; Kashiwagi, M. *J. Electrochem. Soc.* **1980,** *127*, 2243.
45. Fair, R. B. *J. Electrochem. Soc.* **1981,** *128*, 1360.
46. Fahey, P.; Dutton, R. W.; Hu, S. M. *Appl. Phys. Lett.* **1984,** *44*, 777.
47. Harris, R. M.; Antoniadis, D. A. *Appl. Phys. Lett.* **1983,** *43*, 937.
48. Tsai, J. C. C.; Schimmel, D. G.; Ahrens, R. E.; Fair, R. B. *J. Electrochem. Soc.* **1987,** *134*, 2348.
49. Nishi, K.; Sakamoto, K.; Ueda, J. *J. Appl. Phys.* **1986,** *59*, 4177.
50. Marchiando, J. F.; Roitman, P.; Albers, J. *IEEE Trans. Electron Dev.* **1985,** *ED–32*, 2322.
51. Michel, A. E. In *Rapid Thermal Processing*; Sedgwick, T. O.; Seidel, T. E.; Tsaur, B. Y. Eds; Mater. Res. Soc. Symp. Proc.; Pittsburg, 1986; pp 3–13.

52. Hodgson, R. T.; Baglin, J. E. E.; Michel A. E.; Mader, S. M.; Gelpey, J. C. In *Energy Beam–Solid Interactions and Transient Processing*; Fan, J. C. C.; Johnson, N. M., Eds.; North-Holland: Amsterdam, 1984; pp 253–257.
53. Fair, R. B.; Wortman, J. J.; Liu, J. *J. Electrochem. Soc.* **1984,** *131*, 2387.
54. Hofker, W. K. *Phillips Res. Rep.* **1975,** *8*, 1–121.
55. Fair, R. B. In *Energy Beam–Solid Interactions and Transient Thermal Processing*; Biegelsen, D. K.; Rozgonyi, G. A.; Shank, C. V., Eds.; Mater. Res. Soc. Sym. Proc.; Pittsburg, 1985; pp. 381–392.
56. Morehead, F.; Hodgson, R. In *Energy Beam–Solid Interactions and Transient Thermal Processing*; Biegelsen, D. K.; Rozgonyi, G. A.; Shank, C. V., Eds.; Mater. Res. Soc. Sym. Proc.; Pittsburg, 1985; pp. 341–346.
57. Seidel, T.; MacRae, A. U. *Radiat. Eff.* **1971,** *73*, 1.
58. Lervidori, M.; Angelucci, R.; Cembali, F.; Negrini, P.; Solmi, S. in press.
59. Fair, R. B. *IEEE Trans. Electron Dev.* **1988,** *35*, 285.
60. Fair, R. B.; Subrahmanyan, R. In *Advanced Applications of Ion Implantation*; Sadana, D. K.; Current, M. L., Eds.; SPIE: Bellingham, 1985; Vol. 530, pp 88–96.
61. Mele, T. C.; Scarpulla, J.; Nulman, J.; Krusuis, J. P. *J. Vac. Sci. Technol., A.* **1986,** *4*, 832.
62. Ingram, D. C.; Baker, J. A.; Walsh, D. A. In *Ion Implantation Technology*; Current, M. I.; Cheung, N. W.; Weisenberger, W.; Kirby, B., Eds.; North-Holland: Amsterdam, 1987, pp 460–465.
63. Godfrey, D. J.; McMahon, R. A.; Hasko, D. G.; Ahmed, H.; Dowsett, M. G. In *Impurity Diffusion and Gettering in Silicon*; Fair. R. B.; Pearce, C. W.; Washburn, J., Eds.; Mates. Res. Soc. Sym. Proc.; Pittsburgh, 1985; pp 143–149.
64. Crowder, B. L.; Ziegler, J. F.; Cole, G. W. In *Ion Implantation in Semiconductors and Other Materials*; Plenum: New York, 1973; pp 257–266.
65. Cho, K.; Neuman, M.; Finstad, T. G.; Chu, W. K.; Liu, J.; Wortman, J. J. *Appl. Phys. Lett.* **1985,** *47*, 1321.
66. Sedgwick, T. O. Electrochem. Soc. Meeting; Las Vegas, 1985.
67. Oehrlein, G. S.; Cohen, S. A.; Sedgwick, T. O. *Appl. Phys. Lett.* **1984,** *45*, 417.
68. Fair, R. B. *J. Vac. Sci. Technol.* **1986,** *4*, 926.
69. Deal, B. E.; Grove, A S. *J. Appl. Phys.* **1965,** *36*, 3770.
70. Ligenza, J. R.; Spitzer, W. G. *J. Phys. Chem. Solids* **1960,** *14*, 131.
71. Jorgensen, P. J. *J. Chem. Phys.* **1962,** *37*, 874.
72. Reisman, A.; Nicollian, E. H.; Williams, C. K.; Merz, C. J. *J. Electronic Mater.* **1987,** *16*, 45.
73. Nicollian, E. H; Reisman, A. *J. Electronic Mater.* **1988,** *17*, 263.
74. Brawer, S. *Relaxation in Glasses and Composites*; Wiley: New York, 1968; Chapter 1.
75. Cabrera, N. Mott, N. F. *Rep. Prog. Phys.* **1948,** *12*, 163.
76. Ghez, R.; van der Meulen, Y. J. *J. Electrochem. Soc.* **1972,** *119*, 1100.
77. Rayleigh, D. O. *J. Electrochem. Soc.* **1966,** *113*, 782.
78. Hooper, M. A.; Clarke, R. A.; Young, L. *J. Electrochem. Soc.* **1975,** *122*, 1216.
79. Irene, E. A. *Appl Phys. Lett.* **1982,** *40*, 74.
80. Revez, A. G.; Mrstik, B. J.; Hughes, H. L.; McCarthy, H. L. *J. Electrochem. Soc.* **1986,** *133*, 586.
81. Han, C. J.; Helms, C. R. *J. Electrochem. Soc.* **1987,** *134*, 1299.
82. Helms, C. R.; Han, C. J.; de Larios, J. Meeting of the Electrochemical Society, Atlanta, 1988, Extended Abstract No. 213.
83. Han, C. J.; Helms, C. R. *J. Electrochem. Soc.* **1985,** *132*, 402.

84. Ho, C. P.; Plummer, J. D.; Meindl, J. D.; Deal, B. E. *J. Electrochem. Soc.* **1978,** *125*, 665.
85. Taft, E. A. *J. Electrochem. Soc.* **1978,** *125*, 968.
86. Rochet, F.; Aguis, B.; Rigo, S. *J. Electrochem. Soc.* **1984,** *131*, 914.
87. Hetherington, G.; Jack, K. H. *Phys. Chem. Glasses* **1962,** *3*, 129.
88. Bruckner, R. J. *Non-Cryst. Solids* **1971,** *5*, 177.
89. Williams, E. L. *J. Am. Ceram. Soc.* **1965,** *48*, 190.
90. Irene, E. A.; Ghez, R. *J. Electrochem. Soc.* **1977,** *124*, 1757.
91. Helms, C. R. *J. Vac. Sci. Technol.* **1979,** *16*, 608.
92. Krivanek, O. L.; Sheng, T. T.; Tsui, D. C. *Appl. Phys. Lett.* **1978,** *32*, 439.
93. DiStefano, T. H. *J. Vac. Sci. Technol.* **1976,** *13*, 856.
94. Feldman, L. C.; Stensgaard, I.; Silverman, P. J.; Jackson, T. E. In *Physics of SiO_2 and its Interfaces*; Pantelides, S. T., Ed.; Pergamon: New York, 1978; pp 344.
95. Schwettmann, F. N.; Chiang, K. L.; Brown, W. A. Meeting of the Electrochem. Soc., Seattle, 1978, Extended Abstract No. 276.
96. Gould, G.; Irene, E. A.; *J. Electrochem. Soc.* **1987,** *134*, 1031.
97. Revesz, A. G. *J. Electrochem. Soc.* **1977,** *124*, 1811.
98. Hetherington, G.; Jack, K. H. *Phys. Chem. Glasses* **1964,** *5*, 147.
99. Revesz, A. G. *J. Electrochem. Soc.* **1979,** *126*, 122.
100. Nakayama, T.; Collins, F. C. *J. Electrochem. Soc.* **1966,** *113*, 706.
101. Hopper, M. A.; Clarke, R. A.; Young, L. *J. Electrochem. Soc.* **1975,** *122*, 1216.
102. Singh, B. R.; Balk, P. *J. Electrochem. Soc.* **1978,** *125*, 453.
103. Hess, D. W.; Deal, B. E. *J. Electrochem. Soc.* **1977,** *124*, 735.
104. Singh, B. R.; Balk, P. *J. Electrochem. Soc.* **1979,** *126*, 1288.
105. Lewis, E. A.; Irene, E. A. *J. Electrochem. Soc.* **1987,** *134*, 2332.
106. Ligenza, J. R. *J. Phys. Chem.* **1961,** *65*, 2011.
107. Irene, E. A. *J. Electrochem. Soc.* **1974,** *121*, 1613.
108. Mott, N. F. *Proc. Roy. Soc. (London)*, A **1981,** *376*, 201.
109. Hahn, P. O.; Henzler, M. *J. Vac. Sci. Technol., A.* **1984,** *2*, 574.
110. Irene, E. A.; Massoud, H. Z.; Tierney, E. *J. Electrochem. Soc.* **1986,** *133*, 1253.
111. Tiller, W. A. *J. Electrochem. Soc.* **1980,** *127*, 625.
112. Deal, B. E.; Sklar, M. *J. Electrochem. Soc.* **1965, 112, 430.**
113. Dobson, P. S. *Philos. Mag.* **1971,** *24*, 567.
114. Dobson, P. S. *Philos. Mag.* **1972,** *26*, 1301.
115. Ho, C. P. Ph. D. Dissertation, Stanford University, Stanford, California, 1978.
116. Ho, C. P.; Plummer, J. D. *J. Electrochem. Soc.* **1979,** *126*, 1516.
117. Ho, C. P.; Plummer, J. D. *J. Electrochem. Soc.* **1979,** *126*, 1523.
118. Fair, R. B.; In *Semiconductor Materials and Process Technology Handbook*; McGuire, G. E., Ed.; Noyes: Park Ridge, NJ, 1988; pp 455–450.
119. Fair, R. B. In *Applied Solid State Science, Supplement* 2; Kahng, D., Ed.; Academic: New York, 1981; pp 1–108.

Received for review December 30, 1987. Accepted revised manuscript September 26, 1988.

7

Resists in Microlithography

Michael J. O'Brien[1] **and David S. Soane**[2]

[1]**Silicone Products Division, General Electric Company, Waterford, NY 12188**
[2]**Department of Chemical Engineering, University of California, Berkeley, CA 94720**

The drive toward increased circuit density in microelectronic devices has prompted significant efforts aimed at improving the resolution capabilities of lithographic equipment, materials, and processes. This chapter provides an overview of the various microlithographic strategies currently in use, with a special emphasis on resist materials, chemistry, and processing schemes. Emerging technologies are also described, which, although not yet implemented, may hold the key to future progress.

THE DEMAND FOR INCREASED CIRCUIT DENSITY on silicon chips over the last 25 years has continued to push up the level of integration. Photolithography has responded to this demand by improvements in exposure and alignment systems, production of new materials, and innovative fabrication methods. Today, 1–2-μm features are typical of critical geometries for most production devices, whereas in state-of-the-art processes, submicrometer features are becoming more common. In this chapter, various microlithographic strategies and the key role of polymers in this technology will be discussed. Inorganic resist materials, which are still in a highly exploratory stage of development, will not be covered.

Lithographic processes are based on radiation-induced alteration of highly specialized photosensitive polymeric films, which are called *photoresists* or, simply, *resists*. The photoresists used in semiconductor microlithography were originally developed for the printing industry (*1*). For a typical process, the resist is applied onto a substrate to form a thin uniform film. Irradiation through a glass plate or "mask" coated with an opaque material (usually chromium) bearing an array of circuit patterns allows se-

0065–2393/89/0221–0325$13.80/0

lected areas of the photoresist to be exposed (Figure 1). The modified or exposed regions of the polymer exhibit an altered rate of removal or development in certain chemical reagents (developers), which results in the formation of a polymeric relief image of the mask pattern. On the basis of the chemical nature of the photoresist, either a positive or a negative image of the original mask is formed. Resists that produce negative-tone images undergo cross-linking upon irradiation. Cross-linking renders these resists less soluble in the developer solvent. Conversely, positive resists undergo molecular changes that enhance their solubility in the developer such that exposed regions are preferentially removed.

The patterned resist image thus obtained delineates the areas in which subsequent modification or removal of the underlying substrate will take place. Through either chemical or physical processes, the substrate is altered in the unmasked regions, whereas the remaining resist protects the areas

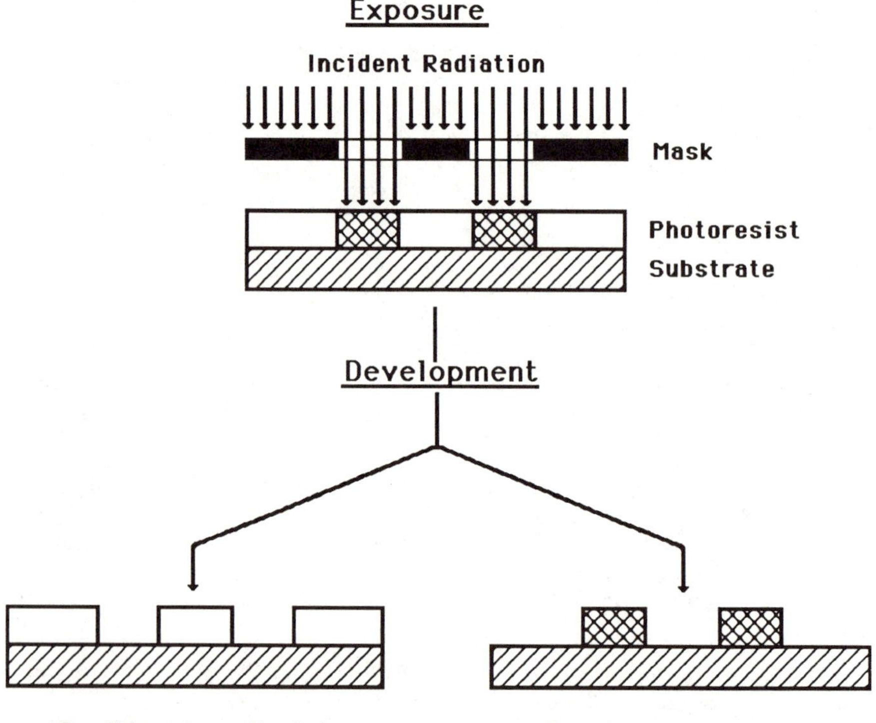

Figure 1. Diagram showing how irradiation through a mask allows selected areas of the photoresist to be exposed. In positive resists, the exposed areas become more soluble in the developer and, therefore, can be selectively removed. In negative resists, the exposed areas become less soluble in the developer, and thus, unexposed material is selectively dissolved.

where minimal change is intended. As the final step in this process, the remaining resist is stripped by wet- or plasma-etching methods. This photolithographic sequence is repeated for every patterned circuit layer on the semiconductor device. Each time, the appropriate mask is precisely aligned to the previous pattern on the wafer. Resist materials used in this application must meet stringent resolution and sensitivity requirements. They must also possess excellent film-forming properties and durability to withstand the highly corrosive chemicals, plasma treatments, and temperature cycles encountered during substrate etching, doping, and deposition processes.

Exposure Techniques

Optical Lithography. Lithographic processes can be classified according to the energy used to expose the resists and the equipment necessary to accomplish the process. Image quality depends on the exposure method, hardware, and resist material. In optical lithography, the resist is exposed to radiation within the near- to deep-UV region (200–450 nm).

Near-UV Lithography. Near-UV lithography, in which resists are exposed to radiation in the 350–450-nm near-UV range, is by far the most commonly used optical lithographic method in production. Exposure systems designed for this purpose are equipped with high-intensity mercury–xenon lamps as radiation sources and a variety of lenses and mirrors for light collimation. The spectral output of the mercury–xenon lamp in the 350–450-nm range has several strong peaks, the most important of which are at 365 nm (i line) and 436 nm (g line).

Several methods are available to image photoresists. In contact printing, the mask and substrate are brought into hard contact under vacuum. Exposure occurs through a mask, with the circuit pattern reproduced many times in an array. This procedure results in a 1:1 image of the entire mask on each wafer. Unfortunately, several major faults of this scheme offset the advantage of excellent resolution and preclude its use in the fabrication of high-density devices. First, scratches resulting from surface contact lead to wear and premature degradation of the mask. Second, unacceptably high levels of resist damage and particulate defects occur. Third, the inherent lack of absolute substrate and mask flatness precludes perfect contact and gives rise to distortion. These problems have prompted the development of more-sophisticated alignment tools, and now, contact printing is relegated primarily to the production of inexpensive chips with large device geometries.

Proximity printing, a variation of contact printing, preserves a minimum gap of approximately 10–30 μm between the silicon wafer and the mask. Although the problem of particulate contamination is avoided, light distortion is enhanced, and a loss in resolution results.

Projection printing uses a series of highly refined reflecting lenses to project the mask image onto the wafer over a distance of many inches. This method allows tight contamination control and prolonged mask life but is prone to optical aberrations. The necessity of using highly sophisticated optical systems and the mechanics required to achieve adequate alignment dramatically increase the cost of projection aligners.

The quality of pattern transfer differs greatly among the three modes of printing. As an example, a mask with parallel bundles of slits and spaces between slits with dimensions comparable with the slits can be considered. In this case, optical interference results in distorted images. The theoretical minimum dimension (for both space and slit) that allows resolvable interference peaks for contact or proximity printing is approximated by:

$$b_{\min} = \frac{3}{2}\left[\lambda\left(s + \frac{1}{2}d\right)\right]^{1/2} \tag{1}$$

In equation 1, $b_{\min}$ is the minimum feature size transferable, λ is the wavelength of light, s is the separation between the mask and the substrate, and d is the thickness of the resist layer. In projection printing, a series of undulating maxima and minima are produced. Because of mutual interference, the dark region is never completely dark, and the maximum brightness does not correspond to 100% transmission. The quality of transfer can be conveniently indicated by the modulation index, M, which is defined as follows:

$$M = \frac{I_{\max} - I_{\min}}{I_{\max} + I_{\min}} \tag{2}$$

In equation 2, $I_{\max}$ and $I_{\min}$ are the peak and trough intensities, respectively. Ideal optics would give an index equal to unity. However, in practice, all exposure systems behave less than ideally (i.e., $M < 1$).

Even though projection optics embodies the inherent limitation of pattern transfer just mentioned, this technique has become a dominant approach in high-resolution work. A key reason for this success is the ability of projection printing to use reduction refraction optics with high numerical apertures. The resolving power of projection systems can be approximated by:

$$W = \frac{k\lambda}{\mathrm{NA}} \tag{3}$$

In equation 3, W is the minimum feature size, k is an empirically determined constant that depends on resist processing, λ is the wavelength of the incident radiation, and NA is the numerical aperture of the optical system. Thus, resolution can be increased by using shorter wavelength radiation or by increasing the numerical aperture. In addition, some improvement in resolution can be achieved by adjusting processing conditions to minimize k.

Unfortunately, resolution gains through the use of high-NA optics or shorter wavelengths have a deleterious effect on the depth of focus (DOF), as shown by equation 4:

$$\mathrm{DOF} = \frac{\lambda}{2(\mathrm{NA})^2} \tag{4}$$

Because DOF is directly proportional to wavelength and inversely proportional to the square of NA, the use of shorter wavelength radiation has less effect than the increase in the numerical aperture.

Despite improvements in projection optics, interference phenomena unrelated to mechanical design continue to limit lithographic resolution. One such example is the standing-wave effect (2). During exposure, the incident light is only partially absorbed by the resist, and unabsorbed radiation can undergo partial reflection at the resist–substrate interface. The resulting reflected beam then sets up an interference pattern with the unabsorbed incident beam. A resist near a constructive node reacts more extensively than does a material near a destructive node. The uneven structural alteration manifests itself as scalloped edge profiles after resist development, which compromises pattern resolution. One solution to this problem is the use of an antireflective coating to reduce reflective waves (3, 4). Another approach involves the use of a postexposure bake step that smoothes the resist edges by diffusion of the reacted species (5). Multilevel resists, which will be discussed in a later section, offer still another remedy to this problem.

The fundamental limitations of optical interference can be suppressed greatly if the wavelength of the source radiation is shortened. Because pattern distortion is severe when feature resolution approaches the exposure wavelength, the use of short-wavelength radiation pushes the resolution towards finer features. Thus, the increasing trend is to explore deep-UV sources and to improve upon the existing near-UV hardware. The desire to reduce feature size has also generated much interest in X-rays and electron beams as alternative radiation sources.

Deep-UV Lithography. The important issues for deep-UV lithography (200–250 nm) are aligner optics and resist materials. Problems in aligner optics stem from the decreased transparency of standard lens materials in this frequency range, which necessitates the use of more-expensive construction materials such as quartz. Typical near-UV positive resists are not useful for deep-UV lithography because of unacceptable absorption at

200–250 nm. Resists tailored for improved performance in the deep-UV region, however, are now becoming available in new products (6). Nevertheless, deep-UV lithography remains in the development phase and is not currently used in integrated-circuit (IC) production.

Deep-UV source brightness is another issue, because the power output of a 1-kW mercury–xenon lamp in the 200–250-nm range is only 30–40 mW. For this reason, excimer lasers (such as KrCl and KrF), which can deliver several watts of power at the required wavelengths, are being considered as alternatives (7). In fact, a deep-UV step-and-repeat projection system with an all-quartz lens and a KrF excimer laser with an output at 248 nm has been reported (8). Even the laser-based systems require resists with a sensitivity of 30–70 mJ/cm^2.

Electron Beam Lithography. The ever-diminishing IC feature size has motivated the development of exposure techniques with high-energy sources. One such radiation source is the electron beam. This technology is routinely used to generate masks for photolithography and is foremost in experimental applications of complex-device fabrication. Its two major drawbacks are low throughput and high capital cost.

For direct wafer writing or mask fabrication, sensitive resists are necessary to ensure a reasonable throughput. A second problem is electron back scattering caused by collisions of electrons with atoms within the resist and substrate. In this situation, the electronic stopping power of organic resists is limited, and a large fraction of the incident electrons is allowed to reach the underlying substrate. Collision with the substrate causes random scattering and secondary and back electron generation. Hence, the resist is showered with electrons from the substrate, and the total energy deposited within the resist has a smeared distribution, with a broad base near the bottom (9). If a low-contrast resist is used, these scattered electrons may have sufficient energy to cause degradation. This effect lowers the line width resolution appreciably. For reasonable throughputs, sensitive resists are required, and for better line width control and resolution, a high-contrast resist is required.

X-ray Lithography. X-ray lithography is similar to optical lithography in that flood exposure of the entire wafer through a patterned mask is possible. Thus, the potential for production applications is greater. X-ray lithography also has the advantages of an essentially infinite depth of field, a high tolerance to dust and contamination, and the absence of standing waves. Because the radiation wavelength varies from about 0.5 to 3 nm, diffraction is not an issue. One challenge of X-ray lithography is the fabrication of high-quality masks. High-atomic-number metals, such as gold, are opaque to X-rays; thus they provide a shadow for pattern definition by the masks. Gold patterns are formed by electron beam (e-beam) lithography on substrates such as boron nitride, silicon carbide, or silicon nitride membranes (*10*). Organic films such as polyimide have been used also (*11, 12*). Because of the mismatch between the thermal expansions of the different materials used in mask making, stress-related pattern distortion is possible.

Other major practical problems must be overcome before X-ray lithography is accepted in production. Foremost is the availability of sensitive X-ray resists. To effect structural changes in the polymer, the incident radiation must be effectively absorbed. Hydrocarbon-based organic resists are often transparent to X-rays, and hence, X-ray resists must be made sensitive by the incorporation of X-ray-absorbing high-atomic-number atoms. This problem is a great challenge in the synthesis or formulation of resists. Alternatively, a higher intensity X-ray source can be developed, so that the total exposure time can be shortened. One such powerful source is synchrotron radiation; however, commercial implementation of this costly source has not been realized yet.

Resist Characterization

To accommodate the diverse needs of lithographic processes and device design specifications, resist properties vary. However, a few primary characteristics common to all resists can be used to gauge their performance. These characteristics include sensitivity, contrast, resolution, and etching resistance. Because resist performance is strongly operation dependent, comparison between materials must be made under identical conditions.

Analysis by Dissolution Curves. Most performance indicators require only an operational definition; these concepts are explained by a film dissolution curve. Figure 2 shows a family of such curves, in which the individual curves correspond to resist behavior in developer solution after exposure to the indicated radiation dose level. Figure 2 is constructed for

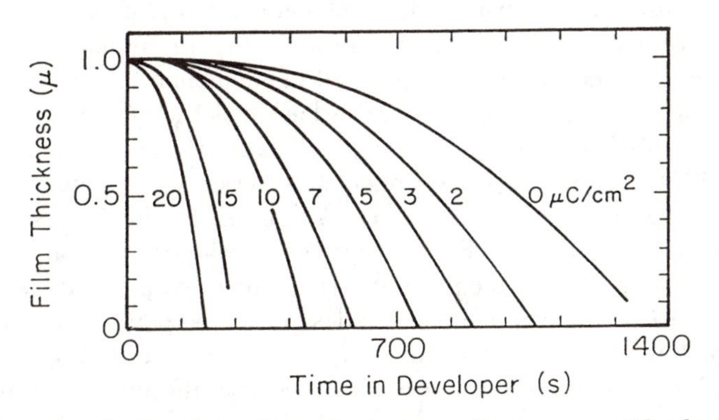

Figure 2. Dissolution curves for positive resists after exposure. The doses are designated by the numbers accompanying the traces. A stronger dose leaves a thinner film at a fixed development time.

positive resists, for which higher doses lead to faster dissolution. With negative resists, high doses tend to decrease the dissolution rate. Furthermore, such a family of curves is highly system specific. The same polymer developed in solutions of different strengths would give different sets of dissolution curves. Similarly, with identical developers and even the same manner of agitation, polymers dissolve at different rates if they are baked (annealed) and cooled at different temperatures and rates (*13*). If the starting material has a slightly different composition or molecular weight distribution, again, these curves would be shifted.

Many experimental techniques exist to determine dissolution curves by in situ monitoring of film development. The simplest technique is the laser end-point-detection system. In this system, monochromatic light from a He–Ne laser is directed at a resist layer from a near-normal direction (*14*). The reflected light is picked up by an adjacent optical fiber, and the intensity is analyzed by a diode detector. The output is a smooth trace with periodic oscillations. The peaks and valleys correspond to successive constructive and destructive interference nodes, which result from film thickness changes as the resist is etched away. From these periodic output traces obtained with resists exposed to varying degrees of radiation, the family of characteristic curves shown in Figure 2 can be constructed. Other more-sophisticated techniques exist for this purpose, including in situ ellipsometry and two-wavelength interferometry. These techniques will be discussed in a later section.

Analysis of Sensitivity. From the characteristic dissolution curves, a cross plot can be made of the normalized remaining film thickness (ratio of current thickness to original thickness) as a function of cumulative dosage. Figure 3 gives such curves for a positive and a negative resist. These curves are referred to as *sensitivity* or *exposure response curves* for resists. For positive resists, the governing phenomenon is film disappearance, whereas for negative resists, the important criterion is the film remaining. The minimum dose needed to cause the relevant phenomenon to emerge, as measured by the development procedure, is known as the *incipient dose* for the particular resist under study. The incipient dose corresponds to the intercept formed by the two extrapolated regions of the curve, denoted as D_p^o and D_x^o in the figure. The *completion dose* is denoted by the same symbols but without the superscript *o*. For positive resists, the completion dose corresponds to the point at which the film is completely dissolved, whereas for negative resists, the completion dose designates the point at which the film is completely intact. These completion doses may be called the resist sensitivity; however, these doses are not necessarily those required to yield a lithographically useful image and are highly dependent on the processing conditions chosen.

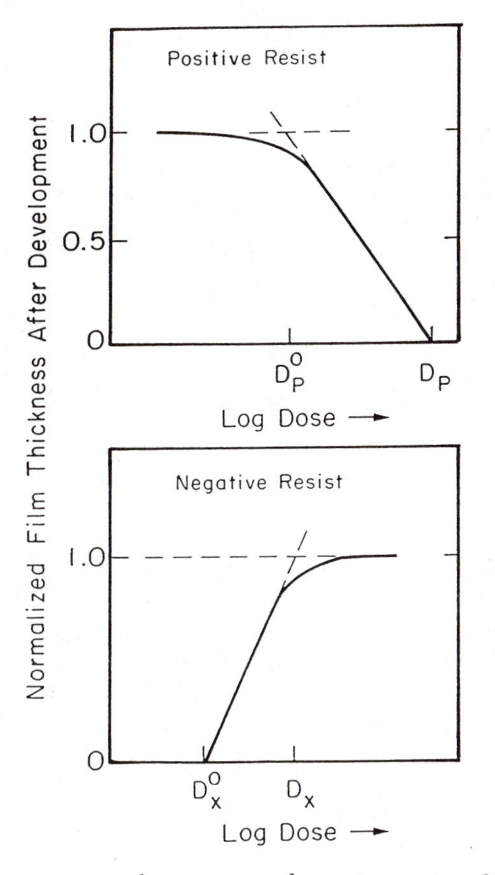

Figure 3. Response curves for positive and negatives resists. Marked on the curves are incipient and completion doses, which indicate the onset and completion of observable events. The variables monitored are film attrition for positive resists and film remaining for negative resists. These traces can be affected by a number of process parameters, particularly development conditions.

Analysis of Contrast. The *contrast*, γ, of a given resist, is defined mathematically for positive and negative resists by equations 5 and 6, respectively.

$$\gamma = \left(\log \frac{D_p}{D_p{}^o}\right)^{-1} \tag{5}$$

$$\gamma = \left(\log \frac{D_x{}^o}{D_x}\right)^{-1} \tag{6}$$

In general, the higher the contrast, the sharper are the edge profiles of developed lines. Contrast is also a resist quality that can be fine tuned by judicious choice of processing parameters.

In the case of polymeric e-beam and X-ray resists, which undergo bond breakage upon irradiation, followed by chain scission (positive resists) or cross-linking (negative resists), resist sensitivity can be represented by a structure-dependent constant called a G value. G_s is a measure of scission efficiency, and G_x is a measure of cross-linking efficiency. G_s values for resists that undergo only chain scission can be determined experimentally by plotting the inverse of the number-average molecular weight (M_n^*) of the polymer versus the exposure dose (D). As shown by equation 7, such a plot gives a straight line with a slope that is directly proportional to G_s.

$$\frac{1}{M_n^*} = \frac{1}{M_n^o} + \left(\frac{G_s}{100}\,\mathrm{AN}\right)D \tag{7}$$

In equation 7, M_n^o is the initial number-average molecular weight and AN is Avogadro's number. When both chain scission and cross-linking occur, the G values for both processes (i.e., G_s and G_x) can be determined. This is accomplished after measuring changes in the number-average molecular weight (M_n) and the weight-average molecular weight (M_w) and then solving the following equations simultaneously:

$$\frac{1}{M_n} = \frac{1}{M_n^o} + (G_s - G_x)D \tag{8}$$

$$\frac{1}{M_w} = \frac{1}{M_w^o} + (G_s - 4G_x)D \tag{9}$$

Again, in equations 8 and 9, D is the cumulative dose and the superscript o reflects the original molecular weights of the unexposed sample (*15*, *16*).

Analysis of Resolution. *Resolution* defines the ability of the resist to resolve fine lines in the final printed pattern. Although resolution is highly dependent on the chemistry of the resist and developer system, it is not determined solely by resist materials. Distortion induced by exposure hardware is one culprit of poor resolution. Another cause of poor resolution is resist deformation because of thermal flow that can occur during resist processing. When a resist is heated above the glass transition temperature (T_g) during a baking or etching step, thermal flow occurs. High-temperature treatments are not normally encountered, but for some processes with poor temperature control, thermal flow is unavoidable. These situations may exist during plasma-etching, doping, or deposition steps with temperatures in excess of 200 °C. Postdevelopment treatments that harden the resist and prevent thermal flow have been formulated.

The resolution of a resist can be determined either optically or electrically by using special line-width-measuring equipment or by examining the resist with a scanning electron microscope (*17*). Correct feature size must be maintained within a wafer and from wafer to wafer, because device performance depends on the absolute size of the patterned structures. The term *critical dimension* (CD) refers to a specific feature size and is a measure of the resolution of a lithographic process.

Analysis of Etching Resistance. After the printed image is formed, resists are often exposed to corrosive or physically abusive environments during subsequent processing steps. For example, solutions used for wet etching usually consist of strong acids or bases. Dry etches often use an oxygen plasma that removes or "ashes" organic materials. Chlorinated plasmas used to etch aluminum and fluorinated plasmas used to etch silicon oxide and silicon nitride are extremely corrosive to resists. *Etching resistance* refers to the ability of the resist to withstand conditions necessary to transfer the printed pattern to the underlying film or substrate. Naturally, this ability is a function of the resist chemistry. In addition to physical and chemical stability, the criteria for etching resistance include adhesion to the substrate. Adhesion is usually monitored qualitatively by visual observation. Resist adhesion must be maintained to a variety of substrates including metals, insulators such as silicon dioxide and silicon nitride, and other semiconductor materials. Typically, adhesion promoters such as hexamethyldisilazane (HMDS) are used prior to resist application (*18*). Loss of adhesion is less of a problem with dry etching than with wet etching, for which it remains a critical concern. However, dry etching places more demands on resist thermal and radiation stabilities.

Resist Materials

Lithography is a central technology that normally uses polymers for semiconductor fabrication. The polymer-based resists must meet rigorous requirements: high sensitivity, high contrast, high T_g, good etching resistance, good resolution, easy processing, purity, long shelf life, minimal solvent use, and reasonable cost. The foregoing list of requirements is formidable. In reality, a particular resist will satisfy these varied stringent requirements only to a certain extent, with the specific lithographic application dictating the acceptable compromises in material properties. As we enter the generation of submicrometer devices, greater demands are placed on each aspect of lithography, including resist material properties. In this section, resist materials that are used traditionally in semiconductor lithographic processes will be examined, as well as the emerging technologies dedicated to material and process improvements for very-large-scale-integration (VLSI) and ultra-large-scale-integration (USLI) applications.

Optical Resists. Photoresists are either negative or positive acting, on the basis of the fundamental chemistry that takes place upon exposure. Both types of resists are exemplified by well-established commercial products currently used in the IC industry (*19*). Except for dry-film resists used in the manufacture of printed circuit boards, photoresists are supplied as premixed solutions that comply with purity standards imposed by this industry. With equipment designed for solution processing, photoresists are dispensed onto wafer substrates and spun to form thin glassy films that are typically 0.5–2.0 μm thick after baking. Subsequently, the resists are exposed in an optical aligner using a chrome-on-glass mask and then developed to form a polymeric relief film that functions as a mask for further processing of the underlying substrate.

Negative-Acting Resists. Historically, negative optical resists were the first to be used in semiconductor device fabrication. The most commonly used negative-acting resists are bis(aryl)azide–rubber resists, whose matrix resin is cyclized poly(*cis*-isoprene), a synthetic rubber. The bis(aryl)azide sensitizers (*20*) lose nitrogen and generate a highly reactive nitrene upon photolysis. The nitrene intermediate then undergoes a series of reactions that result in the cross-linking of the resin and the decrease in the solubility of irradiated areas in organic solvents. For example, the nitrene can add to olefins present in the resin to produce aziridine structures, insert into carbon–hydrogen bonds to give amines, or dimerize to give azobenzene units. Because oxygen can interfere with these reactions, all exposures are done either under the protection of a nitrogen blanket or in vacuum.

The bis(aryl)azide sensitizers must be soluble in the resin, thermally stable, and sensitive to the desired wavelength of light. A commonly used compound is 2,6-bis(4-azidobenzylidene)cyclohexanone, which absorbs at 360 nm (*see* structure). Conjugation extension and other structural changes can shift the absorption maximum to longer wavelengths and allow access to the other mercury lines at 405 and 436 nm. A number of bis(aryl)azides are efficient photosensitizers (*21*). Quantum yields, defined as the number

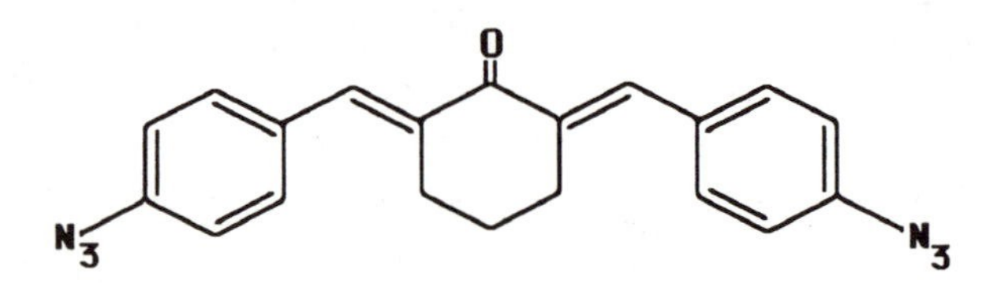

2,6-Bis(4-azidobenzylidene)cyclohexanone, a negative-resist sensitizer

of events triggered by the absorption of each photon, range between 0.4 and 1.0 for these examples.

A major limitation of negative resists is poor resolution resulting from film swelling during development. Swelling is caused by solvent uptake in the cross-linked polymer network. The line width limit of these materials is ~3 μm. This fact, coupled with a growing industrial trend away from organic solvents, favors the use of positive resists, which use water-based developers, in high-density-device applications. However, because they require only small amounts of expensive sensitizers, many negative resists cost substantially less than their positive counterparts. Also, negative resists sometimes have greater process latitude. These facts result in a continuing dominance of negative photoresists in the production of low-cost, high-volume chips (*22*).

More-recent developments have enhanced the resolution capabilities of bis(aryl)azide-based negative resists. For example, a second-generation version that is developable in organic solvent has been reported. This version uses a proprietary polymeric system that greatly diminishes swelling and allows 1.25-μm resolution (*23*). Also, excellent resolution has been achieved with systems combining diazides with aqueous-base-soluble phenolic resins. Again, the enhanced resolution is a result of the use of these nonswelling polymers (*24*).

Positive-Acting Resists. Positive resists have gained popularity in recent years mainly because of their superior resolution potential and also because of their better etching resistance and thermal stability. Diazonaphthoquinone (DNQ)–novolac-based resists represent the workhorse of the industry. These resists are composed of an aqueous-base-soluble novolac resin, which is prepared via the acid-catalyzed polymerization of cresols with formaldehyde, and a sensitizer, DNQ, which is base insoluble (Scheme I). The sensitizer is present in sufficient quantity (usually 15–20% by weight of resin) to drastically inhibit the dissolution of the novolac in aqueous alkali solutions. Upon photolysis, DNQ loses nitrogen to give a carbene, which subsequently undergoes a Wolff rearrangement to yield a highly reactive ketene (*25*). Under normal conditions, the ketene can then react with water present in the resin to form base-soluble indenecarboxylic acid (ICA). Thus, the irradiated areas of the resist are quite soluble in developer solution (typically a 0.05–0.5 N aqueous solution of KOH, NaOH, tetraalkylammonium hydroxide, or other organic bases).

Although DNQ–novolac systems were first used more that 40 years ago in the printing-plate industry, they continue to be the focus of significant interest. For example, several recent studies have addressed the optimization of novolac properties through manipulation of molecular weight, isomeric structure of the phenolic starting materials, and methylene bond position (*26–28*). These changes are reported to influence resist sensitivity, contrast, and process latitude.

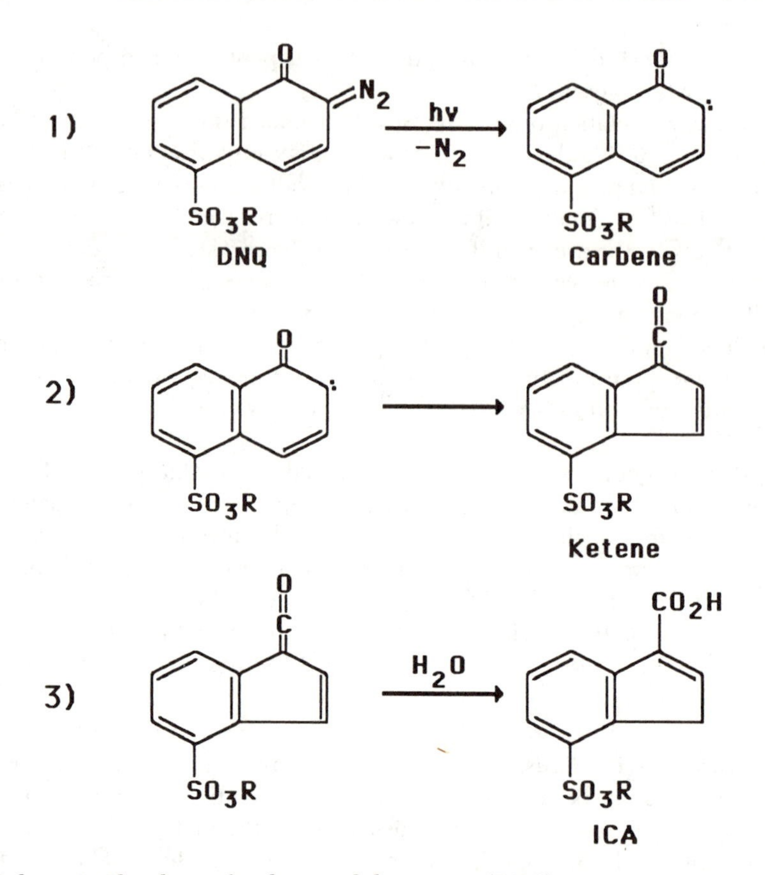

Scheme I. Photolysis of a diazonaphthoquinone (DNQ) positive-resist sensitizer. The reaction leads to a carbene (1), which undergoes a Wolff rearrangement to give a ketene (2). Finally, this ketene can react with water present in the resin to give an indenecarboxylic acid (ICA) (3).

Research has continued also in the area of sensitizer optimization. An example of a recent finding in this area is the so-called "polyphotolysis" phenomenon (*29, 30*). Polyphotolysis refers to enhanced resolution achieved through the attachment of multiple DNQ sensitizer groups to a central ballast molecule. During irradiation, each of the DNQ groups is sequentially and independently converted to an ICA photoproduct group (Scheme II). Maximum resolution enhancement is achieved under conditions in which the totally photolyzed material has a large effect on the dissolution rate, whereas the intermediate forms containing both DNQ and ICA units have very little influence on development.

Another example of variation in DNQ structure was prompted by the advent of i-line (365-nm) steppers. These exposure tools are capable of greater resolution compared with their g-line (436-nm) counterparts because

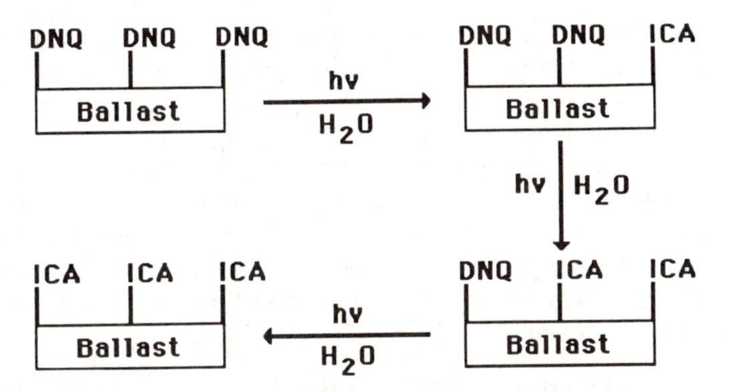

Scheme II. Sequential photolysis of DNQ groups attached to a central ballast molecule during polyphotolysis. Maximum resolution enhancement occurs when the totally photolyzed material controls the dissolution rate.

of the use of shorter wavelength radiation. However, most typical positive resists are not optimized for this wavelength. An improvement in performance at the i line has been accomplished through the use of 2,1,4 isomers of DNQ instead of the more typical 2,1,5 compounds (*31*) (*see* structures).

Further improvements in resist performance have involved the use of special additives. For example, antistriation agents and plasticizers have been used to improve film quality. Other additives that are sometimes used include adhesion promoters, speed enhancers, and nonionic surfactants. Organic dyes are useful in the control of scalloped resist profiles resulting from reflective interference or the so-called standing-wave effect (*32*, *33*). A recent paper (*34*) reported that dyes that contain organic acid groups can also improve the resist side wall angle through formation of a less developer-soluble skin on the surface of the film during soft bake.

Another method of improving the resolution capability of a resist is through optimization of processing conditions. A more in-depth discussion of processing follows; however, some special processing conditions used with DNQ–novolac resists will be mentioned now. As indicated earlier, a postexposure bake step is used sometimes to minimize standing waves. Resist profiles become smooth during heating because of the diffusion of unexposed

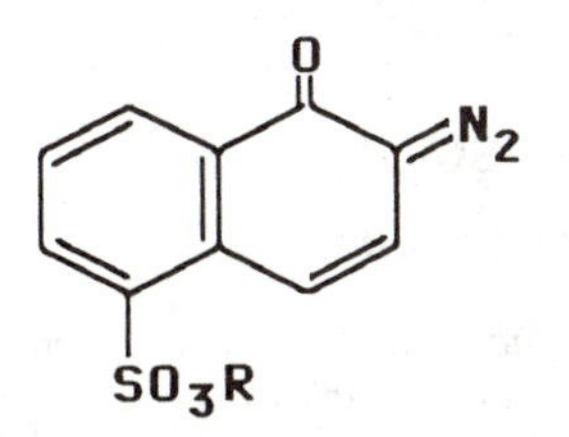

2,1,5 DNQ

2,1,4 DNQ

DNQ from regions of high concentration to regions of low concentration at the edges of the exposed areas (*5*). The application of a deep-UV flood exposure during a postbake step can increase contrast, again by the formation of a less base-soluble surface skin (*35*).

Another method of increasing resist performance through the alteration of processing conditions involves the use of high-contrast developers (*36*). These developers are typically more-dilute aqueous-base solutions that improve selectivity between exposed and unexposed areas of the resist. Unfortunately, the high-contrast developers also tend to increase the time required for development.

Special Modifications of DNQ–Novolac Resists. A technique has been developed that allows DNQ–novolac positive resists to be imaged in the negative mode. Resist image reversal exploits the fact that the ICA photoproducts can be decarboxylated if they are heated in the presence of bases such as imidazole or amines (*37, 38*). Thus, if a positive resist is exposed, treated with a base, and then postbaked, the ICA groups are decarboxylated and converted to aqueous-base-insoluble indenes that act as novolac dissolution inhibitors (Scheme III). A subsequent flood exposure transforms the DNQ sensitizer remaining in the previously unexposed resist to ICA, so that these regions become soluble in developer. The net result is a negative image of the mask. Image reversal of positive resists provides greater resolution, thermal stability, and a reduction of standing-wave effects (*39*).

Recently, other thermally induced image-reversal processes have been

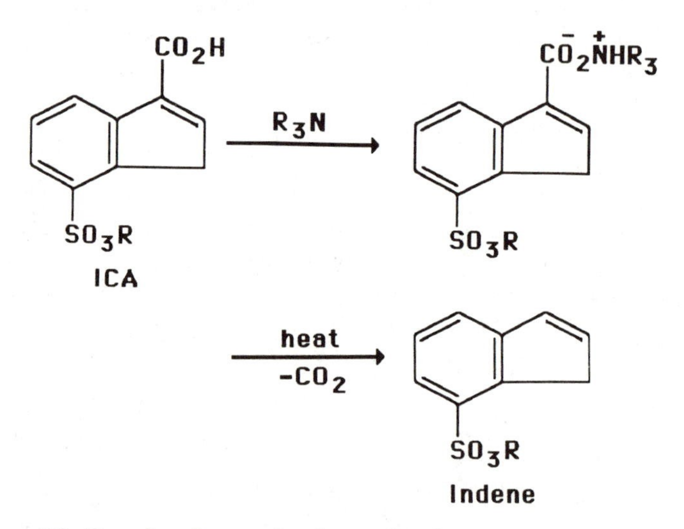

Scheme III. Decarboxylation of indenecarboxylic acid (ICA) photoproducts during image reversal. The process involves treatment with base to form a carboxylate salt, followed by baking to generate ultimately a base-insoluble indene.

described that do not rely on decarboxylation of photolyzed sensitizers (*40*, *41*). Instead, the mechanism of image reversal in these examples involves the acid-catalyzed cross-linking of the resin during postbake. For example, an image-reversal system composed of a 2,1,4 DNQ sensitizer, novolac resin(s), and a thermal cross-linking agent has been reported (*41*). Photolysis of the 2,1,4 DNQ sensitizer produces an acid that is stronger than that generated by the 2,1,5 isomers. Postbaking allows this acid, in combination with the cross-linking agent, to substantially decrease the aqueous-base solubility of the novolac. As in the traditional image-reversal scheme, a flood exposure is then used to generate acid from DNQ in the previously unexposed areas. Thus, development provides a negative image of the mask.

Another technique that has been applied to DNQ–novolac resists is the so-called DESIRE (diffusion-enhanced silylating resist) process (*42*). In this scheme, the resist is exposed and then treated with a silylating agent such as HMDS. Under the appropriate conditions, the silylating agent can react selectively with the novolac resin in the exposed areas. The mechanism by which DNQ inhibits silylation whereas ICA allows it is not well understood. Nonetheless, because the irradiated areas are silylated, they become resistant to etching with an oxygen plasma because of the formation of a silicon oxide layer. Consequently, an oxygen plasma can be used as a developer to etch away the unexposed (nonsilylated) resist. Once again, the net result is the formation of a negative-tone image in a positive resist.

This scheme possesses several distinct advantages over conventional processing. For example, the fact that only the surface needs to be exposed, coupled with the anisotropic nature of plasma etching, allows the use of thick resist layers that can planarize the underlying topography to a higher degree than resist films with typical thicknesses can. Loss of depth of focus because of the use of high-NA lenses is also less of an issue, because only the surface needs to be imaged.

Another innovation, known as *contrast enhancement*, extends the practical resolution of optical lithography (*43–45*). Contrast enhancement uses photobleachable materials in conjunction with standard photoresists to increase the contrast of illumination that reaches the resist. A highly absorbing but photobleachable dye layer (contrast enhancement layer or CEL) is spun on top of a conventional positive resist. This dye layer is typically composed of a diarylnitrone dissolved in a matrix resin. As the system is exposed to the projected mask image, the top layer is gradually converted from an opaque (strongly absorbing) coating to a transparent (nonabsorbing) coating via photochemical conversion of the nitrone to an oxaziridine (Scheme IV). The bleaching rate of the nitrone is governed by the cumulative incident-radiation dose. Edges of the projected mask features where light interference creates intensity shoulders bleach slowly. Meanwhile, the intensity peaks have enough time to bleach locally the entire thickness of the CEL. Hence,

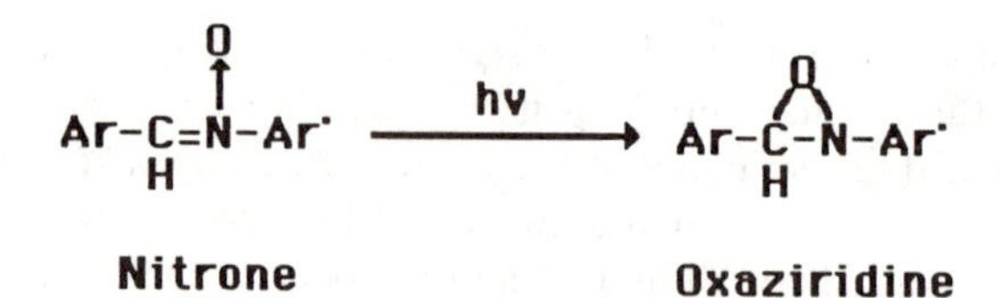

Scheme IV. Photobleaching of a diarylnitrone to an oxaziridine in contrast-enhancement materials.

areas exposed to interference maxima become transparent, whereas areas corresponding to interference fringes remain protected by the CEL. The net result is that the resist layer receives an exposure pattern that is sharper than that transmitted through the aligner optics. Therefore, the contrast is improved over what it would be without the CEL. An example of the successful application of this concept is the recent report that high-contrast 0.5-μm lines and spaces can be printed on an i-line stepper by using a commercial contrast enhancement material (*46*). Trade-offs in the use of CEL materials are longer exposure times, additional processing steps, and, at times, problems of compatibility between dyes and resists.

Deep-UV Photoresists. ***Single-Component Resists.*** Because a variety of polymers can absorb energetic deep-UV photons, many of the resists used in this wavelength region are single component, that is, they are composed of a polymer dissolved in solvent without a sensitizer. Polymeric chain scission (degradation) or cross-linking reactions occur upon UV irradiation and change the solubility of the exposed resist in the developer. As mentioned earlier, resist sensitivity is an important issue because of the low intensity of conventional Hg arc lamps in this wavelength region. In general, e-beam resists, the most common of which is poly(methyl methacrylate) (PMMA), have been used in this capacity (*47*). PMMA shows maximum sensitivity at 220 nm and is insensitive above 260 nm. Development work has focused on producing materials that absorb at 230–280 nm, the range at which lamp output is higher. Attempts to improve PMMA sensitivity include the use of copolymers of methyl methacrylate and the replacement of side chains. All of these changes result in positive-acting resists (*48*). One modified PMMA-type resist is poly(methyl isopropenyl ketone) (PMIPK) (*49*, *50*); its photosensitivity is five times higher than that of PMMA. The poly(olefin sulfone) family also has produced a series of deep-UV resists, despite the fact that standard materials in this class, such as poly(butene-1-sulfone), do not absorb light above 215 nm. Therefore, to make these materials functional in this wavelength range, photosensitizers have been added (*51*); for instance, poly(olefin sulfone)s have been mixed with novolac resins (*51*), or aromatic groups have been attached to the olefinic portion of the polymer (52).

e-Beam epoxy-based resists such as poly(glycidyl methacrylate) (PGMA) have been used also in this application (*53*). Although PGMA normally behaves as a negative e-beam resist, positive-tone imaging is observed under deep-UV radiation. In this case, the chemistry of the methacrylate group is responsible for the photoresponse. As an e-beam resist, PGMA undergoes ring-opening polymerization of the epoxy moiety.

Two-Component Resists. Typical two-component DNQ–novolac photoresists are not well suited for use in the deep UV because of the strong unbleachable absorbance of the novolac and sensitizer photoproducts below 300 nm. Therefore, the optical density of these materials is very high in the deep-UV and it does not decrease (bleach) with exposure. At doses that allow light to penetrate to the bottom of the resist, the top of the film is overexposed, and sloped profiles are produced.

Many modifications of this basic chemistry have been explored to tailor these resists to deep-UV radiation. For example, changes have been made in the sensitizer so that it bleaches in this wavelength region. Early work in this area was performed on diazo-Meldrum's acid (*54*) (*see* structure). This compound functioned as a deep-UV-bleachable dissolution inhibitor; however, it was somewhat volatile and, consequently, could be depleted via evaporation during soft bake. More-recent studies have therefore focused on less-volatile sensitizers incorporating heteroatom substitution (*55*) and on increases in molecular weight (*56*).

Research has also been aimed at the development of more-transparent base-soluble matrix resins. For example, novolacs prepared from pure *p*-cresol absorb less strongly at 250 nm than do typical photoresist novolacs containing a mixture of cresol isomers. Unfortunately, *p*-cresol novolac is only sparingly soluble in aqueous base and has limited usefulness (*28*, *57*). Other examples of more-transparent matrix resins include poly(dimethyl glutarimide) (PMGI) (*58*) and copolymers of methyl methacrylate (MMA) and methacrylic acid (MAA) [P(MMA–MAA)].

Deep-UV resists have also been prepared by changing both sensitizer and matrix resin. For example, materials combining *o*-nitrobenzyl ester derivatives of cholic acid with a P(MMA–MAA) matrix resin (Scheme V) have been reported (*59–61*). Upon photolysis, the nitrobenzyl ester dissolution

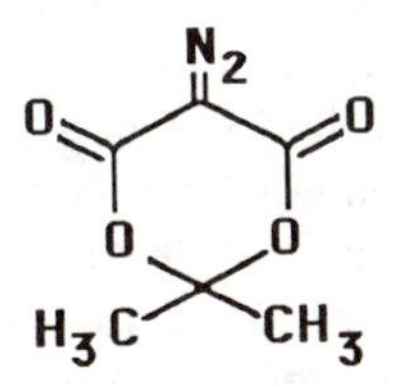

Diazo-Meldrum's acid, a deep-UV bleachable sensitizer

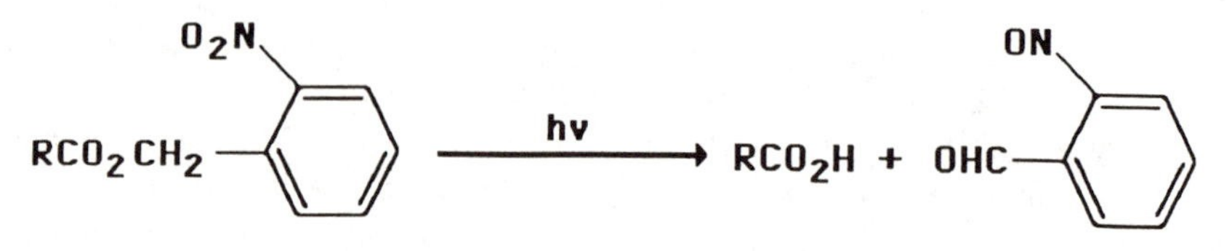

Scheme V. Photolysis of o-*nitrobenyl ester to generate a carboxylic acid.*

inhibitor is converted to a cholic acid derivative, and the irradiated areas of the resist are changed to an aqueous-base-soluble form. The oxygen-plasma-etching resistance of this system has been improved recently by changing the matrix resin to a silicone-substituted P(MMA–MAA) polymer (62).

Resists Based on Chemical Amplification. In a patent issued in 1973, Smith and Bonham reported the preparation of positive resist materials composed of a water-insoluble organic compound containing acid-labile groups (such as acetals) in combination with a material capable of generating an acid upon photolysis (63). The acid-labile functionality of the organic compound was incorporated into a polymeric system by attaching it either directly to the polymer backbone or pendant to the main polymer chain. Alternatively, nonpolymeric acid-labile materials were used. However, in this case, a suitable binder resin was also required to facilitate film formation. The photosensitive component of these resists was the acid generator. For example, trihalomethyl-substituted *s*-triazines, which upon photolysis generate HX, were used. This strong acid then acts as a catalyst in the "dark" or "nonphotochemical" hydrolysis of the acid-labile groups. Thus, the irradiated areas, where catalyst is generated, become more soluble in the developer.

Subsequently, a resist system was reported that is composed of polymers containing recurrent acid-labile pendant groups in combination with an arylonium salt–acid photogenerator (64–67). An example of the type of polymer used in this work is poly(*p-tert*-butyloxycarbonyloxystyrene) (*t*-BOC-styrene).

The chemistry involved in the processing of these materials is illustrated in Scheme VI. First, deep-UV irradiation of either a triarylsulfonium or diaryliodonium salt results in the generation of an extremely strong protonic acid. During a postexposure bake step, this acid can catalytically remove the acid-labile groups to convert the polymer to a much more polar form [for example to poly(4-hydroxystyrene)]. Development with a polar solvent allows selective dissolution of the irradiated areas, and positive-tone images are generated. Alternatively, nonpolar developers can be used to selectively remove unirradiated material and negative-tone images are generated. Therefore, a single resist latent image can be processed to produce a polymeric relief image of either tone by appropriate selection of developer media. The fact that only a catalytic amount of acid needs to be photoge-

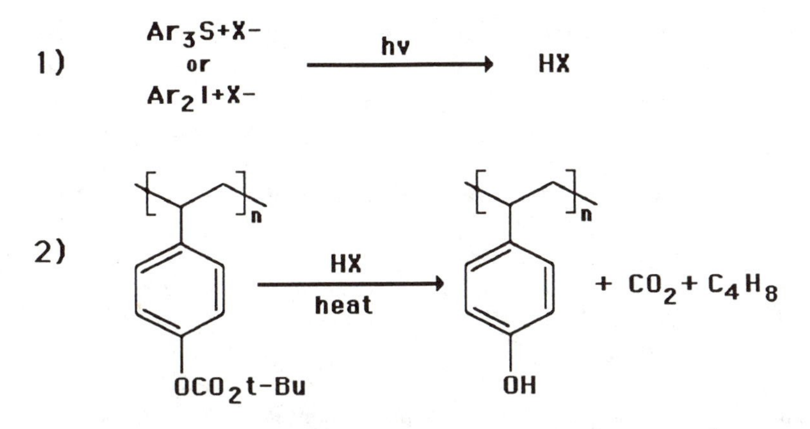

*Scheme VI. Processing of poly(*t-*BOC-styrene)–onium salt resists. The steps are (1) photogeneration of an extremely strong protonic acid from a triarylsulfonium or diaryliodonium salt and (2) baking, which allows acid-catalyzed deblocking of the* t-*BOC groups. Thus, irradiated areas of the polymer are converted to a much more polar form.*

nerated to deblock a large number of acid-labile groups during postbake makes these materials highly photosensitive.

The process by which enhanced photosensitivity is achieved through the use of light to generate a catalyst is *chemical amplification*. Materials based on this chemistry are capable of extremely high resolution. Intuitively, one might predict that the resolution of a resist in which a catalyst is formed and then undergoes a large number of subsequent reactions might be limited because of the migration of the catalyst to unexposed areas. However, recent reports have shown that e-beam exposure of resists based on the acid-catalyzed deblocking of *t*-BOC-styrene polymers are capable of printing line widths as narrow as 18 nm (*68*).

Recently, many similar systems have been reported in the literature. Examples include acid-photogenerating compounds in combination with *t*-BOC-protected maleimide or hydroxyphenyl maleimide copolymers (*69, 70*), *tert*-butyl ethers of phenolic resins (*71*), silylated phenolic resins (*72, 73*), and polycarbonates that contain acid-labile linkages in their backbone (*74, 75*). Onium-salt-photogenerated acid has also been used in another two-component system to catalytically depolymerize polyphthalaldehyde (*76*).

Several groups have investigated three-component systems encompassing both chemical amplification and dissolution inhibition. As stated earlier, Smith and Bonham (*63*) reported resist materials composed of a binder resin (novolac), a nonpolymeric compound containing acid-labile functional groups such as acetals, and a trihalomethyl-substituted *s*-triazine acid photogenerator. The acid-labile compound acts as a novolac dissolution inhibitor in a manner analogous to the action of DNQ in conventional positive resists. However, in this case, the inhibitor is not photochemically active. Instead, irradiation of the *s*-triazine derivative generates an acid that can then catalyze the hydrolysis of the dissolution inhibitor to an aqueous-base-soluble form.

Recently, the use of nonpolymeric *tert*-butyl esters, aryl *tert*-butyl carbonates, and aryl *tert*-butyl ethers as novolac dissolution inhibitors in similar resist materials using onium salt photoinitiators was investigated (*77*). Resist materials based on this chemistry with the di-*tert*-butyl carbonate of bisphenol A as dissolution inhibitor (*78*) have also been reported. Unlike the acetal-type acid-labile compounds, these new materials require a bake step after photogeneration of acid to convert the inhibitors to an aqueous-base-soluble form. Interestingly, these resists function as very sensitive positive-tone materials, despite the fact that previous work has shown that the solubility of novolacs in aqueous base can be substantially decreased by baking them in the presence of onium-salt-photogenerated acids (*79*). Also significant is the fact that although these resists are based on novolac and therefore still possess substantial unbleachable absorbance below 300 nm, chemical amplification of the photochemistry allows high-contrast imaging. In fact, O'Brien and Crivello (*77*) demonstrated that their resists were capable of submicrometer imaging when used with a deep-UV excimer laser imaging tool.

Other three-component systems based on this chemistry have made use of the following acid-labile dissolution inhibitors: polyphthalaldehyde (*80*), ketals of β-ketoesters (*81*), and compounds containing C–O–Si bonds (*82*). Similar resists have also been used with other radiation sources; these will be discussed in subsequent sections.

One final example of the application of onium salt photochemistry in positive resist materials should be mentioned, because it does not include any postexposure acid-catalyzed processes and therefore does not encompass the principle of chemical amplification (*79*). Interestingly, Newman (*79*) has determined that onium salts themselves can inhibit the dissolution of novolac in aqueous base and that irradiation of such an onium salt–novolac resist restores the solubility of the resin in developer and leads to a positive-tone image. In this application, the onium salt behaves like diazonaphthoquinone in a typical positive resist. Recently, Ito (*80*) has reported also the use of onium salts as novolac dissolution inhibitors.

e-Beam Resists. A wide variety of materials that have been investigated as e-beam and X-ray resists have spawned a large amount of literature on these topics (*14*, *83–85*). Currently, numerous commercial e-beam resists are produced for mask making and direct write applications by U.S. and Japanese companies (*85*, *86*).

Bombardment of polymers by electrons causes bond breakage. The free radical or ionic sites thus generated activate subsequent scission or cross-linking reactions. Because all polymers are susceptible to electron-induced reactions, essentially any material can, in principle, function as a resist.

However, the important considerations are sensitivity and tone (positive or negative). When covalent bonds are formed, the increase in molecular weight results in negative-acting resists. A molecular weight decrease because of polymeric backbone scissions produces positive-acting resists.

A large number of polymers have been explored for their suitability as e-beam resists. PMMA, the first reported e-beam resist, has been studied exhaustively (*87, 88*). Its degradation pattern leads to the evolution of small molecular fragments that create voids and microscopic damage, which further enhance the rate of dissolution of the exposed regions. Hence, the development behavior of PMMA cannot be explained merely by molecular weight reduction. PMMA is an inexpensive polymer with a moderate T_g (114 °C) and high resolution capability. Unfortunately, it is not very sensitive (G_s = 1.3 at an acceleration of 20 keV), and its resistance to plasma etching is considerably lower than that of typical novolac-based materials. These deficiences have spawned the search for PMMA variants as resists (*89*).

Generally, the variants can be grouped into three major categories: copolymers, ester-group-substituted variants, and alpha-substituted variants. For example, a terpolymer resist consisting of three monomers, methyl methacrylate, methacrylic acid, and methacrylic anhydride, was developed (*90*). The G_s value for this material is 4.5 and the sensitivity is 10 $\mu C/cm^2$ at 20 kV, an improvement over PMMA. Halogenation of the side chain with fluorine improves some resist properties. For example, poly(hexafluorobutyl methacrylate) demonstrated high sensitivity; however, adhesion was decreased (*91*). An improvement in adhesion and etching resistance was made with poly(dimethyltetrafluoropropyl methacrylate) but, unfortunately, at the expense of sensitivity (*92*). These halogenated variants illustrate some of the unavoidable trade-offs of chemical structuring that plague designers of positive e-beam resists. Another example of these trade-offs is the inverse relationship between dry-etching durability and resist sensitivity that has been observed in methacrylate-based materials (*93*). The incorporation of electron-withdrawing groups, such as nitrile, tends to increase sensitivity, presumably by weakening the main chain bonds and thus facilitating degradation (*94*).

The second class of positive-acting e-beam-sensitive polymers consists of the poly(olefin sulfone)s (*95, 96*). Degradation begins with the generation of radical cations and leads finally to the expulsion of sulfur dioxide (*97, 98*). These materials derive high sensitivity from the selective cleavage of their relatively weak carbon–sulfur bonds. For example, the sensitivity of poly(butene-1-sulfone) (PBS) is 1.6 $\mu C/cm^2$ at 20 kV. Unfortunately, poly(olefin sulfone)s are also sensitive to plasma-etching conditions. However, this deficiency has been greatly diminished by the use of poly(2-methyl-1-pentene sulfone) as a dissolution inhibitor in novolac resins (*99, 100*). The sensitivity of these so-called NPR (novolac positive resist) materials is 3–5 $\mu C/cm^2$ at 20 kV. Consequently, this system maintains the high sensitivity of poly(olefin sulfone)s with the enhanced resistance to plasma etching of the novolac resin.

DNQ–novolac positive resists have been used also with e-beam exposure. The 2,1,4 DNQ isomers give superior performance in these applications (*101*). The e-beam sensitivity of these materials is ~40 $\mu C/cm^2$.

As was mentioned previously, resists based on the acid-catalyzed deblocking of poly(*t*-BOC-styrene) have been used also as e-beam resists (*68*). In fact, these materials are capable of <40-nm resolution in both the positive and negative modes. The sensitivity of these resists is six times that of PMMA.

In the area of negative resists, a copolymer of glycidyl methacrylate and ethyl acrylate (COP) has been developed (*102, 103*). This resist has been used in mask manufacture. Poly(glycidyl methacrylate) homopolymers are also being used as commercial resists (*104*). Both polymers fall into the larger category of epoxy-based resists that have advantages of high sensitivity and thermal stability but lack resistance to plasma etching.

Improved etch resistance has been obtained with another class of negative resists, the substituted polystyrenes. Polystyrene has excellent contrast but low sensitivity (*105*). *para* substitution of halogen-containing segments increases the sensitivity sufficiently to warrant the consideration of these materials as bonafide resists (*106–109*). Copolymers of halogenated styrenes with glycidyl methacrylate or naphthalene-containing polymers show higher reactivity and dry-etch resistance, respectively, compared with styrene homopolymers (*110, 111*).

As with negative UV resists, the resolution of negative e-beam resists is primarily limited by swelling. In addition, high sensitivity and high resolution are often mutually incompatible requirements. For example, resolution usually improves with the use of low-molecular-weight polymers but at the expense of sensitivity. Higher molecular weights benefit sensitivity but adversely affect resolution. However, another parameter that has a marked influence on negative-resist performance is the composition of the developer solution. For example, submicrometer imaging of chloromethylated poly(α-methylstyrene) was accomplished through careful selection of developers based on acetone, methyl ethyl ketone (MEK), or a mixture of these compounds with methanol (*112*).

A new negative resist material based on polystyrene containing a tetrathiafulvalene (TTF) side chain has been reported to have high contrast without swelling (*113*). In the presence of a perhaloaliphatic sensitizer such as carbon tetrabromide, e-beam exposure converts TTF to the radical cation by an electron-transfer process. The difference in solubility of the radical cation and neutral species allows development to form the printed image. Other nonswelling negative e-beam resists based on homopolymers and copolymers of allyl methacrylate (*114*) and an azide–phenolic system (*115*) have been reported recently.

X-ray Resists. The key issues in X-ray resists are source brightness, resist sensitivity, and mask quality. The method used to generate X-rays determines the required resist sensitivity (*116*). For example, synchrotron and plasma sources with moderate to high fluxes can be used with less-sensitive materials. These technologies, however, are unproven and, in the case of synchrotron radiation sources, very expensive. Conventional X-ray sources based on electron bombardment have relatively low radiation intensity, so that high resist sensitivity is critical. Also, decreases in image quality can occur with these X-ray sources because of penumbral blur (*116*). Other factors that affect X-ray lithographic performance are mask absorption, exposure atmosphere, and resist properties, including absorption coefficient and radiation efficiency. An effective method of enhancing the sensitivity of X-ray resists involves matching the resist absorption with the X-ray emission wavelength of the exposure source. Examples of enhanced sensitivity with chlorinated and brominated resists matched to Pd (4.3 Å) and Rh (4.6 Å) sources have been reported (*117–119*). Use of more-reactive groups, higher molecular weight polymers, or both also increase resist sensitivity.

Resist materials that are sensitive to e-beam exposure are also sensitive to X-rays. In fact, a strong correlation exists between resist sensitivities observed with the two radiation sources (*120*). Therefore, the reaction mechanisms responsible for the behavior of e-beam and X-ray resists must be similar for both types of exposure.

In addition to good sensitivity, issues for X-ray resist materials are analogous to those of optical and e-beam resists: resolution, contrast, etch resistance, thermal stability, and adhesion. To stay competitive with e-beam and even optical lithography, X-ray lithography must have a resolution performance better than 0.5 μm. An extensive list of X-ray resist properties has been collected in the literature (*83*, *116*, *121*).

Positive X-ray Resists. PMMA is one of the best-known positive X-ray resists, although it lacks sufficient sensitivity to be of practical use (*122*, *123*). One attempt to increase the sensitivity of methacrylate polymers is the incorporation of more-reactive groups that, upon exposure, produce large amounts of volatile products (*124*). Possibly, dissolution is enhanced in the exposed resist as a result of gas-induced microporosity. Incorporation of metals such as Tl or Cs into MMA–MAA copolymers (*123*) or fluorine atoms (*125*) into PMMA derivatives has also been tried. Often, sensitivity gains have been made at the expense of etching resistance.

Upon X-ray exposure, DNQ–novolac resists undergo unusual chemistry (*126*). Although DNQ sensitizers react with X-rays, very little ICA is formed. This observation was made even when exposure was conducted under ambient conditions in which water vapor was present. The net result is that the irradiated areas of the resist have very poor solubility in aqueous base. In fact, if a UV flood exposure is used after imagewise X-ray irradiation, the areas exposed only to the UV can be selectively removed with a developer, a process that leads to an image-reversal scheme.

Both two-component (*67*) and three-component (*127*, *128*) resists based on the principle of chemical amplification have been used with X-ray irradiation. As for the previously described UV resist, X-ray irradiation is used to generate a strong acid capable of further catalyzing nonphotochemical reactions. The sensitivity of the three-component system is <100 mJ/cm^2, and 0.3-μm gates printed in 0.8 μm of resist have been demonstrated.

Negative X-ray Resists. Negative X-ray resists have inherently higher sensitivities compared with positive X-ray resists, although their resolution capability is limited by swelling. For example, poly(glycidyl methacrylate-*co*-ethyl acrylate) (COP), an e-beam resist, has been used in X-ray lithography. Polystyrene-type negative resists and their halogenated analogs, in particular, have been widely used in this application (*129–131*). The main thrust of current development work in negative resists lies in strategies to reduce swelling. Some apparent success has been achieved by using low-molecular-weight polymers with reactive side groups (*116*). Also, the TTF-substituted polystyrene materials previously described for e-beam lithography have been used (*113*). Finally, aqueous-base-developable novolac-based negative X-ray resists have been reported recently (*132*). These resists are similar to the materials described earlier for optical lithography and are composed of a phenolic resin, an X-ray acid generator, and a cross-linking agent.

Recent Developments in Resist Materials and Processes. ***Dry-Developed Resists.*** During the last decade, a host of dry-developed resists have been described for UV, e-beam, and X-ray lithographic processes (*14*, *83*, *133*, *134*). As mentioned previously, resists developed with conventional solvents or solutions are prone to pattern distortion by polymer swelling or shrinking. Dry development in the absence of liquid solvents eliminates these problems and may offer additional advantages of reduced defect density because of the use of vacuum equipment, reduced organic and chemical waste materials, and better resist side wall angles during plasma development.

Two approaches have been taken to design dry-developed resists. The first technique uses self-developing or ablative resists for direct pattern formation during exposure. This technique eliminates the need for a development step, because the resist is completely removed during irradiation by the action of the incident radiation. This process results only in positive-tone images. Poly(2-methyl-1-pentene sulfone) (PMPS) was the first reported self-developing resist for e-beam exposure (*135*). The relief image was produced by chain scission and depolymerization induced by high exposure doses. Other materials investigated for this purpose include poly(phthalaldehyde) (*136*), sensitized copolymers of methacrylate (*137*), aliphatic aldehyde copolymers (*138*), sensitized poly(methyl isopropenyl ketone) (*139*), and certain charge-transfer complexes such as tetrathiafulvalene bromide (*140*). Although self-developing resists result in a significant reduction in process steps, contamination of the e-beam machine by volatile byproducts remains a problem. Other issues are etching stability of the film and loss of resolution and edge acuity caused by self-propagating depolymerization reactions.

The second approach involves plasma development, during which the latent modifications introduced during exposure are amplified by a plasma treatment. The DESIRE process, which was described earlier, is an example of such a system. In a more general sense, plasma-developed resists incorporate an etch-resistant compound into the polymer matrix. Upon irradiation, the etch-resistant additive is bound to the polymer in the exposed region. In the unexposed areas, the additive is subsequently volatilized during a vacuum bake. Removal of the additive generally increases the plasma-etching rate of unexposed resist relative to the exposed areas. The process is completed by an oxygen plasma or a reactive-ion-etching (RIE) treatment that preferentially removes the unexposed resist. The resulting relief image is negative. Alternatively, irradiation of the resist–additive film may cause the exposed areas to etch faster, and the resulting relief image is positive. Both organic and organometallic etch-resistant compounds have been used.

X-ray plasma-developed resists with silicon bound into the polymer matrix by X-ray-induced polymerization of the metal-containing monomer have been studied (*141, 142*). Vacuum heating removes the moderately volatile monomer in the unexposed areas. During oxygen RIE, a metal oxide forms within the exposed areas of the resist; the metal oxide acts as a partial etch mask. Such a system based on poly(dichloropropyl acrylate) yields 0.5-μm resolution. e-Beam plasma-developed resists have been synthesized by using plasma-polymerized methyl methacrylate (*143*) and poly(methacrylonitrile) and its derivatives (*144*). These polymers are degraded by irradiation. The relief image is ultimately produced by etching in halogen-containing plasmas. e-Beam resists with added etch-resistant components, akin to the X-ray resist example cited earlier, have been investigated also. Deep-UV radiation has been used to expose poly(methyl isopropenyl ketone) containing a bisazide sensitizer, which functions by means of a different mechanism as a negative dry-developed resist (*145*). In this case, the byproducts of bisazide in the exposed resist are believed to inhibit plasma etching of the polymer. These byproducts are formed during the resist postbake rather than during the exposure step.

Organometallic compounds can be incorporated into spun-on resist films either at the time of formation with even distribution or in a separate step after the resist has been applied to the substrate. For example, focused indium (*146*) or gallium (*147*) ion beams have been used to write a pattern directly on the resist surface. Subsequent plasma treatment then produced the etch-resistant oxide. Inorganic halides such as $SiCl_4$ have been introduced by vapor treatment into a UV-exposed bisazide–isoprene-type resist, in which silicon is incorporated predominantly in the unexposed areas (*148*, *149*). Treatment with oxygen plasma selectively removes the exposed resist to produce the positive image. HMDS and other silylating agents have been used in similar schemes to generate positive- and negative-tone resist images developed by oxygen RIE (*150*).

Langmuir–Blodgett Films as Resists. Ultrathin Langmuir–Blodgett (LB) films are prepared by transferring floating organic monolayers onto solid substrates. This technique was first reported about 50 years ago (*151*) and has been reviewed recently for applications in electronic-related fields (*152*). Today, LB films are being investigated for potential use as high-resolution e-beam resists. Because of the thin coatings obtainable with this method, line width broadening, which occurs in conventional 1-μm-thick e-beam resists, can be reduced substantially. For example, ω-tricosenoic acid has been used as a negative resist that is capable of 60-nm line resolution in 30–90-nm-thick films (*153*). LB resists based on α-octadecylacrylic acid have been prepared also (*154*). In this case, on the basis of the extent of UV-induced prepolymerization, either positive or negative images are formed. When prepolymerization is slight, negative-tone images result from further e-beam exposure and development in alcohol. If prepolymerization is more extensive, e-beam exposure causes depolymerization. A resolution of 10 nm has been demonstrated with multilayers of simple-fatty-acid salts (*155*). Irradiation induces sublimation in these films to produce positive images.

Three major drawbacks currently plague LB resists. First, application time is too long (in the order of a few minutes to a few hours), because many coats, each only a few nanometers thick, are required to ensure defect-free etching protection. Second, etching resistance must be improved for adequate resist performance. Third, the substrate and the film preparation bath must be scrupulously clean. Because LB resist technology is still in its infancy, ongoing research may yet provide solutions to these problems.

Resist Processing

Even with the same resist and lithographic equipment, pattern quality can vary considerably, depending on the particular equipment used and the exact processing steps. The proper choice of processing parameters hinges on a firm understanding of the interactions among the various materials in each step. These parameters, in turn, can be grouped into two types. The first type of parameter is hardware related. For these parameters, refinement is likely to be costly, and sometimes, fundamental physics imposes limits. For example, light interference by mask fine structures and energy contour spreading by electron back scattering have been mentioned. In addition, lens imperfections, source stability, beam size and shape, mechanical alignment and focusing abilities are all potential problems. The second type of parameter involves those that can be controlled more directly. Examples include choice of casting solvent, selection of spinner speed, composition of developer solution, duration of exposure and dissolution, baking temperature, etchant formulation, and, if plasma processes are entailed, the various associated parameters, such as gas composition, flow rate, pressure, bias, power level, and radio frequency (rf). Definition of these variables specifies the process train, and most existing fabrication lines have well-tested standard process modules after years of refinement and experience.

The importance of cleanliness in IC processing cannot be overemphasized. Substrates, masks, equipment, human operators, lithographic chemicals, and air and water supplies must be kept as free of chemical and particulate contamination as possible. Contamination in one form or another is probably responsible for a large portion of day-to-day operating problems in fabrication areas. For this reason, environmental control (particle count, air flow and quality, and temperature and humidity levels) account for a major portion of the expense in setting up and maintaining a fabrication environment. Lithographic processes are especially sensitive to particulate levels and fluctuations in temperature and humidity. The rewards of tight environmental control are high, because product yield is directly affected by environmental conditions.

Standard resist processing includes several steps (Figure 4). The major steps are spin coating, baking, exposure, development, and postdevelopment processing (e.g., etching). These five operations will be discussed separately in the following sections. Steps indicated by dashed lines in Figure 4 are not used in all cases.

Substrate Preparation. A clean wafer surface is necessary for defect-free films and good resist adhesion. Cleaning procedures vary according to substrate surface composition and prior processing, but all procedures must, in one way or another, remove organic and inorganic contamination and particles. Substrate surfaces formed by vacuum deposition or thermal oxidation are generally very clean and may not require an additional cleaning step if they are coated immediately with resist. Sources of wafer contamination may be extraneous, as in the case of dirty wafer-handling equipment or impure water. Alternatively, contamination may arise from the interaction of the substrate surface, for example, with ambient air to form oxides and inorganic salts. Particles are the most common contamination, but they constitute a problem only if they cause fatal defects, that is, defects that make the device inoperable.

In lithography, particles can damage the printed pattern either by imbedding themselves in the resist or by casting shadows from the mask

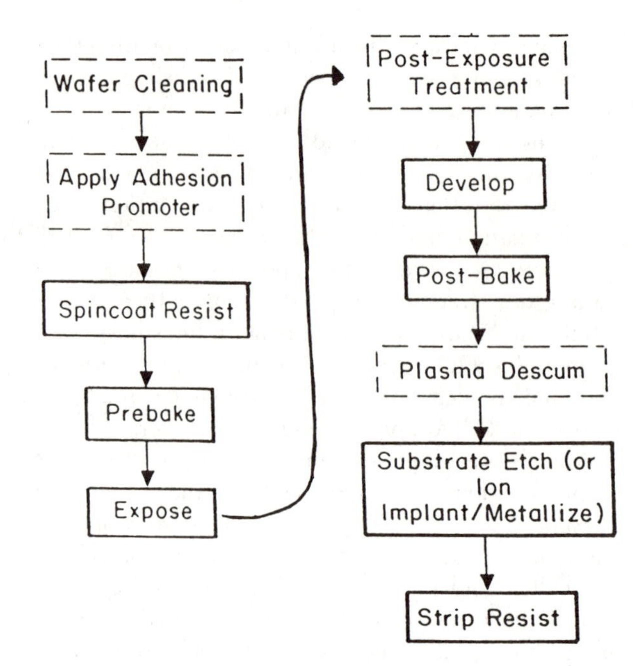

Figure 4. Flow sheet of standard lithographic process. The major steps are outlined in boxes, and optional steps are traced by dashes.

surface. Thus, the resulting resist image can be altered to form gaps or holes or joined to give bridged patterns. These resist irregularities can translate into metal open or short lines or other functional problems. Particulate contamination of the resist can also cause subtle problems that are more difficult to correlate with poor device performance. For example, during resist spin coating, particles can create cometlike shadows that affect film uniformity and planarization.

Wafer-cleaning procedures and reagents have been reviewed in detail (*156*). Cleaning procedures that use solvents are probably most common. Several methods of cleaning with solvents may be used, including dipping, vapor degreasing, spraying, and ultrasonic immersion. Ultrasonic immersion is especially effective in removing particulate contamination. Plasma cleaning is quickly becoming a popular alternative for removing organic and inorganic contamination. A dehydration bake at 200 °C or higher (with optional vacuum) in an oven or on a hot plate typically follows the solvent-cleaning steps. This bake is necessary to remove traces of absorbed water on the wafer surface to promote resist adhesion.

Resist adhesion must be high enough so that the film does not peel off during development. Trapped voids, either in the bulk or at the wafer interface, must be removed completely. Trapped voids in the bulk are eliminated by proper choice of solvent, which must evaporate slowly to avoid creating internal bubbles. Baking conditions are also selected to minimize the evaporation rate of residual solvent and to promote annealing. Interfacial defects are prevented by rigorous cleaning of the substrate surface.

Frequently, adhesion promoters are used. The substrates are primed or coated with a thin layer of an adhesion promoter, typically HMDS, prior to spin coating with resist (*18*). Possibly, the primary action of HMDS is to negate hydrophilic sites such as SiOH and trapped moisture on the substrate surface that would otherwise repel the photoresist. HMDS is applied by dipping, vapor priming in special chambers, or spin coating. Spin coating is most often accomplished with the same equipment that is used to coat resists. No high-temperature bake other than the resist prebake is required.

Spin Coating. Resist layers are deposited by a technique called spin coating. A predetermined amount of the resist solution is poured through a nozzle onto a wafer held on a vacuum chuck. The wafer is accelerated rotationally to the final spin speed of 2000–4000 rpm typically. Much of the original liquid is spun off the wafer edge. A small fraction remains and dries through solvent evaporation. The hardened resist forms a thin film over the wafer; typical layer thickness is on the order of 1 μm. If the wafer is stationary during dispensation, the process is referred to as *static* or *puddle dispense*. Conversely, if a slow rotation is used, the process is called *dynamic dispense*. Following the dispense step, the rotational speed may execute a preprogrammed ramp or simply jump to the final level.

Two important quantities characterize the success of spin coating: targeted film thickness and thickness uniformity. The resist formulation must be established carefully, so that the correct amount of fluid flows off the edge while the solvent evaporates to solidify the remaining film. Flow behavior is dictated by the fraction of solids left, because the radial-convective-loss rate hinges on the concentration-dependent rheology. Residual solvent is removed through an evaporative process. Surface solvent molecules are driven into the ambient by convective mass transfer, a process that is greatly enhanced by the relative motions of air and the resist due to wafer rotation. The solvent gradient thus established induces diffusion through the thickness of the resist from within the bulk fluid. The diffusivity of solvent in the resist is a strong function of its composition. Hence, diffusion (or evaporation) is tightly coupled with fluid flow through concentration-dependent evaporation and rheology. Both mechanisms account for film thickness reduction.

Experimentally, the average film thickness exhibits a power-law dependence on final spin speed (*157*). The relationship can be expressed approximately as $d = kw^{-a}$, in which d is the resulting film thickness, w is the final spin speed, k is a concentration-dependent front factor, and a is the power-law exponent. The power-law exponent a is strictly a function of starting solution composition. The dependence of a on solid content (and in cases in which the polymer molecular weight varies while the total solid weight fraction remains constant) can only be explained by flow-induced non-Newtonian behavior. In these thin liquid layers, radial flow driven by centrifugal forces generates sufficiently high shear rates to cause nonlinearity in the rheological behavior. Hence, the fluid viscosity is lowered by the imposed shear, with the reduction at a given shear rate being greater for more-concentrated solutions and higher molecular weight resins.

Mathematical models that incorporate both flow and evaporative-loss mechanisms have been prepared (*158*). Non-Newtonian features have been introduced recently to refine the modeling efforts and have resulted in models with quantitative predictive abilities (*159*). Hence, data comparison between films spun on different spinners, with nominally identical final rotation speeds, may reveal differences in film uniformity.

Film uniformity becomes a critical issue when the resist layer has to cover an existing topography, such as in applications over partially processed wafers. Step coverage is intimately linked with fluid flow and solvent evaporation. In addition, the exact geometry, such as the space, depth, and width of steps; the orientation of the features relative to the center of the wafer; and the distance of the features from the axis of rotation, all play a role in resist planarization (Figure 5). The existing step is covered with a film, which partially smoothes out the unevenness of the underlying topography. The effectiveness of planarization is measured by the ratio of peak-to-valley distance on the resist top surface to the existing step height of the underlying structure. Normally, planarization improves with a thick coating or multiple applications of resist layers, with each subsequent layer partially screening out the surface undulations of the previous layer. Planarization also depends on the geometrical details of the feature to be covered. Correlations are beginning to appear in the literature, and attempts at their prediction are under way. Resist planarization is an important issue for multilevel-resist processing (*160*), a topic that will be discussed in detail in a later section.

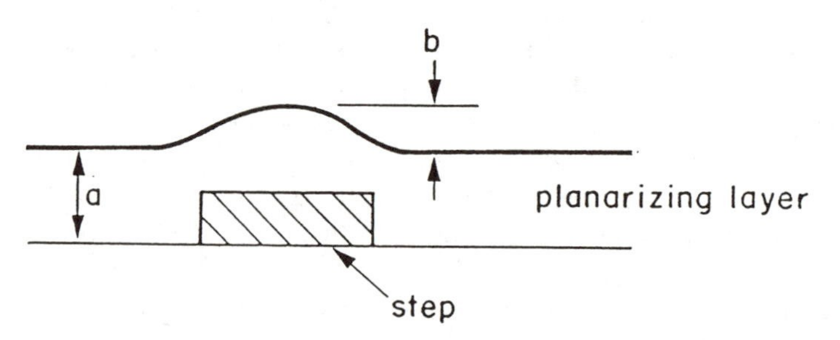

Figure 5. Cross section of a resist film of thickness a *spun on top of a step (or line). The underlying topography is partially obscured by the resist layer, whose surface reveals only a rounded hump with an amplitude* (b) *that is a fraction of the step height below.*

Prebake. After spin coating, the resist is baked to remove residual solvent and to relax locked-in stresses induced by flow. Prebaking also hardens the film for better handling during processing and increases adhesion of the film to the substrate. Although actual prebake conditions vary, 30 min at 90 °C is typical. Forced-air box ovens, conveyorized infrared ovens, and, recently, automated "wafer track" hot plates are all used for this purpose. The temperature at which baking takes place lies below the polymer decomposition point but above its glass transition temperature, because the resist must be soft enough to allow facile stress relaxation and drying. The operating range of annealing temperature is best determined by the combined use of differential thermal analysis (DTA) and thermal gravimetric analysis (TGA). These techniques monitor the enthalpy change (DTA) and weight loss (TGA) as functions of sample temperature. Solvent evaporation is accompanied by an endotherm and sample weight loss, the amount of which is governed by the film composition after spin coating. Baking-temperature selection is not limited by the rate of solvent evaporation alone. Sensitizer degradation and migration are important considerations as well. The annealing temperature must be below the temperature at which the sensitizer is degraded.

The baking process is affected not only by the temperature and duration at which annealing takes place but also by the way the resist-coated wafer is cooled (*161*). If the resist is baked in an oven and then taken out into room air after baking, its temperature drops rather quickly. However, if the oven is shut off while the wafer is left inside, the whole system cools off slowly. The annealed polymer thus experiences different temperature histories, depending on how the wafer is cooled after baking. Rubbery polymers exhibit high glass transition temperatures if the cooling rate is sufficiently fast. Conversely, a slowly cooled sample has much time to contract and thus exhibits a lower glass transition temperature. The slower the rate, the denser is the vitrified polymer. This high-density or, equivalently, low-free-volume state retards solvent penetration. The slower solvent diffusion, in turn, reduces the rate of resist dissolution. Although postbake wafer cooling is not routinely monitored during resist processing, the process should be investigated further.

Exposure. The fundamental principles involved in resist exposure have been discussed in a previous section (*see* Exposure Techniques, pages 327–331) and will not be reiterated.

Development. Resist development is a critical step in lithography, because it exerts great influence on pattern quality. The traditional development method uses a liquid developer solution that preferentially dissolves either the exposed region (positive resists) or the unexposed region (negative resists). Development can be carried out by either spray or immersion techniques. Many operating parameters affect the outcome of resist development, such as developer strength, agitation, temperature and humidity at which development takes place, and the size of developer molecules. All these parameters influence the relative rates of dissolution of both exposed and unexposed regions, as well as the degree of swelling and latent-image distortion.

Novolac–diazonaphthoquinone systems are developed in alkaline solutions. Many commercial developers are available; these vary slightly in pH and counter ions, such as NH_4^+, Na^+, and K^+. The developers are generally buffered, with minor differences in buffer compositions. Investigation of positive resists developed in different alkaline solutions has shown certain correlations between ionic strength, counter ion type and size, and rate of development (*162*). Alkaline solutions neutralize carboxylic acid groups in sensitizers and also react with CO_2 from air. Thus, the pH of these solutions gradually varies, and developers have to be replaced periodically. Cyclized-rubber-type negative resists are developed by using organic solvents.

In e-beam resists, chain scission is often accompanied by the elimination of small volatile fragments. As these fragments leave the polymer, they create microvoids that facilitate solvent penetration during development. Hence, the rate of dissolution is not just a function of the molecular weight of the resist in its postexposure state; rather the dissolution rate also reflects the extent of void formation. The rate would be higher in the exposed region than in the unexposed region, even if the two regions had identical molecular weights, as demonstrated convincingly by Ouano et al. (*163*).

Modeling of glassy-polymer dissolution has been reported in the literature (*164, 165*). In view of the difficulty of correlating the aforementioned gamut of process parameters with the resulting time-dependent resist profiles, most modeling efforts simplify the two-dimensional problem by initially examining samples that have been subjected to flood exposure; this simplification eliminates one spatial variable. Dissolution of resist films conceptually involves two elementary steps: solvent penetration to convert the glassy polymer into a swollen, entangled rubber and coiled dissociation from the gel at the solution interface. Solvent penetration is a stress-relaxation-limited process, and solvent size and matrix free-volume state play a strong role in affecting the rate of this process. Coiled dissociation from the gel is dictated primarily by coil mobility and chain length. Mobility is a function of gel concentration, whereas coil length influences the time scale of coil movement (reptation) needed to free individual polymer molecules from the entangled chains. Experimental evidence to elucidate the detailed mechanisms of film dissolution has been partially collected. Some approaches involve the interruption of development processes by freezing the partially dissolved film at different intervals for later characterization. Other approaches use direct optical means to monitor film thickness as a function of time during development. Through systematic observations, the interdependencies of the film dissolution rate on baking conditions, exposure dosage, initial molecular weight of the resist, developer composition, and size and solubility of solvent molecules may be established. Phenomenological models, which take empirical curve-fit model parameters for one-dimensional situations to describe profile development, have been derived (*166*). Such models would benefit greatly from knowledge of the interrelations just mentioned.

As a first step in the systematic compilation of data on resist dissolution and swelling behavior, a research program was initiated about 4 years ago, which culminated in a number of publications (*167–169*). The swelling and dissolution of thin films of PMMA in methyl isobutyl ketone (MIBK) and in solvent–nonsolvent mixtures of MIBK–methanol and methyl ethyl ketone–isopropyl alcohol (IPA) were investigated. Films were monitored by in situ ellipsometry. Parametric studies of the effects of molecular weight, molecular weight distribution, soft-baking quench rate, solvent size, and temperature were performed with MIBK. These parameters had a significant effect on dissolution. The effects of solvent composition and temperature on swelling and dissolution were studied with binary solvents.

Ternary diagrams based on Flory–Huggins interaction parameters were used to interpret the thermodynamics of swelling and dissolution. A narrow transition region where the developer changed from a swelling to a dissolving agent with a small change in composition or temperature was observed. In the region where the polymer was insoluble, the polymer swelled up to three times the initial thickness. At a 50:50 MEK/IPA ratio, a temperature decrease from 24.8 to 18.4 °C caused a change from complete dissolution to a combined swelling–dissolution behavior and rendered the PMMA film only 68% soluble.

Kinetic effects were determined by measurements of dissolution and penetration rates. A constant penetration velocity was observed for almost all compositions for both binary solvent mixtures. In all studies, case II transport assumptions provided good agreement with experimental results. For MEK–IPA, penetration rates increased with increasing MEK concentration. For MIBK–methanol, however, a maximum penetration rate was observed at a 60:40 MIBK/methanol ratio.

A detailed transport model for resist dissolution has been developed (*169*). In conjunction with standard ellipsometric equations describing multilayer films, the model provides quantitative agreement with the observed traces from the in situ ellipsometer. Model parameters are thus extracted, and their significance in terms of molecular structures of the system can be established. This model can then be extended for predictive purposes in the design and selection of resist materials.

Postbaking. Postbaking, which normally follows the resist development step, is similar to prebaking but uses somewhat higher temperature.

In this step, residual solvent is removed, while adhesion to the substrate is also improved. Postbake conditions vary with the specific resist used, but 30 min at 120 °C is typical. An excessively high bake temperature would initially cause resist flow, and eventually, thermal degradation will occur at temperatures above 200 °C.

Postdevelopment Treatment. ***Descum.*** A photoresist plasma "descum" step is typically but not always used after development. This step removes unwanted resist residues that were not cleared out during development and, in effect, increases the process latitude of the occasionally troublesome exposure–development sequence. Likewise, the descum step can smooth out minor irregularities of the resist side wall. This mild plasma treatment is done with O_2 or O_2–CF_4 gases with low power and pressure settings, short process times, or both. The same plasma under more-vigorous conditions may be used later to strip the resist at the end of the masking operation.

Substrate Treatment. When the desired image is developed in the resist, the pattern created provides a template for substrate modification. The various chemical and physical modifications currently used can be classified into additive and subtractive treatments. Examples of additive treatments include the insertion of dopants (by either diffusion or ion implantation) to alter the semiconductor characteristics and metal deposition (followed by lift-off or electroplating) to complete a conduction network. In most cases, however, the substrate material is etched by a subtractive process.

Wet Etching. Traditionally, pattern transfer (etching) was performed by wet etchants. These substances are generally corrosive liquids (e.g., hydrochloric, hydrofluoric, phosphoric, and nitric acids) and can degrade potentially the remaining resist that serves as a conformable protective mask. Although these processes are simple and inexpensive, wet-etching processes, which are inherently isotropic, form rounded side wall profiles that limit the control of feature size. Process reproducibility is difficult, because many factors, such as temperature, time, degree of agitation, and bath composition, affect the outcome. In addition, "undercutting" of the edges may result because of insufficient resist adhesion to the substrate during wet etching. In this case, highly sloped side walls are formed that lead to a potential failure mechanism. Sloped edges, however, are preferred in some instances. For example, sloped edges are sometimes used to ensure that a subsequently deposited material will cover underlying topography uniformly and continuously. For this purpose, controlled undercutting may be achieved by depositing, prior to resist application, a thin layer of material, which dissolves in the etchant more rapidly than it does in the substrate. Despite many undesirable features, wet etching is still widely used in the industry because of lower capital costs and a large base of experience.

Dry Etching. Both plasma etching and RIE are feasible, with different resulting edge profiles (*170–173*). RIE differs from conventional plasma etching in that the wafers being etched are placed on the rf-driven electrode, which is negatively biased because of differences in mobilities of electrons and positive ions in the plasma. This negative potential accelerates the ions in the plasma toward the substrate and creates etch directionality. The resulting etched profiles are thus anisotropic, with the vertical etch rate dominating over the lateral etch rate. This feature is in sharp contrast to wet etching, which often leaves the pattern overetched (at least near the top). Dry etching also obviates the use of toxic liquid etchants, the disposal of which is a difficult problem. However, the gases used in dry processes are far from benign. Most effluents cause equally difficult disposal problems and can be quite damaging to the pumping and reactor systems. The developed resists must be able to hold up against plasma etchants in order to successfully serve as a mask.

The chemistry and physics of plasmas are extremely complex; Chapter 8 of this book presents detailed information. The use of oxygen RIE has promoted the development of bilevel-resist schemes with silicon-containing top-layer resists (*174–175*). This subject will be deferred to a later section for in-depth discussion.

Additive Processes. For additive processes, the key resist requirement is thermal stability. Most ion implantation or vapor dopant diffusion treatments occur at or induce high temperatures. Bombardment of resists by dopants can also cause degradation. Hence, the resists used as dopant masks must be especially strong, high melting, or hardened. Hardening can be achieved by either plasma treatment or UV exposure, which induces cross-linking. Another additive process is metal layer deposition. Metal deposition can be achieved by condensation of evaporated atoms or, alternatively, by sputtering, in which a shower of ions (generally argon) physically knocks off individual atoms (or clusters) from a metal target (*170*), and these species impinge on the substrate. A typical experimental set up begins with a plasma that serves as an ion source. The colliding ions, which are accelerated toward the target by a direct-current (dc) bias, create morphological damage, as well as eject atoms. The sputtered atoms are physically ejected from the target surface by the momentum carried by the incoming ions. In this collision cascade, some of the energy deposited by the bombarding ions is reflected back toward the surface, and a fraction of the surface atoms acquire enough energy to escape.

Resist Stripping. After the additive or subtractive processes of the substrate are complete, the resist mask must be completely removed by either wet or dry etching. The selection of resist stripper is determined by previous resist history (bakes, exposure to plasma, etc.) that results in chemical alteration and by the underlying substrate stability (*176*). Wet etches are either solvent-based or inorganic reagents such as H_2SO_4, HNO_3 or H_2O_2. Solvent-type strippers are typically acetone for positive resists, trichloroethylene for negative resists, or commercial products developed to remove both types of resists. Commercial organic strippers were initially phenol-based solvents but have been manufactured recently with little or no phenol as a result of health and safety issues associated with the use of this chemical. Plasma stripping or ashing of resist with either O_2 or O_2–CF_4 gases is clearly the method of choice from the standpoint of convenience, cost, and safety. However, the method cannot be used with substrates that are etched by these plasmas.

Auxiliary Process Steps. In addition to the standard process steps, auxiliary processes are sometimes necessary. These steps are not used for all situations but as required and will be considered in this section separately. Certain semiconductor-manufacturing processes are particularly damaging to polymeric films and require an additional step to harden the resist. For example, Al etching with chlorine plasma produces $AlCl_3$, which degrades resists. Ion implantation, in which the chamber temperature and, hence, the wafer temperature increase with increasing implant dose, causes thermal deformation of the resist image. One commonly used method to stabilize novolac-based resists is deep-UV flood exposure after patterning (*177*). With deep-UV exposure, cross-linking of the polymer surface produces a film with increased thermal resistance. With this procedure, positive resists can withstand a 180 °C bake for 30 min. Fluorocarbon plasma treatment also stabilizes resists (*178*), because fluorine insertion impedes subsequent oxidation of the polymers.

As discussed previously, an optional postexposure, predevelopment bake can reduce problems with the standing-wave effect in DNQ–novolac positive resists. However, such a postexposure bake step is indispensable in the image reversal of positive resists (*37–41*) and certain resists based on chemical amplification of a photogenerated catalyst (*64–67, 77, 78*). For both types of resists, the chemistry that differentiates between exposed and unexposed areas does not occur solely during irradiation. Instead, differentiation occurs predominantly during a subsequent bake. Therefore, to obtain acceptable CD control in these systems, the bake conditions must be carefully optimized and monitored.

Rework. Masking steps frequently have the advantage over other IC-manufacturing processes of being able to undergo wafer rework. Rework involves the removal of the original resist layer, cleaning the wafers, and starting over from the spin-coating operation. Rework is necessary when a mistake such as an out-of-specification measurement or misalignment tolerance is discovered prior to the substrate treatment step. However, reworks can be performed only a finite number of times, and each rework operation generally carries the penalty of some yield loss. In some cases, rework is not possible at all because of the sensitivity of the underlying substrate layers.

Multilevel-Resist Processes

Motivation for Multilevel Systems. To date, high-volume IC production is done almost exclusively with one-layer resist microlithography. However, multilevel-resist processes are being developed to contend with inherent lithographic problems, such as reflectivity, back scatter (in the case of e-beam lithography), and uneven topography. In particular, the need for planarization has become important. As device geometries and line pitches shrink to meet the rising demand for higher density circuits, film thicknesses remain relatively constant. This situation creates high, nonuniform steps in multilayer topography and is most apparent in the metallic conductor layers of the circuitry, which usually consist of long, tightly pitched lines. In the metallic conductor layers, resist patterns of high aspect ratio (ratio of height to width) are required for image definition. Inorganic dielectrics also suffer inherent step coverage problems, because they form continuous layers over the highly reflective narrow lines and spaces of underlying metal. As a result, the lithographic process for patterning VLSI geometries has become more challenging, particularly in the case of interconnecting metal and dielectric films. Currently, strategies to planarize conductor and dielectric films are an area of intensive investigation.

The formulation of a single-layer resist that can meet beyond-state-of-the-art demands is an arduous task. To date, very few such materials have been advertized, and their field performance is yet to be proven. The difficulty lies in the fact that requirements of sensitivity, etch resistance, and planarization are mutually exclusive. For example, thinner resists capable of higher resolution sacrifice substrate etching protection and planarization. Consequently, the focus of lithographers lately has centered upon multilevel-resist processes that distribute desirable resist properties among several different organic and inorganic layers.

A typical multilevel structure consists of a thick, planarizing bottom layer, an optional intermediate layer, and a thin top layer of resist. Various etching methods are used to transfer the printed image into the substrate. These layers function synergistically to achieve good resolution, which is otherwise impossible to obtain with single-layer resists. The trade-off for high performance is added process complexity resulting from the incorporation of one or more additional layers for each lithographic step. This added complexity represents a significant impact on throughput and yield. For this reason, multilevel processes have not found their way yet into most production wafer processes (*179*).

Optical, e-beam, and X-ray lithographic processes all benefit from the advantages offered by multilevel resists (*180*). First, in the case of photolithography, line width variations due to varying development times of nonuniform resist film thickness are reduced by the underlying planarizing layer. Width variations due to reflections of topographical features can be suppressed by incorporation of antireflective agents, such as dyes, into the bottom layer (*181*) or by the application of commercially available antireflective coatings spun onto the surface of the bottom resist (*5*). As a result, the top layer is far removed from the reflective surface and sees only the incoming image. The top layer is also relatively smooth, so that the projection optics views it on the same focal plane; thus depth-of-focus problems are reduced. A further advantage of multilayer resists for optical lithography is the increase in sensitivity of the resist imaging layer with thinner films.

Motivations for the use of multilayer systems for e-beam lithography are reductions in proximity and charging effects and an increased sensitivity resulting from the need to image only a thin top-layer resist. e-Beam proximity effects are the result of electron back scattering, which broadens the contours of energy deposition near the back side of the resist. If adjacent printed lines are physically close, the broadened contours begin to overlap, and line width distortions are produced. Charging effects observed during e-beam lithography can be avoided by using a conductive intermediate layer in a trilevel-resist process (*182*).

X-ray lithography also takes advantage of the increased resist sensitivity due to the thinner imaging films of multilayer systems. Thinner imaging films further improve X-ray resolution by minimizing the penumbra effect, a problem associated with an uncollimated X-ray beam. Consequently, the oblique exposure of features near pattern edges are minimized by multilevel resist processes, thereby restoring the desired profile.

Much work has been done in the area of multilayer-resist systems during the last few years, and a variety of schemes have been produced (*180*, *183*, *184*). Typically, however, multilevel resists refer to bilevel and trilevel systems. Both systems incorporate a thicker bottom planarizing layer (typically a novolac resist or PMMA, which is approximately 1–2-μm thick) amenable to pattern transfer with either a wet- or dry-etch process. The degree of planarization depends on resist solution properties, coating thickness, and underlying topographic parameters, such as feature size, aspect ratio, pitch, and location on the wafer. Complete planarization is extremely difficult, and so in reality, underlying steps are only smoothened. Optimal planarization is obtained for narrower lines that are bunched closely, whereas the worst case is that for features far apart enough to be considered practically isolated (*160*, *185*). Planarization is also a function of polymer molecular weight;

shorter polymeric chains are better planarizing agents. The top imaging layer functions as a portable conformal mask (PCM) for patterning the bottom layer.

Trilevel Processes. Multilevel processes were developed during the 1970s. The first reported multilevel process used four different layers of material and two dry-etching steps (*186*). Trilevel schemes use an inorganic intermediate film or transfer layer, which is typically a few hundred angstroms thick, that is deposited (by spinning, sputtering, or chemical vapor deposition [CVD]) on top of the thick planarizing layer. A variety of materials have been used as transfer layers, including Al, Si, Ge, Ti, SiO_x, Si_3N_4, and spin-on glass (SOG) (*180, 183, 184*).

A resist film on which an image is to be patterned is applied directly to the surface of the intermediate transfer layer to form the trilevel structure (Figure 6a). Because the surface is now planar, the top resist layer can be optimized for resolution and can be relieved somewhat from the strict requirements of etch resistance and thermal stability. After the resist is printed, the first pattern transfer into the isolation layer takes place by wet or dry etching (Figure 6b). The second pattern transfer into the planarizing layer is accomplished typically by a dry-etch process (by plasma etching or RIE). The isotropic or anisotropic nature of the etch determines the resist side wall profile. During the second step, the imaging resist is frequently destroyed and the intermediate layer is left to serve as the etch mask (Figure 6c). The pattern of the planarizing layer is then etched into the underlying film. Subsequently, the planarizing layer is stripped (Figure 6d).

Although the trilevel process involves an additional thin-film-deposition step, it is particularly effective for maximum resolution over severe topography and, in the case of RIE, for the generation of straight side walls (*187*). The trilevel process also avoids the formation of resist interfacial layers, a common problem with bilevel systems. Optimization of the image-transfer step enables tailoring of the edge profile for vertical or undercut side walls (*187*). Variations of this process have now been developed for specific applications. As mentioned previously, the addition of dye to the planarizing layer eliminates substrate reflectivity in optical lithography (*180*). On the negative side, trilevel processing is complicated and requires expensive etching and deposition equipment.

Lift-Off Processes. Early trilevel resist schemes were developed for metal lift-off processes (*186*). Lift-off technology is one method of obtaining fine line metallization with a multilevel-resist process. In this case, a resist thickness greater than the final metal thickness is deposited and printed with tapered side walls in the negative image of the metal circuitry, which is accomplished by using an inorganic transfer layer and dry etching. Thus, an overhang of the thin film is produced over the retrograde resist edge

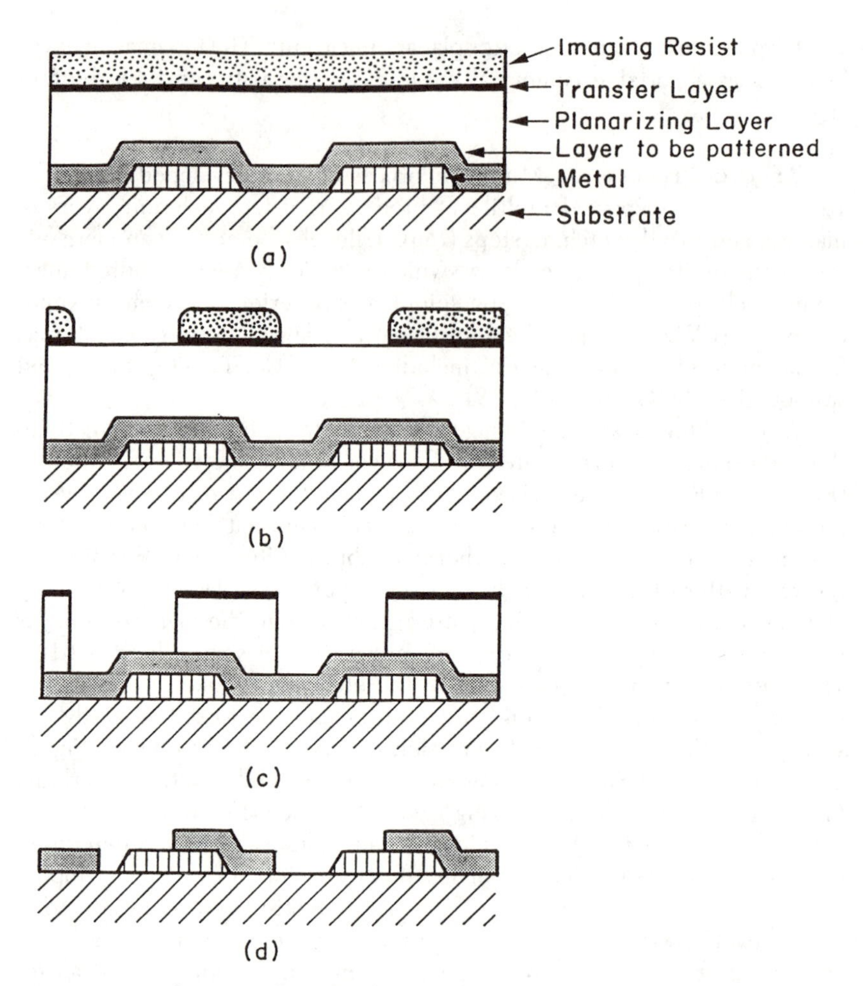

Figure 6. Schematic of a trilevel-resist process. (a) The top imaging layer is separated from the bottom planarizing layer by a transfer (or isolation) layer. (b) The pattern of the top image is transferred into the isolation layer. (c) The top layer is removed, and the pattern is transferred from the isolation layer to the substrate through the planarizing layer. (d) The remaining planarizing layer is stripped to complete the process.

(Figure 7a–b). Metal deposition at low temperature follows, because high-temperature deposition would degrade the resist and make it difficult to remove. A thick layer of metal fills the well created by the resist and the thin-film overhang, and a discontinuous layer is formed that allows penetration of solvent during the lift-off step (Figure 7c). As the resist dissolves and floats away, the inorganic film with its overlying metal is also removed (Figure 7d). Although standard dry etching of some metals can be effective in obtaining fine patterns, lift-off technology offers the advantage of smooth

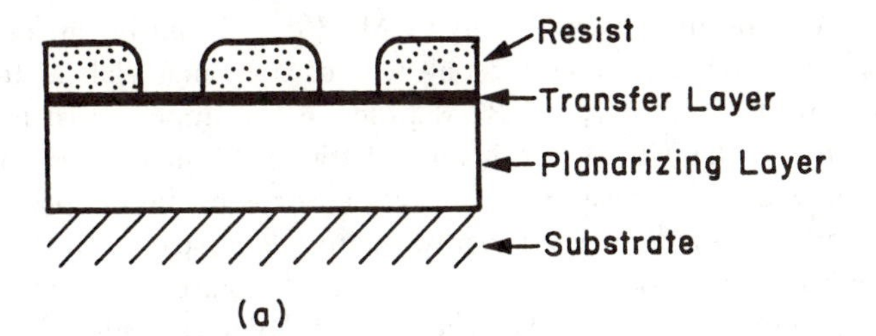

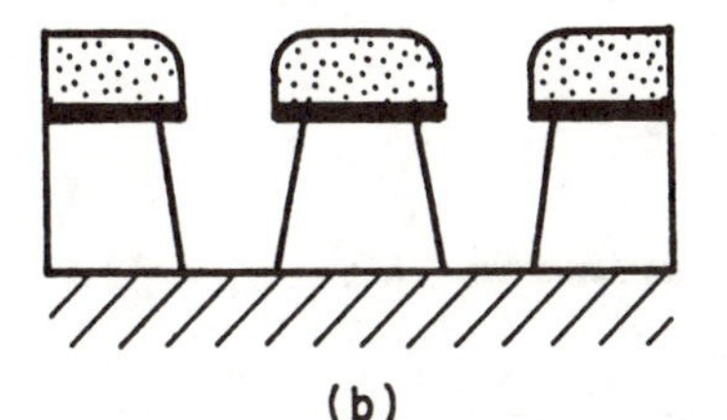

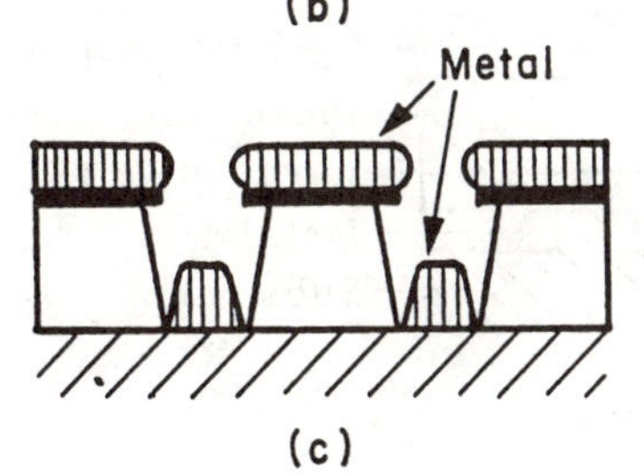

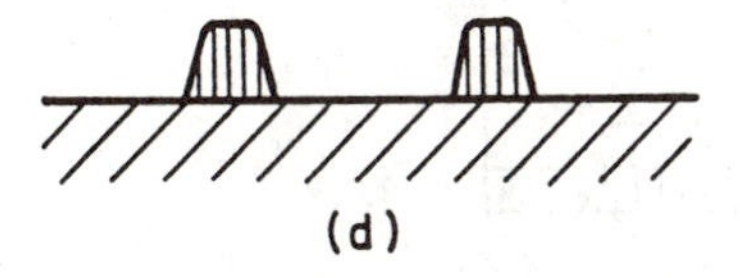

Figure 7. Metal lift-off process using a trilevel-resist scheme. (a and b) The image created in the top-layer resist is transferred via the isolation layer to the bottom planarizing layer by an isotropic etch. (c) The sloped side wall of the planarizing layer has an overhanging transfer layer that breaks up the continuity of the metal film sputter deposited onto the system. (d) Subsequent dissolution of the bottom layer carries off parts of the metal film adhering to the resist layers, and well-defined metal lines are left.

edges of the metallization profile, which precludes poor step coverage during the subsequent deposition step.

Bilevel Processes. A bilevel system consists of a thick resist at the base and a thin imaging resist on top. Many variations have now been reported. Conventional image transfer into the bottom layer is accomplished by deep-UV exposure, as with the PCM (*188*). Recently, an alternative approach using RIE and new oxygen-etch resistant materials as top layers has been reported. In a typical PCM process, the upper resist is exposed and developed with near-UV or e-beam radiation. This material, for instance, may be a novolac-type resist that strongly absorbs in the deep-UV region. The bottom layer is a deep-UV-sensitive material, typically PMMA. Hence, subsequent deep-UV flood exposure replicates the pattern delineated by the top layer, which acts as a PCM. The bottom layer can be developed (by using solvents) with the novolac cap removed or retained.

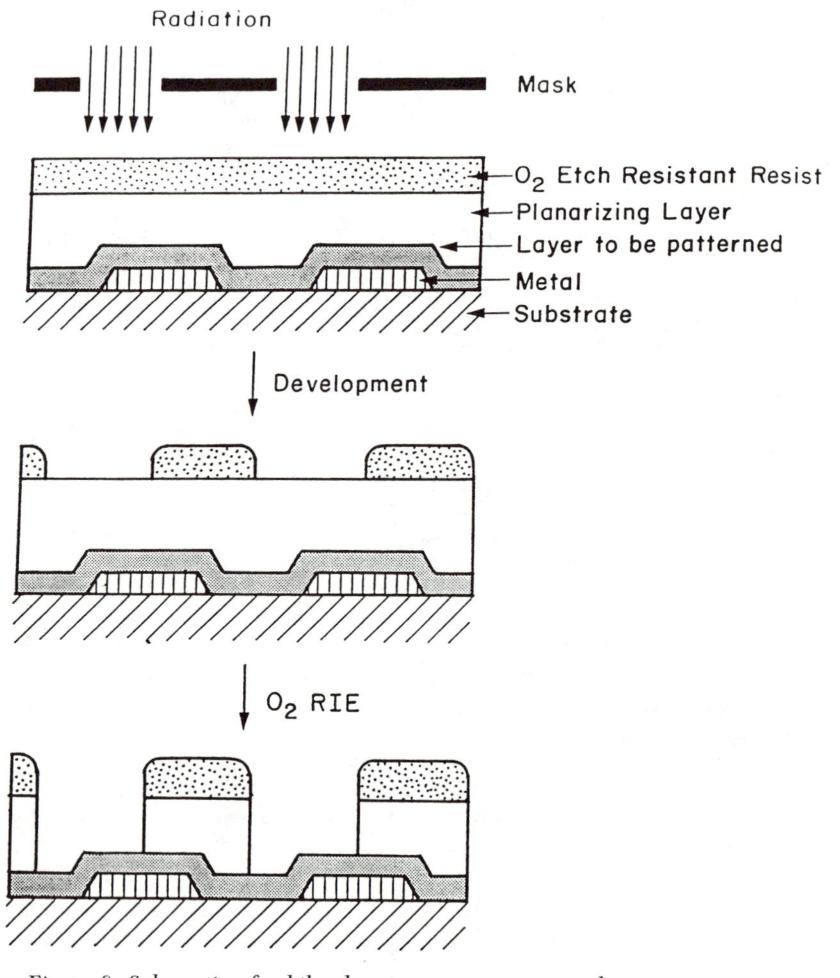

Figure 8. Schematic of a bilevel-resist process using a silicon-containing top layer. Pattern transfer to the bottom planarizing layer is achieved by oxygen RIE.

Problems associated with this system are interfacial mixing resulting in poor image definition, long exposure times for PMMA, reflectivity from the substrate, and the relatively low glass transition temperature of PMMA. Several improvements have been made now on the original system, including dye addition, use of antireflective coatings, and more-powerful deep-UV sources. The development of a new material, PMGI [poly(dimethyl glutarimide)] (58) is noteworthy. When used as the planarizing layer, PMGI provides increased thermal and plasma etch stability and the added advantage of development in aqueous solvents.

Recent developments in organometallic polymers, primarily those containing Si, have enhanced dramatically the interest in bilayer lithography. With these new materials to replace the conventional photoresist or e-beam resist, a single dry-etch step is required for image transfer into the bottom planarizing layer (Figure 8). As a result, the number of publications reporting new Si-containing resist during the last 2 years has mushroomed at a phenomenal rate. An exhaustive review is beyond the scope of this chapter.

Table I. Silicon-Containing Resists

Resist	*Type*	*Reference*
Poly(vinylmethylsiloxane) Poly(dimethylsiloxane)	Negative e-beam, deep UV	189
Poly(trimethylstyrene-*co*-chloromethylstyrene) Poly(trimethylstyrene-*co*-chlorostyrene)	Negative e-beam, deep UV	190, 191
Poly(dimethylsiloxane-*co*-methylphenylsiloxane-*co*-methylvinylsiloxane	Negative photon, e-beam	192
Chloromethylated poly(diphenylsiloxane)	Negative e-beam, deep UV	193
Poly(triallylphenylsilane)–bisazide	Negative deep UV	194
Polysilane	Positive deep UV, mid UV	195, 196
Terpolymer of phenol–trimethylsilylphenol–formaldehyde–*o*-quinonediazide	Positive near UV	197
Trimethylsilyl-substituted PMMA	Positive deep UV	171
Poly(trimethylsilylmethyl methacrylate-*co*-3-oximino-2-butanone methacrylate)	Positive deep UV	198
Poly(ethylphenylstyrene)	Positive deep UV	199
Poly(3-butenyltrimethylsilane sulfone)	Positive e-beam	200
Trimethylsilylmethylated resorcinol–formaldehyde–naphthoquinonediazide	Positive near UV	201
Poly(methyl methacrylate)-*g*-poly(dimethylsiloxane)	Positive e-beam	202
Poly(phenylsilsesquioxane)–phenyl-$T_4(OH)_4$–OFPR-800[a]	Positive near UV	203
Poly(methylstyrene-*b*-dimethylsiloxane)	Negative e-beam	172

[a]Phenyl$T_4(OH)_4$ is *cis*-(1,3,5,7-tetrahydroxy)-1,3,5,7-tetraphenylcyclotetrasiloxane. OFPR-800 is a commercially available positive photoresist.

Silicon-Containing Resists. Photosensitive silicon-containing materials combine the properties of conventional resists with the resistance of silicon to etching by O_2 plasma. In effect, the functions of the imaging resist and the inorganic transfer layer, which was described for trilevel processes, are now served by one film. Even though this area has been the focus of a great deal of research activity, commercial products are not available yet. Some problems encountered are reduced photosensitivity, which results from the high silicon content required for effective O_2 etch resistance, reduced polymer T_g, and hydrophobicity. Ideally, etching-rate ratios greater than 10:1 should exist between the resist and the planarizing layer to minimize line width erosion. The majority of these new resists contain silicon in the polymer backbone (polysiloxanes and polysilanes) or in the side chains. Treatment with an O_2 plasma leads to the formation of silicon oxides, which generate a protective layer on the polymer surface. Negative and positive resists have been reported that are sensitive to both UV and e-beam radiation; Table I gives some examples. The topic of RIE pattern-transfer lithography has been reviewed recently (*184*).

References

1. Deforest, W. S. In *Photoresist Materials and Processes;* McGraw–Hill: New York, 1975.
2. Cuthbert, J. D. *Solid State Technol.* **1977,** *20(10)*, 59.
3. Lin, Y. C.; Purdes, A. J.; Saller, S. A.; Hunter, W. R. *Tech. Digest Int. Electron Devices Meeting,* San Francisco, CA, 1982; p 399.
4. Brewar, T.; Carlson, R.; Arnold, J. *J. Appl. Photographr. Eng.* **1981,** *7(6)*, 184.
5. Walker, E. J. *IEEE Trans. Electron Devices* **1975,** *ED–22*, 464.
6. Willson, G.; Miller, R.; McKean, D.; Clecak N.; Tompkins, T.; Hofer, D. *Proc. SPE Regional Tech. Conf. Photopolym.*, Ellenville, NY, Nov. 1982, p 111.
7. Jain, K.; Willson, C. G.; Lin, B. J. *IEEE Electron Device Lett.* **1982,** *EDL–3*(3), 53.
8. Pol, V.; Bennewitz, J. H.; Escher, G. C.; Feldman, M.; Firtion, V. A.; Jewell, T. E.; Wilcomb, B. E.; Clemens, J. T. *Proc. SPIE Conf. Opt. Microlithogr. V* **1986,** *633*, 6.
9. Kyser, D.; Viswanathan, N. S. *J. Vac. Sci. Technol.* **1975,** *12*, 1305.
10. Luethje, H. *Phillips Tech. Rev.* **1983/84,** *41*(5), 150.
11. Gong, B. M.; Ye, Y. D. *J. Vac. Sci. Technol.* **1981,** *19(4)*, 1204.
12. Wada, T.; Sukuria, S.; Kawabuchi, K. *J. Vac. Sci. Technol.* **1981,** *19(4)*, 1208.
13. Papanu, J.; Manjkow, J.; Soong, D. S.; Hess, D. W.; Bell, A. T. to be published in *J. Electrochem. Soc.* and *J. Appl. Phys.*.
14. Willson, C. G. In *Introduction to Microlithography*; Thompson, L. F.; Willson, C. G.; Bowden, M. J., Eds.; ACS Symposium Series 219, American Chemical Society: Washington, DC, 1983; p 87.
15. *Radiation Chemistry of Macromolecules*; Dole, M., Ed.; Academic: New York, 1973.
16. Kilb, R. *J. Phys. Chem.* **1959,** *63*, 1838.
17. Erasmus, S. R. *Proc. SPE Regional Tech. Conf. Photopolym.*, Ellenville, NY, Oct. 1985, p 297.

18. Collins, R. H.; Deverse, F. T., U. S. Patent 3,549,368, 1970.
19. Singer, P. H. *Semicond. Int.* **1985,** *8(10)*, 68.
20. Thompson, L. F.; Kervin, R. E. *Annu. Rev. Mater. Sci.* **1976,** *6*, 267.
21. Stein, A.; *A Waycoat Tutorial;* P. A. Hunt Chemical Corp.: Palisades Park, NJ.
22. Cox, D. S.; Mills, A. R. *Chem. Eng. Prog.* **1985,** *1*, 11.
23. Benedikt, G. M. *Proc. SPIE Conf., Adv. Resist Technol. Process. II* **1985,** *539*, 242.
24. Iwayanagi, T.; Kohashi, T.; Nonogaki, S.; Matsuzawa, T.; Donta, K.; Yanazawa, H. *IEEE Trans. Electron Devices* **1981,** *ED–28*, 1306.
25. Pacansky, J.; Lyerla, J. R. *IBM J. Res. Develop.* **1979,** *23*, 42.
26. Hanabata, M.; Furuta, A.; Uemura, Y. *Proc. SPIE Conf., Adv. Resist Technol. Process. IV* **1987,** *771*, 85.
27. Hanabata, M.; Furuta, A.; Uemura, Y. *Proc. SPIE Conf., Adv. Resist Technol. Process. III* **1986,** *631*, 76.
28. Templeton, M. K.; Szmanda, C. R.; Zampini, A. *Proc. SPIE Conf., Adv. Resist Technol. Process. IV* **1987,** *771*, 136.
29. Trefonas, P. III; Daniels, B. K. *Proc. SPIE Conf., Adv. Resist Technol. Process. IV* **1987,** *771*, 194.
30. Trefonas, P. III; Daniels, B. K.; Fischer, R. L., Jr. *Solid State Technol.* **1987,** *30*, 131.
31. For example, *see* Lazarus, R. M.; Dixit, S. S. *Proc. SPIE Conf., Electron-Beam, X-Ray, and Ion-Beam Lithographies VI* **1987,** *773*, 68.
32. Watts, M. P. C.; DeBruin, D. *Proc. SPE Regional Tech. Conf. Photopolym.*, Ellenville, NY, Oct. 1985, p 285.
33. Bohland, J. F.; Sandford, H. F, Fine, S. A. *Proc. SPIE Conf., Adv. Resist Technol. Process. II* **1985,** *539*, 267.
34. Pampalone, T. R.; Kuyan, F. A. *J. Electrochem. Soc.* **1988,** *135*, 471.
35. Okuda, Y.; Ohkuma, T.; Takashima, Y.; Miyai, Y.; Inoue, M. *Proc. SPIE Conf., Adv. Resist Technol. Process. IV* **1987,** *771*, 61.
36. Petersen, J. S.; Kozlowski, A. E. *Proc. SPIE Conf., Adv. Resist Technol.* **1984,** *469*, 46.
37. MacDonald, S. A.; Miller, R. D.; Willson, C. G.; Feinberg, G. M.; Gleason, R. T.; Halverson, R. M.; MacIntyre, M. W.; Motsiff, W. T. *Proc. Kodak Interface,* San Diego, CA, 1982; p 114.
38. Alling, E.; Stauffer, C. *Proc. SPIE Conf., Adv. Resist Technol. Process. II* **1985,** *539*, 194.
39. Gijssen, R. M. R.; Kroon, H. J. J.; Vollenbroek, F. A.; Vervoordeldonk, R. *Proc. SPIE Conf., Adv. Resist Technol. Process. III* **1986,** *631*, 108.
40. Spak, M.; Mammato, D.; Jain, S.; Durham, D. *Proc. SPE Regional Tech. Conf. Photopolym.*, Ellenville, NY, Oct. 1985; p 247.
41. Grunwald, J. J.; Cordes, W. F. III; Ben-Shushan, G.; Gal, C.; Harding, K.; Spencer, A. C.; Shalom, E. *Proc. SPIE Conf., Adv. Resist Technol. Process. IV* **1987,** *771*, 317.
42. Roland, B.; Vandendriessche, J.; Lombaerts, R.; Denturck, B.; Jakus, C. *Proc. SPIE Conf., Adv. Resist Technol. Process. V* **1988,** *920*, 120 and references cited therein.
43. Griffing, B. F.; West, P. R. *Polym. Eng. Sci.* **1983,** *23*, 947.
44. Griffing, B. F.; West, P. R.; Balch, E. W. *Proc. SPIE Conf., Adv. Resist Technol.* **1984,** *469*, 94.
45. Griffing, B. F.; Lornenson, W. E. *Proc. SPIE Conf., Adv. Resist Technol.* **1984,** *469*, 102.
46. Petrillo, K. E.; Smyth, M. J.; Hall, D. R. *Proc. SPIE Conf., Adv. in Resist Technol. Process. V* **1988,** *920*, 82.

47. Lin, B. J. *J. Vac. Sci. Technol.* **1975,** *12*(*6*), 1317.
48. Chandross, E. A.; Reichmanis E.; Wilkins, C. W., Jr.; Hartless, R. L. *Solid State Technol.* **1981,** *24*(*8*), 81.
49. Kaplan, M.; Levine, A. W.; Poliniak, E. S. *Polym. Eng. Sci.* **1974,** *14,* 518.
50. Tsuda, M.; Oikawa, S.; Nakamura, Y.; Nagata, H.; Yokota, A.; Nakane, H.; Tsumori, T.; Nakane, Y.; Mifune, T. *Photogr. Sci. Tech.* **1979,** *23,* 290.
51. Hiraoka, H.; Welsh, L. W., Jr. In *Polymers in Electronics*; Davidson, T.; Ed.; ACS Symposium Series 242, American Chemical Society: Washington, DC; 1984, p 55.
52. Bowden, M. J.; Chandross, E. A. *J. Electrochem. Soc.* **1975,** *122*(*10*), 1371.
53. Yamashita, Y.; Ogura, Kunishi, K.; M.; Kawazu, R.; Ohno, S.; Mizokami, Y. *J. Vac. Sci. Technol.* **1979,** *16*(*6*), 2026.
54. Grant, B. D.; Clecak, N. J.; Twieg, R. J.; Willson, C. G. *IEEE Trans. Electron Devices* **1981, ED-28, 1300.**
55. Willson, C. G.; Miller, R. D.; McKean, D. R.; Pederson, L. A.; Regitz, M. *Proc. SPIE Conf., Adv. Resist Technol. Process. IV* **1987,** *771,* 2.
56. Schwartzkopf, G. *Proc. SPIE Conf., Adv. Resist Technol. Process. V* **1988,** *920,* 51.
57. Gipstein, E.; Ouano, A. C.; Tompkins, T. *J. Electrochem. Soc.* **1982,** *129,* 201.
58. Legenza, M. W.; Vidusek, D. A.; de Grandpre, M. *Proc. SPIE Conf., Adv. Resist Technol. Process. II* **1985,** *539,* 250.
59. Reichmanis, E.; Wilkins, C. W., Jr.; Chandross, E. A. *J. Vac. Sci. Technol.* **1981,** *19,* 1338.
60. Reichmanis, E.; Wilkins, C. W., Jr., Price, D. A.; Chandross, E. A. *J. Electrochem. Soc.* **1983,** *130,* 1433.
61. Reichmanis, E.; Gooden, R.; Wilkins, C. W., Jr.; Schonhorn, H. *J. Polym. Sci. : Polym. Chem. Ed.* **1983,** *21,* 1075.
62. Reichmanis, E.; Smith, B. C.; Smolinsky, G.; Wilkins, C. W., Jr., *J. Electrochem. Soc.* **1987,** *134,* 653.
63. Smith, G. H.; Bonham, J. A.; U. S. Patent 3,779,778, 1973.
64. Frechet, J. M. J.; Ito, H.; Willson, C. G. *Proc. Microcircuit Eng.* **1982,** *82,* 260.
65. Frechet, J. M. J.; Eichler, E.; Willson, C. G.; Ito, H. *Polymer* **1983,** *24,* 995.
66. Ito, H.; Willson, C. G.; Frechet, J. M. J.; Farrall, M. J.; Eichler, E. *Macromolecules* **1983,** *16,* 510.
67. Ito, H.; Willson, C. G.; Frechet, J. M. J.; U. S. Patent 4,491,628, 1985.
68. Umbach, C. P.; Broers, A. N.; Willson, C. G.; Koch, R.; Laibowitz, R. B. *J. Vac. Sci. Technol.* **1988,** *6,* 319.
69. Osuch, C. E.; Brahim, K.; Hopf, F. R.; McFarland, M. J.; Mooring, A.; Wu, C. J. *Proc. SPIE Conf., Adv. Resist Technol. Process. III* **1986,** *631,* 68.
70. Turner, S. R.; Willson, C. G.; In *Polymers for High Technology: Electronics and Photonics;* Bowden, M. J.; Turner, S. R., Eds.; ACS Symposium Series 346, American Chemical Society: Washington, DC; 1987, p 200.
71. Conlon, D. A.; Crivello, J. V.; Lee, J. L, O'Brien, M. J. *Macromolecules* **1989,** *22,* 509. See also Crivello, J. V.; U. S. Patent 4,603,101, 1986.
72. Buiguez, F.; Guibert, J. Ch.; Tacussel, M. Ch.; Rosilio, C.; Rosilio, A. *Proc. Microcircuit Eng.* **1984, 471.**
73. McFarland, J. C.; Orvek, K. J.; Ditmer, G. A. *Proc. SPIE Conf., Adv. Resist Technol. Process. V* **1988,** *920,* 162.
74. Narang, S. C.; Attarwala, S. T. *Polym. Prepr.* **1985,** *26,* 323 and U. S. Patent 4,663,269, 1987.
75. Frechet, J. M. J.; Bouchard, F.; Houlihan, F. M.; Kryczka, B.; Eichler, E.; Clecak, N.; Willson, C. G. *J. Imaging Sci.* **1986,** *30,* 59.

76. Willson, C. G.; Ito, H.; Frechet, J. M. J.; Houlihan, F. M. *Proc. IUPAC 28th Macromol. Symp.* **1982, 448.**
77. O'Brien, M. J.; Crivello, J. V. *Proc. SPIE Conf., Adv. Resist Technol. Process. V* **1988,** *920,* 42.
78. McKean, D. R.; MacDonald, S. A.; Clecak, N. J.; Willson, C. G. *Proc. SPIE Conf., Adv. Resist Technol. Process. V* **1988,** *920,* 60.
79. Newman, S.; U. S. Patent 4,708,925, 1987.
80. Ito, H. *Proc. SPIE Conf., Adv. Resist Technol. Process. V* **1988,** *920,* 33.
81. Roth, M.; Eur. Pat. Appl. EP 202,196; 1986, *Chem. Abstr.* **1987,** *106,* 205254q.
82. Azuma, T.; Aoso, T.; Kamiya, A.; Kita, N., *Jpn. Kokai, Tokkyo Koho* Japanese Patent 61,169,835, 1986.
83. Bowden, M. J. In *Materials for Microlithography*; Thompson, L. F.; Willson, C. G.; and Bowden, M. J., Eds.; ACS Symposium Series 266, American Chemical Society: Washington, DC; 1984, p 39.
84. Toshiaki, T.; Imamura, S.; Sugawara, S. In *Polymers in Electronics*; Davidson, T.; Ed.; ACS Symposium Series 242, American Chemical Society: Washington DC; 1984, p 103.
85. Watts, M. P. C. *Solid State Technol.* **1984,** *27*(*2*), 111.
86. Takahashi, Y. *Semicond. Int.* **1984,** *7*(*12*), 91.
87. Hatzakis, M. *J. Electrochem. Soc.* **1969,** *116,* 1033.
88. Hiraoka, H. *IBM J. Res. Develop.* **1977,** *21,* 121.
89. Moreau, W. M. *Proc. SPIE Conf.* **1982,** *333,* 2.
90. Moreau, W.; Merrit, D.; Moyer, W.; Hatzakis, M.; Johnson, D.; Pederson, L. *J. Vac. Sci. Technol.* **1979,** *16,* 1989.
91. Kakuchi, M.; Sugawara, S.; Murase, K.; Matsuyama, K. *J. Electrochem. Soc.* **1977,** *124,* 1648.
92. Sakakibara, Y.; Ogawa, T.; Komatsu, K.; Moriya, S.; Kobayashi, M.; Kobayashi, T. *IEEE Trans.* **1981,** *ED-28,* 1279.
93. Harada, K. *J. Appl. Polym. Sci.* **1981,** *26,* 3395.
94. Clemens, S. In *Plastics for Electronics*; Goosey, M. T.; Ed.; Elsevier: London, 1985, p 207.
95. Himics, R. J.; Kaplan, M.; Desai, N. V.; Poliniak, E. S. *Polym. Eng. Sci.* **1977,** *17,* 406.
96. Thompson, L. F.; Bowden, M. J. *J. Electrochem. Soc.* **1973,** *120,* 1722.
97. Broun, J. R.; O'Donnell, J. H. *Macromolecules,* **1972,** *5,* 109.
98. Bowmer, T. N.; O'Donnell, J. H.; *Radiat. Phys. Chem.* **1973,** *17,* 177.
99. Bowden, M. J.; Thompson, L. F.; Fahrenhold, S. R.; Doerries, E. M. *J. Electrochem. Soc.* **1981,** *128,* 1304.
100. Ito, H.; Pederson, L. A.; MacDonald, S. A.; Cheng, Y. Y.; Lyerla, J. L.; Willson, C. G. *Proc. SPE Regional Tech. Conf. Photopolym.*, Ellenville, NY, Oct. 1986, p 127.
101. Tanigaki, K. *J. Vac. Sci. Technol. B* **1988,** *6,* 91 and references cited therein.
102. Thompson, L. F.; Feit, E. D.; Heidenreich, R. D. *Polym. Eng. Sci.* **1974,** *14,* 529.
103. Thompson, L. F.; Ballantyne, J. P.; Feit, E. D. *J. Vac. Sci. Technol.* **1975,** *12,* 1280.
104. Taniguchi, Y.; Hatano, Y.; Shiraishi, H.; Horigami, S.; Nonogaki, S.; Naraoka, K.; *Jpn. J. Appl. Phys.* **1979,** *18,* 1143.
105. Lai, J. H.; Shepard, L. T. *J. Electrochem. Soc.* **1979,** *126,* 696.
106. Imamura, S.; Tamamura, T.; Harada, K.; Sugawara, S. *J. Appl. Polym. Sci.* **1982,** *27,* 937.
107. Shiraishi, H.; Taniguchi, Y.; Horigami, S.; Nonogaki, S. *Polym. Eng. Sci.* **1980,** *20*(*16*), 1054.

108. Feit, E.; Stillwagon, L. *Polym. Eng. Sci.* **1980,** *20(16)*, 1058.
109. Liutkis, J.; Paraszczak, J.; Shaw, J.; Hatzakis, M. *Proc. SPE Regional Tech. Conf. Photopolym.*, Ellenville, NY, Nov. 1982, p 223.
110. Thompson, L. F.; Stillwagon, L. E.; Doerries, E. M. *J. Vac. Sci. Technol.* **1978,** *15*, 938.
111. Ohnishi, Y. *J. Vac. Sci. Technol.* **1981,** *14*, 1136.
112. Sukegawa, K.; Sugawara, S.; *Jpn. J. Appl. Phys.* **1981,** *20*, L583.
113. Hofer, D. C.; Kaufman, F. B.; Kramer, S. R.; Aviram, A.; *Appl. Phys. Lett.* **1980,** *37*(3), 314.
114. Daly, R. C.; Hahrahan, M. J.; Blevins, R. W. *Proc. SPIE Conf., Adv. Resist Technol. Process. II* **1985,** *539*, 138.
115. Shiraishi, H.; Hayashi, N.; Ueno, T.; Suga, O.; Murai, F. *Proc. PMSE, ACS,* **1986,** *55*, 279.
116. Taylor, G. N. *Solid State Technol.* **1984,** *27*(6), 124.
117. Yamaoka, T.; Tsunoda, T.; Goto, Y.; Photogr. Sci. Eng. **1979,** *23*, 196.
118. Taylor, G. N.; Coquin, G. A., Someku, S. *Polym. Eng. Sci.* **1977,** *17*, 420.
119. Taylor, G. N.; Wolf, T. M. *J. Electrochem. Soc.* **1980,** *127*, 2665.
120. Murase, K.; Kakuchi, M.; Sugawara, S.; *Int. Conf. Microlithogr.* Paris, June 1977.
121. Taylor, G. N.; *Solid State Technol.* **1980,** *23*(5), 73.
122. Smith, H. I.; Flanders, D. C. *J. Vac. Sci. Technol.* **1980,** *17*, 533.
123. Haller, I.; Feder, R.; Hatzakis, M.; Spiller, E. *J. Electrochem. Soc.* **1979,** *126*, 154.
124. Ouano, A. C. *Polym. Eng. Sci.* **1978,** *18*, 306.
125. Kakuchi, M.; Sugawara, S.; Murase, K.; Matsuyama, K. *J. Electrochem. Soc.* **1977,** *124*, 1648.
126. Mochiji, K.; Kimura, T.; *Microelectron. Eng.* **1986,** *4*, 251.
127. Dossel, K.; Huber, H. L.; Oertel, H.; *Microelectron. Eng.* **1986,** *5*, 97.
128. Dammel, R.; Dossel, K.; Lingnau, J.; Theis, J.; Huber, H. L.; Oertel, H. *Microelectron. Eng.* **1987,** *6*, 503.
129. Tarascon, R.; Hartney, M.; Bowden, M. J. In *Materials for Microlithography*; Thompson, L. F.; Willson, C. G.; Bowden, M. J., Eds.; ACS Symposium Series 266, American Chemical Society: Washington, DC; 1984, p 39.
130. Choong, H. S.; Kahn, F. J. *J. Vac. Sci. Technol.* **1981,** *19*, 1121.
131. Tamamura, T.; Sukegawa, K.; Sugawara, S. *J. Electrochem. Soc.* **1982,** *129*, 831.
132. Bruns, A.; Luethje, H.; Vollenbroek, F. A.; Spiertz, E. J. *Microelectron. Eng.* **1987,** *6*, 467.
133. Taylor, G. N.; Wolf, T. M.; Stillwagon, L. E. *Solid State Technol.* **1984,** *27*(2), 145.
134. Roberts, E. D. *Solid State Technol.* **1984,** *27*(6), 135.
135. Bowden, M. J.; Thompson, L. F. *ACS Appl. Polym. Symp.* **1974,** *23*, 99.
136. Ito, H.; Willson, C. G. *Proc. SPE Regional Tech. Conf. Photopolym.*, Ellenville, NY, Nov. 1982, p 331.
137. Yamada, M.; Tamano, J.; Yoneda, K.; Moritu, S.; Hattori, S.; *Jpn. J. Appl. Phys.* **1982,** *12*, 768.
138. Hatada, K.; Kitiyama, T.; Danjo, S.; Yuki, H.; Aritome, H.; Namaba, S.; Nate, K.; Yokono, H. *Polym. Bull.* **1982,** *8*, 469.
139. Tsuda, M.; Oikawa, S.; Yabuta, M.; Yokota, A.; Nakane, H.; Atoda, N.; Hoh, K.; Gamo, K.; Namba, S. *Proc. SPE Regional Tech. Conf. Photopolym.*, Ellenville, NY, Oct. 1985, p 369.
140. Tomkiewicz, Y.; Engler, E. M.; Kuptsis, J. D.; Schad, R. G.; Patel, V. V.; Hatzakis, M. *Appl. Phys. Lett.* **1982,** *40*, 90.
141. Taylor, G. N.; Wolf, T. M.; Moran, J. M. *J. Vac. Sci. Technol.* **1981,** *19*, 872.

142. Taylor, G. N.; Wolf, T. M. *Proc. Microcircuit Eng. 81, Lausanne, Switzerland* Sept. 1981, p 381.
143. Morita, S.; Tamano, J.; Hattori, S.; Ieda, M. *J. Appl. Phys.* **1980,** *51*, 3938.
144. Hiraoka, H. *J. Electrochem. Soc.* **1981,** *128*, 1065.
145. Tsuda, M.; Oikawa, O.; Kanai, W.; Hashimoto, K.; Yokota, A.; Nuino, K.; Hijikata, I.; Uehara, A.; Nakane, H. *J. Vac. Sci. Technol.* **1981,** *19*, 1351.
146. Venkatesan, T.; Taylor, G. N.; Wagner, A.; Wilkens, B.; Bar, D. *J. Vac. Sci. Technol.* **1981,** *19*, 1379.
147. Kuwano, H. *J. Appl. Phys.* **1984,** *55*, 1149.
148. Taylor, G. N.; Stillwagon, L. E.; Venkatesan, T. *J. Electrochem. Soc.* **1984,** *131*, 1658.
149. Wolf, E. D.; Taylor, G. N.; Venkatesan, T.; Kretsch, R. T. *J. Electrochem. Soc.* **1984,** *131*, 1664.
150. MacDonald, S. A.; Ito, H.; Willson, C. G. *Proc. SPE Regional Tech. Conf. Photopolym.*, Ellenville, NY, Oct. 1985, p 177.
151. Blodgett, K. L.; Langmuir, I. *Phys. Rev.* **1937,** *51*, 964.
152. Roberts, G. G. *Adv. Phys.* **1985,** *34(4)*, 475.
153. Barrand, A. *Thin Solid Films* **1983,** *99*, 317.
154. Fariss, G.; Lando, J.; Rickert, S. *Thin Solid Films* **1983,** *99*, 305.
155. Boers, A. N.; Promerantz, M. *Thin Solid Films* **1983,** 99, 323.
156. Thompson, L. F.; Bowden, M. J. In *Introduction to Microlithography*; Thompson, L. F.; Willson, C. G.; Bowden, M. J., Eds.; ACS Symposium Series 219, American Chemical Society: Washington DC; 1983, p 160.
157. Jenekhe, S. A. *Polym. Eng. Sci.* **1983,** *23, 713*, 830.
158. Meyerhofer, D. *J. Appl. Phys.* **1978,** *49*, 3993.
159. Flack, W. W.; Soong, D. S.; Bell, A. T.; Hess, D. W., *J. Appl. Phys.* **1984,** *56*, 1199.
160. White, L. K. *Proc. SPIE Conf., Adv. Resist Technol. Process. II* **1985,** 539, 29.
161. Majkow, J.; M. S. Thesis, University of California, Berkeley, 1986.
162. Hinsberg, W. D.; Gutierrez, M. L. *Proc. Kodak Microelectron. Seminar*, **1983, 52.**
163. Onano, A. C. In *Polymers in Electronics*; Davidson, T., Ed.; ACS Symposium Series 242, American Chemical Society: Washington DC; 1984, p 79.
164. Tu, Y. O.; Ouano, A. C. *IBM J. Res. Develop.* **1977,** *21:2*, 131.
165. Soong, D. S. *Proc. SPIE Conf., Adv. Resist Technol. Process. II* **1985,** 539, 2.
166. Exterkamp, M.; Wong, W.; Damar, H.; Neureuther, A. R.; Ting, C. W.; Oldham, W. G. *Proc. SPIE Conf., Opt. Microlithogr.* **1982,** *334*, 182.
167. Manjkow, J.; Papanu, J. S.; Soong, D. S.; Hess, D. W.; Bell, A. T. *J. Appl. Phys.* **1987,** *62*, 682.
168. Manjkow, J.; Papanu, J. S.; Hess, D. W.; Soane (Soong), D. S.; Bell, A. T. *J. Electrochem. Soc.* **1987,** *134*, 2003.
169. Papanu, J. S.; Ph. D. Dissertation, University of California, Berkeley, 1987.
170. Chapman, B. *Glow Discharge Processes*; Wiley: New York, 1980.
171. Sawin, H. H. *Solid State Technol.* **1985,** *28(4)*, 211.
172. Mucha, J. A.; Hess, D. W. In *Introduction to Microlithography*; Thompson, L. F.; Willson, C. G.; Bowden, M. J., Eds.; ACS Symposium Series 219, American Chemical Society: Washington DC; 1983, p 215.
173. Coburn, J. W.; Kay, E. *IBM J. Res. Develop.* **1979,** *23*, 33.
174. Reichmanis, E.; Smolinsky, G. *Proc. SPIE Conf., Adv. Resist Technol.* **1984,** *469*, 38.
175. Hartney, M. A.; Novembre, A. E. *Proc. SPIE Conf., Adv. Resist Technol. Process. II* **1985,** *539*, 90.
176. Kaplan, L.; Bergin, B. *J. Electrochem. Soc.* **1980,** *127*(2), 386.

177. Allen, R.; Foster, M.; Yen, Y. T. *J. Electrochem. Soc.* **1982,** *129(6)*, 1379.
178. Ma, W. H-L. *Proc. SPIE Conf., Submicron Lithogr.* **1982,** *333*, 19.
179. Burggraaf, P. *Semicond. Int.* **1985,** *28(8)*, 88.
180. Lin, B. J. In *Introduction to Microlithography*; Thompson, L. F.; Willson, C. G.; Bowden, M. J., Eds.; ACS Symposium Series 219, American Chemical Society: Washington DC; 1983, p 279.
181. O'Toole, M. M.; Liu, E. D.; Chang, M. S. *Proc. SPIE Conf., Develop. Semicond. Microlithogr. IV* **1981, 128.**
182. Suzuki, M.; Namamatsu, H.; Yashikawa, A. *J. Vac. Sci. Technol.* **1984,** *B,2:6*, 665.
183. Hatzakis, M. *Solid State Technol.* **1981,** *24(8)*, 74.
184. McDonnell Bushnell, L. P.; Gregor, L. V.; Lyons, C. F. *Solid State Technol.* **1986,** *29(6)*, 133.
185. Bassons, E.; Pepper, G. IBM Research Report RC 9480, 1982.
186. Franeo, J. R.; Havas, J. R.; Levine, H. A.; U. S. Patent 3,873,361, 1973.
187. Hatzakis, M.; Hofer, D.; Chang, T. H. *J. Vac. Sci. Technol.* **1984,** *16(6)*, 1631.
188. Lin, B. J. *Proc. SPIE Conf* **1979,** *174*, 114.
189. Hatzakis, M.; Paraszczak, J.; Shaw, J. *Int. Conf. Microlith.; Microcircuit Eng.* **1981,** *81*, 386.
190. Suzuki, M.; Saigo, K.; Golan, H.; Ohnishi, Y. *J. Electrochem. Soc.* **1983,** *130*, 1962.
191. MacDonald, S. A.; Steinmann, A. S.; Ito, H.; Hatzakis, M.; Lee, W.; Hiraoka H.; Willson, C. G. *Int. Symp. Electron, Ion, Photon Beams* Los Angeles, CA, May 1983.
192. Shaw, J. M.; Hatzakis, M.; Paraszczak, J.; Liutkus, J.; Babich, E. *Proc. SPE Regional Tech. Conf. Photopolym.*, Ellenville, NY, Nov. 1982, p 285.
193. Tanaka, A.; Morita, M.; Imamura, A.; Tamamura, T.; Koyure, O. *Polym. Prepr.* **1984,** *25*, 309.
194. Saigo, K.; Ohnishi, Y.; Suzuki, M.; Goka, H. *Int. Symp. Electron, Ion, Photon Beams* Tarrytown, NY, May 1984.
195. Hofer, D. C.; Miller, R. D.; Willson, C. G. *Proc. SPIE Conf., Adv. Resist Technol.* **1984,** *469*, 16.
196. Miller, R. D.; Hofer, D. C.; Willson, C. G. *Polym. Prepr.* **1984,** *25*, 307.
197. Wilkins, C. W., Jr.; Reichmanis, E.; Wolf, T. M.; Smith, B. C. *J. Vac. Sci. Technol.* **1985,** *3*, 306.
198. Reichmanis, E.; Wilkins, C. W., Jr. In *Polymer Materials for Electronic Applications*; Feit, E. D.; Wilkins, C. W., Jr., Eds.; ACS Symposium Series 184, American Chemical Society: Washington, DC; 1982, p 29.
199. Nate, K.; Sugiyama, H.; Inoue, T. *Electrochem. Soc. Ext. Abs.* 94:2, New Orleans, LA, Abstract 530 (Oct. 1984).
200. Gozda, A. S.; Craighead, H. G.; Bowden, M. J. *J. Electrochem. Soc.* **1985,** *132*, 2809.
201. Saotome, Y.; Goken, H.; Saigo, K.; Suzuki, M.; Ohnishi, Y. *J. Electrochem. Soc.* **1985,** *132*, 909.
202. Bowden, M. J.; Gazdz, A. S.; Klausner, C.; McGrath, J. G.; Smith, S.; *Proc. Polym. Mater. Sci. Eng.* **1986,** *55*, 298.
203. Hayashi, Ueno, T.; Shiraishi, H.; Nishida, T.; Toriumi, M.; Nonogaki, S.; *Proc. Polym. Mater. Sci. Eng.* **1986,** *55*, 611.

RECEIVED for review December 30, 1987. ACCEPTED revised manuscript March 15, 1989.

8

Plasma-Enhanced Etching and Deposition

Dennis W. Hess and David B. Graves

Department of Chemical Engineering, University of California, Berkeley, CA 94720

Chemical and chemical engineering principles involved in plasma-enhanced etching and deposition are reviewed, modeling approaches to describe and predict plasma behavior are indicated, and specific examples of plasma-enhanced etching and deposition of thin-film materials of interest to the fabrication of microelectronic and optical devices are discussed.

THE INCREASING COMPLEXITY OF SOLID-STATE electronic and optical devices places stringent demands upon the control of thin-film processes. For example, as device geometries drop below the 1-μm level, previously standard processing techniques for thin-film etching and deposition become inadequate. For etching, the control of film etch rate, uniformity, and selectivity is no longer sufficient; the establishment of film cross sections or profiles is crucial to achieving overall reliability and high-density circuits. Low-temperature deposition methods are required to minimize defect formation and solid-state diffusion and to be compatible with low-melting-point substrates or films. Therefore, the established techniques of liquid etching and, to some extent, chemical vapor deposition (CVD) are being replaced by plasma-assisted methods. Plasma-assisted etching and plasma-enhanced CVD (PECVD) take advantage of the high-energy electrons present in glow discharges to dissociate and ionize gaseous molecules to form chemically reactive radicals and ions. Because thermal energy is not needed to break chemical bonds, reactions can be promoted at low temperatures (<200 °C).

Although the chemistry and physics of a glow discharge are extraordinarily complex, the plasma performs only two basic functions. First, reactive

0065–2393/89/0221–0377$15.60/0

chemical species are generated by electron-impact collisions; thus they overcome kinetic limitations that may exist in thermally activated processes. Second, the discharge supplies energetic radiation (e.g., positive ions, neutral species, metastable species, electrons, and photons) that bombard surfaces immersed in the plasma and thus alter the surface chemistry during etching and deposition. The combination of these physical processes with the strictly chemical reactions due primarily to atoms, radicals, or molecules yields etch rates, etch profiles, and material properties unattainable with either process individually.

Dry Processing

Liquid etching has been the preferred method for pattern delineation for thin films for many years (*1*). Its pervasive use has been due primarily to two considerations. First, although the exact chemistry is often poorly understood, the technology of liquid etching is firmly established. Second, the selectivity (ratio of film etch rate to the etch rate of the underlying film or substrate) can be essentially infinite with the proper choice of etchant solution.

Despite these advantages, several critical problems arise for micrometer and submicrometer pattern sizes. Resist materials often lose adhesion in the acid solutions used for most etch processes and thereby alter pattern dimensions and prevent line width control. As etching proceeds downward into the film, it proceeds laterally at an approximately equal rate. The mask is undercut, and an isotropic profile (Figure 1) results. Because film thickness and etch rate are often nonuniform across a substrate, overetching is required to ensure complete film removal. Overetching generates a decrease in pattern size because of the continued lateral etching and thus affects process control. When the film thickness is small relative to the minimum pattern dimension, undercutting is insignificant. But when the film thickness is comparable with the pattern size, as is the case for current and future devices, undercutting is intolerable. Finally, as device geometries decrease, spacings between resist stripes also decrease. With micrometer and submicrometer patterns, the surface tension of etch solutions can cause the liquid to bridge the space between resist stripes. Because the etch solution does not contact the film, etching is precluded.

The limitations encountered with solution etching can be overcome by plasma-enhanced etching. Adhesion is not a major problem with dry-etch methods. Undercutting can be controlled by varying the plasma chemistry, gas pressure, and electrode potentials (*2–6*) and thereby generate directional or anisotropic profiles.

Numerous techniques have been developed for the formation of thin-film materials (*7–9*). Because of the versatility and throughput capability of CVD, this method has gained wide acceptance for a variety of film materials.

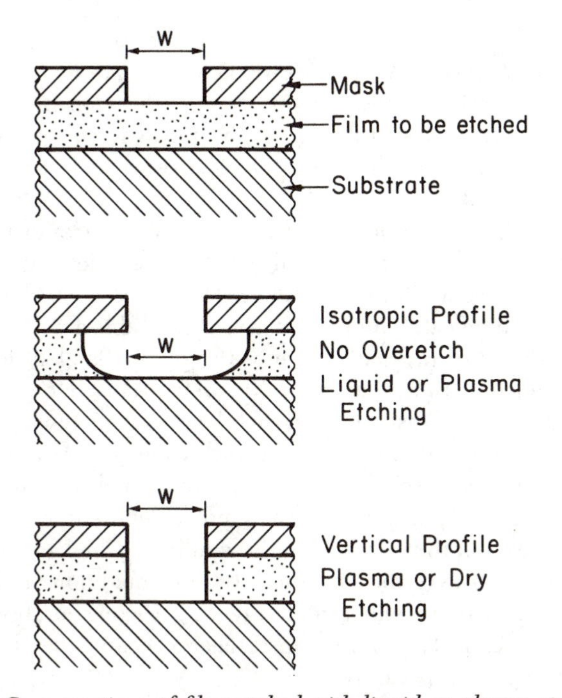

Figure 1. Cross sections of films etched with liquid or plasma etchants. The isotropic profile is the result of zero overetch and can be generated with liquid or plasma etch techniques. The anisotropic (vertical) profile requires plasma or dry-etch processes. W is the width of the resist pattern. (Reproduced from reference 2. Copyright 1983 American Chemical Society.)

However, deposition rates are often low with CVD, and the presence of temperature-sensitive substrates or films (e.g., polymers or low-melting-point metals) prior to deposition, along with the possibility of generating defects (e.g., vacancies, interstitials, stacking faults, and dislocations) often precludes the use of elevated temperatures (>300 °C) for film growth. In such cases, deposition rates can be enhanced by using high-energy electrons in a discharge rather than thermal means to supply the energy for bond breaking (*10–14*).

rf Glow Discharges

The rf (radio frequency) glow discharges (*2*) or plasmas used for plasma etching or PECVD are partially ionized gases composed of ions, electrons, and a host of neutral species in both ground and excited states. Typically, the plasma is formed by applying an electric field across a volume of gas. Many types of plasmas exist (*15*); they differ primarily in electron concentration n_e and average electron energy kT_e. A quantity that is useful in characterizing the average electron energy is the ratio of the electric field to the pressure, E/p (*15*). As the electric field strength increases, free electrons, whose velocities increase because of acceleration by the field, gain energy. The electrons lose this energy by inelastic collisions, so that an increase in pressure, which decreases the electron mean free path, decreases the electron energy.

In thin-film processes for the fabrication of electronic materials and devices, rf glow discharges are primarily used. The application of an rf voltage at frequencies between 50 kHz and 40 MHz to a low-pressure (6–600 Pa) gas results in a chemically unique environment (Table I.)

Electron densities (n_e) and, because the plasma is electrically neutral, positive-ion densities (n_i) range from 10^8 to $10^{12}/cm^3$. However, the ratio of the neutral-species density (n_N) to the electron density is usually greater than 10^3, so that these plasmas are only weakly ionized. As a result, radicals and molecules in the discharge are primarily responsible for etching and deposition reactions. That is, radicals and molecules are not inherently more reactive than ions, but they are present in significantly higher concentrations. The glow discharges described by Table I are termed nonequilibrium plasmas, because the average electron energy (kT_e) is considerably higher than the ion energy (kT_i). Therefore, the discharge cannot be described adequately by a single temperature.

Physical and Electrical Characteristics. The electrical potentials established in the reaction chamber determine the energy of ions and electrons striking the surfaces immersed in a discharge. Etching and deposition of thin films are usually performed in a capacitively coupled parallel-plate rf reactor (*see* Plasma Reactors). Therefore, the following discussion will be directed toward this configuration.

The important potentials in rf glow discharge systems (*16, 17*) are the plasma potential (potential of the glow region), the floating potential (potential assumed by a surface within the plasma that is not externally biased or grounded and thus draws no net current), and the potential of the powered or externally biased electrode. When the plasma contacts a surface, that surface, even if grounded, is usually at a negative potential with respect to the plasma (*16, 18, 19*). Therefore, positive-ion bombardment occurs. The energy of the bombarding ions is established by the difference in potential

Table I. Properties of rf Glow Discharges (Plasmas) Used for Thin-Film Etching and Deposition

Parameter	*Value*
$n_e = n_i$	10^8–$10^{12}/cm^3$
n_N	~10^{15}–$10^{16}/cm^3$
kT_e	1–10 eV
kT_i	~0.04 eV

between the plasma and the surface that the ion strikes, the rf frequency (because of mobility considerations), and the gas pressure (because of collisions). Because ion energies may range from a few volts to more than 500 V, surface bonds can be broken, and in certain instances, sputtering of film or electrode material may occur (*16*).

The reason for the different potentials within a plasma system becomes obvious when electron and ion mobilities are considered (*19a*). Imagine applying an rf field between two plates (electrodes) positioned within a low-pressure gas. On the first half-cycle of the field, one electrode is negative and attracts positive ions; the other electrode is positive and attracts electrons. Because of the frequencies used and because the mobility of electrons is considerably greater than that of positive ions, the flux (current) of electrons is much larger than that of positive ions. This situation causes a depletion of electrons in the plasma and results in a positive plasma potential.

On the second half-cycle, a large flux of electrons flows to the electrode that previously received the small flux of ions. Because plasma-etching systems generally have a dielectric coating on the electrodes or a series (blocking) capacitor between the power supply and the electrode, no direct current (dc) can be passed. Therefore, on each subsequent half-cycle, negative charge continues to build on the electrodes and on other surfaces in contact with the plasma, and so electrons are repelled and positive ions are attracted to the surface. This transient situation ceases when a sufficient negative bias is achieved on the electrodes such that the fluxes of electrons and positive ions striking these surfaces are equal. At this point, time-average (positive) plasma and (negative) electrode potentials are established.

A plasma potential that is positive with respect to electrode potentials is primarily a consequence of the greater mobility of electrons compared with positive ions. When there are many more negative ions than electrons in the plasma (e.g., in highly electronegative gases), plasma potentials are below electrode potentials, at least during part of the rf cycle (*19b*).

The plasma potential is nearly uniform throughout the observed glow volume in an rf discharge, although a small electric field directed from the discharge toward the edge of the glow region exists. Between the glow and the electrode is a narrow region (typically 0.01–1 cm, depending primarily upon pressure, power, and frequency) wherein a change from the plasma potential to the electrode potential occurs. This region is called a sheath or dark space and can be likened to a depletion layer in a semiconductor device in that most of the voltage is dropped across this region.

Positive ions drift to the sheath edge where they encounter the strong field. The ions are then accelerated across the potential drop and strike the electrode or substrate surface. Because of the series capacitor or the dielectric coating of the electrodes, the negative potentials established on the two electrodes in a plasma system may not be the same. For instance, the ratio of the voltages on the electrodes depends upon the relative electrode areas

(*20*). The theoretical dependence is given by equation 1, where V is the voltage and A is the electrode area (*20*).

$$V_1/V_2 = (A_2/A_1)^4 \tag{1}$$

If V_1 is the voltage on the powered electrode and V_2 is the voltage on the grounded electrode, then the voltage ratio is the inverse ratio of the electrode areas raised to the fourth power. However, for typical etch systems, the exponent of the area ratio is generally less than 4 and may be less than 1.2 (*16*). This apparent deviation from theory is in part due to the reactor configuration. Although the physical electrodes in a plasma reactor often have the same area, A_2 represents the grounded electrode area, that is, the area of all grounded surfaces in contact with the plasma. Because this area usually includes the chamber walls, the area ratio can be quite large. Because of such considerations, the average potential distribution in a typical commercial plasma reactor with two parallel electrodes immersed in the plasma is similar to that shown in Figure 2 (*16*). In this case, the energy of ions striking the powered electrode or substrates on this electrode will be higher than that of ions reaching the grounded electrode. Indeed, equation 1 can be used to design electrode areas for reactors such that a particular voltage can be established on an electrode surface.

In addition to the ratio of electrode areas, other plasma parameters can

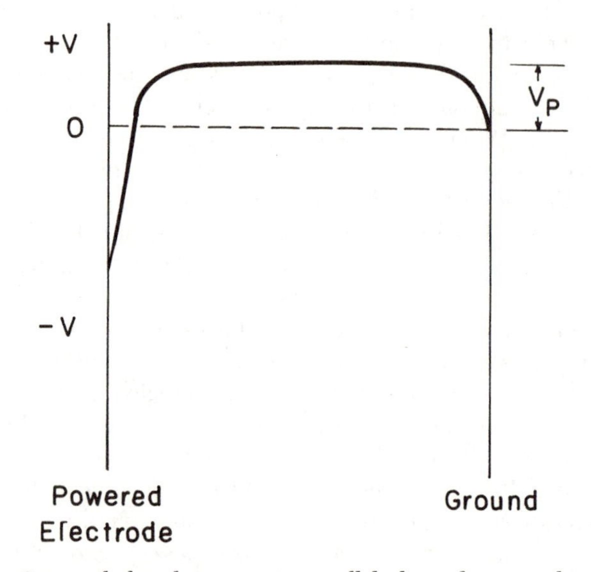

Figure 2. Potential distribution in a parallel-plate plasma etcher with the grounded surface area larger than the powered electrode area. V *is the potential, and* V_p *is the plasma potential. (Reproduced with permission from reference 16. Copyright 1979 The Electrochemical Society, Inc.)*

affect the electrical characteristics of the discharge. Varying the rf power input will alter plasma and electrode potentials, as well as ion concentrations, and thereby change ion energies and fluxes. Also, radio frequency affects the kinetic energy of ions that strike surfaces in contact with the plasma. This effect can be readily understood by considering the behavior of an ion experiencing an oscillating plasma potential caused by applied rf voltages (*21, 22*). Depending upon the ion mobility, some frequency exists above which the ion can no longer follow the alternating voltage. Therefore, the ion cannot traverse the sheath in one half-cycle. Above this frequency, ions experience an accelerating field (the difference between the plasma and electrode potentials divided by the sheath thickness) that is an average over a number of half-cycles. At lower frequencies, where the ions can respond directly to the oscillating field, they are accelerated by instantaneous fields. Thus, the ions can attain the maximum energy corresponding to the maximum instantaneous field across the sheath. As a result, for a constant sheath potential, ion bombardment energies and fluxes are higher at lower frequencies.

Chemical Characteristics. Because etching or deposition processes are merely chemical reactions that yield a volatile or involatile product, respectively, the overall process can be broken down into the following six primary steps:

1. Generation of reactive species
2. Diffusion to the surface
3. Adsorption
4. Reaction
5. Desorption of volatile products
6. Diffusion of volatile products away from the surface

First, reactive atoms, molecules, and ions must be generated by electron–molecule collisions. Because most of the reactant gases or vapors used for plasma-enhanced etching and deposition do not spontaneously undergo reaction at the low temperatures involved, radicals or atoms must be formed so that heterogeneous chemical reactions can proceed at reasonable rates. The reactive species thus generated diffuse to surfaces where they can adsorb onto a surface site. Sticking coefficients are believed to be large for free radicals, such that chemisorption and surface reactions occur readily (*23*). Surface diffusion of physically adsorbed species or volatile product molecules can occur.

The nature of the primary reaction product differentiates plasma-enhanced etching from deposition. In etching, the volatility of reaction products is crucial to film removal. Although the principal reaction product in deposition processes is not volatile, secondary products (e.g., hydrogen or halide molecules) must desorb to avoid incorporation into, and thus contamination of, the growing film. Complete elimination of such contamination is difficult, because particle bombardment of adsorbed species can assist incorporation.

As indicated previously, the chemical reactions taking place in glow discharges are exceedingly complex. However, two general types of chemical processes can be categorized: homogeneous gas-phase collisions and heterogeneous surface interactions. To completely understand and characterize plasma processes, the fundamental principles of both processes must be understood.

Homogeneous Processes. Homogeneous gas-phase collisions generate reactive free radicals, metastable species, and ions. Therefore, chemical dissociation and ionization are independent of the thermodynamic temperature. Electron impact can result in a number of different reactions depending upon the electron energy. The following list indicates these reaction types in order of increasing energy requirement (*24–26*).

- Excitation (rotational, vibrational, or electronic)

$$e + X_2 \rightarrow X_2^* + e$$

- Dissociative attachment

$$e + X_2 \rightarrow X^- + X^+ + e$$

- Dissociation

$$e + X_2 \rightarrow 2X + e$$

- Ionization

$$e + X_2 \rightarrow X_2^+ + 2e$$

- Dissociative ionization

$$e + X_2 \rightarrow X^+ + X + 2e$$

Excitation and dissociation processes can occur with mean electron energies below a few electronvolts. Thus, the discharge is extremely effective in producing large quantities of free radicals. Many of these species are generated by direct dissociation, although if attachment of an electron to a molecule results in the formation of a repulsive excited state, the molecule can dissociate by dissociative attachment. These attachment processes are prevalent at low electron energies (<1 eV) when electronegative gases or vapors are used. By comparison, the ionization of many molecules or atoms requires energies greater than ~8 eV, so that relatively few ions exist. The generation of reactive species is balanced by losses due to recombination processes at surfaces (electrodes and chamber walls) and in the gas phase, along with diffusion out of the plasma.

Electron-impact reactions occur at a rate (R) determined by the concentrations of both electrons (n_e) and a particular reactant (N) species (*24*).

$$R = kn_eN \tag{2}$$

The proportionality constant k is the rate coefficient, which can be expressed by

$$k = \int_{E_{\mathrm{thres}}}^{\infty} (2\epsilon/m)\sigma(\epsilon)f(\epsilon)d\epsilon \tag{3}$$

where ϵ and m are the impinging electron energy and mass, respectively; $\sigma(\epsilon)$ is the cross section for the specific reaction; and $f(\epsilon)$ is the electron energy distribution function. The limits of the integral run from the threshold energy for the impact reaction to infinity. If an accurate expression for $f(\epsilon)$ and electron collision cross sections for the various gas-phase species present are known, k can be calculated. Unfortunately, such information is generally unavailable for many of the molecules used in plasma etching and deposition.

Because of the highly nonequilibrium conditions experienced by electrons in the plasma, $f(\epsilon)$ almost never follows the Maxwell–Boltzmann distribution. In general, the distribution function is determined by the electric field that accelerates electrons and collisions that cause electrons to change energy. Very few direct measurements of $f(\epsilon)$ have been made under conditions of interest to plasma etching or deposition; consequently, the current understanding of $f(\epsilon)$ is limited, at best. This fact impedes the ability to make quantitative predictions of electron-impact rates. As previously described, ionization due to electron impact occurs through the action of the most energetic electrons in the distribution. The number of electrons in the high-energy tail of the distribution that are capable of ionizing neutral species in the discharge is considerably less than the number of electrons capable of molecular dissociation. As a result, the degree of ionization is usually much less than the degree of molecular dissociation.

A second type of homogeneous impact reaction is that occurring between the various heavy species generated by electron collisions, as well as between these species and unreacted gas-phase molecules (*27, 28*). Again, dissociation and ionization processes occur, but in addition, recombination and molecular rearrangements are prevalent. Particularly important inelastic collisions are those called Penning processes (*29*). In these collisions, metastable species (species in excited states where quantum mechanical selection rules forbid transition to the ground state and thus have long lifetimes) collide with neutral species, transfer their excess energy, and thereby cause dissociation or ionization. These processes are particularly important with gases, such as argon and helium, that have available a number of long-lifetime metastable states. Furthermore, Penning ionization has a large cross section, which enhances the probability of this process.

Heterogeneous Processes. A variety of heterogeneous processes can occur at solid surfaces exposed to a glow discharge (*28*, *30–32*). The primary processes of interest in plasma etching and deposition are summarized in the following list (*23*). These interactions result from the bombardment of surfaces by particles.

- Ion–surface interactions
  1. Neutralization and secondary electron emission
  2. Sputtering
  3. Ion-induced chemistry
- Electron–surface interactions
  1. Secondary electron emission
  2. Electron-induced chemistry
- Radical– or atom–surface interactions
  1. Surface etching
  2. Film deposition

Although vacuum-UV photons and soft X-rays present in the plasma are sufficiently energetic to break chemical bonds, electron and, particularly, ion bombardments are the most effective methods of promoting surface reactions (*33*).

Several theoretical investigations (*23*, *34*, *35*) indicate that nearly all incident ions will be neutralized within a few atomic radii of a surface, presumably because of electrons arising from Auger emission processes. These results suggest that the particles ultimately striking surfaces in contact with a glow discharge are neutral species rather than ions. To a first approximation, effects due to energetic ions and neutral species should be similar, provided that the particle energies are the same.

Auger emission to neutralize incoming ions leaves the solid surface in an excited state; relaxation of the surface results in secondary electron generation (*23*, *24*). Secondary electrons are ejected when high-energy ions, electrons, or neutral species strike the solid surface. These electrons enhance the electron density in the plasma and can alter the plasma chemistry near a solid surface. Radiation impingement on a surface can induce a number of phenomena that depend upon the bombardment flux and energy.

As noted previously (*33*), positive ions (or fast neutral species) are extremely efficient in enhancing surface processes; thus this chapter will concentrate on ion bombardment effects. The various surface, thin-film, and bulk phenomena affected by bombarding species are indicated in Figure 3 (*36*). The specific processes taking place are designated above the labeled abscissa in Figure 3, along with the range of particle energies that cause such effects.

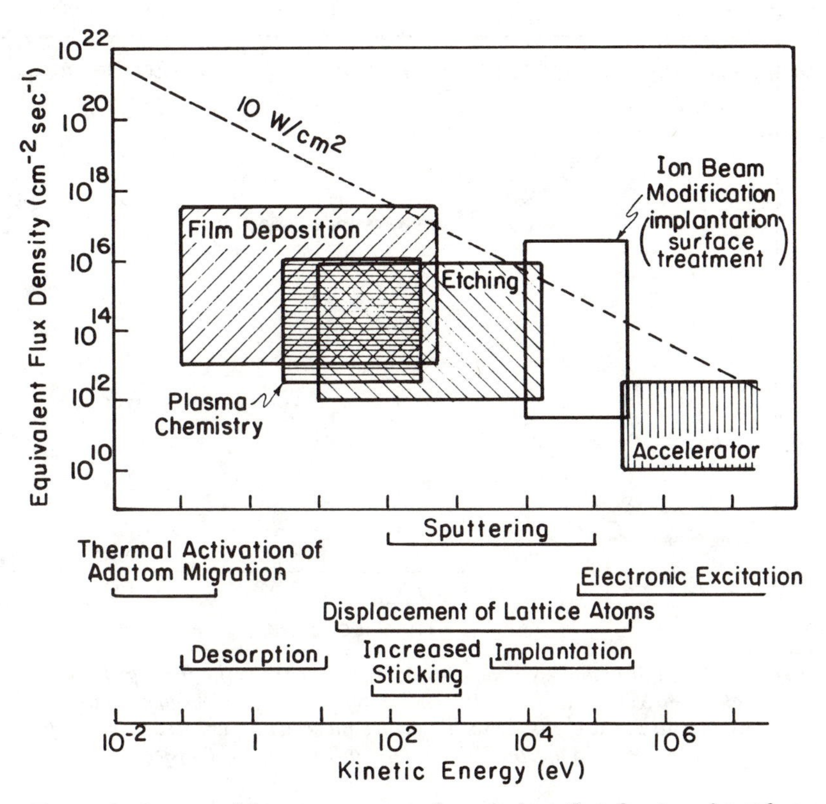

Figure 3. Ranges of kinetic energy and equivalent flux density of incident species for various engineering applications for ion–surface and gas–surface interactions. Kinetic energy ranges of particles in which significant interactions occur are also shown. (Reproduced with permission from reference 36. Copyright 1984 American Institute of Physics.)

Synergistic Phenomena. The enhancement of etch rates and the alteration of deposited film properties due to particle bombardment are well-known phenomena whose generic origins can be simply envisioned by referring to the process steps listed on page 383. Steps 3–5 are heterogeneous processes whose kinetics are temperature dependent. However, temperature is merely one method of increasing the energy of surface bonds. Particle bombardment is another means of imparting energy to a surface. Specifically, ion (or electron) bombardment can break surface bonds and thereby create crystal damage and adsorption sites (*37*), as well as assist product desorption (*38*). Also, chemical reactions on the solid surface can be promoted by such bombardment (*33*). Which of these steps is primarily responsible for enhanced etch rates and film property alteration is not yet clear. Nevertheless, particle bombardment obviously promotes etch processes, as demonstrated by the beam experiments described by Figure 4.

In this study (*39*), a beam of XeF_2 molecules and a beam of argon ions

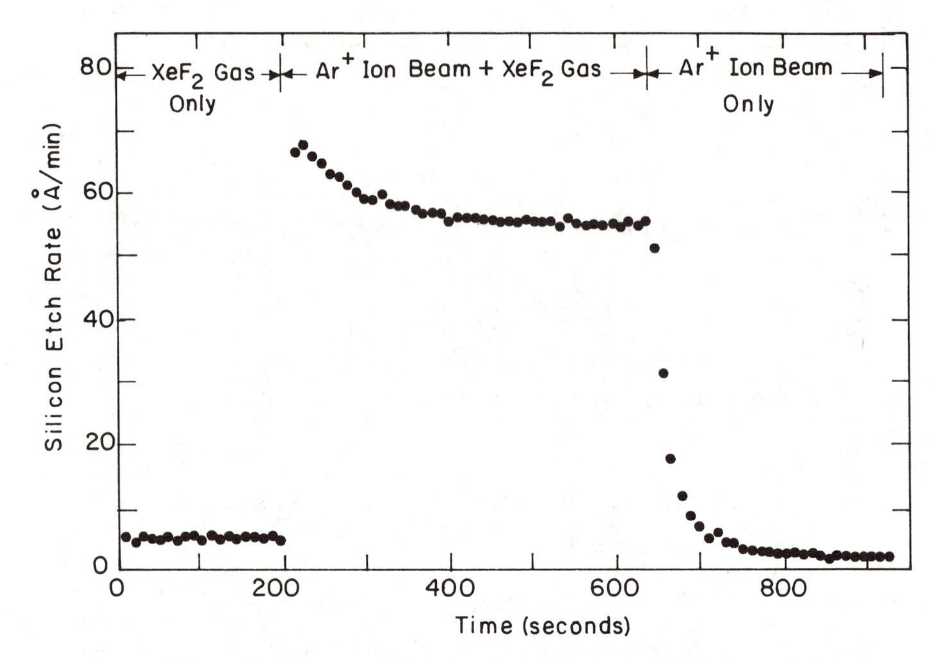

Figure 4. Example of ion-assisted gas–surface chemistry in the etching of silicon with XeF_2. The XeF_2 flow is 2×10^{15} mol/s, and the argon energy and current are 450 eV and 2.5 μA, respectively. (*Reproduced with permission from reference 39. Copyright 1979 American Institute of Physics.*)

were directed at a silicon film. Xenon difluoride was used because it afforded a method of depositing fluorine atoms onto the silicon surface. Measurement of the silicon etch rate as a function of XeF_2 or Ar^+ exposure allowed a comparison of chemical and physical etch processes. When only XeF_2 contacted the silicon surface, a small (~5 Å/min) purely chemical etch reaction was observed. Likewise, when Ar^+ impinged on the surface, pure sputtering (~2 Å/min) was noted. However, when the beams were simultaneously directed at the silicon surface, a relatively large etch rate (~55 Å/min) was observed; the measured rate was approximately an order of magnitude greater than the sum of the chemical and physical components. Therefore, synergistic effects due to ion bombardment are crucial to this chemical etch process. However, the exact nature of these effects is at present not well understood. Regardless of the precise cause, particle-bombardment-induced chemistry clearly results in directional etching by promoting higher etch rates in the vertical direction (where ions strike) compared with the lateral direction.

Synergistic phenomena similar to those described for etching are expected during film formation processes. In particular, the creation of adsorption and nucleation sites, along with the promotion of chemical reactions and the dissociation of adsorbed species because of particle bombardment, should be prevalent with particle energies above 20 eV. However, the various effects of ion bombardment (Figure 3) on the primary processes occurring during growth are extremely difficult to separate (*32*). Furthermore, although the basic plasma chemistry, physics, and synergistic effects for both etching and deposition are analogous, PECVD introduces one additional complication: Film bonding configurations must be controlled if films with specified and reproducible properties are to be formed.

The previous discussions indicate that a fundamental understanding of gas-phase plasma chemistry and physics, along with surface chemistry modified by radiation effects, is needed in order to define film etch and growth mechanisms. These phenomena ultimately establish etch rates and profiles, as well as film deposition rates and properties. The complex interactions involved in PECVD are outlined in Figure 5 (*14*, *40*). If the basic or microscopic plasma parameters (neutral-species, ion, and electron densities; electron energy distribution; and residence time) can be controlled, the gas-phase chemistry can be defined. Many macroscopic plasma variables (gas flow, discharge gas, pumping speed, rf power, frequency, etc.) can be changed to alter the basic plasma conditions. However, the precise manner in which a change in any of these variables affects basic plasma parameters is currently unknown.

The variation of a macroscopic variable usually results in a change in two or more basic gas-phase parameters, as well as surface potential, particle flux, and surface temperature. For instance, rf power determines the current and voltage between the electrodes in a parallel-plate plasma reactor. Varying the rf frequency changes the number and energy of ions (because of mobility considerations) that can follow the alternating field; thus, bombardment flux and energy are affected. The gas flow rate, the pump speed, and the pressure are interrelated, and two ways of changing the gas pressure can be envisioned. The gas flow rate can be varied at constant pump speed, or the pump speed can be varied (by throttling the pump) at constant gas flow rate. These two methods of pressure variation yield different residence times for the chemical species in the reactor, so that the precise chemistry is altered.

The particular reactant gas and the surface temperature (not necessarily equal to the electrode temperature) are critical parameters because of the dependence of the process on the type and concentration of reactive species and because of the observation that most deposition and etching processes follow an Arrhenius rate expression. Electrode and chamber materials can alter the chemistry occurring in glow discharges because of chemical reactions (adsorption, recombination, etc.) on or with the surfaces. Electrode potential and reactor configuration (equation 1) determine the energy of ions and electrons that strike the surfaces in contact with the discharge. Synergism between these numerous processes results in specific film growth (and etch) mechanisms. Ultimately, these factors establish film composition, bonding structure, and thus film properties.

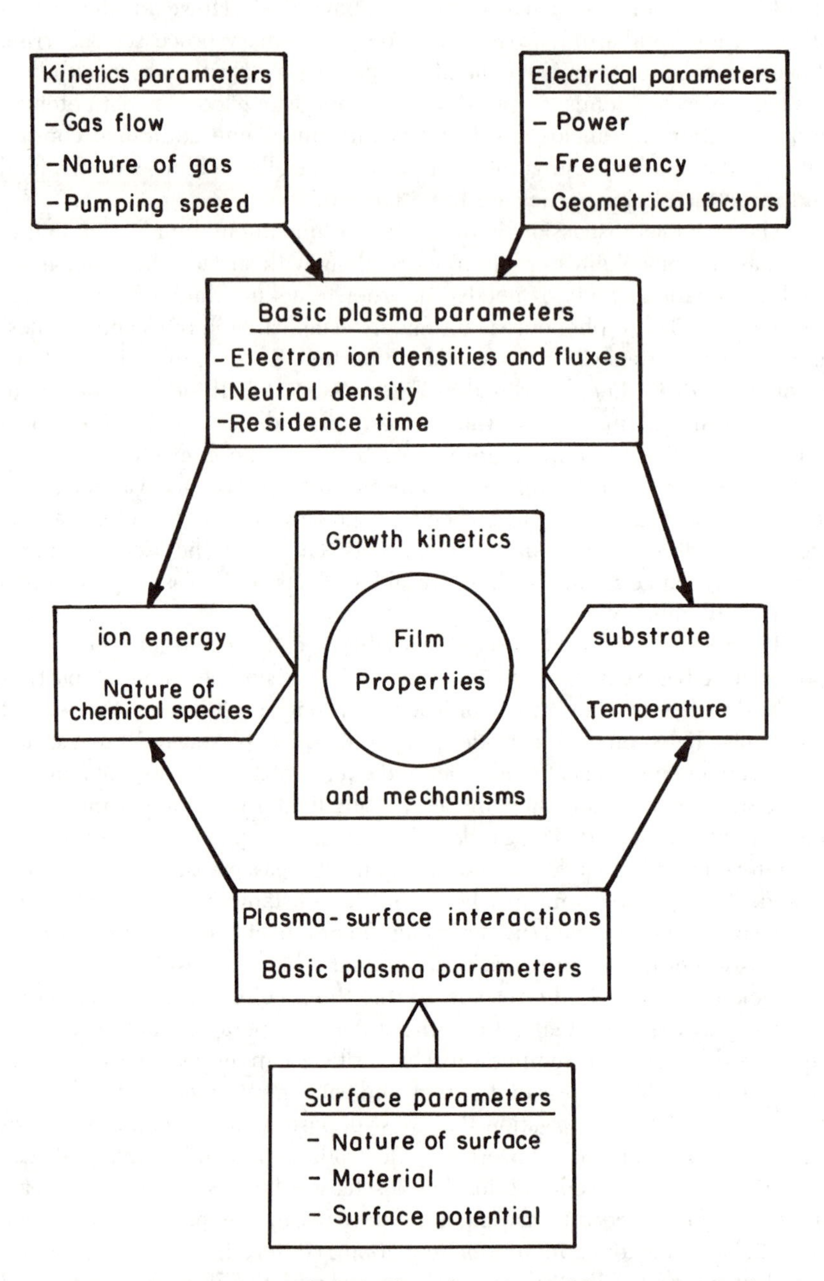

Figure 5. Interaction complexity of homogeneous and heterogeneous plasma processes that determine film properties in PECVD. (Reproduced with permission from reference 14. Copyright 1985 The Electrochemical Society, Inc.)

Radiation Damage. The particle bombardment that occurs during plasma-enhanced film etching and deposition is responsible for etch directionality and for the alteration of film properties. However, if the particle (primarily ion) energies or fluxes are sufficiently high, damage can be created in existing films or substrates. The displacement of atoms generates vacancies, interstitials, dislocation loops, and stacking faults. Even at relatively low bombardment energies, incorporation of light ions can occur, which results in crystal damage.

Considerable efforts have been expended recently to characterize the damage incurred during plasma-enhanced etching (*41–44*). These studies indicate that dislocation loops and impurity incorporation in substrate materials and films exposed to the plasma are prevalent. Furthermore, the damage can extend to more than 30 nm into substrates even at fairly low ion energies (~ 450 eV). An example of these effects for the case of CF_4–H_2 etching of SiO_2 on a silicon substrate is shown in Figure 6, for which X-ray photoelectron spectroscopy (XPS), hydrogen depth profiling, and Rutherford

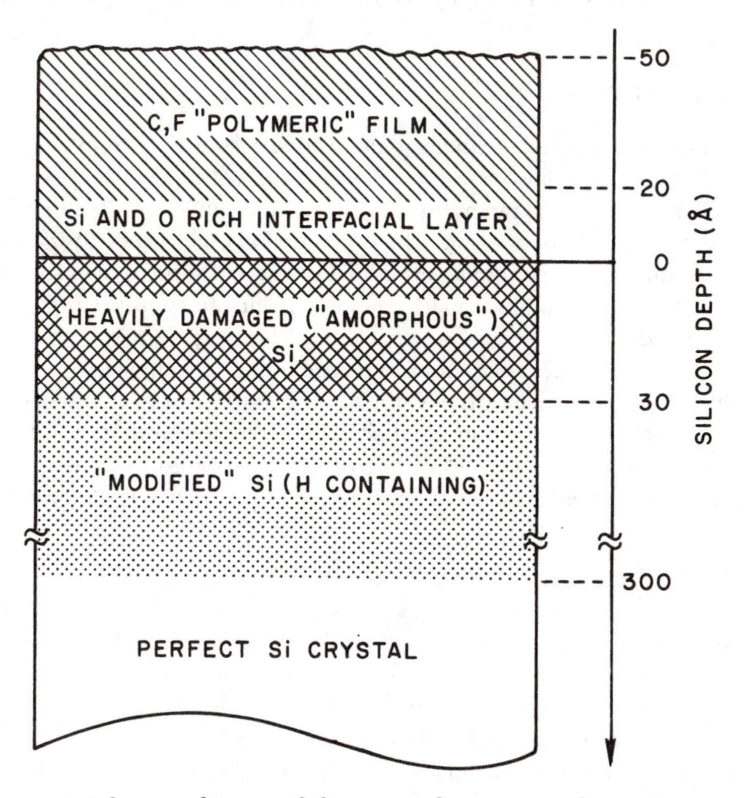

Figure 6. Schematic diagram of changes in silicon near-surface region caused by CF_4–H_2 reactive-ion etching. (Reproduced with permission from reference 43. Copyright 1985 The Electrochemical Society, Inc.)

back-scattering spectrometry (RBS) were used to characterize the carbonaceous residue and silicon surface region (*43*). Such damage degrades material properties and alters the characteristics of fabricated devices. In addition, charge formation and accumulation in insulators subjected to radiation during plasma etching can lower the dielectric breakdown strength of device structures (*44*).

Few studies have been reported that address radiation damage in PECVD processes. Recent work comparing sputter deposition and PECVD for dielectric film deposition indicates that structural damage is minimal in PECVD, although substrate damage is noted for sputtered coatings (*45*). These differences are probably due to lower ion energies resulting from the higher pressures and lower power densities used in PECVD compared with sputter deposition. Also, substrate temperatures above 200 °C are generally used in PECVD, so that any damage incurred may be annealed during deposition. Finally, in plasma etching, the underlying film or substrate is exposed to the discharge at the end of the etch cycle. With PECVD, underlying surfaces are only briefly exposed at the start of the deposition cycle.

Plasma Reactors

Like CVD units, plasma etching and deposition systems are simply chemical reactors. Therefore, flow rates and flow patterns of reactant vapors, along with substrate or film temperature, must be precisely controlled to achieve uniform etching and deposition. The prediction of etch and deposition rates and uniformity require a detailed understanding of thermodynamics, kinetics, fluid flow, and mass-transport phenomena for the appropriate reactions and reactor designs.

For the most part, plasma-enhanced etching and deposition are performed in four basic reactor types (Figure 7; *2*, *46*). Each reactor has several basic components: a vacuum chamber and pumping system to maintain reduced pressures, a power supply to create the discharge, and gas- or vapor-handling capabilities to meter and control the flow of reactants and products.

Barrel Reactor. Chronologically, the first and still the simplest etching system is the barrel reactor. This configuration generally uses a cylindrical chamber with rf power applied to external coils or external electrodes. Substrates are placed in a holder or "boat" within the chamber. To improve temperature uniformity along the length of the holder and to minimize particle bombardment of substrates, a perforated cylindrical "etch tunnel" is often inserted in the reactor. This metal cylinder acts as a Faraday cage and confines the glow region to the annulus between the etch tunnel and the chamber wall. Substrates are thereby shielded from the plasma and are subjected to little, if any, ion or electron bombardment. However, neutral species diffuse through the perforations and reach film surfaces. In this case,

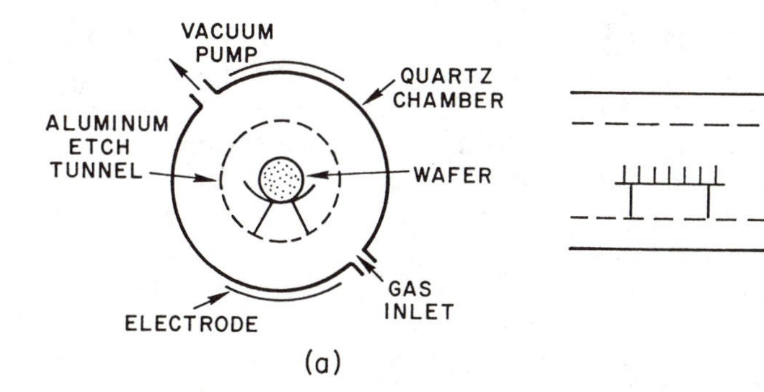

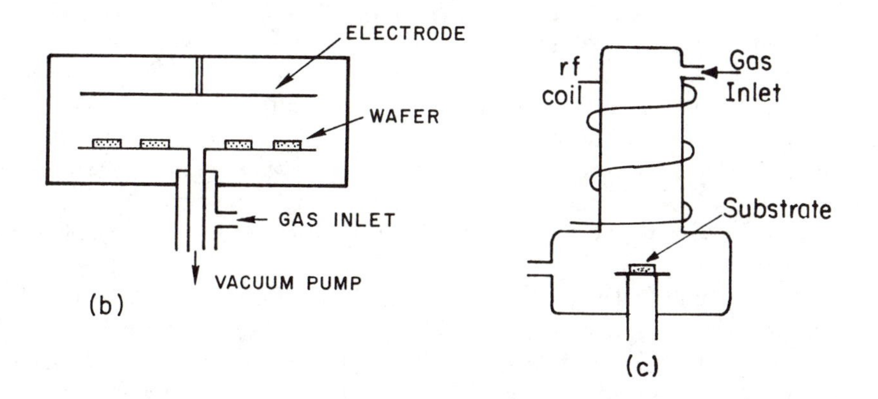

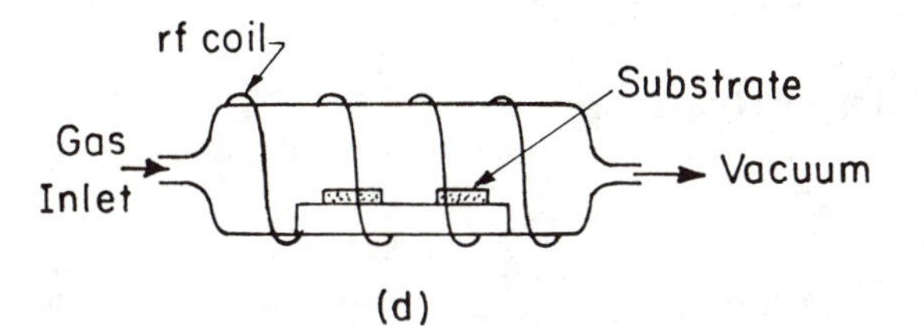

Figure 7. Configuration for plasma etch and deposition reactors. (a) parallel-plate or surface-loaded design with wafers positioned horizontally; (b) parallel-plate design with vertical electrodes with a furnace tube; (c) external coupling, downstream; (d) external coupling. (a: Reproduced from reference 2. Copyright 1983 American Chemical Society. b–d: Reproduced with permission from reference 46. Copyright 1983 American Society for Testing and Materials.)

the etch reaction is almost purely chemical (like liquid etching) and results in an isotropic etch profile. Therefore, barrel systems are usually used for resist stripping or in noncritical etching steps where undercutting can be tolerated.

Planar or Parallel-Plate Reactor. Because very-large-scale integration (VLSI) demands nearly vertical etch profiles, planar or parallel-plate reactors are favored for many etch processes. In this configuration (Figure 7a), which is similar to that used in rf sputtering, the substrates lie on an electrode within the plasma. The substrates are subjected to energetic radiation that imparts both physical and chemical components to the etch (or deposition) process. The physical component can induce directionality during film etching and can alter the properties of the deposited film by promoting surface reaction steps.

In PECVD, substrates are placed on the grounded electrode, and the opposite electrode is powered. Etching that is performed in this mode is generally called plasma etching. However, when substrates are positioned on the powered electrode for etching purposes, a reactive-ion-etching (RIE) or reactive-sputter-etching configuration results (*38*, *47*, *48*). The RIE arrangement generally causes the wafers to be subjected to higher energy ion bombardment than does the plasma-etching mode because of the potentials established on the grounded electrode versus the powered electrode (*see* Physical and Electrical Characteristics of rf Glow Discharges). Another reason for higher energy ions in RIE is the lower operating pressure. No exact pressure demarcation exists; however, RIE is typically carried out at pressures below 13 Pa, whereas plasma etching is performed at higher pressures.

When parallel-plate reactors are used for PECVD, substrates can be positioned horizontally or vertically (Figure 7b). The vertical position is often used to enhance throughput. PECVD processes are usually performed at electrode temperatures below 400 °C. If higher temperatures are desired (e.g., for epitaxial deposition), inductive coupling using external coils is often invoked to eliminate problems of film contamination arising from the electrode material. In this approach, the substrates may be positioned outside of (Figure 7c) or within (Figure 7d) the plasma volume.

The "downstream" configuration (Figure 7c) is particularly interesting because of its flexibility in controlling or modifying the specific chemistry that occurs (*49*, *50*). All feed gases can be passed through the discharge, and the fragments can be allowed to react at the heated (or cooled) substrate surface. Alternatively, one (or more) of the reactants (or diluents) can be excited or dissociated by the plasma, and the reactive fragments can be brought into contact with other reactant molecules at the substrate surface. In this way, specific chemical bonds can be broken or specific molecules can be excited, and improved discrimination may result in the reaction chemistry and thus in the film-bonding structure.

Finally, downstream configurations are also used for etching processes. In these applications, problems of temperature control and radiation damage (bond breaking because of radiation impingement on solid surfaces) are minimized or eliminated. To enhance the etch rate, microwave radiation has been used to generate long-lived chemical species that are then transported to the wafers (*51*, *52*). Such approaches often permit independent control of ion flux and energy to etching surfaces.

Modeling Plasma Processes

Modeling Requirements. To understand plasma processes in a systematic and comprehensive way, a mathematical model is useful. Because of the enormous complexity of chemically reacting plasmas, the relationships between different physical and chemical processes in the discharge are difficult to quantify and, sometimes, even to identify. To date, only a few preliminary attempts to model discharges in any detail have been made. Even for chemically inert gases, few efforts have been made to model the complete discharge (sheath-and-glow) region. Much of the modeling work has addressed the positive column of a direct-current (dc) discharge in a cylindrical tube. The positive column has received more attention because of its importance in gas discharge lighting and gas discharge lasers and also, undoubtedly, because of the relative simplicity of the analysis for this region. Under the proper conditions, the properties of the positive column vary only slightly in the axial direction, and therefore, discharge properties are assumed to vary only in the radial direction. A comprehensive review of the state of the art in modeling the positive column is available (*53*).

Discharge Behavior. A chemical reactor model of discharge behavior must take into account both discharge physics and discharge chemistry. Discharge physics includes the charged-particle densities and energies and the electric field strength. These quantities are important in creating chemically active radicals (primarily through electron-impact dissociation of reactants) and in influencing surface reactions (mainly through ion bombardment). Discharge chemistry includes gas-phase and surface reactions between ion and neutral species, as well as the transport of these species by convection and diffusion. Plasma reactors are generally not isothermal, although temperature gradients in plasma reactors are relatively small compared with those in CVD. As a result, an energy balance must be solved in order to determine the temperatures in the system.

The physical structure of a discharge is complicated by the fact that the discharge is far from local thermodynamic equilibrium (LTE). The existence of LTE implies that all species at a given point in space and time have the same temperature. Indeed, the term "temperature" has no meaning in the absence of LTE. In discharges of interest in plasma processing, charged species (electrons and ions) are in general at elevated energies compared with the energy of neutral species because of the electric field. Usually, ions are near the neutral gas temperature, but electrons have a much higher mean energy, because most collisions between charged and neutral species are elastic (i.e., there is no change in internal energy; only kinetic energy is exchanged between particles). The fraction of energy exchanged in an elastic collision is a function of the relative masses of colliding particles. For particles of similar mass, kinetic energy is exchanged readily (e.g., one billiard ball can lose all of its momentum to another in a collision). However, a small particle loses only a small fraction of its kinetic energy when colliding with a much heavier particle. The fraction of kinetic energy transferred is proportional to twice the smaller mass divided by the sum of the small and large masses. However, if the relative energy of colliding particles is sufficiently high, then inelastic collisions can occur. These processes have been described in the list on page 384.

Solutions to the Boltzmann Equation for Discharges. In conventional chemically reacting systems, all species are assumed to have the same distribution of energies, namely the well-known Maxwell–Boltzmann distribution. This assumption is incorrect for discharges at low pressure because of the large deviation from LTE. Because the rates of important electron–neutral species inelastic collision processes such as molecular dissociation depend on the form of the electron energy distribution function rather than on mean energy (equation 3), such assumptions can create serious problems in the modeling of discharges. The calculation of the distribution function for electrons and ions is not a simple task, particularly under the extreme conditions of a discharge. The normal procedure for obtaining the electron velocity distribution function in a weakly ionized gas is to solve the Boltzmann equation (*15, 54–56*):

$$\frac{\partial f}{\partial t} + \frac{\mathbf{v} \cdot \partial f}{\partial \mathbf{x}} + \frac{e\mathbf{E}}{m} \cdot \frac{\partial f}{\partial \mathbf{v}} = \left(\frac{\partial f}{\partial t}\right)_{\text{collisions}} \tag{4}$$

where f is the velocity distribution function, t is time, $\mathbf{x}$ is position, $\mathbf{v}$ is velocity, $\mathbf{E}$ is the electric field, e is the charge on the electron, and m is electron mass. Equation 4 gives the rate at which the distribution function changes because of collisions.

The Boltzmann equation works reasonably well when electrons collide mainly with neutral species. Electron–electron or electron–ion collisions involve coulombic interactions that have a longer range than that of electron–neutral species interactions. Coulombic-interaction potentials vary inversely with separation, but electron–neutral species interaction potentials vary inversely with the fifth or sixth power of separation.

Most of the work in solving the Boltzmann equation for electrons has been for the relatively simple conditions of electron swarm experiments. In these experiments, electrons are released from a cathode in low concentrations and drift under the influence of a uniform applied electric field in a low-pressure gas towards an anode at which the electrons are collected. If the electrons are released as a pulse, then the time for electrons to drift to the anode is a measure of electron mobility. If the radial and axial spreads of the electron pulse can be measured at the anode (with a radially segmented electrode, for example), then a measure of the electron longitudinal and transverse (with respect to the direction of drift) diffusivities can be obtained. Further, if electrons are created through ionization or lost through electron attachment, then rate coefficients for these processes can be determined. Such experiments provide rate and transport coefficients but not collision cross sections.

To deduce cross sections, the following iterative process is used. A trial set of cross sections is chosen, the Boltzmann equation is solved numerically, and the corresponding rate and transport coefficients are calculated. If the measured and predicted coefficients fail to agree, a revised set of cross sections is chosen, and the process is repeated until satisfactory agreement is reached.

A great deal of research has gone into improving the procedures for solving the Boltzmann equation (57) in the context of electron swarm experiments. Rather than attempt to review this large area, suffice it to say that a considerable amount of work will be necessary to adapt these approaches to conditions in actual discharges. For example, spatial variations, which are neglected or minimized in swarm analyses, are of considerable importance in discharges. In addition, solving for ion characteristics will generally require approaches that are different from those taken for electrons, because ion velocity distribution functions are characterized by large anisotropies in velocity space. Net ion velocity tends to be the same order of magnitude as random ion velocity, a fact that has implications for the techniques used to solve the Boltzmann equation. The most common approach for ions is to solve for velocity moments directly rather than for the distribution function.

An alternative to solving the Boltzmann equation directly is the use of particle simulation techniques, sometimes referred to as Monte Carlo methods (58, 59). Major difficulties with the Monte Carlo approach include self-consistency, inclusion of ions, and extension to two spatial dimensions. However, these difficulties are probably not insurmountable, and the Monte Carlo approach may well turn out to be a very powerful tool for discharge analysis.

Fluid Model of Discharges. An important question is whether it makes sense to attempt to solve for distribution functions or moments in the absence of a commensurate accuracy in the treatment of neutral-species chemistry. As already stated, modeling of the chemically reacting plasma requires solutions to the bulk gas momentum and energy balance equations and continuity equations for each reacting neutral species. Surface chemistry is incorporated through boundary conditions on species continuity equations, which is analogous to the approach used for thermal CVD (*60, 61*).

As an alternative to the massive computational effort in coupling Boltzmann or moment equations for each charged species to CVD equations, it may make sense to assume a velocity distribution function for charged species and deal with mean energies. This technique is sometimes referred to as a "fluid" approach, because charged species are treated as if they formed a continuous fluid. In this approach, charged species are treated in nearly the same way as neutral species, except that momentum and energy balance equations are solved for each species. Strictly speaking, this treatment is not correct, because a mean energy does not uniquely determine the value of a rate or transport coefficient. However, this approach is a useful approximation. The major advantage of this approach is that it is much less expensive computationally than solving the Boltzmann equation for each charged species in two velocity dimensions, one or two spatial dimensions, and time. Probably the major disadvantage of the fluid approach is that charged particles may not always behave like a continuous fluid, for example, in high-field sheath regions or where electrons and ions act more like directed beams rather than fluids.

Another point in favor of the simpler, but less accurate, fluid approach is that discharge diagnostics are still quite primitive. In studies of electron and ion swarms, experimentalists routinely measure mobilities and diffusivities with a precision in the order of a few percent. A sophisticated model must be used to properly interpret such experiments. However, for discharges, even relative concentration profiles for a few of the dozens of important neutral and charged species are difficult to attain. Thus, an overly complex and expensive model is probably inappropriate, and the fluid model is a good compromise at present.

Validation of Models. An important goal in establishing the validity of large-scale models of reacting plasmas is to demonstrate agreement between predicted and experimentally measured quantities. This effort has barely begun for plasma processing. Ideally, all fundamental quantities predicted in equations describing discharge physics and chemistry must be measured. Electron and ion densities, net velocity and energy profiles, and the electric field strength profile are all predicted by the discharge physics model, and measurement of these quantities as a function of position and time is helpful. For discharge chemistry models, important variables include neutral-species densities, fluid velocity and temperature profiles, and rates of reaction at surfaces. Many of these quantities are difficult to measure, particularly under realistic etch or deposition conditions.

Some of the most powerful tools for in situ discharge diagnostics are optical (*62*). Plasma-induced emission spectroscopy, laser-induced fluorescence, laser absorption, and laser optogalvanic spectroscopy have all been recently applied in discharge diagnostics. Considerable work remains in the development of these tools so that systematic and comprehensive measurements can be compared with model predictions.

Plasma Processes versus Combustion. To model discharge chemistry, detailed mechanisms and rate coefficients for all elementary reactions are required. This requirement represents a large amount of information even for simple systems. By analogy, for the combustion of hydrogen, a typical model would include eight species (H_2, O_2, H_2O, O, OH, H, HO_2, and H_2O_2) and about 20 elementary reactions. This case is probably the simplest nontrivial combustion problem. Computationally, the major expense is in the number of species, because a continuity equation for each species must be solved. As the complexity of the set of reactants increases, the number of intermediate reactants and elementary reactions increases, and the computational costs rise correspondingly. The inclusion of all important species is not always clear, and values for rate coefficient parameters can be difficult to obtain. In particular, the measurement of reaction rates is not easy when intermediate radical species are involved. Intermediate species are not stable and will sooner or later react to form other species. Sampling of these species is difficult, and estimates of rate coefficients can vary widely.

Combustion modeling is a good paradigm for plasma process modeling because the goals and problems involved are similar. For both, interest is in predicting the detailed chemistry of many different reactive species. Both gas-phase and surface reactions are important, and physical processes must be included in models. For combustion, chemistry interacts strongly with fluid mechanics and heat transfer. One area of combustion research that the plasma analyst, thankfully, can ignore is the study of turbulent reacting flows. At the low pressures encountered in plasma systems, flows are virtually always laminar. The plasma modeler has many other unique difficulties, however, including the possibility that at sufficiently low pressures (i.e., <6.7 Pa), the system may no longer be treated as a continuum, because mean free paths can be in the same order as reactor dimensions.

A perspective on the status and prospects of detailed plasma modeling may be gained by examining the present situation in combustion modeling (63). Currently, the only fuels that can be modeled with real confidence are pure hydrocarbons (no nitrogen or sulfur) that have at most two carbon atoms. When particulates form during combustion or when combustion occurs at solid surfaces, confidence in the model drops substantially. The presence of trace species such as nitrogen and sulfur complicates matters further. This state of relative ignorance in combustion is particularly striking because combustion chemistry has been studied fairly intensely for decades. In contrast, the present status of plasma chemistry research indicates that only the first stumbling steps have been taken; a great deal of work is yet to be done. The plasma-process-modeling community can, however, take advantage of progress made in related areas such as combustion and CVD by initially adopting approaches and techniques that have proved successful.

System of Model Equations. Because of the tremendous complexity of the processes involved in plasma chemical reactors, the number of possible sets of model equations is large. In the following treatment, one set of equations is presented, but this set is by no means exhaustive.

Qualitatively, to model the chemically reacting plasma, neutral-species chemistry and transport and the electrical or plasma physics aspects of the discharge must be considered. The chemical and physical properties included in the model of the discharge must at least reflect the current understanding of the most important processes. The conventional view of discharge behavior has been described previously, but will now be reviewed briefly to set the stage for equation formulation.

Current Understanding of Discharge Behavior. Electron-impact dissociation of neutral reactant molecules creates highly chemically reactive radicals that can react further in the gas phase and diffuse and convect to surfaces where they react to form or etch films. The surface chemistry is substantially influenced by the action of ions bombarding the surface. To model the chemistry of neutral reactants, a continuity equation for each chemically important species must be solved. These continuity equations must include important transport terms (convection of bulk gas and diffusion) and creation-and-loss terms (electron-impact creation or loss and neutral species–neutral species gas-phase reactions). To account for bulk gas convection, equations for bulk gas momentum and overall mass continuity must be solved. In addition, because surfaces in plasma deposition reactors are often heated to temperatures of 300 °C or higher and because both gas-phase chemistry and gas density depend on temperature, an energy balance for gas temperature must be solved. In plasma etching systems, surfaces may not be heated, and so the energy balance may not be required. This set of equations—continuity for each species, momentum, mass, and energy—are generally required for any reacting flow.

Often, to simplify the analysis, the gas is assumed to be well mixed or to flow as a plug with no diffusion. In the well-mixed system, no gradients exist, and the set of coupled partial differential equations becomes sets of coupled algebraic equations, which is an enormous simplification. In general, however, spatial variations must be considered.

Formulation of Equations. Discharge structure influences chemistry primarily through electron-impact dissociation and surface ion bombardment. To predict the rate of electron-impact dissociation, local electron number density and energy must be known. These quantities are obtained from equations for electron continuity and electron energy, respectively.

Ion bombardment rate is determined from ion momentum or continuity equations, depending upon assumptions made in the model. To solve equations for ion and electron momentum and energy balances, the electric field profile must be known. This profile is obtained from the governing Maxwell equation, which is usually Poisson's equation.

Discharges of interest in plasma processing are usually sustained through the application of radio frequency power to an electrode in the reactor. In this case, a grounded counter electrode is present in the discharge or the entire chamber wall serves as the grounded electrode. Voltages and currents at the powered electrode vary with time, and therefore, quantities in the discharge such as electron density and energy also vary with time. Thus, equations for discharge structure must, in general, include terms for the rate of change of quantities with time, in addition to terms for spatial variations. However, because neutral species respond on a time scale that is much longer than the time scale for charged species, equations describing neutral species do not usually require time derivatives.

Equations for Neutral Species. The balanced equations for neutral species can be expressed in the following form:

$$\nabla \cdot (\mathbf{v} n_s + \mathbf{J}_s) = \sum r_j^s \tag{5}$$

where v is the gas velocity, n_s is the mass density of species s, $\mathbf{J}_s$ is the diffusion flux of species s, and r_s^j is the net rate of creation of species s in reaction j. A sum is made over all such reactions. Gas velocity is obtained from the momentum balance (*see* equation 7), and reaction rates r_s^j are in general functions of the concentrations of other species and gas temperature, which are obtained from the energy balance (*see* equation 8). Coupling to gas discharge equations occurs primarily in equation 5 through the role of electron-impact reaction-and-loss terms. In addition, boundary conditions for species continuity equations will include terms for ion-bombardment-enhanced chemical reactions. The equation for total mass continuity is

$$\nabla \cdot (\rho \mathbf{v}) = 0 \tag{6}$$

where ρ is the total gas mass density. The gas momentum balance equation can be expressed as

$$\nabla \cdot (\rho \mathbf{v}\mathbf{v}) = -\nabla p - \nabla \cdot \boldsymbol{\tau} + \rho \mathbf{g} \tag{7}$$

where p is gas pressure, $\boldsymbol{\tau}$ is the viscous stress tensor, and $\mathbf{g}$ is the gravitational force. One form of the energy balance is presented in equation 8

$$\nabla \cdot [(\rho U + p)\mathbf{v} - K\nabla T] = 0 \tag{8}$$

where U is the internal energy of the gas, K is thermal conductivity, and T is temperature. Equation 8 implicitly assumes that gas-phase reactions do not contribute significantly to the energy balance either because rates of reactions are too low or because the enthalpy of reaction is too low.

The equation of state for the gas is given by equation 9, which relates mass density, pressure, molecular weight, and temperature. The molecular weight, M, is assumed to be a mean quantity, because its value depends upon the exact composition of the gas.

$$\rho = pM/kT \tag{9}$$

To be complete, equations 5–9 require boundary conditions. The basis for choosing appropriate boundary conditions is difficult to generalize, partly because the choice of boundary conditions is what differentiates one system from another.

Typically, there are two types of boundaries in reacting flows. The first is a solid surface at which a reaction may be occurring, where the flow velocity is usually set to zero (the no-slip condition) and where either a temperature or a heat flux is specified or a balance between heat generated and lost is made. The second type of boundary is an inflow or outflow boundary. Generally, either the species concentration is specified or the Dankwerts boundary condition is used wherein a flux balance is made across the inflow boundary (*64*). The gas temperature and gas velocity profile are usually specified at an inflow boundary. At outflow boundaries, choices often become more difficult. If the outflow boundary is far away from the reaction zone, the species concentration gradient and temperature gradient in the direction of flow are often assumed to be zero. In addition, the outflow boundary condition on the momentum balance is usually that normal or shear stresses are also zero (*64*).

Equation for Electric Field Strength. To include electron-impact source terms in continuity equations for neutral species and to include the effect of ion bombardment on the rate of surface reactions, equations that predict electron and ion densities, momentums, and energy profiles are required. The profiles require an equation for the electric field strength. In general equations for electron and ion continuities (these equations yield electron density and ion density, respectively), electron and ion momentums (for electron and ion net or directed velocity), and electron and ion energies (for electron and ion random or thermal energy) must be solved. Finally, the electric field profile is obtained from Poisson's equation.

This set of equations can be simplified somewhat, depending on discharge operating conditions and on the nature of the discharge gas. For example, in the following equations, electron and ion motions are assumed to be collisionally dominated so electron and ion momentum balance equations can be simplified. In addition, the positive ions are assumed to exchange energy rapidly in elastic collisions and are therefore at the bulk, neutral gas temperature. This assumption eliminates the need for an ion energy balance. The following equations neglect negative ions, and the gas considered is assumed to be electropositive. This restriction can be easily generalized to include negative ions by incorporating equations for negative ion continuity and momentum.

Electron Continuity Equation. The electron continuity equation is

$$\partial n_e/\partial t + \nabla \cdot \mathbf{j}_e = \sum r_j \tag{10}$$

where n_e is the electron density and $\mathbf{j}_e$ is the electron flux (*see* equation 14). The right side represents the sum of all processes that create electrons through ionization, primarily electron-impact ionization of ground-state molecules. For electron-impact processes, the rate is linearly dependent on electron density and the neutral collision partner (equation 2) and usually exponentially dependent on electron temperature. The continuity equation for all other charged species will have a similar form, as shown by the continuity equation for positive ions:

$$\partial n_+/\partial t + \nabla \cdot \mathbf{j}_+ = \sum r_j \tag{11}$$

where n_+ is the positive-ion density and j_+ is the positive-ion flux. The right side represents the sum of all processes that create positive ions through ionization.

In ionization processes, a positive ion is created every time an electron is created, and the net rate of creation of positive ions equals that of electrons. The equation for mean electron energy (referred to as electron temperature) is

$$\partial(3/2 n_e kT_e)/\partial t + \nabla \cdot \mathbf{q}_e + \mathbf{j}_e \cdot e\mathbf{E} + \sum H_j r_j = 0 \tag{12}$$

where $3/2\ n_e kT_e$ is the mean electron random or thermal energy, $\mathbf{q}_e$ is the electron energy flux (*see* equation 16), $\mathbf{j}_e$ is the electron flux, e is the charge on the electron, $\mathbf{E}$ is the electric field strength, and H_j is the energy lost in inelastic electron–molecule collisions. The r_j in equation 12 is the rate of all inelastic collisions experienced by electrons. Most of the electron energy is lost via these collisions. Equation 12 is coupled to equations 10 and 11 through the dependence of r_j on mean electron energy.

Electric Field Profile and Electron Energy Flux. The equation that determines the electric field profile through the reactor is Poisson's equation:

$$\nabla \cdot \mathbf{E} = e/\epsilon_0(n_+ - n_e) \tag{13}$$

In equation 13, ϵ_0 is the permittivity of free space. Poisson's equation is the governing Maxwell equation, and use of this equation is justified by the fact that the wavelength of the radio frequency field is large compared with discharge dimensions. If this situation were not true (as in the case of microwave discharges), then the complication of electromagnetic wave propagation in the discharge has to be treated. The electron and ion momentum balance equations are simplified by the assumption that charged-particle motion is dominated by collisions. In this case, momentum balance equations for electrons and ions reduce to the following:

$$\mathbf{j}_e = -D_e \nabla n_e - \mu_e n_e \mathbf{E} \tag{14}$$

$$\mathbf{j}_+ = -D_+ \nabla n_+ + \mu_+ n_+ \mathbf{E} \tag{15}$$

Both equations 14 and 15 contain a diffusion term, D, for diffusivity and a drift term, μ, for mobility. For positive ions, the diffusion term is nearly always negligible compared with the drift term, but both terms are retained for reasons that have to do with numerical stability.

Finally, the expression for electron energy flux is given by

$$\mathbf{q}_e = -(5/2 n_e D_e)\nabla T_e + 5/2 n_e k T_e \mathbf{j}_e \tag{16}$$

where the first term on the right side of the equation accounts for electron random motion (thermal conduction) and the second term accounts for energy transported by electron-directed motion.

Boundary Conditions. To complete the set of equations for the discharge physical structure, boundary conditions are needed. One possible set of boundary conditions is presented in Table II.

Boundary conditions for electron density, ion density, and electron energy are all "flux" boundary conditions, wherein an expression is given for the flux of the quantity over which the balance is made. The net flux of electrons to the surface (assumed to be conducting) is the difference between the rate of recombination and the rate of creation through secondary electron

Table II. Boundary Conditions for Fluid Equations

Boundary Condition	*Physical Significance*
$\mathbf{n} \cdot \mathbf{j}_e = n_e(kT_e/2\pi m_e)^{1/2} - \gamma \mathbf{n} \cdot \mathbf{j}_+$	Electron flux balance[a]
$\mathbf{n} \cdot \mathbf{j}_+ = \mu_+ n_+ \mathbf{E} \cdot \mathbf{n}$	Ion flux balance
$V = V_{dc} + V_{rf} \sin(2\pi \upsilon t)$	Voltage boundary value
$\mathbf{n} \cdot \mathbf{q}_e = n_e (kT_e/2\pi m_e)^{1/2} (3/2kT_e) - \gamma \mathbf{n} \cdot \mathbf{j}_+ (3/2kT_{ec})$	Electron energy flux balance
$F(x, t) = F(x, t_p)$	Periodicity conditions on all solutions

[a]The electron flux balance is equal to rate of recombination – rate of emission.

emission. The boundary condition for positive ions is that the flux is due only to drift motion. The flux of electron energy is assumed to result from a balance between energy lost through recombination and energy gained through secondary emission. This boundary condition is subject to criticism, because effects due to the work function for the surface and energy of impinging ions have been neglected. However, preliminary solutions show that the important features of the electron energy profile are not strongly affected by the simplified form of the boundary condition.

The boundary condition for Poisson's equation is to specify voltage at the conducting surfaces. Some workers have suggested that a more stable numerical solution results from a specification of current at the boundary rather than voltage. Typically, one electrode is grounded and the other is powered. In the model, the grounded electrode is assigned zero voltage, and the powered electrode can have a dc bias voltage (as discussed earlier), in addition to the rf voltage. In fact, the actual voltage wave form at the powered electrode is a consequence of the external circuit, including the power supply, matching network, and cables. Simple specification of the voltage at the powered electrode is an approximation, but in many cases, this approximation can be done with acceptable error. An obvious extension would be to solve an equation representing the external circuit at the electrode for the voltage.

The solutions sought in the discharge simulation are time periodic and match the normal stable operating conditions in the discharge. Time periodicity in the solutions is driven by the time-varying voltage at the powered electrode. The time periodicity condition is expressed as

$$\mathbf{u}(x, t) = \mathbf{u}(x, t + t_p) \tag{17}$$

where $\mathbf{u}$ is any function in the simulation, x is position in the discharge, t is time, and t_p is the rf period ($t_p = 1/\nu$, where ν is frequency in hertz).

The numerical solution of the discharge equations and boundary conditions can be approached by using either finite-difference or finite-element methods to make the spatial derivatives discrete. The resulting set of coupled, nonlinear ordinary differential equations can be solved by using relatively standard methods. Typically, the equations are integrated with respect to time until the desired time-periodic solution is obtained. In fact, obtaining a numerical solution is by far the most difficult part of the entire modeling exercise. The set of coupled, nonlinear partial differential equations is stiff, with widely varying temporal and spatial scales in the solution. The size and nonlinearity of the set of coupled equations require either a dedicated minicomputer or a supercomputer for reasonable solution times. Preliminary solutions have been obtained (65) but a great deal of work remains before this modeling approach becomes well established.

Other Modeling Approaches. The modeling just outlined should be viewed in the context of previous models of plasma chemical reactors. A representative list of models of plasma chemical reactors is presented in Table III (*66–84*), with brief descriptions of the following characteristics of each model: reactor type, chemistry studied, transport model for neutral species, and treatment of discharge physics. One of the reasons for the detailed discussion of the discharge fluid equations is that discharge physics seems to be the part of plasma chemical reactor models that is the least well understood and is probably the major source of mystery concerning reactor behavior. Other than the fluid model just described, the only treatment of spatially and temporally dependent discharge behavior involves the use of a (non-self-consistent) Monte Carlo technique (*79*) to describe electron dynamics.

Most of the models assume that neutral-species transport can be represented with either a well-mixed model or a plug flow model. The major drawback to these assumptions is that important inelastic rate processes such as molecular dissociation are usually localized in space in the reactor and are often fast compared with rates of diffusion or convection. As a result, the spatial variation of fluid flow in the reactor must be accounted for. This variation introduces a major complication in the model, because the solution of the nonisothermal Navier–Stokes equations in multidimensional geometries is expensive and difficult.

The most glaring drawback of most of the models presented heretofore is the lack of proper experimental verification. Model predictions must be compared with experimental measurements in order to verify the model. If the model assumes that concentration gradients are negligible, then this assumption should be tested. If the model predicts detailed spatial- and time-dependent profiles, then these profiles must be measured and compared with the predictions. Only when such comparisons have been made and measurements and predictions agree over a wide range of conditions can the model be considered accurate.

Detailed models of plasma chemical reactors are just beginning to be developed. The combination of neutral-species chemistry, hydrodynamics, and discharge physics is sufficiently complex that modeling lags behind both industrial practice and even certain diagnostics such as laser spectroscopy and molecular-beam mass spectrometry. However, the widespread availability of large-scale computers, coupled with an increasing interest in a fundamental understanding of plasma processes, will spur the continued development of mathematical modeling and numerical solutions of model equations.

Etching of Thin-Film Materials

The basic principle behind etching a solid material with a reactive gas discharge is inherently simple. A gas is chosen that dissociates into reactive species that can form a volatile product with the material to be etched. However, a viable etch process must also display reasonable etch rates, good selectivity, and directional (anisotropic) etching. All of these criteria can be met by judicious choice of reactant gases and plasma conditions. For instance, because etch rates usually follow an Arrhenius dependence on substrate temperature and because the effective activation energy is material dependent, etch rate and, thus, selectivity vary exponentially with temperature. Also, ion bombardment generally imparts different degrees of damage to various materials; thus, etch selectivity can be altered to tailor a specific etch process by changing the ion bombardment energy.

Most isotropic etchants exhibit a loading effect, wherein a measurable depletion of the active etchant results from consumption in the etch process. In these cases, the overall etch rate depends upon the area of film to be etched. Under extreme circumstances, with carbon-based etch gases, etchant depletion can be so severe that polymer deposition occurs instead of etching. An analysis (*85*) of the loading effect, which has been extended (*86*) to include multiple etchant loading and other etchant loss processes, indicates that the etch rate (R) for N wafers each of area A is given by

$$R(N) = \frac{(k_{\text{etch}}/k_{\text{loss}})G}{1 + (k_{\text{etch}}\rho NA/k_{\text{loss}}V)} \tag{18}$$

where k_{etch} and k_{loss} are the rate constants (first order) for etching and etchant loss in an "empty" reactor, respectively; ρ is the number density of substrate molecules per unit wafer area in the reactor; G is the generation rate of the active etchant species; and V is the volume of the reactor. This formalism indicates that as the number of wafers (or wafer area) increases to a point where $k_{\text{etch}}\rho NA/k_{\text{loss}}V >> 1$, the etch rate

$$R(N) = GV/\rho NA \tag{19}$$

varies inversely with the number of wafers. This dependence is characteristic of a marked loading effect and shows that loading effects can be controlled to some extent by using large-volume reactors. Further, these effects can be eliminated when homogeneous or heterogeneous etchant loss dominates ($k_{\text{loss}} >> k_{\text{etch}}$). When ion bombardment controls the reaction, equation 18 is inapplicable, because the etch rate is determined by bombardment flux rather than etchant supply.

Other transport limitations, such as diffusion-controlled reactions, can lead to localized depletion of etchant, which results in a number of observable etch effects. The size and density of features can influence the etch rate at different locations on a single wafer and thus produce pattern sensitivity. Depletion across a wafer produces a bull's-eye effect, whereas depletion across a reactor is indicated by the fact that the leading wafer edge etches faster than the trailing edge. Similar effects are noted when product removal

Table III. Discharge Reactor Models

Reactor Type	*Chemistry*	*Transport*	*Discharge Physics*	*Reference*
Parallel-plate electrodes	Ethane polymerization	Plug flow with axial dispersion	Assumed average electron density and energy	66
Parallel-plate with electric barriers	Helium discharge	Not applicable	Ambipolar diffusion model	67
Flow through plasma then afterglow	CO oxidation; CO_2 dissociation	Axial plug flow	Mean electron density and energy	68
Parallel-plate pyrex pillbox	Oxygen dissociation	Plug flow	Cosine electron density between parallel-plate electrodes	69
Radio frequency diode reactor	Silane decomposition	Well mixed	Mean electron density and energy	70
Cylindrical tube with ring electrodes	Silane decomposition and dispersion	Plug flow with axial dispersion	Mean electron density measured with interferometry	71
Inductive coupling to quartz tube	Silicon nitride from silane–N_2	Plug flow	Mean electron density measured with microwave interferometry	72
Tubular and radial	No specific chemistry	Parabolic flow	Dissociation rate coefficient treated as a parameter	73
Arbitrary	No specific chemistry	Well mixed	Not treated	74
Parallel-plate electrodes	C_nF_m in H_2 and O_2 to etch Si and SiO_2	Well mixed	Mean electron density from power balance; electron energy distribution from Boltzmann code	75
Tubular reactor with afterglow	CF_4 etching Si	Plug flow	Assumed mean density and energy in discharge	76

Flow between parallel-plate electrodes	SF_6 etching silicon	Plug flow	Boltzmann equation and Monte Carlo for electron energy; electron density power balance	77
Flow between parallel plate electrodes	SF_6–O_2 etching silicon	Well mixed	Boltzmann equation for electron energy; power balance for electron density	78
Parallel-plate electrodes	SiH_4 dissociation in silane–Ar mixtures	Not treated	Spatially and temporally dependent Monte Carlo treatment for electrons; electric field is assumed	79
Parallel-plate electrodes; downstream plasma	SiH_4 dissociation for Si films; both thermal and plasma dissociation methods	Two-dimensional nonisothermal Navier–Stokes	Assumed electron density and energy	80
Hollow cathode	NF_3 etching	Well mixed	Electron density and energy treated as adjustable parameters	81
Parallel-plate reactor	Dissociation of CH_4 for amorphous carbon	Well mixed	Mean electron density; energy distribution measured with Langmuir probe	82
Barrel etcher with etch tunnel	Arbitrary between wafers	Diffusion	Not applicable	83
Parallel-plate radial flow	CF_4 etching Si	Two-dimensional isothermal Navier–Stokes	Ambipolar diffusion with assumed electron density and energy	84

is transport limited. Pressure and flow interact by determining the residence time relative to diffusion, convection, and reaction rates. Residence times that are short compared with reaction times will reduce diffusion limitations, whereas long residence times will enhance them.

Most of the effects just discussed can be reduced by the judicious choice of pressure and flow rates. However, good or even adequate process control requires precise termination of the etch sequence by using suitable end-point detection or plasma-diagnostic schemes. Numerous methods have been invoked to follow plasma processes either by monitoring film thickness in situ or by detecting changes in plasma composition or impedance. Discussion of these optical, chemical, electrical, or physical techniques and their application to plasma etching (or deposition) is beyond the scope of this chapter; however, recent publications cover this important area in great detail (*2, 62, 87–89*). Finally, if process design rather than diagnostics is desired, response surface methodology (statistical design) can be invoked (*90, 91*). This technique permits the determination of a parametric operating window for a particular complex process with the execution of a minimal number of experiments. Unfortunately, no fundamental insight into the effect of specific variables on the etch process is generated.

Etch Models. As has been discussed, an extensive parameter space is associated with plasma techniques. Therefore, if the development of etch processes is to proceed efficiently, some means of data assimilation and prediction must be available. Two general schemes have been proposed to organize chemical and physical information on plasma etching. Both schemes deal primarily with carbon-containing gases, but with a slight modification, they can be easily applied to other etchants. Conceptually, the two models are similar, although they emphasize different aspects of plasma etching.

F/C Ratio Model. The fluorine-to-carbon ratio (F/C) of the active species can be used (*92*) to explain the observed etch results (Figure 8). This model does not consider the specific chemistry occurring in a glow discharge but, rather, views the plasma as a ratio of fluorine species to carbon species that can react with a silicon surface. The generation or elimination of the active species by various processes or gas additions then modifies the initial F/C ratio of the inlet gas.

The F/C ratio model accounts for the fact that for carbon-containing gases etching and polymerization occur simultaneously. The process that dominates depends upon etch gas stoichiometry, reactive-gas additions, amount of material to be etched, and electrode potential and upon how these factors affect the F/C ratio. For instance, as described in Figure 8, the F/C ratio of the etchant gas determines whether etching or polymerization is favored. If the primary etchant species for silicon (F atoms) is consumed either by a loading effect or by reaction with hydrogen to form HF, the F/

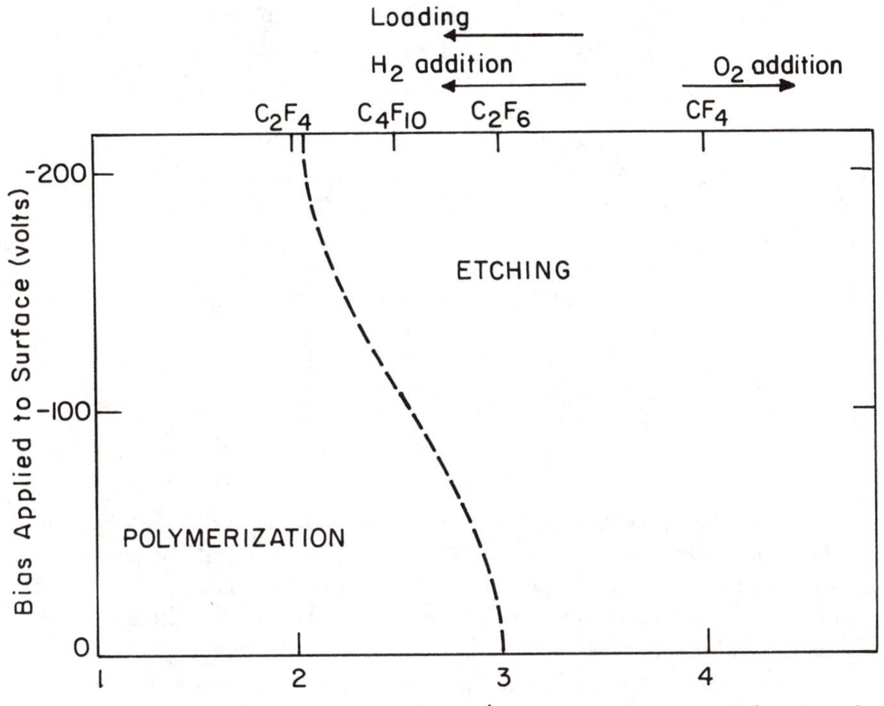

Figure 8. Schematic diagram of the influence of fluorine-to-carbon ratio and electrode bias on etching versus polymerization processes in fluorocarbon plasmas. (Reproduced with permission from reference 92. Copyright 1980 Springer–Verlag New York, Inc.)

C ratio decreases, thereby enhancing polymerization. However, if oxygen is added to the etchant gas, reaction with carbon-containing fragments to form CO or CO_2 can occur, thus increasing the F/C ratio and favoring etching. Further, as the negative bias of a surface exposed to the plasma increases at a constant F/C ratio, etching of this surface becomes more important relative to polymerization. Such effects are caused primarily by the enhanced energies of the ions striking these surfaces that result in polymer sputtering.

Etchant–Unsaturated Species Model. In the etchant–unsaturated species model described by equations 20–23 (93), specific chemical species derived from electron collisions with etchant gases are considered.

$$\text{e} + \text{halocarbon} \rightarrow \begin{matrix}\text{saturated}\\\text{radicals}\end{matrix} + \begin{matrix}\text{unsaturated}\\\text{radicals}\end{matrix} + \text{atoms} \qquad (20)$$

$$\text{reactive atoms and molecules} + \text{unsaturated species} \rightarrow \text{saturated species} \quad (21)$$

$$\text{atoms} + \text{surfaces} \rightarrow \text{chemisorbed layer} + \text{volatile products} \quad (22)$$

$$\text{unsaturated species} + \text{surfaces} \rightarrow \text{films} \quad (23)$$

Application of this model to a CF_4 plasma results in the chemical scheme indicated by equations 24–27.

$$2e + 2CF_4 \rightarrow CF_3 + CF_2 + 3F + 2e \quad (24)$$

$$F + CF_2 \rightarrow CF_3 \quad (25)$$

$$4F + Si \rightarrow SiF_4 \quad (26)$$

$$nCF_2 + \text{surface} \rightarrow (CF_2)_n \quad (27)$$

Depending upon the particular precursors generated in the gas phase, etching, recombination, or film formation (i.e., polymerization) can occur. Also, gas-phase oxidant additives (O_2, F_2, etc.) can dissociate and react with unsaturated species. Mass spectrometric studies of oxidant additions to fluorocarbon and chlorocarbon gases have demonstrated that the relative reactivity of atoms with unsaturated species in a glow discharge follows the sequence $F \sim O > Cl > Br$ (*93*). The most reactive species present will preferentially undergo saturation reactions that reduce polymer formation and that may increase halogen atom concentration. Ultimately, the determination of the relative reactivity of the plasma species allows the prediction of the primary atomic etchants in a plasma of specific composition.

Specific Materials. Because the fundamental physics and chemistry of reactive gas discharges used for etching is not yet fully understood, empirical approaches to the design of etch processes abound. As indicated earlier, etchant selection begins with considerations of product volatility. Virtually any gas or vapor that can dissociate to form etchant species has been considered or studied. An abbreviated list of the most common reactant molecules used to etch films of interest in electronic materials processing is given in Table IV.

Generally, a single etchant species is not used in production processes; rather, mixtures of gases or vapors are used to tailor the etch process to establish a particular rate, selectivity, anisotropy, etc. Often, the additives include inert gases such as Ar or He. These gases are responsible for Penning processes that enhance dissociation and ionization. Further, these gases can stabilize and homogenize discharges that oscillate between multiple states (*94*). In the following sections, etch considerations for individual materials are discussed.

Table IV. Etch Gases Used for Various IC Film Materials

Film	*Gases*
Si	CF_4, CF_4–O_2, CF_3Cl, CCl_4, SF_6–O_2, SiF_4–O_2, NF_3, ClF_3
SiO_2	C_2F_6, C_3F_8, CF_4–H_2, CHF_3
Si_3N_4	CF_4–O_2, C_2F_6, C_3F_8, CF_4–H_2
Organic materials	O_2, O_2–CF_4, O_2–SF_6
Al	CCl_4, CCl_4–Cl_2, BCl_3, BCl_3–Cl_2, $SiCl_4$
W, WSi_2, Mo	CF_4, C_2F_6, SF_6
$MoSi_2$	NF_3
Au	$C_2Cl_2F_4$, Cl_2
Cr	Cl_2–O_2, CCl_4–O_2

NOTE: Often, inert gases such as Ar or He are added to the mixtures.

Silicon and Polysilicon. The isotropic etching of silicon and polycrystalline silicon (poly-Si) by atomic fluorine (F) is probably the most completely understood of all etch processes, particularly for the cases in which F atoms are produced in discharges of F_2 (95) and CF_4–O_2 (96). Fluorine atoms etch <100>-oriented Si at a rate (in angstroms per minute) given by (95)

$$R_F(\text{Si}) = 2.91 \times 10^{-12} T^{1/2} n_F \exp(-0.108 \text{ eV}/kT) \tag{28}$$

where n_F is the F atom concentration per cubic centimeter, T is the temperature (in kelvin units), and k is Boltzmann's constant. The combination of these and other studies (97, 98) has generated a reasonably detailed mechanism of silicon etching in F-containing discharges, in which the major etch products are SiF_2 and SiF_4 (2, 99). Fluorine atoms chemisorb onto clean silicon surfaces to form a stable SiF_2-like steady-state surface under low-pressure (0.013 Pa) conditions (98, *100a*).

However, during conditions of higher pressure, SiF_3 moieties are the primary species detected on Si surfaces (*100b*). Continued reaction requires that the fluorinated surface layer be penetrated by impinging F atoms. The reaction probability for this step is low (<0.01) at normal substrate temperatures (95); thus, the penetration is rate limiting and proceeds at a rate defined by equation 28. Further, most F-containing etchant gases display similar apparent activation energies (~0.1 eV per atom) so that the same mechanism and etchant (F atoms) are probably operative in these somewhat different chemical systems.

Gases such as CF_4 and SF_6 offer advantages because of their low toxicity; however, the formation of unsaturated species such as C_xF_{2x} and S_xF_{2x} in the discharge can scavenge free F atoms and, in extreme cases, can lead to significant polymer or residue formation. The role of oxygen in these plasmas is twofold. First, in accord with the etchant–unsaturated species model, O atoms react with unsaturated species to enhance F atom generation and to eliminate polymerization. Second, with sufficient O_2 present, O_2 or O can occupy film adsorption sites and thus inhibit etching (*96, 101*). Gases such as NF_3 and ClF_3 (*81, 86, 102*) are interesting, because they do not contain atoms that form residues. Therefore, high concentrations of F result without the addition of O_2, which can attack resist materials.

Plasmas that produce chlorine and bromine atoms are excellent for Si etching, because they can generate a high degree of anisotropy. The most commonly used gases have been Cl_2, CCl_4, CF_2Cl_2, CF_3Cl, Br_2, and CF_3Br, along with mixtures such as Cl_2–C_2F_6, Cl_2–CCl_4, HCl–CCl_4, and C_2F_6–CF_3Cl (*103–106*). High etch rates (500–6300 Å/min for undoped and doped poly-Si) and selectivities (Si:SiO_2 ~ 10–50:1) have been attained. The active etchants in these plasmas are likely to be Cl and Br atoms; however, ion bombardment plays a significant role in achieving high etch rates and anisotropy control (*107–110*). The high degree of anisotropy that is readily achieved indicates that ion bombardment tends to dominate the etch mechanism by enhancing either the reaction with the chemisorbed $SiCl_2$ (penetration of Cl into the Si surface) or product volatility. These conclusions are also consistent with the fact that only a small loading effect is observed in Cl-based etching (*105*), whereas F source plasmas (*85, 86*) exhibit strong loading effects.

Heavily doped ($>10^{18}/cm^3$) n-type Si and poly-Si etch faster in Cl- and F-containing plasmas than do their boron-doped or undoped counterparts (*103a, 105, 111, 112*). Because ion bombardment is apparently not required in these cases, isotropic etch profiles (undercutting) in n+ poly-Si etching often occur. Although the exact mechanisms behind these observations are not completely understood, enhanced chemisorption (*103b, 111*) and space charge effects on reactant diffusion (*112*) have been proposed.

Silicon Dioxide and Silicon Nitride. Silicon dioxide can also be etched by F atoms in a downstream discharge configuration. However, because of the strength of the Si–O bond, etch rates (equation 29) are low without particle bombardment (*95*).

$$R_F(SiO_2) = 6.14 \times 10^{-13} n_F T^{1/2} \exp(-0.163 \text{ eV}/kT) \quad (29)$$

XeF_2 does not etch SiO_2 in the absence of bombardment (*39*). Thus, F and XeF_2 are not equivalent sources of F atoms, probably because of differences in sticking coefficients and the Xe–F bond energy. Because ion-bombardment-assisted etching of SiO_2 occurs in F-containing gases, directional etching can be achieved. This fact suggests that the etchants described for Si are suitable for etching oxide and nitride; however, they can be used only in the absence of silicon. Therefore, selective etching of these materials represents an important application of the chemical models presented earlier (*see* Etch Models). The earliest reported etchant gases for selectively etching SiO_2 and Si_3N_4 in the presence of silicon were CF_4–H_2, CHF_3, C_3F_8, and C_2F_6 (*113*). Selectivities as high as 15:1 (Si_3N_4:Si) and 10:1 (SiO_2:Si) were achieved. However, these selectivities are lower than one would like for a production process. All of these plasmas tend to be fluorine deficient, a fact suggesting that CF, CF_2, or CF_3 might be active etchants (*114*). In any case, carbon-containing species are clearly involved in the etch mechanism, because in addition to SiF_4, products such as CO and COF_x have been observed by mass spectrometry (*114–116*) during SiO_2 and Si_3N_4 etching.

To achieve selective oxide and nitride etching, additives to F source plasmas are chosen to make a F-deficient chemical environment. Additives include H_2, C_2H_4, and CH_4, which are efficient F scavengers. Alternatively, molecules that contain F, C, and H are used, such as CHF_3. Deciding the amount of additive necessary remains more of an art than a science, because oxide and nitride selectivity requires operation in a chemical environment very close to the demarcation between etching and polymerization shown in Figure 8. In fact, polymer deposition on Si generally occurs while active etching of oxide proceeds (*103c, 117, 118*). Thus, the mechanism for selective oxide and nitride etching may not involve CF_x as primary etchants but as film formers that inhibit Si etching by passivating chemisorptive sites.

An example of the improvement in selectivity with CF_4–H_2 mixtures is shown in Figure 9 (*118*). With no added H_2, the selectivity of silicon dioxide over silicon (SiO_2:Si $\sim$ 1.3) is unacceptable for a controllable process. However, as H_2 is added, F atoms are scavenged (to form HF), and thus the C/F ratio of unsaturated species in the gas phase is increased (Figure 8 or equations 24–27), and polymerization is promoted. Because oxygen is released from SiO_2 as it etches, carbonaceous residues can be removed (as CO, CO_2, or COF_2) from this material. On Si, no mechanism other than sputtering is available for carbon removal; thus, the carbon deposits inhibit etching even with low H_2 additions. The small decrease in SiO_2 etch rate with increasing H_2 is probably a result of etchant dilution. At 40% H_2, a selectivity of ~40 (SiO_2:Si) is achieved under the designated conditions of pressure, power, etc. (*118*). If more H_2 is added, polymerization on SiO_2 surfaces retards etching still further.

The qualitative role of ions in promoting oxide and nitride selectivity has also been established. Discharge conditions (i.e., pressure and voltages) are generally those that favor ion-enhanced reactivity, and ion bombardment is usually required to initiate etching (*117*). The absence of a noticeable dependence on substrate temperature (*119*), minimal loading effects (*120*), and the sensitivity of etch rate with respect to electrode (*121*) and sheath potential (*104*) reinforces the importance of ion bombardment. In one study (*121*), selective etching of SiO_2 over Si by $>$50:1 was obtained in a CHF_3 plasma by using cathode (powered-electrode) coupling, whereas selectivities of $<$10:1 were observed at the anode.

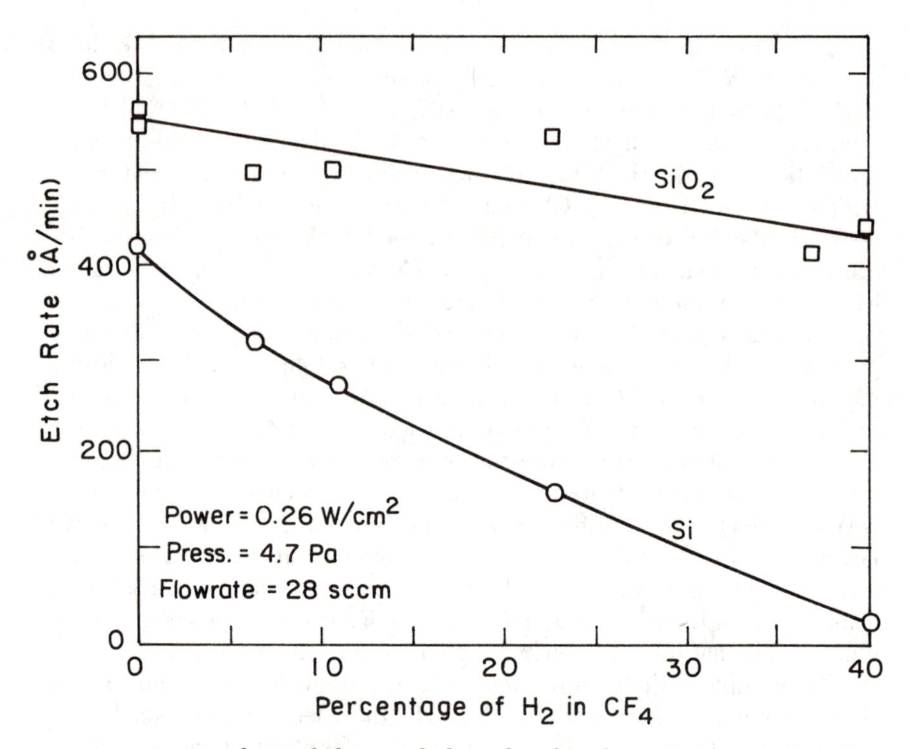

Figure 9. Dependence of silicon and silicon dioxide etch rates on the percentage of H_2 in CF_4–H_2 plasmas. (sccm is standard cubic centimeters per minute.) (Reproduced with permission from reference 118. Copyright 1979 The Electrochemical Society, Inc.)

Group III–V Materials. Group III–V semiconductor compounds such as GaAs, GaP, InP and GaAlAs form the basis for many new electronic applications, particularly high-speed integrated circuits (ICs) and microwave and optoelectronic devices. The development of plasma pattern delineation methods is an area of active research because of the parallelism with the more highly developed silicon technology; however, these systems are somewhat more complicated. Because most of the applications involve binary, ternary, and even quaternary alloys, suitable etchants must be capable of selective etching when stoichiometric changes are made to obtain specific device characteristics.

At present, primarily the binary compounds have been studied, and considerations of product volatility are the most significant driving force behind much of this research. The group III–V compounds are particularly difficult, because the group V elements form volatile halides, whereas group III halides, particularly the fluorides, tend to be involatile. As a result, F source plasmas, which have been the mainstay of silicon technology, are generally not practical for the etching of group III–V materials unless high ion bombardment energies are used. H_2 discharges have been used to etch group III–V materials (*122*). Temperatures above 100 °C were used, selectivities of GaAs oxide to GaAs were low (~2), and etch rates of the oxide were reasonable (<20 Å/s). However, GaAs etching displayed undercut profiles.

Because of the problems just described, most studies have used chlorine-containing plasmas with elevated substrate temperatures to take advantage of the volatility (albeit limited) of the group III chlorides. The importance of the volatility concept in the etching of group III–V materials relative to the reaction concepts that dominate Si etching is evident from results of chlorine-rich vapors. When gases such as CCl_xF_{4-x} (where x = 2, 3, or 4), $COCl_2$, PCl_3, BCl_3, $SiCl_4$, HCl or Cl_2 are used in rf (usually RIE configurations) or dc plasmas, the etch rates of GaAs and other group III-V compounds increase substantially (*123–130*).

Br_2 discharges have also been used to etch GaAs (*127*). These studies were performed under conditions in which isotropic etch profiles or chemical etching occurred. Because the chemical compositions (Ga/As ratio) of the different atomic planes in GaAs vary, crystallographic etch patterns are observed under etch conditions in which chemical processes dominate. Similar results are noted for Cl_2 plasmas. Furthermore, pure Br_2 vapor (no plasma and, thus, no Br atoms or particle or photon bombardment) etches GaAs, and smooth surfaces are observed (*127*). By using absorption spectroscopy to determine absolute Br concentrations in discharges, the etch rate of GaAs is found to be proportional to the Br atom concentration (*127*). In addition, the etch rate increases with frequency (from 0.1 to 14 MHz), in agreement with an increase in Br. Such results are consistent with plasma-etching studies with Cl_2 at 40 Pa, in which the etch rate decreases with decreasing frequency (*131*).

Temperature plays an important role in the etching of GaAs and InP. Both materials display an Arrhenius dependence on substrate temperature between 200 and 300 °C in Cl_2 discharges (*132*). The apparent activation energies (34.5 and 10.5 kcal/mol for InP and GaAs, respectively at 40 Pa) are in reasonable agreement with the heats of vaporization of $InCl_3$ and $GaCl_3$. Estimates of InP etch rates from kinetic molecular theory, assuming that volatilization of the product ($InCl_3$) is rate limiting, are in good agreement with the observed rates. However, similar estimates for GaAs are higher than the observed etch rate (*132*). The investigators conclude that a chemical reaction on the surface must be rate limiting in the case of GaAs, perhaps because of dimerization to form Ga_2Cl_6, which desorbs, or a slow chemical reaction between etchant (presumably Cl) and GaAs (*132*). Indeed, surface reactions with chlorine have been postulated recently to be rate controlling during the etching of GaAs with a modulated Ar^+ beam in the presence of Cl_2 molecules (*133*).

Metals. Because of the high reactivity of most metals with oxygen and water vapor, plasma etching of metals may require more attention to reactor design and process details than is necessary with other materials. Unless the metal forms a volatile oxyhalide or an unstable oxide, water vapor and oxygen must be excluded from or scavenged in the plasma reactor. In addition, the metal–oxygen bond may be extremely strong and therefore ion bombardment is required to assist native-oxide removal.

As a result of ion bombardment effects and of their ability to reduce native oxides chemically, chlorocarbon or fluorocarbon gases, rather than pure halogens, are used typically to etch metal films. However, halocarbon vapors are particularly susceptible to polymerization, which causes residue formation that can interface with etch processes (*134*).

Aluminum. Aluminum is widely used as an interconnect layer for integrated circuits, and its plasma etch characteristics have been extensively studied (*134, 135*). Because AlF_3 is not volatile under normal (nonsputtering) plasma etch conditions, chlorine-containing gases (CCl_4, BCl_3, $SiCl_4$, and $CHCl_3$) have been the preferred etchants for aluminum (*136–139*). A few studies have also investigated the use of brominated gases (HBr, Br_2, and BBr_3) and, consistent with volatility considerations, find results similar to those of chlorinated vapors (*140, 141*).

An initiation period or a lag time exists at the start of aluminum etching because of scavenging or removal of oxygen and water vapor present in the reactor and etching of the thin (~30 Å) native aluminum oxide layer always present on the aluminum surface (*136*). The removal of oxygen and water vapor can be minimized by using a load lock so that the chamber is not exposed to air or water vapor between etch runs (*137*) or by using an etch gas (BCl_3 or $SiCl_4$) that effectively scavenges water and oxygen. Native aluminum oxide can be etched by enhancing the ion bombardment of the surface and by supplying chemical species capable of reacting directly with the oxide, such as CCl_x, BCl_x, or $SiCl_x$.

After the removal of the native aluminum oxide layer, molecular chlorine (Cl_2) can etch pure, clean aluminum without a plasma (*142–144*). Indeed, Cl_2 rather than Cl appears to be the primary etchant species for aluminum in a glow discharge (*145*). At least at low temperature (<200 °C), the main product of the etch reaction seems to be Al_2Cl_6 rather than $AlCl_3$ (*146*).

To prevent aluminum from spiking through shallow junctions, 1–2% silicon is often added to the film. Because $SiCl_4$ is volatile at room temperature, aluminum–silicon films can be etched in chlorine-containing discharges.

Copper additions to aluminum films enhance electromigration resistance. However, copper does not form volatile chlorides or other halides, and therefore its removal during aluminum plasma etching is difficult. Two methods can be used to promote copper chloride desorption: increase the substrate temperature (consistent with the resist material being used) or enhance the ion bombardment so that significant sputtering or surface heating is attained.

After plasma etching is completed, aluminum films often corrode upon exposure to atmospheric conditions. The corrosion is a result of the hydrolysis of chlorine-containing residues (mostly $AlCl_3$) remaining on the film side walls, on the substrate, or in the photoresist. Because the passivating native oxide film normally present on the aluminum surface is removed during etching, chlorine species are left in contact with aluminum and ultimately cause corrosion. Further, contamination with carbon and radiation damage caused by particle bombardment may enhance corrosion susceptibility (*147*).

When copper is present in aluminum films, accelerated postetch corrosion is observed. This phenomenon occurs for at least two reasons. Because of the low vapor pressure, hygroscopic $CuCl_2$ is probably left in contact with the aluminum film, and because most of the copper is present in the grain boundaries as $CuAl_2$, the grain boundaries have a cathodic potential relative to the aluminum grains (*147*). Electrolytic corrosion then takes place upon adsorption of water vapor, followed by hydrolysis of $AlCl_3$ and $CuCl_3$ to generate HCl and chlorine ions.

Significant efforts have been made to eliminate or at least minimize corrosion. A water rinse or an oxygen plasma treatment after etching reduces the amount of chlorine left on the etched surfaces, but this step is usually not adequate to preclude corrosion. Low-temperature thermal oxidation in dry oxygen appears effective in restoring a passivating native aluminum oxide film (*147*). Another method of preventing postetch corrosion is to expose the aluminum film to fluorocarbon plasmas such as CF_4 or CHF_3 (*37*, *148*). This treatment converts the chloride residues into nonhygroscopic fluorides and deposits a fluorocarbon polymer film onto the Al surface so that the Al film can be exposed to ambient conditions without immediate corrosion. Subsequently, a nitric acid rinse can be used to remove the fluoride layer and to regrow the protective oxide.

Chlorine-based plasma etching of aluminum films causes serious degradation of photoresist materials. To some extent, these effects are a result of the etch product, Al_2Cl_6 or $AlCl_3$. Aluminum trichloride is a Lewis acid used extensively as a Friedel–Crafts catalyst. Therefore, this material reacts with and severely degrades photoresists (*149*).

Other Metals. Numerous other metal films have been etched in glow discharges. The following is a brief summary that gives specific information on the etching of the metal films commonly used in VLSI.

After aluminum, the refractory metals and their silicides have been the subject of the most extensive efforts in metal etching (*150–155*). Because the fluorides and chlorides of the transition metals and silicon are volatile in the presence of ion bombardment, etch studies have been performed with nearly any fluorine- or chlorine-containing gas. Mixtures of these gases (e.g., SF_6–Cl_2) have also been used. When C or S is present in the reactant gas, oxygen is often necessary to prevent polymer and residue formation and to increase the concentration of fluorine atoms. Oxygen can be added to enhance the transition-metal etch rate, because some oxyfluorides (e.g., those of Mo and W) have reasonable vapor pressures. Layered films (silicide–poly-Si; termed *polycides*) are frequently used to meet specific device requirements. These structures present problems in profile control, and selectivity to underlying SiO_2 or Si materials can be difficult to achieve.

Gold can be etched effectively with $C_2Cl_2F_4$ (*156, 157*) or with $CClF_3$ (*158*), whereas CF_4–O_2 etching causes staining. The observed staining is believed to be caused by gold oxides, whose formation is enhanced by the presence of atomic fluorine (*158*). Chromium is etched readily in plasmas containing chlorine and oxygen (*159*) because of the high volatility of the oxychloride (CrO_2Cl_2). Indeed, the high boiling point of $CrCl_2$ (1300 °C) results in significantly reduced etch rates of chromium in chlorine plasmas without oxygen.

Titanium can be etched in fluorine-, chlorine-, or bromine-containing gases, because all the titanium halides are volatile. Chlorides and bromides have been studied to a great extent, because they result in high selectivity over silicon-containing films and do not promote staining on gold (*158, 160*).

Organic Films. Organic films are present during the plasma processing of all materials discussed in the preceding sections, because polymeric resist masks are the primary method of pattern transfer. An ideal mask should be highly resistant to the reactive species, ion bombardment, and the UV radiation produced in the glow discharge and should be readily removed after pattern delineation is complete. However, very high selectivities of the etching film to resist material are not often achieved. As a result, line width loss in pattern delineation frequently occurs as the thinner mask edges erode during etching.

Although CF_4–O_2 plasmas often severely degrade resist materials, resist durability in some plasmas is quite high. For example, conditions favorable for the selective etching of SiO_2 and Si_3N_4 are not as conducive to resist degradation as are CF_4–O_2 plasmas. Thus, CF_4–C_2H_4 (*161*), CF_4–H_2 (*118*), CHF_3 (*48, 161–163*), and C_2F_6–C_2H_4 (*119*) plasmas exhibit excellent selectivity of oxide over resist even when ion bombardment is present. In these cases, selective oxide etching occurs in a saturated-species-rich plasma near the borderline of polymer deposition, and any degradation of polymeric resist material is likely to be compensated by condensation reactions of saturated species at these sites.

Increased durability can be designed into polymer resists. For instance, for the relative etch rates of a variety of polymers in an O_2 plasma, a high correlation exists between the structural properties of the polymer and their stability when exposed to the plasma (*164*). More importantly, the results indicate a synergistic degradation involving atomic oxygen from the plasma and halogen present in either the polymer or the plasma. These studies explain why CF_4–O_2 discharges produce high mask erosion rates. Atomic fluorine (or chlorine) abstracts hydrogen from the polymer and produces sites that react more readily with molecular oxygen. The etch rate of organic materials in oxygen discharges can be significantly enhanced by the addition of fluorine-containing gases (*165, 166*). Conversely, oxygen-free plasmas generally result in polymer stabilization. Halogen abstraction of polymer hydrogens followed by reaction with halogen or halocarbon radicals leads to halocarbon groups in the polymer that can make the mask more resistant to plasma degradation.

During the patterning of thin-film materials, two inconsistent demands are imposed on resists. Initially, resists must be highly sensitive to radiation so that exposure times are short. However, after development, the remaining resist should be stable to radiation (plasma) environments. Ion, electron, or photon bombardment from the plasma atmosphere causes heating, sputtering, and the degradation of resists. Thus, efforts to improve temperature control (wafer cooling) and minimize ion bombardment energy have been mounted. Because these approaches are sometimes not effective or are incompatible with profile considerations, methods to toughen resists (particularly AZ [phenol–formaldehyde-based resin] materials) against plasma exposure have been developed.

UV exposure (at $\lambda < 300$ nm) of the AZ resist prior to plasma etching causes polymer cross-linking (*167, 168*) or decomposition (*169*) of the resist photosensitizer near the surface. Thus, a hardened shell or case is formed that permits a higher bake temperature without resist flow and also reduces the etch rate due to plasma exposure. Exposure to inert plasma (e.g., N_2) causes similar effects (*170*), possibly because of ion and electron, as well as UV, bombardment of the resist surface. When F-containing discharges are used, fluorination of the resist surface occurs that strengthens the resist (because of the formation of C–F bonds) and minimizes reactivity (*171*).

Oxygen plasmas provide a highly selective medium for the removal of organic materials and have been used extensively for stripping photoresists (*172–176*), removing epoxy smears from other electronic components, and etching printed circuit boards (*177*). In addition, the use of oxygen plasmas in delineating the original mask pattern is of interest. Termed "plasma-developable resists", these resists are designed such that, upon exposure, they can be selectively etched in either the exposed or unexposed regions by plasma techniques (*178–180*).

Ion bombardment can be used to enhance resist etch rates and thus achieve anisotropic resist profiles. Reactive sites produced by bombardment permit more rapid attack by oxygen species in the plasma. Multilevel processing (*181, 182*), in which an etch-resistant layer serves as a mask to pattern a polymer or other planarizing film, relies heavily on such anisotropic resist development (*183*). This approach can be implemented by using a thin organic resist on top of the masking layer (trilevel process) or by combining the function of the resist layer with the intermediate (masking) layer (bilevel process). In the bilevel process, silicon-containing resists such as polysilanes and polysiloxanes have been used (*184*). Alternatively, a resist can be exposed but functionalized prior to development. The functionalization involves selective incorporation (because of different reactivities) of a silicon-containing vapor such as hexamethyldisilazane into the exposed resist areas (*185*).

Ultraviolet light, which is present in all glow discharges, also enhances the purely chemical etchant activity and is particularly important for organic film materials. The photochemistry of organic molecules is a well-established field (*186*). Although polymeric films are a more complex photochemical system, many of the chromophoric groups in the polymer react similarly under exposure to UV radiation. The most serious degradation results when scission of backbone or side-chain (near the backbone) bonds of the polymer occurs. The predominant degradation mechanism in polymers exposed to plasma environments is random chain scission (*187*). Active sites then become available for reaction with atomic and molecular species in the plasma. To stabilize these bonds, polymer scientists design side groups far removed from the backbone to reduce the impact of photochemical attack on the polymer chain.

After the resist has served its purpose as an etch mask, it must be removed prior to subsequent processing. Exposure to plasma atmospheres (or ion implantation) often renders the resist material impervious to complete removal by liquid etch baths. As a result, glow discharge methods are invoked to strip resists. With these methods, the most important criteria are etch rate and the minimization of radiation damage to or attack of underlying films or substrates. Etch rates are typically low in O_2 plasmas unless ion bombardment or elevated temperatures ($>$150 °C) are used, and recent efforts have concentrated on downstream reactors using microwave source gas (O_2 or O_2–CF_4) excitation (*188–190*). This configuration takes advantage of the efficiency of microwave generation of atoms and radicals, as well as the controlled rise in substrate temperature without high-energy-particle bombardment. These effects can lead to high resist stripping rates with little or no radiation damage. However, device degradation can still occur if contaminants (e.g., iron or lead) are present in the resist. These species are diffused into SiO_2 surfaces at elevated temperatures ($>$150 °C) even during downstream stripping (*191*).

Profile Control

The goal of any pattern-etching process is to transfer an exact replica of the mask features to the underlying film. However, this transfer establishes only a two-dimensional criterion for the quality of the replication. The third dimension relates to the cross section or edge profile of the etched feature (*2, 192*).

The simplest and perhaps the most useful measure of anisotropy is the ratio of lateral or horizontal undercut distance (d_h) to the vertical etch distance (d_v). This ratio is inversely related to the quality of replication. In anisotropic etching, $d_h/d_v = 0$, and exact dimensional transfer is achieved. A low-quality transfer is obtained with isotropic etching, in which $d_h/d_v = 1$. Anisotropic etching is imperative for high-density-device fabrication.

The desired edge profile depends upon the specific application of the film in the final device. For instance, if a metal line is being defined, a steep-walled profile is desirable to maximize the conductor cross-sectional area. Anisotropic etching can yield such profiles with line dimensions of ~0.1 μm. However, if good step coverage is desired, a tapered or sloped profile is required so that subsequent film deposition will uniformly cover the step, that is, without thinning of the film because of shadowing during deposition. Some control of the taper can be achieved by the proper choice of resist process and plasma conditions. If isotropic etching is needed, for example, to clean the side walls after a sputter etch (physical ablation of material) or to produce a slight undercut, a purely chemical etch process is required.

Isotropic etching is achieved by using a barrel reactor with an etch tunnel or, better yet, with a downstream reactor configuration so that radiation bombardment of etching surfaces is eliminated (*193, 194*). Anisotropic or directional etching (in which the taper is 90° or less) can be achieved in several ways. Sputtering generates a high degree of anisotropy because of the directional nature of the inert gas ions that physically ablate (via momentum transfer) the film material. However, sputter etching (*94, 192*) redeposits etched material, is only weakly selective, and exhibits low etch rates. Ion bombardment in the presence of chemically reactive species (plasma etching) can alleviate these problems by forming volatile etch products, invoking chemical reactions for specificity, and taking advantage of the density of highly reactive neutral species in glow discharges. Although a fundamental understanding of the generation of anisotropic profiles has not been established (*see* Synergistic Phenomena), phenomenological models that present guidelines for variations in anisotropic edge profiles do exist (*94, 99, 195*).

Ion bombardment promotes surface bond breaking (*see* Synergistic Phenomena) and causes sputtering; therefore, etch rates are generally enhanced where bombardment occurs. Qualitatively, these phenomena can lead to anisotropy in two ways (*94, 95, 105, 195*), as depicted in Figure 10 (*94*). In the ion-induced-damage mechanism, energetic ions break chemical bonds on the film surface, thereby making the film more re-

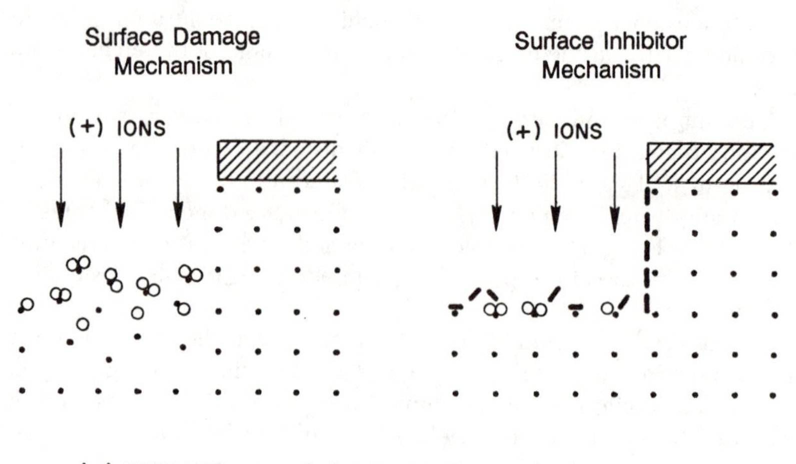

Figure 10. Surface damage and surface inhibitor mechanisms for ion-assisted anisotropic etching. (Reproduced with permission from reference 94. Copyright 1981 Plenum Publishing Corporation.)

active to chemical etchant species. However, the side walls remain relatively unperturbed, because ions primarily impinge perpendicularly to the film surface. Thus, etching proceeds at the nominal chemical etch rate. Material removal is more rapid in the direction of the ion flux and results in anisotropy.

In actual fact, the surfaces exposed to the plasma are usually composed of a chemisorbed coating of etchant radicals, as well as unsaturated species and resist fragments that inhibit the etch rate. Ion bombardment can stimulate the desorption of these species so that surface coverage caused by inhibitors is smallest in regions of high ion flux (Figure 10). Again, increased anisotropy is the net result. The ion-induced-damage mechanism requires considerably more energy than the surface inhibitor mechanism for anisotropy, and both are likely to play a role in plasma-etching processes. The degree of anisotropy will depend on ion flux, ion energy, and chemistry in the plasma.

As noted earlier, ion flux, energy, and plasma chemistry depend strongly on reactor design and gas composition and are therefore virtually impossible to translate directly from one reactor to another. Such possibilities await the development of accurate plasma models (*see* Modeling Plasma Processes). However, the important parameters can be identified to some extent. Ion bombardment is enhanced by decreasing pressure in a high-frequency plasma (~5 MHz) or by decreasing the frequency of the plasma discharge (22); anisotropic profiles thereby result (*196*). Anisotropies induced by surface damage and surface inhibitor (from photoresist as well as other fragments) are possible; however, as pressure increases and ion energies become moderate because of collisions, the inhibitor mechanism becomes more favorable. Under these conditions, anisotropy can be achieved by adding film-forming precursors (CHF_3, C_2H_2, C_2F_6, etc.) to the plasma (*105, 197*). At low frequencies (1 kHz–1 MHz), ions are strongly accelerated because of the longer duration of the accelerating potential, and both anisotropic mechanisms operate at much higher gas pressures. Even at high pressures (133 Pa) and frequencies, however, anisotropy can be achieved if the electric field across the sheath is sufficiently large (*198*).

An interesting demonstration of profile control via alteration of the specific chemistry is that of silicon etching in ClF_3 mixtures (*86*). Because a pure chemical (isotropic) etchant (F atoms) is combined with an ion-bombardment-controlled (anisotropic) etchant (Cl atoms), a continuous spectrum of profiles with varying anisotropies is generated by changing the gas composition.

Etch profiles can also be altered by controlling the susceptibility to erosion of the masking layer as shown in Figure 11 (*199*). If a masking layer that does not erode is used (e.g., MgO and Al_2O_3), a vertical (perfectly anisotropic) etch profile results. However, if the masking layer is attacked by chemical reaction or physical ablation (e.g., organic resist materials), the edges of the layer at the opening are removed. This removal exposes the edges of the underlying film to the plasma atmosphere. Further removal of the resist exposes additional film surface for etching. In this way, a tapered profile can be achieved. Such procedures require close control of resist processes as well as plasma conditions.

Deposition of Specific Film Materials

Like the literature of plasma-assisted etching, the literature on the PECVD of specific materials is considerable. Because film properties are ultimately determined by chemical reaction mechanisms, reactor design, and film structure (Figure 5), the determination of the exact relationships between properties and processing is difficult. At present, the fundamental understanding of such relationships is limited, and thus, empirical efforts have been the norm. In this chapter, the more widely studied film materials deposited by PECVD will be briefly discussed. More extensive information on these and other films can be found in a number of review articles (*9–14, 32, 50, 200–203*) and references therein.

Silicon. The most widely studied and perhaps the best understood PECVD film is that of amorphous silicon (a-Si). Glow discharge a-Si is an "alloy" of silicon and hydrogen, with the hydrogen content ranging from ~5 to 35 atom percent (atom %), depending upon the deposition conditions (temperature, rf power, rf frequency, etc.) and the resulting film structure.

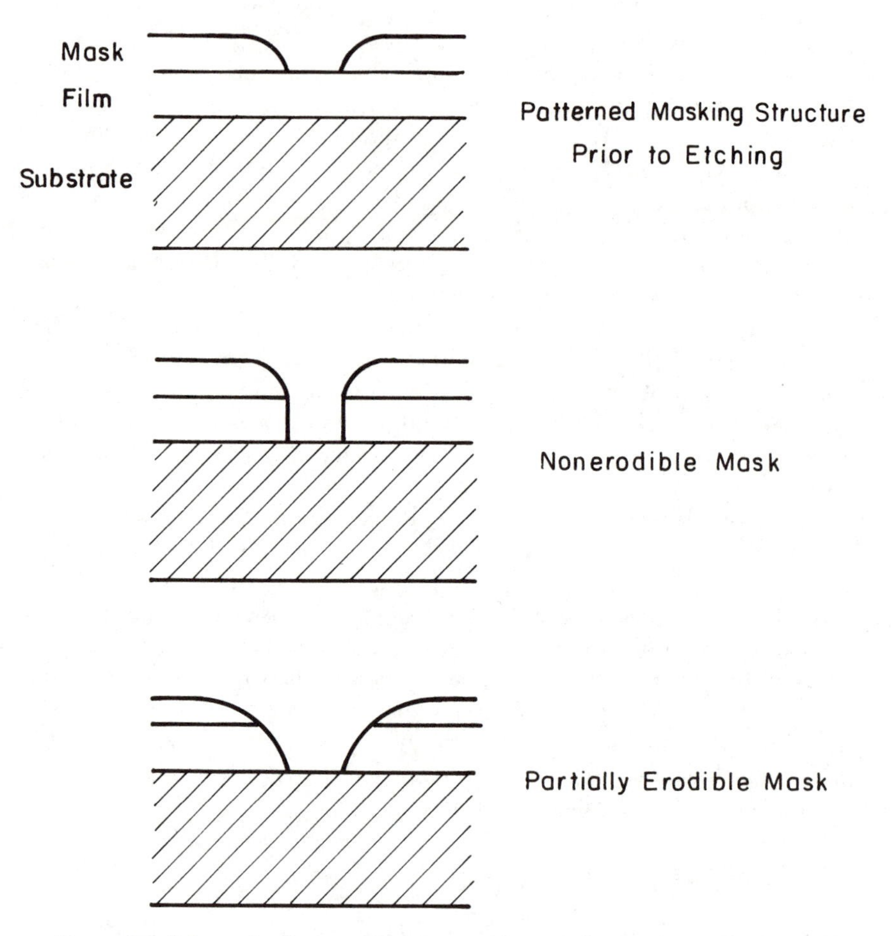

Figure 11. Schematic diagram comparing the control of etch profiles by the use of erodible and nonerodible mask materials. (Reproduced with permission from reference 199. Copyright 1980 John Wiley.)

The great interest in this material stems from its applications in solar-energy conversion, thin-film transistors, memory switches, and electrophotography (*204*).

Although the production of a-Si from SiH_4 (or from mixtures of SiH_4 with inert gases or H_2) is chemically one of the simplest reactions imaginable, the fundamental reactions involved in film formation and the structure–property relationships are not yet clearly defined (*70, 200, 205–207*). In particular, the role of hydrogen in film structure is complex and involves the reduction of silicon dangling bonds that degrade device properties. Infrared (*208*) and nuclear magnetic resonance (NMR) (*209*) spectra of deposited films have established that hydrogen exists in several bonding configurations: monohydride ($\equiv$Si–H), dihydride ($=SiH_2$), trihydride ($-SiH_3$), and polymeric $(-SiH_2-)_n$. Furthermore, NMR studies have suggested that Si–H can be randomly distributed or clustered in the structure (*207, 209*).

For the most part, even qualitative relationships between such structural considerations and film properties are lacking. Recently, however, two important correlations have been presented: Device quality a-Si contains only monohydride bonding structures (*207*), and the intrinsic film stress rises drastically as a morphological transition from columnar to noncolumnar growth occurs (*210*). Both studies point to the importance of defects in a-Si, although the precise role of these defects is still unclear.

The gas-phase and surface reactions in a-Si deposition by PECVD are equally nebulous. Reactive species, primarily H, SiH, SiH_2, SiH_3, and their positive ions (*70, 205, 206*), are produced by electron-impact decomposition of silane. Secondary reactions between these precursors to form species such as disilane (Si_2H_6) and higher molecular weight compounds are important in establishing reactive-species concentrations. Laser-light-scattering studies suggest that particle growth occurs in the plasma near the sheath edges (*211, 212*). Radicals, atoms, and ions diffuse or drift to the growing film surface, where they are adsorbed and undergo various reactions yielding solid and gaseous (e.g., SiH_4, Si_2H_6, and H_2) products. The surface reactions are generally modified by ion and photon bombardment. Adsorption sites can be created by bond breaking due to ion impingement or by atom (e.g., H) abstraction due to reaction with H atoms, SiH_2, or SiH_3 from the gas phase. Finally, bond rearrangement and an increase in film density due to loss of hydrogen occur.

Polycrystalline silicon (poly-Si) has been formed by the plasma-enhanced decomposition of dichlorosilane in argon at temperatures above 625 °C, a frequency of 450 kHz, and a total pressure of 27 Pa. Doped films have been deposited by the addition of phosphine to the deposition atmospheres (*213*). Approximately 1 atom % of chlorine was found in the as-deposited films. Annealing in nitrogen at temperatures above ~750 °C caused chlorine to diffuse from the film surface, grain growth to occur, and the film resistivity to drop. Such heat treatments were necessary to achieve integrated-circuit-quality films.

The drive to reduce process temperatures has led to an interest in growing epitaxial (crystalline) silicon films at temperatures below 800 °C. In addition to increasing deposition rates, glow discharges can assist surface cleaning so that epitaxy can be achieved. Indeed, crystalline silicon films have been produced by PECVD from SiH_4 atmospheres by using external coil excitation (*214*) or by downstream configurations (*215, 216*). In all cases, in situ surface cleaning, including the removal of the native silicon oxide layer, was imperative for epitaxy. If adequate cleaning (usually by ion bombardment) is performed, crystalline layers of silicon can be grown by low-pressure CVD, albeit with a low deposition rate (relative to that of PECVD). Thus, the primary role of a plasma in epitaxial growth may be surface cleaning, in addition to enhancement of reaction kinetics.

Silicon Nitride. Silicon nitride produced by high-temperature (>700 °C) CVD is a dense, stable, adherent dielectric that is useful as a passivation or protective coating, interlevel metal dielectric layer, and antireflection coating in solar cells and photodetectors. However, these applications often demand low deposition temperatures (<400 °C) so that low-melting-point substrates or films (e.g., Al or polymers) can be coated. Therefore, considerable effort has been expended to form high-quality silicon nitride films by PECVD.

A critical aspect of silicon nitride deposition by PECVD is that a significant concentration of hydrogen ($>10^{21}/cm^3$) is present in the deposited films (*217–223*). For this reason, silicon nitride will be referred to as SiN_xH_y. The hydrogen originates from SiH_4 or NH_3, the typical reactants for PECVD. For the most part, hydrogen is bonded to either silicon (~75%) or nitrogen (~25%) (*217, 218, 220, 223*). The exact concentration and chemical distribution of hydrogen greatly affect film properties such as refractive index, etch rate, optical absorption edge, stress, and electrical conductivity. The concentrations and bonding configurations of Si, N, and H depend upon deposition and plasma conditions such as pressure, reactant ratio, rf power, rf frequency, and substrate temperature (*217–224*). In addition, rearrangement of hydrogen bonds in the film or diffusion of hydrogen out of the film can cause instabilities in devices fabricated with SiN_xH_y (*225, 226*).

In general, an increase in rf power density decreases the Si/N ratio in the film (*227–229*). Because the binding energy of the Si–H bond is less than those of the N–H and N–N bonds, an increase in rf power should increase the concentration of reactive nitrogen species relative to the number of reactive silicon species and thereby decrease the Si/H ratio in the film. At high power densities and at high temperatures, the Si/N ratio approaches 0.75, which is the stoichiometric ratio for Si_3N_4.

Much of the data on film etch rate, density, refractive index, and conductivity can be correlated to the concentration and bonding configuration of hydrogen in the films. The total hydrogen concentration decreases with increasing temperature and decreasing frequency (*219, 221, 227, 230, 231*). With increasing substrate temperature, adsorbed surface species have more energy, can preferentially form Si–N bonds, and thereby release hydrogen. Similar effects are operative at low frequency (<4 MHz), where adatom mobility and hydrogen removal by sputtering are favored. The correlation between the effects of deposition temperature and ion bombardment in SiN_xH_y is shown in Figure 12 (*221*). When adequate thermal energy is available (deposition temperature > 300 °C), the total hydrogen content in the film is controlled by thermally activated desorption. Below 300 °C, the hydrogen concentration is determined by ion bombardment.

The hydrogen content of SiN_xH_y produced by PECVD can be reduced but not eliminated in several ways. The use of N_2 rather than NH_3 as the

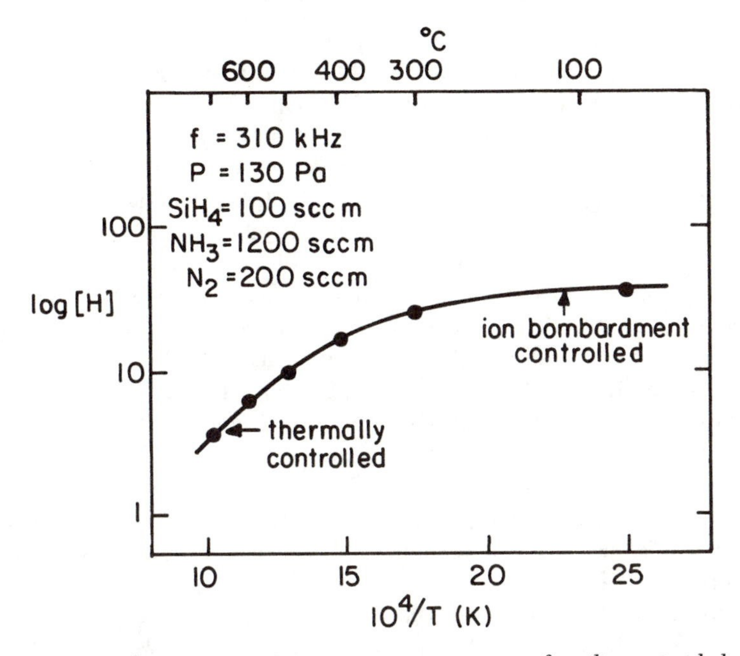

Figure 12. Hydrogen concentration versus temperature for silicon nitride layers deposited by PECVD. (sccm is standard cubic centimeters per minute.) (Reproduced with permission from reference 221. Copyright 1985 The Electrochemical Society, Inc.)

nitrogen source reduces the amount of H in the film, but difficulty is often encountered with silicon-rich films because of the strong N–N bond (*232*). Even with N_2, high temperatures and power levels are needed to effectively minimize H (*228*). Hydrogen can also be scavenged by incorporating a fluorine-containing reactant so that H content is reduced by formation of HF that is removed from the chamber. However, when SiF_4 or SiF_2 is used as a reactant with N_2–H_2 mixtures, large concentrations of fluorine (10–25 atom %) are incorporated into the films (*233*). The Si–F bonds in such films can hydrolyze upon postdeposition exposure to air (*234*) to form silicon oxynitride films.

An alternative but effective way to scavenge hydrogen is to use NF_3–NH_3 mixtures as the nitrogen source (*234*). With this nitrogen source, lower concentrations of F (<8 atom %) can be incorporated in the films by altering the NF_3/NH_3 ratio. Finally, the use of downstream PECVD in which only NH_3–He or NH_3–Ar mixtures are plasma excited has shown promise in minimizing the formation of Si–H bonds (or N–H bonds with N_2–He mixtures) (*50*).

Hydrogen is not the only impurity frequently incorporated into SiN_xH_y films produced by PECVD. Oxygen is generally found in the films and is believed to come from desorbed water vapor from reactor walls or from small vacuum system leaks. Even small quantities of oxygen reduce the refractive index, increase the etch rate, and decrease the film stress (*221, 235*).

SiN_xH_y can be deposited by PECVD with either a compressive or a tensile stress, depending upon the deposition conditions. Although rf power, pressure, and temperature affect stress, rf frequency is a critical parameter at low temperatures (<300 °C) (*221, 236*). At low frequencies (<4 MHz), the stress is generally compressive, whereas at higher frequencies (especially at elevated pressures), tensile stress is often observed. Upon postdeposition, high-temperature (>500 °C) heat treatments, the stress varies inversely with hydrogen concentration; in this study (*221*), the stress was independent of Si–H content but increased linearly with decreasing N–H concentration. If the stress level of SiN_xH_y deposited by PECVD is too high, the films crack either during or after deposition or after subsequent heat treatments (*235, 237*). Film stress can significantly affect adhesion.

Silicon Dioxide. SiO_2 layers produced by PECVD are useful for intermetal dielectric layers and mechanical or chemical protection and as diffusion masks and gate oxides on compound-semiconductor devices. The films are generally formed by the plasma-enhanced reaction of SiH_4 at 200–300 °C with nitrous oxide (N_2O), but CO, CO_2, or O_2 have also been used (*238–241*). Other silicon sources including tetramethoxysilane, methyl dimethoxysilane, and tetramethylsilane have also been investigated (*202*). Diborane or phosphine can be added to the deposition atmosphere to form doped oxide layers.

Like SiN_xH_y deposited by PECVD, SiO_2 films deposited by PECVD contain hydrogen (*203, 240, 242*). Because of the enhanced reactivity of oxygen with SiH_4 fragments compared with nitrogen species, lower concentrations of hydrogen are present in SiO_2 (2–9 atom %) than in SiN_xH_y (15–30 atom %) films. The primary bonding configurations for H are SiH, SiOH, and H_2O (*240, 242*). The distribution of hydrogen between these moieties depends upon deposition conditions. The electrical properties of SiO_2 films are improved if SiOH bonds are minimized (*203*). Because of the use of N_2O, a small amount of N (<5 atom %) is also incorporated into the deposited films.

The properties of silicon dioxide films also depend upon all plasma deposition parameters. Temperature is the critical parameter (*240*), although the compressive stress level varies with rf frequency (*237, 240*). Film topography can be varied during deposition by altering ion bombardment conditions (*242, 243*). In particular, the incorporation of Ar in the deposition atmosphere enhances sputtering and thus promotes conformal step coverage during film formation (*243*).

Conducting Films. To improve adhesion, grain structure, and step coverage of metal films at low temperatures, much interest has been centered recently in PECVD. One of the major limitations of this approach to metal deposition is the availability of suitable source vapors. Generally, reactant species are halides, carbonyls, or alkyls. In the reactive plasma atmosphere, halogens, carbon, or oxygen are therefore often incorporated into the deposited film. Because metallic-film properties are highly sensitive to small concentrations of impurities present in the film, the incorporation of these atoms, as well as oxygen from leaks and chamber outgassing, must be minimized in order to produce low-resistivity films. In addition, the halogen atoms are often etchants for the depositing film material. This situation results in a dynamic equilibrium between etching and deposition so that deposition rates may be low. Nevertheless, a number of film materials have been formed, albeit generally with higher than bulk (or even evaporated, sputtered, or deposited by CVD) resistivities. In certain cases, the elevated resistivities may also be due to metastable-phase formation (*244*). A summary of these film materials, their deposition conditions, and their resistivities is shown in Table V (*201*, *245–253*).

Carbon. Carbon films have recently generated considerable interest because of their potentially useful properties: electrical insulation, thermal conductivity, optical transparency, chemical resistance, and mechanical hardness. Much of the effort has centered around amorphous carbon (a-C), which is a metastable phase of carbon that contains hydrogen and displays properties that are intermediate between those of graphite and diamond (*254–256*). Numerous hydrocarbons have been used as source gases, including methane, acetylene, ethylene, propane, butane, and benzene (*254–259*). Although these films are often referred to as diamond-like, at least at PECVD deposition temperatures below ~400 °C, at least one hydrogen is bonded to each carbon atom (*256*). At ~425 °C, deposited films have properties similar to those of graphite, a fact suggesting that a-C is metastable with respect to graphite, perhaps because of the removal of

Table V. Summary of Conducting Films Deposited in rf Glow Discharges

Film	*Reactants*	*Electrode Temperature (°C)*	*Pressure (Pascal)*	*Frequency (MHz)*	*As-Deposited Sheet Resistivity (Ω/□)*[a]	*Reference*
W	$WF_6 + H_2$	350	27	4.5	2	246
Mo	$MoF_6 + H_2$	350	27	4.5	400	246
	$MoCl_5 + H_2$	430	–	–	(500)	247
WSi_x	$WF_6 + SiH_4$	230	80	13.56	(~500)	248
$MoSi_x$	$MoCl_5 + SiH_4$	400	–	–	(800)	247
$TiSi_x$	$TiCl_4 + SiH_4$	450	267	0.05	15–20	249
	$TiSi_x + SiH_4$	350	133	0.3	–	250
$TaSi_x$	$TaCl_5 + SiH_2Cl_2 + H_2$	650	200	0.6	(70)	251
TiN_x	$TiCl_4 + NH_3 + H_2$	600	27	13.56	30	252
TiB_x	$TiCl_4 + BCl_3 + H_2$	600	40	15	(200)	253

[a]Numbers in parentheses are resistivities in microohm-centimeter.

hydrogen with the subsequent collapse of the C–H structure (*256*). Polycrystalline and crystalline diamond films have been formed (*260–262*), although the exact nature of the reactions needed to form crystallites is not yet clear. Theories include hydrogen atom generation or the specific energy of certain plasma-excited hydrocarbon fragments.

Summary and Conclusions

Glow discharges or plasmas have been used extensively to promote chemical reactions for thin-film etching and deposition in a variety of technologically important areas. The reactive chemical atmosphere and complex discharge–surface interactions in these systems permit the attainment of unique etch profiles and film properties.

At present, reproducibility and control are the primary limitations to the implementation of plasma processes; clearly, the large number of interacting parameters accounts for such problems. In addition, deposition and etching processes are inordinately sensitive to small (part-per-billion) concentrations of impurities. These difficulties can be overcome only by careful investigation of the complex chemistry and physics of glow discharges.

Much progress has been made in recent years. Many of the required gas-phase parameters (reaction rates, cross sections, species concentrations, etc.) can be measured, even though the necessary attempts have not been mounted in all cases.

The principal impediment to effective process design and analysis is the limited understanding of synergistic effects due to ion, photon, and electron bombardment of solid surfaces during etching and deposition. Fundamental relationships must be established between the gas-phase chemistry; the surface chemistry as modified by radiation; and etch profiles, rates, selectivities, and film properties.

This lack of fundamental understanding of the science and engineering of plasma processing is reminiscent of the situation in the catalytic field a few decades ago. With the proper research efforts in surface- and gas-phase chemistry, engineering, and reactor design, most of the current problems can be overcome, and the ultimate capabilities of plasma processing can be realized.

References

1. Kern, W.; Deckert, C. A. In *Thin Film Processes*; Vossen, J. L.; Kern, W., Eds.; Academic: New York, 1978; p 401.
2. Mucha, J. A; Hess, D. W. In *Introduction to Microlithography*; Thompson, L. F.; Willson, C. G.; Bowden, M. J., Eds.; ACS Symposium Series 219; American Chemical Society: Washington, DC, 1983; p 215.
3. Flamm, D. L.; Donnelly, V. M.; Ibbotson, D. E. In *VLSI Electronics: Microstructure Science*; Einspruch, N. G.; Brown, D. M., Eds.; Academic: New York, 1984; Vol. 8, p 189.

4. Smith, D. L. In *VLSI Electronics: Microstructure Science*; Einspruch, N. G.; Brown, D. M., Eds.; Academic: New York, 1984; Vol. 8, p 253.
5. Goworwitz, B.; Saia, R. J. In *VLSI Electronics: Microstructure Science*; Einspruch, N. G.; Brown, D. M., Eds.; Academic: New York 1984; Vol. 8, p 297.
6. *Dry Etching for Microelctronics*; Powell, R. A., Ed.; North-Holland: Amsterdam, 1984.
7. *Handbook of Thin Film Technology*; Maissel L. I.; Glang, R., Eds.; McGraw–Hill, New York, 1970.
8. Vossen, J. L.; Kern, W. *Thin Film Processes*; Academic: New York, 1978.
9. *Deposition Technologies for Films and Coatings*; Bunshah, R. F., Ed.; Noyes: Park Ridge, NJ, 1982.
10. Ojha, S. M. In *Physics of Thin Films*; Haas, G.; Francombe, M. H.; Vossen, J. L., Eds.; Academic: New York, 1982; p 237.
11. Thornton, J. A. *Thin Solid Films* **1983,** *107*, 3.
12. Hess, D. W. *J. Vac. Sci. Technol., A* **1984,** *2*, 244.
13. Reif, R. *J. Vac. Sci. Technol., A* **1984,** *2*, 429.
14. Catherine, Y. In *Plasma Processing*; Mathad, G. S.; Schwartz, G. C.; Smolinsky, G., Eds.; Electrochemical Society: Pennington, 1985; p 317.
15. Bell, A. T. In *Techniques and Applications of Plasma Chemistry*; Hollahan, J. R.; Bell, A. T., Eds.; Wiley: New York, 1974; p 1.
16. Vossen, V. L. *J. Electrochem. Soc.* **1979,** *126*, 319.
17. Chapman, B. N. *Glow Discharge Processes*; Wiley: New York, 1980.
18. Butler, H. S.; Kino, G. S. *Phys. Fluids* **1963,** *6*,1346.
19. a. Anderson, G. S.; Mayer, W. B.; Wehner, G. K. *J. Appl. Phys.* **1962,** *33*, 2991. b. Gottscho, R. A.; Gaebe, C. E. *IEEE Trans. Plasma Sci.* **1986,** *PS–14*, 92.
20. Koenig, H. R.; Maissel, L. I. *IBM Res. Dev.* **1970,** *14*, 168.
21. Coburn, J. W.; Kay, E. *J. Appl. Phys.* **1972,** *43*, 4965.
22. Bruce, R. H. *J. Appl. Phys.* **1981,** *52*, 7064.
23. Winters, H. F. In *Plasma Chemistry III*; Veprek S.; Venugopalan M., Eds.; Springer–Verlag: New York, 1980; p 69.
24. Bell, A. T. *Solid State Technol.* **1978,** *214*, 89.
25. McDaniel, E. W. *Collision Phenomena In Ionized Gases*; Wiley: New York, 1964.
26. Massey, H. S. W.; Burhop, E. H. S.; Gilbody, H. B. *Electronic and Ionic Impact Phenomena*; Oxford: New York, 1971.
27. McDavid, E. W.; Cermak, V.; Dalgarno, A.; Ferguson, E. E.; Friedman, L. *Ion–Molecule Reactions*; Wiley: New York, 1970.
28. Kondratiev, V. N. *Chemical Kinetics of Gas Reactions*; Addison–Wesley: Boston, 1964.
29. Muschlitz, E. E., Jr. *Science (Washington, D.C.)* **1968,** *159*, 599.
30. Carter, G.; Colligan, J. S. *Ion Bombardment of Solids*; Elsevier: New York, 1969.
31. *Ion Bombardment Modification of Surfaces*; Aucellio, O.; Kelly, R., Eds.; Elsevier: Amsterdam, 1984.
32. Hess, D. W. *Annu. Rev. Mat. Sci.* **1986,** *16*, 163.
33. Gerlach-Meyer, U.; Coburn, J. W.; Kay, E. *Surf. Sci.* **1981,** *103*, 177.
34. Hagstrum, H. D. *Phys. Rev.* **1961,** *396*, 36.
35. Greene, J. E.; Barrett, S. A. *J. Vac. Sci. Technol.* **1982,** *21*, 285.
36. Takagi, T. *J. Vac. Sci. Technol., A* **1984,** *2*, 382.
37. Donnelly, V. M.; Flamm, D. L. *Solid State Technol.* **1981,** *244*, 161.
38. Coburn, J. W. In *Proceedings of the Tutorial Symposium on Semiconductor Technology*; Doane, D. A.; Fraser, D. B.; D. W. Hess, Eds.; Electrochemical Society: Pennington, 1982; p 177.

39. Coburn, J. W.; Winters, H. F. *J. Appl. Phys.* **1979,** *50,* 3189.
40. Kay, E.; Coburn, J. W.; Dilks, A. In *Plasma Chemistry III*; Veprek, V.; Vengugopalan, V., Eds; Springer–Verlag: New York, 1980; p 1.
41. Frieser, R. G.; Montillo, F. J.; Zingerman, N. B.; Chu, W. K.; Mader, S. R. *J. Electrochem. Soc.* **1983,** *130,* 2237.
42. a. Pang, S. *Solid State Technol.* **1984,** *274,* 249. b. Fonash, S. J. *Solid State Technol.* **1985,** *284,* 201.
43. Oehrlein, G. S.; Tromp, R. M.; Tsang, J. C.; Lee, Y. H.; Petrillo, E. J. *J. Electrochem. Soc.* **1985,** *132,* 1441.
44. Ryden, K.-H.; Norstrom, H.; Nender, C.; Berg, S. *J. Electrochem. Soc.* **1987,** *134,* 3113.
45. Dautremont-Smith, W. C.; Feldman, L. C. *J. Vac. Sci. Technol., A* **1985,** *3,* 873.
46. Hess, D. W. In *Silicon Processing*; Gupta, D. C., Ed.; Am. Soc. Testing and Materials: Washington, DC, 1983; p 218.
47. Bondur, J. A. *J. Vac. Sci. Technol.* **1978,** *13,* 1023.
48. Lehmann, H. W.; Widmer, R. *J. Vac. Sci. Technol.* **1978,** *15,* 319.
49. Meiners, L. G. *J. Vac. Sci. Technol.* **1982,** *21,* 655.
50. Lucovsky, G.; Tsu, D. V.; Markunas, R. J. In *Plasma Processing*; Coburn, J. W.; Gottscho, R. A.; Hess, D. W., Eds.; Materials Research Society: Pittsburgh, 1986; Vol. 68, p 323.
51. Horiike, Y.; Shibagaki, M. In *Semiconductor Silicon 1977*; Huff, H. R.; Sirtl, E., Eds.; Electrochemical Society: Pennington, 1977; p 1071.
52. Spencer, J. E.; Borel, R. A.; Hoff, A. *J. Electrochem. Soc.* **1986,** *133,* 1922.
53. von Engel, A. *Ionized Gases*; Oxford University Press: London, 1965.
54. Pitchford, L. C.; O'Neil, S. V.; Rumble, J. R. *Phys. Rev.* **1981,** *23,* 294.
55. Winkler, R.; Deutsch, H.; Wilhelm, J.; Wilke, C. *Beitr. Plasmaphys.* **1984,** *243,* 285–316.
56. Kitamori, K.; Tagashira, H.; Sakai, Y. *J. Phys. D* **1980,** *13,* 535.
57. Penetrante, B. M.; Bardsley, J. N. *J. Phys. D* **1984,** *17,* 1971.
58. Hunter, S. R. *Austr. J. Phys.* **1977,** *30,* 83.
59. Blevin, H. A.; Fletcher, J.; Hunter, S. R. *Austr. J. Phys.* **1978,** *31,* 299.
60. Coltrin, M. E.; Kee, R. J.; Miller, J. A. *J. Electrochem. Soc.* **1984,** *131,* 425.
61. Wahl, G. *Thin Solid Films* **1977,** *40,* 13.
62. Gottscho, R. A.; Miller, T. A. *Pure Appl. Chem.* **1984,** *56,* 189.
63. *Combustion Chemistry*; Gardiner, W. C., Ed.; Springer–Verlag, New York, 1984.
64. Hess, D. W.; Jensen, K. F.; Anderson, T. J. *Rev. Chem. Eng.* **1985,** *3,* 97.
65. Jensen, R. J.; Bell, A. T.; Soong, D. S. *Plasma Chem. Plasma Proc.* **1983,** *3,* 163.
66. a. Graves, D. B. In *Plasma Processing*; Mathad, G. S.; Schwartz, G. C.; Gottscho, R. A., Eds.; Electrochemical Society: Pennington, 1987; p 267. b. Boeuf, J. P. *Phys. Rev. A* **1987,** *36,* 2782. c. Barnes, M. S.; Cotler, T. J.; Elta, M. E. *J. Appl. Phys.* **1987,** *61,* 81. d. Richards, A. D.; Thompson, B. E.; Sawin, H. H. *Appl. Phys. Lett.* **1987,** *50,* 492.
67. Bell, A. T. *I&EC Fundamentals* **1970,** *9,* 160.
68. Brown, L. C.; Bell, A. T. *I&EC Fundamentals* **1970,** *13,* 210.
69. Bell, A. T.; Kwong, K. *AIChE J.* **1972,** *18,* 990.
70. Turban, G.; Catherine, Y.; Grolleau, B. *Plasma Chem. Plasma Proc.* **1982,** *2,* 61.
71. Turban, G.; Catherine, Y.; Grolleau, B. *Thin Solid Films* **1979,** *60,* 147.
72. Turban, G.; Catherine, Y. *Thin Solid Films* **1978,** *48,* 57.
73. Chen, I. *Thin Solid Films* **1983,** *101,* 41.

74. Winters, H. F.; Coburn, J. W.; Kay, E. *J. Appl. Phys.* **1977,** *48,* 4973.
75. Kushner, M. J. *J. Appl. Phys.* **1982,** *53,* 2923.
76. Edelson, D.; Flamm, D. L. *J. Appl. Phys.* **1984,** *56,* 1522.
77. Kline, L. *IEEE Trans. Plasma Sci.* **1986,** *PS–14,* 145.
78. Anderson, H. M.; Merson, J. A.; Light, R. W. *IEEE Trans. Plasma Sci.* **1986,** *PS–14,* 156.
79. Kushner, M. J. *IEEE Trans. Plasma Sci.* **1986,** *PS–14,* 188.
80. Rhee, S.; Szekely, J. *J. Electrochem. Soc.* **1986,** *133,* 2194.
81. Greenberg, K. E.; Verdeyen, J. T. *J. Appl. Phys.* **1985,** *57,* 1596.
82. Tachibana, K.; Nishida, M.; Harima, H.; Urano, Y. *J. Phys. D* **1984,** *17,* 1727.
83. Alkire, R. C.; Economou, D. F. *J. Electrochem. Soc.* **1985,** *132,* 648.
84. Dalvie, M.; Jensen, K. F.; Graves, D. B. *Chem. Eng. Sci.* **1986,** *41,* 653.
85. Mogab, C. J. *J. Electrochem. Soc.* **1977,** *124,* 1262.
86. Flamm, D. L.; Wang, D. N.; Maydan, D. *J. Electrochem. Soc.* **1982,** *129,* 2755.
87. Marcoux, P. J.; Foo, P. W. *Solid State Technol.* **1981,** *244,* 115.
88. Dreyfus, R. W.; Jasinski, J. M.; Walkup, R. E.; Selwyn, G. S. *Pure Appl. Chem.* **1985,** *57,* 1265.
89. Roland, J. P.; Marcoux, P. J.; Ray, G. W.; Rankin, G. H. *J. Vac. Sci. Technol., A* **1985,** *3,* 631.
90. Bergeron, S. F.; Duncan, B. F. *Solid State Technol.* **1982,** *258,* 98.
91. Sawin, H. H.; Allen, K. D.; Jenkins, M. W. In *Eleventh Tegal Plasma Seminar Proceedings,* 1985; p 17.
92. Kay, E.; Coburn, J. W.; Dilks, In *Topics in Current Chemistry;* Veprek, S.; Venugopalan, M., Eds.; Springer–Verlag: New York, 1980; p 94, 1.
93. Flamm, D. L. *Plasma Chem. Plasma Proc.* **1981,** *1,* 37.
94. Flamm, D. L.; Donnelly, V. M. *Plasma Chem. Plasma Proc.* **1981,** *1,* 317.
95. Flamm, D. L.; Donnelly, V. M.; Mucha, J. A. *J. Appl. Phys.* **1981,** *52,* 3633.
96. Mogab, C. J.; Adams, A. C.; Flamm, D. L. *J. Appl. Phys.* **1979,** *49,* 3796.
97. Mucha, J. A.; Flamm, D. L.; Donnelly, V. M. *J. Appl. Phys.* **1982,** *53,* 4553.
98. Chuang, T. J. *J. Appl. Phys.* **1980,** *51,* 2614.
99. Flamm, D. L.; Donnelly, V. M.; Ibbotson, D. E. In *VLSI Electronics: Microstructure Science;* Einspruch, N. G.; Brown, D. M.; Eds.; Academic: Orlando, FL, 1984; Vol. 8, p 189.
100. a. Stinespring, C. D.; Freedman, A. *Appl. Phys. Lett.* **1986,** *48,* 718. b. McFeely, F. R.; Morar, J. F.; Himpsel, F. J. *Surf. Sci.* **1986,** *165,* 277.
101. d'Agostino, R.; Flamm, D. L. *J. Appl. Phys.* **1981,** *52,* 162.
102. Picard, A.; Turban, G. *Plasma Chem. Plasma Proc.* **1985,** *5,* 333.
103. a. Schwartz, G. C.; Schaible, P. M. *J. Vac. Sci. Technol.* **1979,** *16,* 410. b. Mogab, C. J.; Levenstein, H. J. *J. Vac. Sci. Technol.* **1980,** *17,* 721. c. Flamm, D. L.; Cowen, P. L.; Golovchenko, J. A. *J. Vac. Sci. Tech.* **1980,** *17,* 1341.
104. Chow, T. P.; Maciel, P. A.; Fanelli, G. M. *J. Electrochem. Soc.* **1987,** *134,* 1281.
105. Engelhardt, M.; Schwarzl, S. *J. Electrochem. Soc.* **1987,** *134,* 1985.
106. Adams, A. C.; Capio, C. D. *J. Electrochem. Soc.* **1981,** *128,* 366.
107. Barker, R. A.; Mayer, T. M.; Pearson, W. C. *J. Vac. Sci. Technol., B* **1983,** *1,* 37.
108. Sanders, F. H. M.; Kolfschoten, A. W.; Dieleman, J.; Haring, R. A.; de Vries, A. E. *J. Vac. Sci. Technol., A* **1984,** *2,* 487.
109. Rossen, R. A.; Sawin, H. H. *J. Vac. Sci. Technol., A* **1985,** *3,* 881.
110. McNevin, S. C.; Becker, G. E. *J. Vac. Sci. Technol., B* **1985,** *3,* 485.
111. Baldi, L.; Beards, D. *J. Appl. Phys.* **1985,** *57,* 2221.
112. Lee, Y. H.; Chen, M.-M. *J. Vac. Sci. Technol., B* **1986,** *4,* 468.

113. Heinicke, R. A. H. *Solid-State Electron.* **1975,** *18,* 1146; **1976,** *19,* 1039.
114. Clarke, P. E.; Field, D.; Hydes, A. J.; Klemperer, D. F.; Seakins, M. J. *J. Vac. Sci. Technol., B* **1985,** *3,* 1614.
115. Coburn, J. W.; Winters, H. F. *J. Vac. Sci. Technol.* **1979,** *16,* 391.
116. Turban, G.; Grolleau B.; Launay, P.; Briaud, P. *Rev. Appl. Phys.* **1985,** *20,* 609.
117. Coburn, J. W.; Winters, H. F. *Solid State Technol.* **1979,** *224,* 117.
118. Ephrath, L. M. *J. Electrochem. Soc.* **1979,** *126,* 1419.
119. Matsuo, S. *J. Vac. Sci. Technol.* **1980,** *17,* 587.
120. Mayer, T. M. J. *Electron. Mat.* **1980,** *9,* 513.
121. Toyoda, H.; Komiya, H.; Itakura, H. J. *Electron. Mat.* **1980,** *9,* 569.
122. Chang, R. H. P.; Chang, C. C.; Durack, S. *J. Vac. Sci. Technol.* **1982,** *20,* 45.
123. Klinger, R. E.; Greene, J. E. *J. Appl. Phys.* **1983,** *54,* 1595.
124. Smolinsky, S.; Chang, R. P. H.; Mayer, T. M. *J. Vac. Sci. Technol.* **1981,** *18,* 12.
125. Gottscho, R. A.; Smolinsky, G.; Burton, R. H. *J. Appl. Phys.* **1982,** *53,* 5908.
126. Stern, M. B.; Liao, P. F. *J. Vac. Sci. Technol., B* **1983,** *1,* 1053.
127. Ibbotson, D. E.; Flamm, D. L.; Donnelly, V. M. *J. Appl. Phys.* **1983,** *54,* 5974.
128. Li, J. Z.; Adesida, J.; Wolf, E. D. *J. Vac. Sci. Technol.* **1985,** *B3,* 406.
129. Burton, R. H.; Gottscho, R. A.; Smolinsky, G. In *Dry Etching for Microelectronics*; Powell, R. A., Ed.; Elsevier: Amsterdam, 1984; p 79.
130. Hipwood, L. G.; Wood, P. N. *J. Vac. Sci. Technol., B* **1985,** *3,* 395.
131. Donnelly, V. M.; Flamm, D. L.; Collins, G. J. *J. Vac. Sci. Technol.* **1982,** *21,* 817.
132. Donnelly, V. M.; Flamm, D. L.; Tu, C. W.; Ibbotson, D. E. *J. Electrochem. Soc.* **1982,** *129,* 2533.
133. McNevin, S. C.; Becker, G. E. *J. Appl. Phys.* **1985,** *58,* 4670.
134. Hess, D. W.; Bruce, R. H. In *Dry Etching for Microelectronics*; Powell, R. A., Ed.; Elsevier: Amsterdam, 1984, p 1.
135. Schwartz, G. C. In *Plasma Processing*; Mathad, G. S.; Schwartz, G. C.; Smolinsky, G., Eds.; Electrochemical Society: Pennington, 1985; p 26.
136. Tokunaga, K.; Redeker, F. C.; Danner, D. A.; Hess, D. W. *J. Electrochem. Soc.* **1981,** *128,* 851.
137. Winkler, U.; Schmidt, F.; Hoffman, N. In *Plasma Processing*; Frieser, R. G.; Mobab, C. J., Eds.; Electrochemical Society: Pennington, 1981; p 253.
138. Herb, G. K.; Porter, R. A.; Cruzan, P. D.; Agraz-Guerena, K.; Soller, B. R. *Electrochem. Soc. Ext. Abstr.*; **1981,** *81–2,* 710.
139. Purdes, A. J. *J. Vac. Sci. Technol., A* **1983,** *1,* 712.
140. Schaible, P. M.; Metzger, W. C.; Anderson, J. P. *J. Vac. Sci. Technol.* **1978,** *15,* 334.
141. a. Keaton, A. L.; Hess, D. W. *J. Vac. Sci. Technol., A* **1985,** *3,* 962. b. Keaton, A. L.; Hess, D. W. *J. Appl. Phys.* **1988,** *63,* 533. c. Keaton, A. L.; Hess, D. W. *J. Vac. Sci. Technol., B* **1988,** *6,* 72. d. Bell, H. B.; Light, R. W.; Anderson, H. M. In *Plasma Processing*; Mathad, G. S.; Schwartz, G. C.; Gottscho, R. A., Eds.; Electrochemical Society: Pennington, 1987; p 35.
142. Poulsen, R. G.; Nentwich, H.; Ingrey, S. In Proc. of the Int. Elec. Dev. Mtg.; Washington, DC, 1976; p 205.
143. Smith, D. L.; Bruce, R. H. *J. Electrochem. Soc.* **1985,** *129,* 2045.
144. Danner, D. A.; Hess, D. W. *J. Electrochem. Soc.* **1986,** *133,* 151.
145. Danner, D. A.; Hess, D. W. *J. Appl. Phys.* **1986,** *59,* 940.
146. Winters, H. F. *J. Vac. Sci. Technol., B* **1985,** *3,* 9.
147. Lee, W. Y.; Eldridge, J. M.; Schwartz, G. C. *J. Appl. Phys.* **1981,** *52,* 2994.

148. Fok, Y. T. *Electrochem. Soc. Ext. Abstr.* **1980,** *80–1,* 301.
149. Hess, D. W. *Plasma Chem. Plasma Proc.* **1982,** *2,* 141.
150. Chow, T. P.; Saxena, A. N.; Ephrath, L. M.; Bennett, R. S. In *Dry Etching for Microelectronics*; Powell, R. A., Ed.; Elsevier: Amsterdam, 1984; p 9.
151. Chow, T. P.; Steckl, A. J. *J. Electrochem. Soc.* **1984,** *131,* 2325.
152. a. Tang, C. C.; Hess, D. W. *J. Electrochem. Soc.* **1984,** *132,* 115. b. Fischl, D. S.; Hess, D. W. *J. Electrochem. Soc.* **1987,** *134,* 2265. c. Hess, D. W. *Solid State Technol.* **1988,** *314,* 97.
153. Robb, F. Y. In *Plasma Processing*; Mathad, G. S., Schwartz, G. C.; Smolinsky, G., Eds.; Electrochemical Society: Pennington, 1985; p 1.
154. Mattausch, H. J.; Hasler, B.; Beinvogl, W. *J. Vac. Sci. Technol., B* **1983,** *1,* 15.
155. Cadien, J. C.; Sivaram, S.; Reintsema, C. D. *J. Vac. Sci. Technol., A* **1986,** *4,* 739.
156. Legat, W. H.; Schilling, H. *Electrochem. Soc. Ext. Abstr.* **1975,** *75–2,* 336.
157. Poulsen, R. G. *J. Vac. Sci. Technol.* **1977,** *14,* 266.
158. Mogab, C. J.; Shankoff, T. A. *J. Electrochem. Soc.* **1977,** *124,* 1766.
159. Nakata, H.; Nishioka, K.; Abe, H. *J. Vac. Sci. Technol.* **1980,** *17,* 1351.
160. Harada, T.; Gamo, K.; Namba, S. *Jpn. J. Appl. Phys.* **1981,** *20,* 259.
161. Matsuo, S.; Takehara, Y. *Jpn. J. Appl. Phys.* **1977,** *16,* 175.
162. Lehmann, H. W.; Widmer, R. *Appl. Phys. Lett.* **1978,** *32,* 163.
163. Toyoda, H.; Komiya, H.; Itakura, H. J. *Electron. Mat.* **1980,** *9,* 569.
164. Taylor, G. N.; Wolf, T. M. *Polym. Eng. Sci.* **1980,** *20,* 1086.
165. Goldstein, I. S.; Kalk, F. *J. Vac. Sci. Technol.* **1981,** *19,* 743.
166. Turban, G.; Rapeaux, M. *J. Electrochem. Soc.* **1983,** *130,* 2231.
167. Lin, B. J. *J. Electrochem. Soc.* **1980,** *127,* 202.
168. Allen, R.; Foster, M.; Yen, Y.-T. *J. Electrochem. Soc.* **1982,** *129,* 1379.
169. Hiraoka, H.; Pacansky, J. *J. Electrochem. Soc.* **1981,** *128,* 2645.
170. Moran, J. M.; Taylor, G. N. *J. Vac. Sci. Technol.* **1981,** *19,* 1127.
171. Dobkin, D. M.; Cantos, B. D. *IEEE Electron Device Lett.* **1981,** *EDL–2,* 222.
172. a. Cook, J. M.; Benson, B. W. *J. Electrochem. Soc.* **1983,** *130,* 2459. b. Cook, J. M. *Solid State Technol.* **1987,** *304,* 147.
173. Degenkolb, E. O.; Mogab, C. J.; Goldrick, M. R.; Griffiths, J. E. *Appl. Spectrosc.* **1976,** *30,* 520.
174. Stafford, B. B.; Gorin, G. J. *Solid State Technol.* **1977,** *209,* 51.
175. Reichelderfer, R. F.; Welty, J. M.; Battey, J. F. *J. Electrochem. Soc.* **1977,** *124,* 1926.
176. Szekeres, A.; Kirov, K.; Alexandrova, S. *Phys. Status Solidi A* **1981,** *63,* 371.
177. a. Kegel, B. *Circuits Mfg.* **1981,** *21,* 27. b. Rust, R. D.; Rhodes, R. J.; Parker, A. A. *Solid State Technol.* **1984,** *274,* 270. c. Lu, N. H.; Nielsen, C.; Welsh, J. A.; Babu, S. V.; Rembetski, J. F. In *Plasma Processing*; Mathad, G. S.; Schwartz G. C.; Smolinksy, G., Eds.; Electrochemical Society: Pennington, 1985; p 175.
178. a. Taylor, G. N.; Wolf, T. M. *J. Electrochem. Soc.* **1980,** *127,* 2665. b. Taylor, G. N.; Wolf, T. M.; Moran, J. M. *J. Vac. Sci. Technol.* **1981,** *19,* 872.
179. Geis, M. W.; Randall, J. N., Deutsch, T. F., DeGraff, P. D., Krohn, K. E.; Stern, L. A. *Appl. Phys. Lett.* **1983,** *43,* 74.
180. Hori, M.; Miwa, T.; Hattori, S.; Morita, S. *Plasma Chem. Plasma Proc.* **1984,** *4,* 119.
181. Lin, B. J. In *Introduction to Microlithography*; Thompson, L. F., Willson, C. G.; Bowden, M. J., Eds.; ACS Symposium Series 219; American Chemical Society: Washington, DC, 1983; p 287.
182. Kruger, J. B.; O'Toole, M. M.; Rissman, P. In *VLSI Electronic Microstructure*

Science: Plasma Processing for VLSI; Einspruch, N. G.; Brown, D. M., Eds.; Academic: Orlando, FL, 1984; Vol. 8.
183. Chow, N. J.; Tang, C. H.; Paraszczak, J.; Babich, E. *Appl. Phys. Lett.* **1985,** *46*, 31.
184. a. Hatzakis, M.; Paraszczak, J.; Shaw, J. In *Proc. Microcircuit Eng. 81*; Lausanne, 1981. b. Saotome, Y.; Gokan, H.; Saigo, K.; Suzuki, M.; Ohnisahi, Y. *J. Electrochem. Soc.* **1985,** *132*, 919. c. Reichmanis, E.; Smolinsky, G.; Wilkins, C. W., Jr. *Solid State Technol.* **1985,** *288*, 130.
185. a. Coopmans, F.; Roland, B. *SPIE Proc. Advances In Resist Technology III* **1986,** *633*, 126. b. Roland, B.; Lombaerts, R.; Jakus, C.; Coopmans, F. *SPIE Proc. Advances In Resist Technology IV,* **1988,** *771*, 69.
186. Calvert, J. G.; Pitts, J. N., Jr. *Photochemistry*; Wiley: New York, 1967.
187. Wu, B. J.; Hess, D. W.; Soong, D. S.; Bell, A. T. *J. Appl. Phys.* **1983,** *54*, 1725.
188. Dzioba, S.; Este, G.; Naguib, H. M. *J. Electrochem. Soc.* **1982,** *129*, 2537.
189. Charlet, B.; Peccoud, L. In *Plasma Processing*; Mathad; G. S.; Schwartz, C. G.; Smolinsky, G., Eds.; Electrochemical Society: Pennington, 1985; p 227.
190. Robinson, B.; Shivashankar, In *Plasma Processing*; Mathad, G. S, Schwartz, C. G.; Smolinsky, G., Eds.; Electrochemical Society: Pennington, 1985; p 206.
191. Fujima, S.; Yano, H. *Electrochem. Soc. Ext. Abst.* **1986,** *86–2*, 456; *J. Electrochem. Soc.* **1988,** *135*, 1195.
192. Melliar-Smith, C. M.; Mogab, C. J. In *Thin Film Processes*; Vossen, J. L.; Kern, W., Eds.; Academic: New York, 1978; p 497.
193. Robb, F. *Semicond. Int.* **1979,** *2(11)*, 60.
194. Dieleman, J.; Sanders, F. H. M. *Solid State Technol.* **1984,** *274*, 191.
195. a. Tu, Y. Y.; Chuang, T. J.; Winters, H. F. *Phys. Rev. B* **1981,** *23*, 23. b. Gerlach-Meyer, U.; Coburn, J. W.; Kay, E. *Surf. Sci.* **1981,** *103*, 177. c. Coburn, J. W. *Solid State Technol.* **1984,** *294*, 117.
196. Donnelly, V. M.; Flamm, D. L.; Bruce, R. H. *J. Appl. Phys.* **1985,** *58*, 2135.
197. Adams, A. C.; Capio, C. D. *J. Electrochem. Soc.* **1981,** *128*, 366.
198. Zarowin, C. B. *J. Vac. Sci. Technol., A* **1984,** *2*, 1537.
199. Chapman, B. N. *Glow Discharge Processes*; Wiley: New York, 1980; pp 299–305.
200. Hirose, M. In *Semiconductors and Semimetals*; Pankove, J. I., Ed.; Academic: New York 1984; Vol. 21A, p 9.
201. Hess, D. W. In *Reduced Temperature Processing for VLSI*; Reif, R.; Srinivasan, G. R., Eds.; Electrochemical Society: Pennington, 1986; p 3.
202. Adams, A. C. In *Reduced Temperature Processing for VLSI*; Reif, R.; Srinivasan, G. R., Eds.; Electrochemical Society: Pennington, 1986; p 111.
203. Nguyen, S. V. *J. Vac. Sci. Technol., B* **1986,** *4*, 1159
204. Stuke, J. *Annu. Rev. Mat. Sci.* **1985,** *15*, 79.
205. Kampas, F. J. In *Semiconductors and Semimetals*; Pankove, J. I., Ed.; Academic: New York 1984; Vol. 21A, p 153.
206. Longeway, P. A. In *Semiconductors and Semimetals*; Pankove, J. I., Ed.; Academic: New York 1984; Vol. 21A, p 179.
207. Reimer, J. A. In *Plasma Processing*; Coburn, J. W., Gottscho, R. A.; Hess, D. W., Eds.; Materials Research Society: Pittsburgh, 1986; Vol. 68, p. 157.
208. Zanzucchi, P. J. In *Semiconductors and Semimetals*; Pankove, J. I., Ed.; Academic: New York, 1984; Vol. 21B, p 113.
209. Reimer, J. A., Vaughan, R. W.; Knights, J. C. *Phys. Rev. Lett.* **1980,** *44*, 193; *Phys. Rev. B* **1981,** *124*, 3360.
210. Harbison, J. P. *J. Non-Cryst. Solids* **1984,** *66*, 87.
211. Roth, R. M.; Spears, K. G.; Stein, G. D.; Wong, G. *Appl. Phys. Lett.* **1985,** *46*, 253.

212. Spears, K. G.; Robinson, T. J. In *Plasma Processing*; Coburn, J. W.; Gottscho, R. A.; Hess, D. W., Eds.; Materials Research Society: Pittsburgh **1986,** *68,* 121.
213. Kamins, T. I.; Chiang, K. L. *J. Electrochem. Soc.* **1982,** *129,* 2326.
214. Townsend, W. G.; Uddin, M. E. *Solid-State Electron.* **1963,** *16,* 39.
215. Suzuki, S.; Itoh, T. *J. Appl. Phys.* **1983,** *54,* 1466.
216. Donahue, T. J.; Burger, W. R.; Reif, R. *Appl. Phys. Lett.* **1984,** *44,* 346.
217. Lanford, W. A.; Rand, M. J. *J. Appl. Phys.* **1978,** *49,* 2473.
218. Stein, H. J.; Wells, V. A.; Hampy, R. E. *J. Electrochem. Soc.* **1979,** *126,* 1750.
219. Chow, R.; Landford, W. A.; Ke-Ming, W.; Rosler, R. S. *J. Appl. Phys.* **1982,** *53,* 5630.
220. Paduschek, P.; Hopfl, C.; Mitlehner, H. *Thin Solid Films* **1983,** *110,* 291.
221. Claassen, W. A. P.; Valkenburg, W. G. J. N.; Willemsen, M. F. C.; Wijert, W. M. v. d. *J. Electrochem. Soc.* **1985,** *132,* 893.
222. Maeda, M.; Nakamura, H. *J. Appl. Phys.* **1985,** *58,* 484.
223. Knolle, W. R.; Osenbach, J. W. *J. Appl. Phys.* **1985,** *58,* 1248.
224. Samuelson, G.; Mar, K. M. *J. Electrochem. Soc.* **1982,** *129,* 1773.
225. Sun, R. C.; Clemens, J. T.; Nelson, J. T. In *Proc. IEEE Rel. Phys. Symps.*; 1980; p 244.
226. Fair, R. B.; Sun, R. C. *IEEE Trans. Electron Devices* **1981,** *ED–28,* 83.
227. Dun, W.; Pan, P.; White, F. R.; Douse, R. W. *J. Electrochem. Soc.* **1981,** *128,* 1555.
228. Maeda, M.; Arita, Y. *J. Appl. Phys.* **1982,** *53,* 6852.
229. Katoh, K.; Yasui, M.; Watanabe, H. *Jpn. J. Appl. Phys.* **1983,** *22,* L321.
230. Cheng, H. S.; Zhou, Z. Y.; Yang, F. C.; Xu, Z. W.; Ren, Y. H. *Nucl. Instrum. Methods Phys. Res.* **1983,** *218,* 601.
231. Watanabe, H.; Katoh K.; Yasui, M. *Jpn. J. Appl. Phys.* **1984,** *21,* 1.
232. Reinberg, A. R. *Annu. Rev. Mat. Sci.* **1979,** *9,* 341.
233. Fujita, S.; Toyoshima, H.; Ohsihi, T.; Sasaki, A. *Jpn. J. Appl. Phys.* **1984,** *23,* L144; *23,* L268.
234. a. Livengood, R. E.; Hess, D. W. *Appl. Phys. Lett.* **1987,** *50,* 560. b. Flamm, D. L.; Chang, C.-P.; Ibbotson, D. E.; Mucha, J. A. *Solid State Technol.* **1987,** *30,* 43.
235. Livengood, R. E.; Petrich, M. A.; Hess, D. W.; Reimer, J. A. *J. Appl. Phys.* **1988,** *63,* 2651.
236. Sinha, A. K.; Levinstein, H. J.; Smith, T. E.; Quintana, G.; Haszako, S. E. *J. Electrochem Soc.* **1978,** *125,* 601.
237. Koyama, K.; Takasaki, K.; Maeda, M.; Takgi, M. In *Plasma Processing*; Dieleman, J.; Freiser, R. G.; Mathad, G. S., Eds.; Electrochemical Society: Pennington, 1982; p 478.
238. van de Ven, E. P. G. T. *Solid State Technol.* **1981,** *244,* 167.
239. Hollahan, J. R. *J. Electrochem. Soc.* **1974,** *126,* 930.
240. Reinberg, A. R. *J. Electron. Mat.* **1979,** *8,* 345.
241. Adams, A. C. *Solid State Technol.* **1983,** *264,* 135.
242. Gorczyca, T. B.; Gorowitz, B. In *VLSI Electronics: Microstructure Science*; Einspruch, N. G.; Brown, D. M., Eds.; Academic: New York, 1984; Vol. 8, p 69.
243. Adams, A. C.; Alexander, F. B.; Capio, C. D.; Smith, T. E. *J. Electrochem. Soc.* **1981,** *128,* 1545.
244. Smith, G. C; Purdes, A. J. *J. Electrochem. Soc.* **1985,** *132,* 2721.
245. a. Tang, C. C.; Hess, D. W. *Appl. Phys. Lett.* **1984,** *45,* 633. b. Greene, W. M.; Oldham, W. G.; Hess, D. W. *Appl. Phys. Lett.* **1988,** *52,* 1133.
246. Hess, D. W. In *VLSI Electronics: Microstructure Science*; Einspruch, N. G.; Brown, D. M., Eds.; Academic: New York, 1984; Vol. 8, p 55.

247. Tang, C. C.; Chu, J. K.; Hess, D. W. *Solid State Technol.* **1983,** *263*, 125.
248. Tabuchi, A.; Inoue, S.; Maeda, M.; Takagi, M. *Jpn. Semicond. Technol. News*, **1983,** *February*, 43.
249. Akimoto, K.; Watanabe, K. *Appl. Phys. Lett.* **1981,** *39*, 445.
250. Rosler, R. S.; Engle, G. M. *J. Vac. Sci. Technol., B* **1984,** *2*, 733.
251. Kemper, M. J. H.; Koo, S. W.; Huizinga, F. *Electrochem. Soc. Ext. Abstracts*; New Orleans, October 7–12, 1984; Abst. No. 377.
252. Hieber, K.; Stolz, M.; Wieczorek, In *Proceedings, Ninth International Conference on CVD*; Robinson, M.; Cullen, G. W.; van den Brekel, C. H. J.; Blocker, J. M., Jr.; Rai-Choudhury, P., Eds.; Electrochemical Society: Pennington, 1979; p 205.
253. Gleason, E. F.; Hess, D. W. In *Plasma Processing*; Coburn, J. W.; Gottscho, R. A.; Hess, D. W., Eds.; Materials Research Society: Pittsburgh, 1986; Vol. 68, p 343.
254. Williams, L. M. *Appl. Phys. Lett.* **1985,** *46*, 43.
255. Natarajan, V.; Lamb, J. D.; Woollam, J. A.; Liu, D. C.; Gulino, D. A. *J. Vac. Sci. Technol., A* **1985,** *3*, 681.
256. Warner, J. D.; Pouch, J. J.; Alterovitz, S. A.; Liu, D. C.; Landford, W. A. *J. Vac. Sci. Technol., A* **1985,** *3*, 900.
257. Meyerson, B. S. In *Plasma Processing*; Coburn, J. W.; Gottscho, R. A.; Hess, D. W., Eds.; Materials Research Society: Pittsburgh, 1986; Vol. 68, p 191.
258. Park, S. C.; Bodart, J. R.; Han, He.-X.; Feldman, B. J. In *Plasma Processing*; Coburn, J. W., Gottscho, R. A.; Hess, D. W., Eds.; Materials Research Society: Pittsburgh, 1986; Vol. 68, p 199.
259. Wagner, J.; Wild, Ch.; Bubenzer, A.; Koidl, P. In *Plasma Processing*; Coburn, J. W., Gottscho, R. A.; Hess, D. W., Eds.; Materials Research Society: Pittsburgh, 1986; Vol. 68, p 205.
260. Pouch, J. J.; Alterovitz, S. A.; Warner, J. D. In *Plasma Processing*; Coburn, J. W.; Gottscho, R. A.; Hess, D. W., Eds.; Materials Research Society: Pittsburgh, 1986; Vol. 68, p 211.
261. Fedoseev, D. V.; Varnin, V. P.; Deryagin, B. V. *Russ. Chem. Rev.* **1984,** *53*, 435.
262. Fujimori, N.; Imai, T.; Doi, A. *Vacuum* **1986,** *36*, 99.
263. Robinson, A. L. *Science (Washington, D.C.)* **1986,** *234*, 1074.

RECEIVED for review December 30, 1987. ACCEPTED revised manuscript October 27, 1988.

9

Interconnection and Packaging of High-Performance Integrated Circuits

Ronald J. Jensen

Sensors and Signal Processing Laboratory, Honeywell, Bloomington, MN 55420

The increasing density and speed of advanced integrated circuits (ICs) have created a need for new materials, processes, and designs for packaging and interconnecting the chips. Conventional single-chip packaging frequently limits the overall density and performance of electronic systems. These limitations are overcome by a variety of customized multichip-packaging approaches that provide short and dense chip-to-chip interconnections. Single-chip-packaging approaches and their limitations in the packaging of high-performance ICs are reviewed in this chapter. The advantages and the geometrical, electrical, and thermal design considerations for multichip packaging are described. Current multichip-packaging technologies are reviewed, with emphasis on those technologies that use thin-film processes to achieve high-density interconnections in multiple layers of a thin-film conductor and a polymer dielectric. Material and process options for these thin-film multilayer interconnections are discussed, followed by several recent demonstrations of the technology. Finally, some emerging technologies for even higher performance packaging and interconnection are examined.

THE INVENTION OF THE INTEGRATED CIRCUIT (IC) in 1958–1959 was motivated by the interconnection problem. The high cost, large size, and poor reliability of discrete solid-state components severely limited the performance of digital electronic systems (*1*). Since the introduction of the IC, system performance has been enhanced primarily through increased levels of integration and faster devices on the chip, whereas IC-packaging and interconnection technologies have been relatively unchanged.

0065–2393/89/0221–0441$15.60/0

With the conventional technology, ICs are mounted individually in plastic or ceramic single-chip packages (SCPs), such as dual-in-line packages (DIPs) or chip carriers, and the SCPs are interconnected on printed wiring boards (PWBs). The number of pins on SCPs has increased significantly, and line widths on PWBs, like IC feature sizes, have followed a historical downward trend (*2*). However, the basic SCP-on-PWB approach has remained predominant.

Interconnection and Packaging Requirements

The high levels of integration and fast switching speeds of advanced ICs, such as very-large-scale ICs (VLSICs) and GaAs digital ICs, impose unique demands on the package and interconnections between chips. Some typical characteristics anticipated for ICs in the next 5–10 years (*3*) are the following:

- Maximum die size, 2.5 by 2.5 cm
- Input–output pins, 1000
- Signal rise time, 100 ps
- Power supply voltage, 1.5 V
- Noise tolerance, 0.5 V
- Simultaneously switching I/Os, 256
- Power dissipation, 10–100 W

The large number of inputs and outputs (I/Os) on VLSICs (currently 200–300 and up to 1000 anticipated in the 1990s) requires high-density bonding from the chip to the package and dense interconnections between chips. Chips must be closely spaced to minimize interconnection lengths and, thus, the propagation delay for signals between chips. The propagation delay can become a significant fraction of the clock cycle time. Short signal rise times (<100 ps for GaAs digital ICs) require interconnections that are designed and fabricated as transmission lines, with controlled characteristic impedance to prevent reflections, low resistance to minimize signal attenuation and dispersion, and sufficient spacing between adjacent lines to minimize crosstalk.

Because of low power supply voltages and low noise margins, power distribution lines or planes must have low resistance to minimize voltage drops and low inductance to minimize the noise induced by a large number of simultaneously switching I/Os. High power dissipation and chip packing densities require materials and designs that can efficiently transfer heat away from the chip. Finally, the high cost of advanced ICs and the severe environments under which they must perform place a premium on the stability of the packaging materials and the reliability of the assembled system.

Conventional single-chip-packaging technologies are often unable to meet these combined requirements, and the performance of advanced systems is again becoming limited by the packaging and interconnection technologies, as it was before the advent of the IC. Multichip packages (MCPs) containing a number of bare ICs interconnected on a common substrate overcome many of the limitations of single-chip packaging, but a number of new challenges are created by this new packaging approach.

New materials and processes are needed to provide the high-density, multilayer interconnections between chips in MCPs. High-thermal-conductivity substrates, low-dielectric-constant dielectrics, and high-conductivity conductors are required for optimum performance. New models and computer-aided design (CAD) tools are being developed to route the dense interconnections and to simulate the electrical, thermal, and mechanical performance of the package before fabrication. Advanced testing and assembly methods are needed to populate the MCPs with functional ICs and to interface the packages with the rest of the system.

The development of new packaging technologies, which is a highly interdisciplinary effort, involves many trade-offs between design requirements and material and process limitations and is gaining increasing importance and recognition within the microelectronic industry.

Conventional IC Packaging

Function and Scope of IC Packaging. The package of an electronic system performs several functions. The package must mechanically support the ICs, protect the ICs from adverse environmental effects, distribute electrical power, remove heat from the chips, and provide signal interconnections between the ICs and the rest of the system.

In large systems such as mainframe computers, packaging is typically divided into three levels (Figure 1) (*4*). At the first level of packaging, the IC is mounted on a single- or multichip carrier, bonds are made to provide electrical contact between the chip and carrier, and the package is sealed. At the second level, a number of IC packages and other discrete components are interconnected by mounting them onto a PWB or card, which is made usually from laminated layers of fiberglass-reinforced epoxy with printed conductor patterns. The package is connected to the PWB by inserting pins into holes in the board or by soldering the package to the surface of the PWB. At the third level of packaging, the cards are connected to a back plane or card rack, which interfaces with the rest of the system through multiconductor cables.

The focus of this chapter is the first level of IC packaging, that is, the SCP or the MCP. However, many important advancements are still being made at the second and third levels of packaging. These advances include improved materials and fabrication technology for PWBs (*5–7*) and new

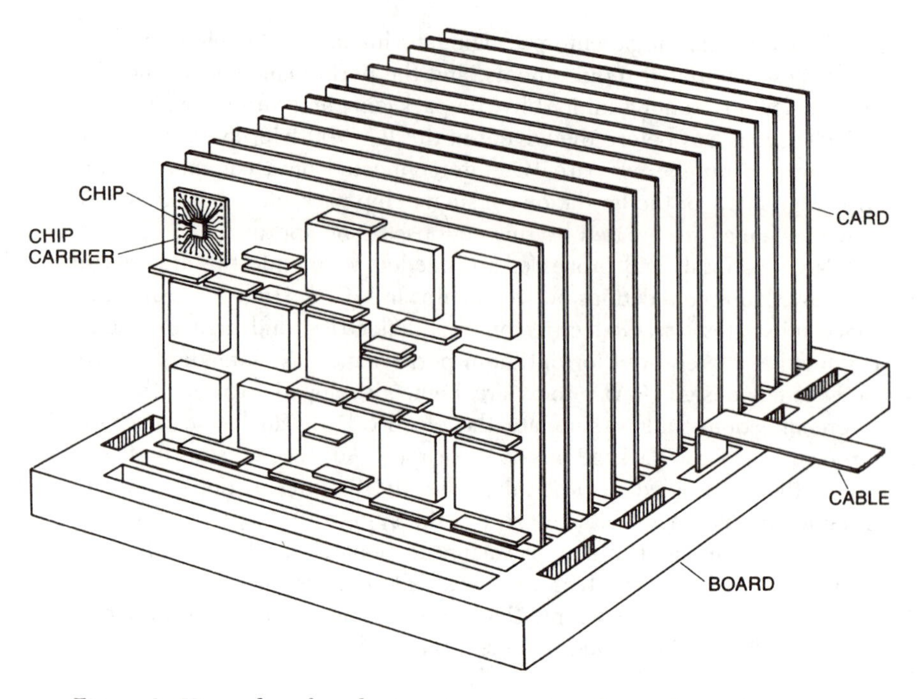

Figure 1. Hierarchy of packaging structures in a large electronic system. (Reproduced with permission from reference 4. Copyright 1983 Scientific American, Inc.)

assembly processes for mounting IC packages to PWBs. In particular, surface mount technology (SMT), probably the most significant current development in conventional packaging, has been extensively investigated (*8–11*).

Also outside the scope of this chapter but of great importance in first-level packaging technology are the materials and processes used to attach (*12*) and electrically bond the chips to packages by techniques such as wire bonding (*13*, *14*), tape-automated bonding (TAB) (*15–17*), and controlled-collapse flip-chip bonding (*18–20*). Comprehensive reviews of packaging and assembly technology (*4*, *21–25*) are available, and updates of recent trends in packaging appear periodically (*26–28*). Reference 25 is a comprehensive, up-to-date handbook on microelectronic packaging.

Single-Chip Packaging. ***Dual-In-Line Package.*** The predominant form of IC packaging throughout the 1960s and the 1970s was the dual-in-line package (DIP), a rectangular ceramic or plastic housing with a row of metal leads along each of the two long sides (Figure 2). Molded plastic DIPs are used for low-cost applications, whereas glass-sealed ceramic DIPs are used for high-reliability applications that require hermetic sealing of the chip. The DIP is attached to a PWB by inserting and soldering the leads

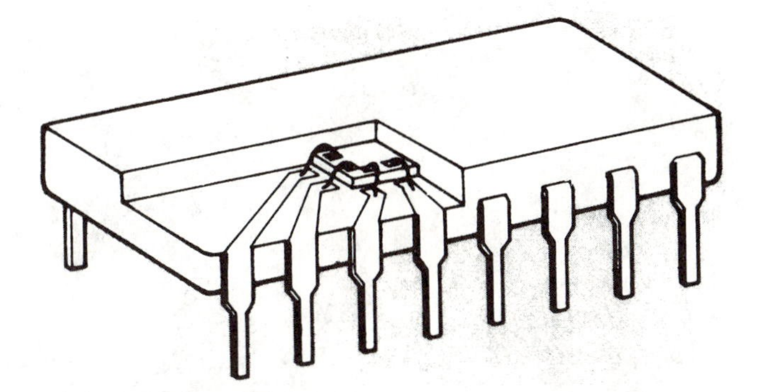

Figure 2. Cutaway view of a dual-in-line package. (Reproduced with permission from reference 27. Copyright 1986 Technical Publishing.)

into holes on the board. The holes are on a standard pitch (center-to-center spacing) of 0.100 in. (0.254 cm). The standard sizes and footprints of DIPs have contributed greatly to the standardization of PWBs and of assembly and testing equipment in the electronic industry (27).

The DIP still accounted for 80% of all IC packaging in 1985 (28). Although the use of DIPs is decreasing, the DIP will continue to be an inexpensive format for packaging small- and medium scale ICs with 10–20 leads. For chips with higher lead counts, the DIP consumes too much board area, and the large distance from the chip to the outer pins causes a degradation in electrical performance. The largest DIP, with 64 leads, requires nearly 3 in.^{2} (19.4 cm^2) of board area. A chip carrier with leads along four sides can accommodate 128 leads in the same area, and a pin grid array package can have 256 leads.

Surface-Mounted Packages. The most significant trend in conventional packaging technology during the 1980s has been the replacement of insertion-mounted packages with components that are soldered to the surface of the PWB. Surface mounting eliminates through-holes in the PWB and thus reduces its cost and increases the wiring density. Surface mounting also permits a finer lead pitch and allows packages to be mounted closely together on both sides of the PWB. Figure 3 shows a section of a PWB populated with surface-mounted leadless ceramic chip carriers.

The most prevalent surface-mounted packages are the small-outline ICs (SOIC) and leaded and leadless chip carriers. The SOIC (Figure 4) is similar to a plastic DIP but is flatter, shorter, and contains 8–28 gull-wing leads on 0.050-in. (0.13-cm) centers. Because of the finer lead pitch, the SOIC has a size advantage over the DIP of about 3:1 and will continue to replace much of the DIP market for small ICs throughout the 1990s (28).

Chip carriers are square ceramic or plastic packages with leads on four

Figure 3. Printed wiring board with surface-mounted leadless ceramic chip carriers.

sides. The leads are typically on 0.050-in. (0.13-cm) centers, but they can also be on 0.040- and 0.025-in. (0.102- and 0.064-cm) centers and even on 0.011-in. (0.028-cm) centers for some customized packages (*29*). Plastic-leaded chip carriers (PLCCs) with J-shaped leads (Figure 5), initially the most common chip carrier (*28*), are available with up to 124 leads. However, problems with solder joint reliability on J-shaped leads have led to the greater use of gull-wing leads, like those on the SOIC.

Leaded and leadless ceramic chip carriers can be hermetically sealed. Because of this property, these packages are important in medical, space, and military applications. Leadless ceramic chip carriers (LCCCs) are attached to the PWB with solder fillets rather than metal leads. The solder bond supports the package and provides a standoff between the package and board. Leadless packages have the highest possible packing density on a PWB (Figure 3). However, the biggest technical problem with the surface mounting of LCCCs (and large leaded chip carriers) is fatigue of the solder bonds because of the mismatch between the thermal expansion of ceramic packages and that of the glass-filled polymer PWB. Extensive work has been done to analyze these failures and to develop soldering and PWB materials and designs that improve the reliability of surface mounting (*30, 31*).

Figure 4. Small-outline IC package with 16 gull-wing leads on a 50-mil pitch (1 mil = 0.001 in. = 25.4 μm). (Reproduced with permission from reference 28. Copyright 1985 Institute of Electrical and Electronics Engineers.)

Figure 5. Plastic-leaded chip carrier with J-shaped leads. (Reproduced with permission from reference 28. Copyright 1985 Institute of Electrical and Electronics Engineers.)

Pin Grid Arrays. Standard chip carriers are limited to ICs with 124 or fewer leads. Many VLSICs with higher lead counts are now being developed, and the predominant package for these chips is the ceramic pin grid array (PGA) (*27*, *28*). Plastic PGAs have also been introduced recently (*26*, *32*). Typically, the PGA is a multilayer ceramic package with pins brazed to the bottom in an area grid array. The standard pin pitch is 0.100 in. (0.254 cm), but a 0.050-in. (0.013-cm)-pitch PGA has been developed for some packages with very high lead counts. The chip cavity may be on top of the package or on the same side as the pins to permit full contact to the top of the package for cooling.

A 270-pin cofired ceramic PGA containing a tape-automated-bonded IC is shown in Figure 6. In this PGA, the decoupling capacitors are on top of the package, and the chip cavity is surrounded by a seal ring to which a metal lid can be welded. An area area at the bottom of the package is cleared of pins to contact a heat-dissipating structure on the PWB. This package represents the state of the art in high-lead-count single-chip packaging.

Although the PGA achieves the highest possible number of I/O interconnections for a given board area, it greatly increases the cost and complexity of the PWB because of the large number of leads and the need for

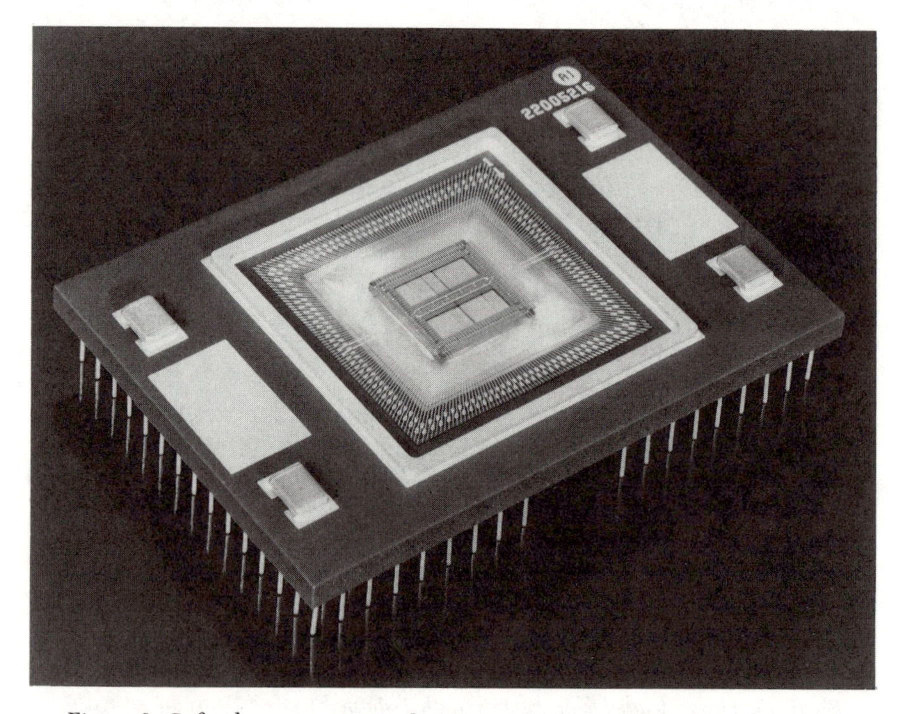

Figure 6. Cofired ceramic pin grid array package with 270 pins and a tape-automated-bonded IC.

through-holes in the board. For this reason, surface-mounted SCPs with fine-pitch peripheral leads are beginning to replace high-lead-count PGAs. Also, the ceramic PGA is generally the most expensive type of SCP because of the high tooling charges for fabricating multilayer ceramics with brazed and plated pins.

Limitations of Single-Chip Packaging. All of the single-chip packaging approaches just discussed impose fundamental limitations to chip density and system performance. The SCP is basically a pitch extender; it provides a fan-out from the fine pitch of I/O pads on the chip (typically 0.005–0.010 in. [0.013–0.025 cm]) to the larger pitch of interconnection lines on a PWB (minimum of 0.010 in. [0.025 cm]). The package size is usually determined by the density of SCP-to-PWB connections. Packages with very high I/Os have a high ratio of package area to chip area and a large number of connections, because there are two connections for each lead, one from the chip to the package and one from the package to the board. The reliability of a system is inversely related to the total number of connections (*21*); thus maintaining the reliability of high-lead-count SCPs presents a significant challenge.

Single-chip packaging can also limit system performance. The large distance between SCPs results in large signal propagation delays, which can limit the clock speed in high-performance systems. The large resistance and capacitance of the long interconnection lines increase signal delay and degradation and consume additional power from IC output drivers. Many impedance discontinuities and reflection points exist for signals passing through the bonds, leads, and PWB interconnections between SCPs. Controlling impedances and providing adequate grounding for high-speed signals throughout this path are very difficult tasks.

Multichip Packaging. A solution to these problems is to mount a number of ICs closely together on a common substrate within an MCP. The obvious advantages of the MCP are the reduction in the overall size and weight of the package and, thus, the reduction in the size and cost of the system. The reliability of the system is also improved, because the chip-to-chip interconnections can be accomplished within the package, which reduces the number of package-to-board connections.

The MCP significantly enhances the speed and reduces the power consumption of the system. Because the ICs are spaced closely together, the interconnection length and propagation delay are greatly reduced, and faster clock speeds are possible. The short interconnections also reduce the need for line termination to prevent reflections. Characteristic impedance is better controlled within the MCP, and fewer signal reflection points exist. Finally, the power dissipation of output drivers can be reduced because of the lower resistive losses and capacitive load of the interconnection.

Increasing the chip density on an MCP presents several new problems. First, the high-density interconnections between chips require processes that are capable of patterning high-resolution, high-aspect-ratio features in multiple conductor layers on relatively large substrates. The cross section of the conductor lines must be large to maintain low resistive losses, and the dielectric layers must be thick to minimize capacitance. However, the routing pitch must be kept small.

The distribution of power to a large number of ICs is another crucial issue; the power must be distributed through planes that are tapped at many points to minimize voltage drops and the noise induced by a large number of simultaneously switching circuits. Another potential source of electrical noise that must be controlled is crosstalk between the closely spaced interconnection lines carrying high-frequency signals. Finally, the removal of heat from closely spaced, high-power ICs requires efficient heat transfer through the package and to the ambient. The next section will describe in more detail the geometrical, electrical, and thermal design requirements of MCPs and their impact on material properties and physical dimensions.

Design Requirements for High-Performance Packaging

Geometrical Requirements. The maximum packing density of chips on an MCP is determined by one of the following factors: (1) the number and spacing of off-package connections, (2) the area required for chip bonding and other accessible features on the surface of the package, (3) the maximum density of interconnection lines, or (4) the maximum power density. In addition, processing or assembly capabilities can place an upper limit on the package size. The first three factors, which pertain to the interconnection geometries of an MCP, are discussed in the following sections.

Off-Package Connections. The number of terminals needed to connect randomly a cluster of logic circuits (such as a chip or an MCP) to the rest of a system is predicted by a well-known empirical power law relationship known as Rent's rule:

$$t = AC^p \tag{1}$$

where t is the number of terminals, C is the number of circuits in the cluster, A is an empirical parameter related to the number of terminals per circuit, and p is the Rent exponent (<1.0), which is related to the architecture of the system, with more-parallel architectures having higher values of p.

Rent's rule can be applied at many levels of packaging, including chips, MCPs, cards, or boards. Values of 2.5 for A and 0.6 for p have been empirically derived for large data-processing systems (*4*). Thus the number of signal I/Os on a chip (t_{chip}) is given by

$$t_{chip} = 2.5C_{chip}^{0.6} \quad (2)$$

where C_{chip} is the number of circuits (i.e., gates) on the chip. The number of off-package I/Os on an MCP (t_{MCP}) containing N chips is given by

$$t_{MCP} = 2.5(C_{chip}N)^{0.6} = t_{chip}N^{0.6} \quad (3)$$

Compared with N SCPs, the MCP reduces the total number of interconnections by a factor of $N^{0.4}$.

The number of I/Os predicted by Rent's rule is often large for an MCP. However, Rent's rule assumes random interconnection to the rest of the system. Many systems can be partitioned into functional blocks (such as a microprocessor) that require significantly fewer interconnections to the rest of the system and, thus, break Rent's rule. This partitioning of the system into functional blocks containing a feasible number of chips is the first crucial decision in designing an MCP and often determines the number of chips and, thus, the size of the MCP.

In addition to signal I/Os, off-package connections are required to supply power and ground to the MCP. The number of power and ground pins depends on the circuit technology. Bipolar circuits generally require more power than do MOS (metal–oxide–semiconductor) circuits and thus require more parallel pins to reduce voltage drops. As many as 30% of the pins on an MCP may be required for off-package power and ground connections.

The minimum size of the MCP may be limited by the spacing of the off-package connections. If the connections are on a square grid array with a center-to-center spacing of P_{grid}, the side length of the package must be at least $t_{MCP}^{0.5}P_{grid}$. If the I/Os are on the perimeter of a square package with a pitch of P_{perim}, the side length of the package must be $0.25t_{MCP}P_{perim}$. For SCPs, the off-package I/Os often limit the size of the package; however, the size of a well-partitioned MCP will often be limited by other factors such as interconnection density.

Surface Features. A number of features must be accessible on the top surface of the package. These features may be used to attach components or to test and repair defects and include chip attach pads; bonding pads; testing points; pads for repair or design changes; pads for discrete components, such as termination resistors and decoupling capacitors; off-package I/O pads; and seal rings. The size and spacing of most of these features depend on the specific bonding and assembly technologies. In general, bonding footprints have not decreased in size as much as chip or package interconnections have and thus may limit the chip packing density for high-I/O chips.

The largest features on an MCP are the chip footprints, which include the die attach pads and bonding pads. The bonding pads are usually placed on a single or a double row around the perimeter of the chips. Currently, the highest density perimeter-bonding technology is tape-automated bonding (TAB), which can achieve a 100-mm pitch for inner lead bonds to the chip (*19*) and a somewhat larger pitch for outer lead bonds to the package or PWB (*29*). Bonding techniques that mount the chip face down ("flip-chip") and that electrically connect the chip through a grid array of solder bumps (termed "controlled-collapse chip connections" or C4 bonding) can achieve the highest bonding and chip packing density, with chips nearly touching one another (*18*, *19*). However, if engineering change pads or repair pads are required, the chip footprint for flip-chip bonding will be expanded (*see* Figure 14).

Figure 7 shows an example of the top surface features of an MCP designed for electrooptical-signal-processing applications (*33*). The MCP has 18 chip attach pads surrounded by dumbbell-shaped pads for wire bonding and repair. The top surface also contains off-package I/Os along two sides, wide power distribution lines, and sites for decoupling capacitors. In this design, the package size of 2.25 by 2.25 in. (5.7 by 5.7 cm) was determined by the top-layer features rather than by the maximum interconnection density.

Wirability. Signal interconnections on an MCP are normally routed on multiple orthogonal layers of conductor lines. Wirability refers to the extent to which all of the interconnections in a package are routed on a given number of layers and with a given line pitch. CAD tools are normally required to route the interconnections on a complex MCP, but an estimate of wirability can be obtained by using statistical relationships such as Rent's rule combined with geometrical arguments.

If no branching is assumed such that each interconnection contains two end points (a worst-case assumption), the total number of interconnections (I_{tot}) to be routed on an MCP is one-half the number of terminals:

$$I_{tot} = 0.5(Nt_{chip} + t_{MCP}) \quad (4)$$

Each interconnection is broken up into an average number of segments, s_i, that will be routed on orthogonal layers. Each routing layer contains a number of wiring channels determined by the package side length, L, divided by the routing pitch, d. Each channel contains an average number of segments, s_t. The total number of segments available on n layers must at least equal the total number of segments required.

$$ns_tL/d = s_iI_{tot} \quad (5)$$

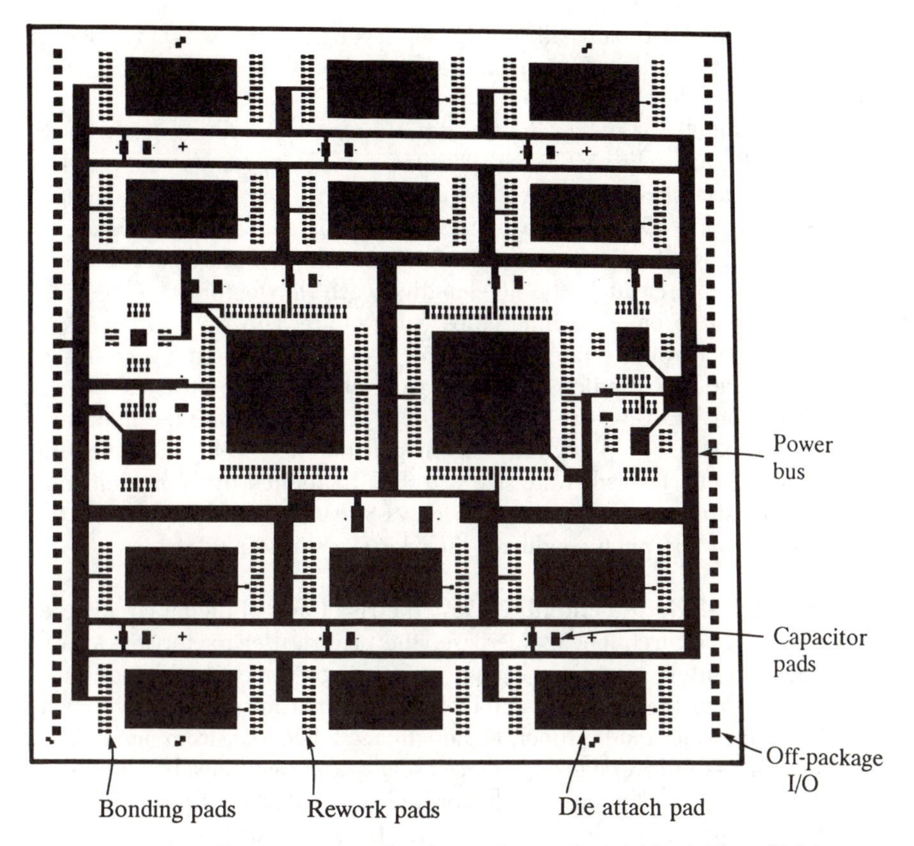

Figure 7. Metallization pattern for the top layer of a 2.25 by 2.25 in. (5.1 by 5.1 cm) multichip package, with chip footprints, I/O pads, and other surface features.

By substituting equation 4 into equation 5, the equation for the package side length is derived.

$$L = (s_i d/2ns_t)(Nt_{\text{chip}} + t_{\text{MCP}}) \qquad (6)$$

Typical values for s_i and s_t in an MCP computer module are three segments per interconnection and four segments per channel, respectively, and thus

$$L = (3d/8n)(Nt_{\text{chip}} + t_{\text{MCP}}) \qquad (7)$$

Equation 7 shows that the package size is directly proportional to the conductor line pitch (d) and the number of chip and package I/Os and inversely proportional to the number of layers (n).

For chips that are flip-chip bonded, additional conductor layers may be required to redistribute the I/Os from underneath the chip to the engineering change or repair pads on the top layer (*see* Figure 14). The layers required for this redistribution can be derived also from geometrical arguments (*34*).

The geometrical factors limiting MCP size may be illustrated with an example of a nine-chip MCP with 200 I/Os per chip. Rent's rule predicts 747 off-package signal I/Os. Additional pins for power and ground might bring the total to 1000 package pins. With pins on a 0.25-cm-grid array, an 8 by 8 cm substrate would be required. If the chip bonding pads are on a 0.025-cm-perimeter pitch, the side length of each chip footprint will be (200/4)(0.025) = 1.25 cm, and nine chips will require 3.75 by 3.75 cm. If an interconnection routing pitch of 0.0125 cm and two routing layers are assumed, equation 7 predicts a package side length of 6.0 cm for signal interconnections. In this case, the off-package I/Os determine the minimum size of the package, although intelligent partitioning will likely reduce the number of I/Os. The substrate size will then be limited by interconnection routing density, a case that is typical for VLSI MCPs and that indicates the importance of fine-pitch conductor lines for signal routing.

Electrical Requirements. As the rise times of digital circuits decrease and clock frequencies increase, the electrical characteristics of the package and interconnections become increasingly important. Signal interconnections must be designed and fabricated as transmission lines with controlled impedance and proper termination. Reflections from impedance discontinuities and crosstalk between closely spaced lines must be minimized to prevent false switching. Conductor and dielectric losses must be minimized to limit signal attenuation and rise time degradation. Power distribution networks must have low resistance to minimize voltage drops and low inductance and capacitive decoupling to reduce voltage spikes caused by sudden current demands. The electrical design of high-performance packages has been widely discussed in the literature (*21, 35–45*). The important electrical characteristics of a high-performance package and the effects of material properties and physical geometries on these characteristics are summarized in the following sections.

Electrical Length. The electrical length of an interconnection determines how it is designed and analyzed. A time-varying voltage imposed on one end of an interconnection will propagate along the line with a finite velocity. For electrically short lines, the delay due to the interconnection can be ignored, and the voltage can be assumed to appear simultaneously at all points along the line. Short interconnections can be treated as a lumped electrical element, such as resistance, capacitance, or both. However, for electrically long lines, the current and voltage depend on position and time, and the distributed characteristics of the interconnection must be considered. From a design standpoint, long interconnections must be designed and fabricated as transmission lines with controlled characteristic impedance and proper terminations to avoid reflections.

For digital circuits, the electrical length of the line can be defined relative to the signal rise time. A common rule of thumb for simple lines with no branches is that interconnections must be treated as transmission lines when the round-trip transit time of the signal, $2\tau_{pd}l$, exceeds the signal rise time, t_r (37). This rule defines a critical line length (l_c) given by

$$l_c \geq t_r/2\tau_{pd} \tag{8}$$

in which l is line length and τ_{pd} is the propagation delay per unit length of line, which is equal to the inverse of the phase velocity, v_p. When the critical line length is exceeded, reflections from the end of the line will arrive at the receiver after the steady-state voltage is reached and may cause false switching.

The critical line length may also be defined relative to the wavelength of the highest frequency components of the signal. In the frequency domain, a digital signal can be represented by a discrete series of Fourier harmonics (for a periodic signal) or by a continuous Fourier function (for a single pulse). To accurately transmit a signal of rise time t_r, the bandwidth of the interconnection must include frequencies from 0 to the maximum frequency, $f_{max} = 0.5/t_r$ (45–47). Thus the shortest wavelength (λ_{min}) components are

$$\lambda_{min} = v_p/f_{max} = 2t_r/\tau_{pd} \tag{9}$$

and by comparison with equation 8

$$l_c \geq 0.25\lambda_{min} \tag{10}$$

In other words, transmission lines are needed when the interconnection length is greater than one-fourth of the wavelength of the highest frequency harmonic.

Figure 8 shows the required bandwidth and critical line length as a function of signal rise time or clock frequency, in which the clock cycle time is assumed to be $10t_r$. Short signal rise times or high clock frequencies require larger bandwidths and have shorter critical line lengths, whereas materials with lower dielectric constants permit longer critical line lengths because of the higher phase velocity of electromagnetic waves in these materials.

Transmission Line Structures and Characteristics. When interconnections exceed the critical line length, transmission lines with proper terminations are used to effectively couple the signal into the line, to minimize reflections along the line and from its ends, and to minimize crosstalk with adjacent lines. Transmission line structures provide a controlled environment

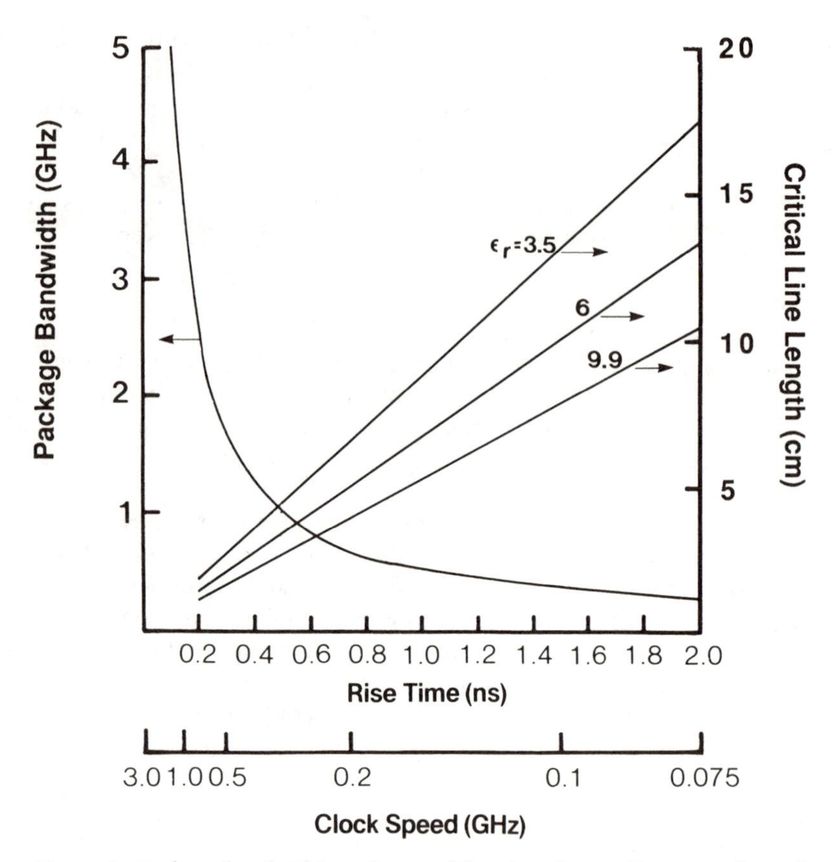

Figure 8. Package bandwidth and critical line length as a function of signal rise time or system clock speed.

for signal propagation by confining the electric and magnetic fields between the signal line and a reference voltage or ground plane.

A variety of transmission line structures can be fabricated in planar layers of conductor and dielectric (Figure 9). The stripline and offset stripline are best suited for multilayer structures. The offset stripline, with two orthogonal signal layers between a pair of reference voltage planes, eliminates one intermediate plane and achieves higher characteristic impedance for a given dielectric thickness than do two stripline layers but increases the possibility for crosstalk between layers.

The two fundamental parameters that describe a transmission line are the characteristic impedance, Z_o, defined as the ratio of the voltage to the current at any point on an infinitely long line, and the propagation constant, γ, which describes the phase shift and attenuation of the signal as it propagates down the line. For signals that propagate in a transverse electromagnetic (TEM) mode (i.e., the electromagnetic fields are transverse to the

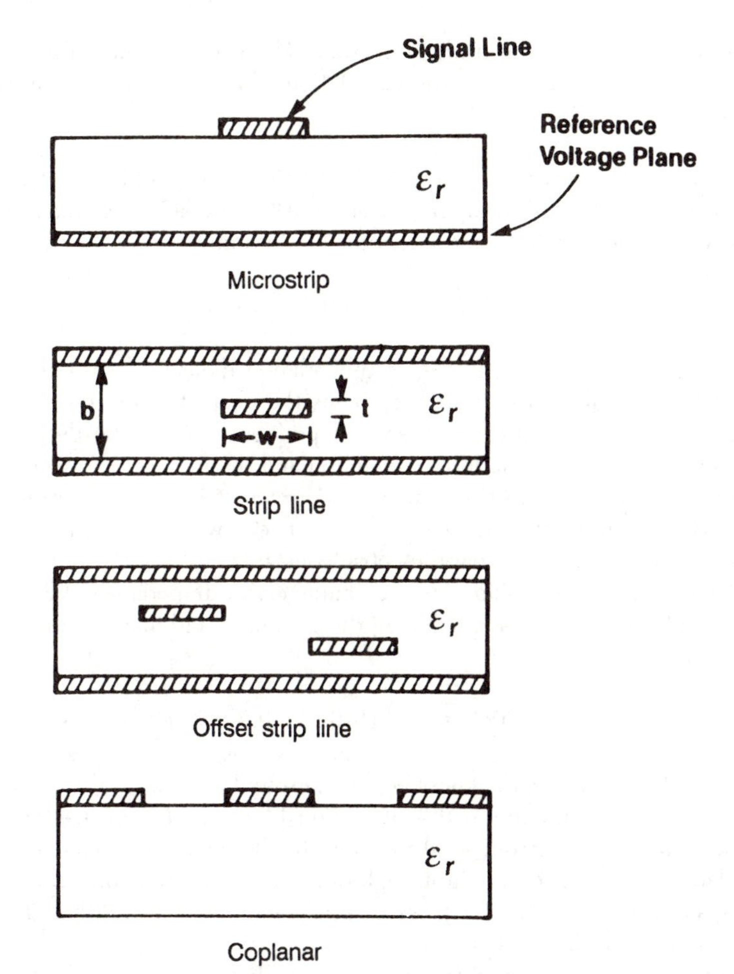

Figure 9. Examples of planar transmission line structures. Dimensions are as follows: b *is dielectric thickness,* w *is line width, and* t *is conductor thickness.*

direction of propagation), Z_o and γ at a given sinusoidal frequency $\omega = 2\pi f$ (where f is frequency in reciprocal second) can be related to the series resistance (R), series inductance (L), shunt capacitance (C), and shunt dielectric conductance (G) of an infinitesimal (i.e., electrically short) length of line by the following expressions (*42*):

$$Z_o = \sqrt{(R + j\omega L)/(G + j\omega C)} \qquad (11)$$

$$\gamma = \sqrt{(R + j\omega L)(G + j\omega C)} \qquad (12)$$

The parameters R, L, C, and G can be calculated from the geometries and material properties of the interconnection. The extensive literature of microwave analysis is useful for the calculation of Z_o and other transmission line parameters (*48–50*).

Lossless Transmission Lines. If the line resistance and dielectric conductance are both small (i.e., $R << \omega L$ and $G << \omega C$), the transmission line is considered lossless, and equation 11 simplifies to

$$Z_o = \sqrt{L/C} = 1/v_p C = \sqrt{\epsilon_r}/cC \tag{13}$$

where c is the speed of light, ϵ_r is the relative dielectric constant of the dielectric material, and $v_p = c/\sqrt{\epsilon_r}$. Thus the characteristic impedance of a lossless line is real and frequency independent and can be determined from the capacitance of the interconnection and the dielectric constant. For structures with two dielectrics, such as the microstrip, ϵ_r in equation 13 is replaced with an effective dielectric constant, ϵ_{eff}, which is calculated from the geometries and individual dielectric constants.

A closed-form expression for the characteristic impedance of a lossless stripline illustrates its dependence on the geometry and material properties of the interconnection (*51*)

$$Z_o = (60/\sqrt{\epsilon_r}) \ln [4b/0.67\pi(0.8w + t)] \tag{14}$$

where b is the dielectric thickness between ground planes, t is the conductor thickness, and w is the line width. Equation 14 is valid for $w/(b - t) < 0.35$ and for $t/b < 0.25$. From a fabrication standpoint, a high characteristic impedance is difficult to obtain in high-density interconnections, because it requires a large dielectric thickness, a small conductor line width and thickness, and a small dielectric constant.

The optimum characteristic impedance is dictated by a combination of factors. Interconnections with low characteristic impedance (<40 Ω) cause high power dissipation and delay in driver circuits, increased switching noise, and reduced receiver noise tolerance (35). High characteristic impedance causes increased coupling noise and usually has higher loss. Generally, a characteristic impedance of 50–100 Ω is optimal for most systems (35), and a Z_o of 50 Ω has become standard for a variety of cables, connectors, and PWBs. For a polyimide dielectric with $\epsilon_r = 3.5$, a 50-Ω stripline can be obtained with $b = 50$ μm, $w = 25$ μm, and $t = 5$ μm.

Lossy Transmission Lines. For lossy transmission lines, the conductor resistance, R, and dielectric conductance, G, must be considered. The assumption $G << \omega C$ is usually valid, because the dissipation factor, $\tan \delta = G/\omega C$, is usually less than 0.01 for most packaging dielectrics (although the dissipation factor may become larger at very high frequencies). For high line resistance or low frequency, the lossless assumption $R << \omega L$ is often not valid, and the characteristic impedance becomes frequency dependent. Resistive losses cause a degradation in signal rise time and an attenuation of the signal. Further analysis of lossy transmission lines is given by Ho (*21*, *36*).

Conductor Resistance. The conductor resistance depends on frequency. The dc (direct-current) resistance per unit length (R_{dc}) is given by the simple expression

$$R_{dc} = \rho / wt \tag{15}$$

where ρ is the conductor resistivity. The large cross-sectional area required for low dc resistance, combined with the narrow line widths required for high characteristic impedance and high wiring density, means that a large conductor aspect ratio, t/w, is required for low dc resistance.

As frequency increases, the current is forced out of the center of the conductor toward its periphery, a phenomenon known as the "skin effect". A measure of the depth of penetration of the current into the conductor is the skin depth, defined as $\delta = \sqrt{(\rho/\pi f \mu)}$, where f is the frequency and μ is the conductor permeability (1.26×10^{-6} H/m for nonmagnetic conductors). For copper, the skin depth is 2 μm at 1 GHz. When the skin depth is less than the conductor thickness, the line resistance becomes greater than the dc resistance.

The skin effect resistance of a rectangular-cross-section line also depends on its aspect ratio. For a given cross-sectional area, as the ratio t/w approaches 1, the skin depth perimeter decreases and the resistance increases, as shown in Figure 10, in which the measured resistance is plotted as a function of frequency for lines of fixed cross-sectional area with different aspect ratios (*52*, *53*). Unfortunately, the lines with high aspect ratios that are desirable for high wiring density and low dc resistance have a higher skin effect resistance compared with thin, wide lines.

Effects of Transmission Line Characteristics on Digital Signals. The effects of propagation delay and resistive losses on an idealized input signal are illustrated in Figure 11, which shows output signals on both lossless and lossy transmission lines, with no reflections or sources of noise. The input voltage first appears at the output after a minimum propagation delay $\tau_{pd}l$, which is determined by the speed of light in the dielectric medium and the line length l. If the transmission line is lossless and properly terminated, the input signal will be reproduced exactly at the receiving end after the $\tau_{pd}l$ delay. Resistive losses will increase the signal rise time, attenuate the steady-state voltage, and smooth out the sharp transitions at the beginning and end of the signal (Figure 11). Additional noise (not shown in Figure 11)

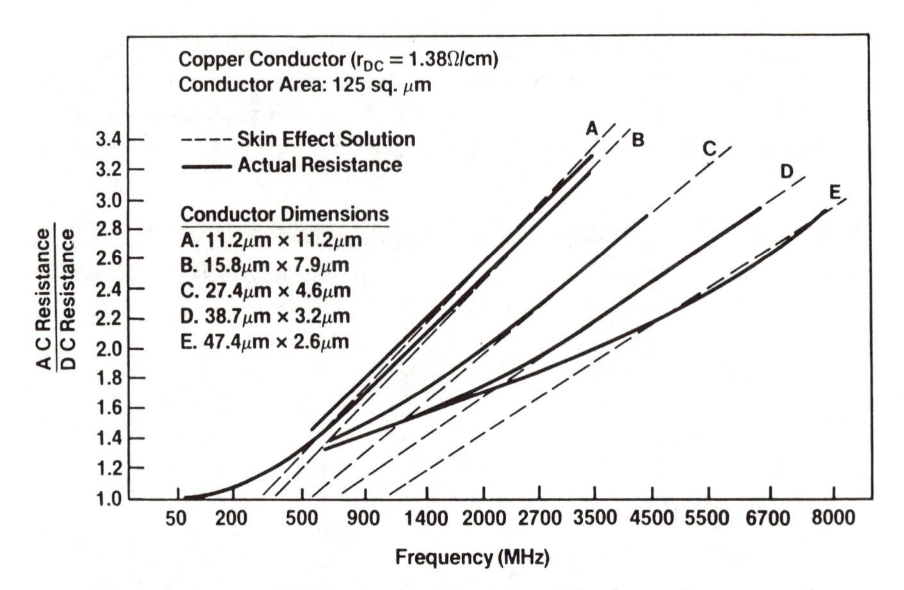

Figure 10. Measured resistance as a function of frequency for rectangular-cross-section lines with different cross sections (product of line width and thickness). Abbreviations are as follows: AC is alternating current and DC is direct current. (Data were derived from reference 53.)

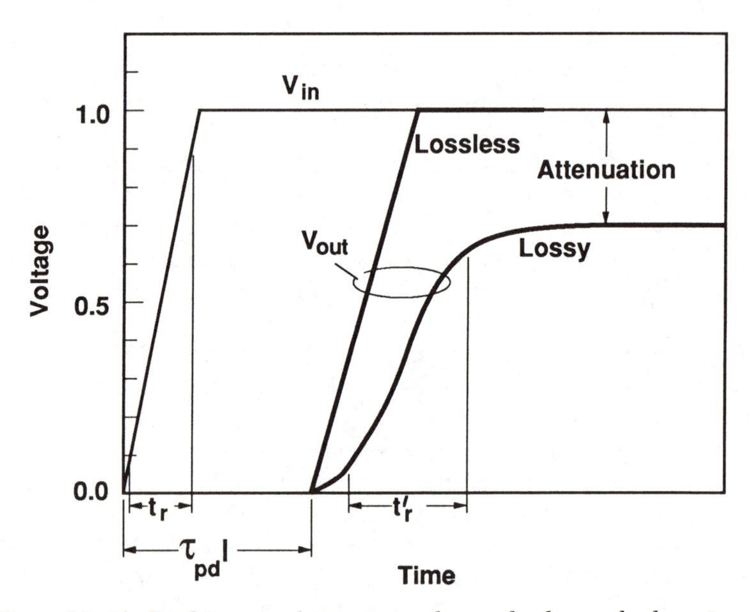

Figure 11. Idealized input and output signals on a lossless and a lossy transmission line of length l. Rise time (t_r) is defined as the time between the 10 and 90% voltage levels. Other abbreviations are defined in the text.

will be caused by reflections from impedance discontinuities, crosstalk from neighboring signal lines, and noise from the power source.

Propagation Delay of Lossless Lines. The minimum propagation delay (τ_{pd}; expressed in picoseconds per centimeter) for a unit length of lossless line is the inverse of the phase velocity (v_p) of the electromagnetic waves propagating through the dielectric medium surrounding the conductor line:

$$\tau_{pd} = 1/v_p = \sqrt{\epsilon_r}/c = 33\sqrt{\epsilon_r} \quad (16)$$

Propagation delay is minimized by keeping lines short and by using low-dielectric-constant materials. In general, organic dielectrics have lower dielectric constants compared with inorganic dielectrics. For example, Al_2O_3, a typical ceramic dielectric, has an ϵ_r of 9.5 and a propagation delay of 102 ps/cm, whereas polyimide, a common polymer dielectric, has an ϵ_r of 3.5 and a propagation delay of 62 ps/cm.

Even electrically short lossless interconnections may have additional delay due to charging of the interconnection capacitance by the output driver's resistance ($R_{driver}C_{line}$). Such a delay increases the signal rise time. Thus low interconnection capacitance is required even for long-rise-time signals to minimize delay due to rise time degradation.

Additional delay may be imposed by the capacitive loading of branch lines, receivers, and bonding pads attached to the interconnection or by the series inductance of vias or wire bonds. If the additional capacitance can be assumed to be uniformly distributed over the interconnection with electrically short spacing (an assumption that is not always valid), the modified propagation delay (τ_{pd}') is given by (*37*)

$$\tau_{pd'} = \tau_{pd}\sqrt{1 + c_d/c_o} \quad (17)$$

in which c_o is the isolated line capacitance and c_d is the capacitance of the distributed loads.

Delay Due to Resistive Losses. On electrically long, lossy lines, the signal rise time is degraded by dispersion in the interconnection. Dispersion delays and attenuates the high-frequency components of the signal more than the low-frequency components because of the frequency-dependent resistance of the interconnection. The rise time degradation contributes additional delay before the switching threshold is reached at the end of the line.

For electrically short interconnections with high resistance, such as on-chip interconnections, signal delays are dominated by the rise time degradation due to the charging of the receiver capacitance by the interconnection resistance ($R_{line}C_{rec}$) (*54*, *55*). On PWBs in which resistance is negligible, the time-of-flight or propagation delay given by equation 16 is dominant. In the intermediate case of high-density multichip packages, both propagation delay and rise time degradation must be considered.

Noise in Interconnections. Real interconnections will not behave as shown in Figure 11 but will have several sources of noise. Ringing on the transmission line can be caused by signal reflections from any point of impedance discontinuity, such as vias, bends, branch points, wire bonds, pins, or the end of an unterminated line. The amount of reflection depends on the amount of mismatch between the impedance of the discontinuity and that of the transmission line. If the line is electrically short, as defined by equation 8, the reflections will affect the signal only during its rise time. However, if the line is electrically long, reflections may cause an overshoot or false switching after the signal has reached its steady-state value. Therefore, long, high-speed signal lines must be terminated in their characteristic impedance to avoid reflections, and other sources of impedance discontinuity must be minimized. Interconnections are normally terminated by a series resistor at the driving end of the line or by a shunt resistor to ground at the receiver end, with either resistor matching the line's characteristic impedance. One disadvantage of these termination resistors is that they increase the power requirements of driver circuits.

Even properly terminated lines can have reflections from impedance discontinuities along the line, and these reflections can degrade the signal rise time. Some of the transmitted signal is lost at each reflection point, and higher frequency components tend to be reflected (and thus attenuated) more than low-frequency components; thus, the interconnection behaves like a low-pass filter and causes additional degradation of the signal rise time.

Crosstalk. Crosstalk between adjacent, parallel signal lines is another source of noise that can cause false switching. Crosstalk is caused by capacitive and inductive coupling between lines (*42*, *56*, *57*). The coupled voltage propagates in both directions along the line, and its amplitude varies with line length as a function of frequency. Forward crosstalk propagates in the same direction as the active signal, and backward crosstalk travels back toward the driver end. For true TEM transmission lines, such as striplines, the backward crosstalk is the only significant component and reaches maxima at line lengths of $\lambda(n + 1)/4$ and minima at line lengths of $\lambda(n/2)$ (where n is an integer). The forward crosstalk, which appears in quasi TEM structures such as microstriplines, increases with the length of coupled lines. The magnitude of crosstalk increases with a decrease in the spacing between lines and with increases in the distance to the ground plane and the dielectric constant. Thus the design requirements for high characteristic impedance and high line density are in conflict with those for low crosstalk.

Modeling of High-Speed Interconnections. Modeling the electrical behavior of an interconnection involves two steps. First, the transmission line characteristics, such as the characteristic impedance, propagation constant, capacitance, resistance, dielectric conductance, and coupling parameters, must be calculated from the physical dimensions and material properties of the interconnection. In addition, structures, such as wire bonds, vias, and pins, must be represented by lumped resistance (R), inductance (L), and capacitance (C) elements.

Second, the response of the circuit to a time-varying input signal is obtained by using one of two models. In the distributed-circuit model, the input signal is transformed to the frequency domain, the transmission line equations are solved, and an inverse Laplace or Fourier transform is performed to obtain the response in the time domain (*42*). In the lumped-element model, the transmission line can be approximated by a series of lumped elements containing a series resistance and inductance and a shunt capacitance and resistance (inverse of dielectric conductance) to ground such that each lumped element represents an electrically short segment of the interconnection. The lumped-element model can then be solved directly in the time domain by using nonlinear-circuit-simulation software packages such as SPICE (*38*). The distributed-circuit model is computationally more efficient and can handle frequency-dependent parameters such as resistance and crosstalk. However, the lumped-element model can be more easily coupled to nonlinear driver and receiver circuit models. Both approaches have been used to to predict signal waveforms at the receiver, forward and backward crosstalks, and the power delivered to various loads.

Power Distribution. The electrical design of a package must also take into account the distribution of power to and from the ICs. The design requirements for power distribution depend on the circuit technology. For example, MOS circuits require power only during switching, whereas bipolar circuits require continuous power (*39*). In general, power distribution structures in the package must have low resistance to minimize voltage drops and low inductance to minimize voltage spikes (defined by $LN\ di/dt$, where L is the effective inductance of the power distribution system, N is the number of simultaneously switching driver circuits, and di/dt is the rate of current change [25]) that can occur when a large number of circuits switch simultaneously at high speeds.

Power is usually fed into the package and chip through multiple pins to distribute the current and reduce voltage drops. Within the package, power is distributed on solid or mesh planes or wide lines that present low inductance and resistance. The high inductance of small-diameter wire bonds can be a significant source of $L\ di/dt$ noise and can provide a motivation for short and flat bonding leads. Decoupling capacitors are frequently provided on power input lines to shunt the high-frequency noise to ground. More-detailed discussions of power distribution requirements can be found elsewhere (*21*, *25*, *35*, *36*, *39–41*).

Thermal Design. Because both the performance and reliability of ICs degrade with increasing temperature (*58*), an important function of the package is to dissipate the heat generated by the chip by providing a path of low thermal resistance to the ambient air or heat sink. The ultimate level of integration of the IC or the packing density of chips on an MCP can be limited by the package's ability to dissipate heat. Conventional forced-air cooling can dissipate power densities of up to 1 W/cm^2, whereas advanced cooling technologies based on boiling heat transfer to liquid Freon may extend power densities to 20 W/cm^2 (*59*).

In almost all packages, heat is transferred from the chip to the external surfaces of the package by thermal conduction. The heat may then be conducted through the PWB to an external heat sink, or it may be transferred by forced or natural convection to the surrounding air or cooling fluids. At steady state, the heat generated by the chip equals the heat flux through the package, and the conductive heat transfer is described by Fourier's law (*60*)

$$\vec{q} = -k\vec{\nabla}T \tag{18}$$

where $\vec{q}$ is the conductive heat flux (typically in watts per square centimeter [W/cm^2]), k is the thermal conductivity (in watts per centimeter per degree Celcius or kelvin [W/cm-°C or W/cm-K]), and $\vec{\nabla}T$ is the temperature gradient. For the simplest case of one-dimensional heat flow through a layer of thickness t and cross-sectional area A, the heat transfer Q is given by

$$Q = kA\Delta T/t \tag{19}$$

where ΔT is the temperature difference across the layer. For more-complex two- and three-dimensional designs, finite-element methods can be used to solve equation 18 (*61*).

The convective heat transfer to a fluid is described by using a heat-transfer coefficient, h

$$Q = hA\Delta T \tag{20}$$

where ΔT is the difference between the surface temperature and the fluid temperature and A is the total surface area in contact with the fluid. Heat-transfer coefficients must be determined experimentally, or they can be obtained from the extensive literature on heat transfer.

In the analysis of thermal performance, an electrical analogy is often applied, and heat transfer is described in terms of a thermal resistance, θ, defined as

$$\theta = \Delta T/Q \tag{21}$$

and expressed in units of degrees Celsius or kelvins per watt (°C/W or K/W). Thermal resistances can be determined for various elements of the package, and these resistances can be combined in series or parallel to estimate the overall thermal resistance.

The overall temperature difference ($T_j - T_a$) between the chip junction (the point on the chip where the heat is generated) and the ambient is given by

$$T_j - T_a = \theta_{ja} Q \tag{22}$$

where Q is the total heat generated by the chip or chips and θ_{ja} is the thermal resistance between the junction and the ambient. θ_{ja} is often separated into two components, a conductive thermal resistance between the junction and case (or external surface of the package), θ_{jc}, and a convective resistance between the case and the ambient, θ_{ca}.

Although ICs often can operate at temperatures as high as 150 °C, the maximum junction temperature, T_j, is usually held below 75–85 °C for high-performance ICs (*4*) or below 125 °C for lower reliability applications (*62*). The ambient temperature, T_a, is typically 30 °C for air cooling (although it may increase by 10–15 °C in passing through the chassis of the machine) or 24 °C for water cooling. Thus $T_j - T_a$ for high-performance ICs must often be kept below 50 °C. Measured values of θ_{ja} for a 64-pin DIP are 36.5 °C/W in still air and 22 °C/W in air moving at 600 ft^3/min (~17 m^3/min) (*63*). Therefore, to hold $T_j - T_a$ to less than 50 °C, the maximum power generated by a still-air-cooled DIP IC must be less than 1.4 W (50 ° C ÷ 36.5 °C/W).

Emerging VLSI bipolar ICs commonly generate 5–10 W, a fact that clearly precludes the use of air-cooled DIPs from a thermal standpoint. A wide variety of materials and designs have been used to reduce θ_{jc} and θ_{ca}. Equations 16–18 indicate that θ is equal to t/kA for one-dimensional heat conduction and to $1/hA$ for thermal convection. The conductive thermal resistance is reduced by using high-thermal-conductivity (k) materials and by providing large conductive cross sections (A) with short thermal path lengths (t). Examples of high-conductivity materials used in high-performance packages are metal-filled die attach materials; high-conductivity metals, such as Cu and Al; and high-conductivity ceramics, such as BeO, AlN, or SiC, in place of the standard Al_2O_3.

Convective heat transfer to the ambient is enhanced by (1) increasing the heat-transfer area through the use of fins or cooling channels, (2) increasing the heat-transfer coefficient, or (3) reducing the temperature of the cooling fluid. Heat-transfer coefficients for natural convection in air, typically in the order of 10^{-3} W/cm^2-°C, are improved by an order of magnitude by forced-convection cooling, by another order of magnitude by jet-impingement cooling, and by still another order of magnitude, to 1 W/cm^2-°C, by forced-convection liquid cooling (*62*). Still higher convective heat transfer is obtained through boiling heat transfer or by direct immersion of packages in fluorocarbon liquids or in liquid nitrogen (*64–66*).

Overall Design Strategy. The preceding discussion indicates that the package design involves trade-offs among a number of competing requirements for interconnectability, electrical performance, and thermal performance. Because of the complexity of these design trade-offs, CAD tools are becoming increasingly important in designing packages and simulating performance. Widely used software packages include ANSYS, NASTRAN, and NISA, for finite-element modeling of heat transfer and mechanical stress, and SPICE, for modeling nonlinear electrical circuits and transmission lines (*38*). In addition, new models are being developed to predict the static and frequency-dependent electrical characteristics of various interconnection structures and their effect on the time–domain behavior of high-speed signals (*42, 44, 58*). The ultimate goal is to integrate these individual models and data bases into a knowledge-based system that can perform the electrical, thermal, and mechanical design trade-offs and that can be highly interactive with the package designer.

Summary of Package Design Requirements. The impact of design requirements on the materials and geometries of multichip packages can be summarized as follows: The number of ICs in an MCP will often be determined by the partitioning of the system architecture into functional blocks having a reduced number of interconnections to the rest of the system. The package size and the density of chips on the package are then determined by one of several factors: (1) the area required for off-package connections, (2) the area required for bonding chips and other components on the top layer, (3) the signal interconnection density, or (4) the thermal or power density.

Technological advancements should be aimed at the limiting factor in order to increase the chip density and reduce interconnection lengths. Shorter interconnections will reduce propagation delay, resistive losses, and the need to terminate signal lines. High interconnection density requires a small conductor line routing pitch and narrow line widths. The conductor line cross section should be large to maintain low resistive losses. A large dielectric thickness between signal lines and reference planes is required for high characteristic impedance and low interconnection capacitance. However, crosstalk becomes more of a concern as the dielectric thickness increases and line spacing decreases. A high-conductivity material is desirable for low resistive losses, and a low-dielectric-constant material is desirable because it results in low propagation delay, low interconnection capacitance, and low crosstalk. High-thermal-conductivity materials and large heat-transfer areas are needed for effective thermal dissipation. Finally, the material system must be stable against a variety of environmental stresses, such as temperature cycling, thermal and mechanical shocks, humidity, radiation, and chemical attack, to ensure high system reliability.

Multichip-Packaging and Interconnection Technologies

Three multichip-packaging technologies that use fundamentally different materials and processes to obtain multilayer interconnections have been developed or are in production: (1) multilayer thick-film systems with screened conductor pastes and glass–refractory dielectrics, (2) cofired multilayer ceramic, and (3) multilayer thin-film metallization with polymer dielectrics. The typical material properties, process geometries, and electrical characteristics for these three technologies are summarized in Table I. The following sections describe each technology in relation to the design requirements discussed in the previous sections.

Thick-Film Multilayer. Thick-film multilayer technology has been used for many years to fabricate hybrid circuits that interconnect small-scale ICs or discrete components on a ceramic or metal substrate (*67–70*). This technology has also been used for multichip packaging of more highly integrated ICs for large computer applications.

The top of Figure 12 shows a thick-film multilayer multichip package used in a mainframe computer (Honeywell DPS 88) (*71*), and the bottom shows the signal lines on an internal metal layer in this package. The package

Table I. Comparison of Multichip–Packaging Technologies

Feature	*Cofired Ceramic*	*Thick Film*	*Thin Film*
Conductor material	W (Mo)	Cu (Au)	Cu (Au, Al)
Sheet resistance (mΩ/□)	10	3	3.5
Thickness (μm)	15	15	5
Linewidth (μm)	100	100–150	10–25
Pitch with vias (μm)	250–750	250–300	50–125
Maximum number of layers	30+	5–8	5
Dielectric material	Al_2O_3	Glass–ceramic	Polyimide
Dielectric constant	9.5	6–9	3.5
Thickness/layer (μm)	250–500	35–65	25
Minimum via diameter (μm)	100–200	200	25
Propagation delay (ps/cm)	102	90	62
Minimum stripline capacitance (pF/cm)	2.0	4.3	1.2
Minimum line resistance (Ω/cm)	1.0	0.2	1.35

NOTE: Standard conductor materials are given, with alternative materials in parentheses.

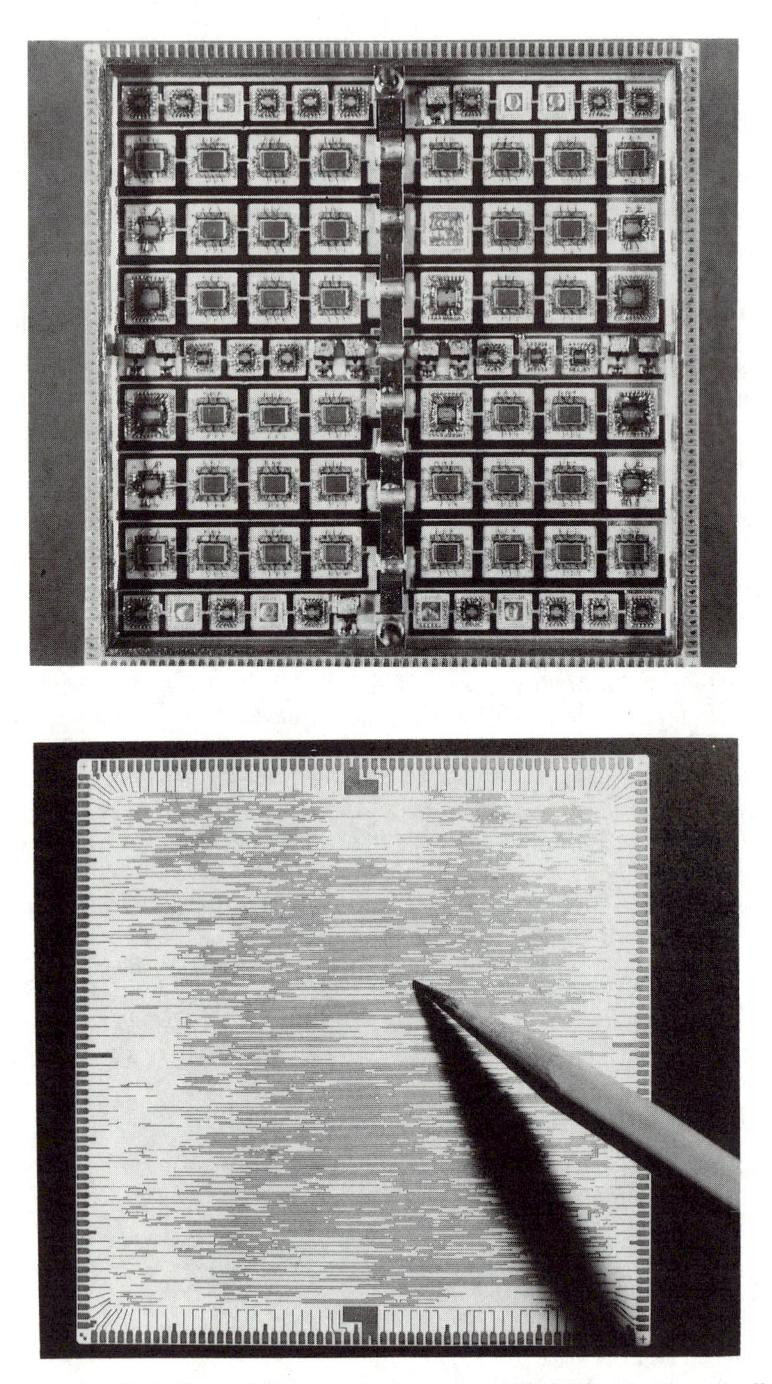

Figure 12. Thick-film multilayer multichip micropackage for the Honeywell DPS 88 mainframe computer. Top, completed package populated with chips; bottom, internal conductor layer with signal lines.

accommodates 60–110 tape-automated-bonded CML (current-mode logic) small- and medium- scale ICs on an 80 by 80 mm ceramic substrate with five layers of thick-film metallization: two layers for signal interconnection, two layers for voltage distribution, and a top layer for ground and chip bonding. This MCP contains circuitry equivalent to that of two 12 by 12 in. (30.5 by 30.5 cm) PWBs with dual-in-line-packaged TTL (transistor–transistor logic) ICs.

Thick-film multilayer structures are fabricated by sequentially screen printing and firing pastes of conductor and dielectric materials. The conductor pastes are a complex mixture of metal or alloy powders, organic binders and solvents, and inorganic glasses. The organic components are adjusted to control the flow and wetting of the pastes during printing, and the inorganic additives promote the heat transfer and densification of the film during firing.

The conductor materials may be noble metals, such as gold, silver, palladium, or platinum, or nonnoble metals, such as copper, nickel, or aluminum. Copper is a desirable material for its low cost and low resistivity (3×10^{-6}–5×10^{-6} Ω-cm, or two to three times the resistivity of elemental copper); however, it requires an oxygen-free atmosphere for firing.

The dielectric pastes are roughly equal mixtures of ceramic powders, such as Al_2O_3, and glasses, such as SiO_2, with organic additives for control of printing properties. After the firing step, the ceramic forms an aggregate in the the glass matrix. Dielectric constants vary from 6 to 9. Thick-film pastes of resistor materials such as ruthenium oxide, indium oxide, and tantalum nitride are also used to incorporate resistors into the package.

The conductor and dielectric pastes are patterned by forcing them with a squeegee through openings in an emulsion or metal mask patterned on a screen. Each conductor and dielectric layer is then fired at high temperatures (550–1000 °C) in inert or oxidizing atmospheres. The maximum number of conductor layers is only five to eight layers because of the increasing topography over underlying patterns and the stress developed in the thick films. The resolution of screen-printing processes limits conductor line widths and pitches to about 100 and 250 μm, respectively. Conductor thicknesses are typically 12 μm. Multiple printings are used to obtain a maximum dielectric thickness of about 60 μm. The wide conductor lines and thin dielectric layers result in a relatively large interconnection capacitance, which inhibits electrical performance.

In summary, the main advantages of thick-film technology are the high-conductivity conductors and the relatively inexpensive process equipment, whereas the drawbacks are the limited number of layers, the multiple printing and firing steps, the low conductor line resolution, and the limited dielectric thickness.

Cofired Multilayer Ceramic. Cofired multilayer ceramic technology (Figure 13) is used to fabricate most ceramic SCPs (ceramic DIPs,

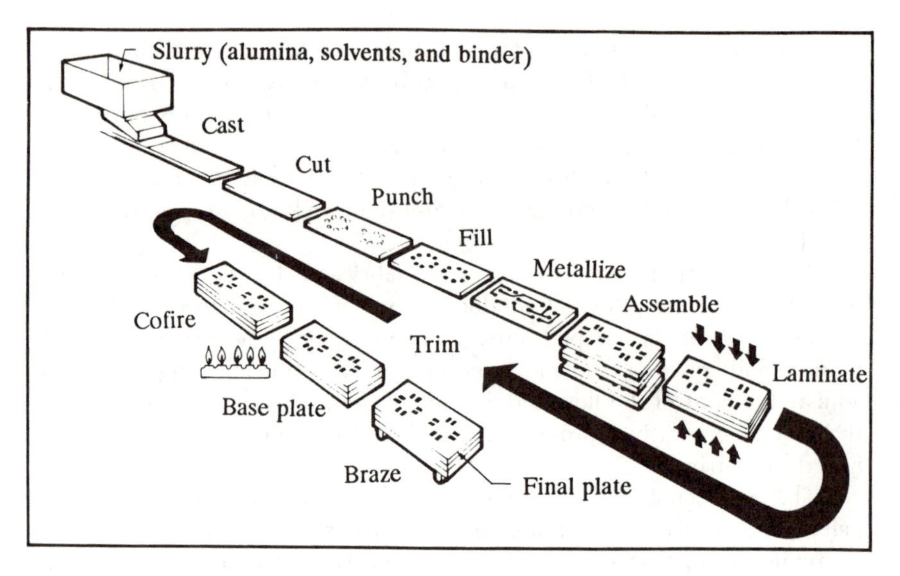

Figure 13. Process flow for cofired multilayer ceramic packages. (Used by courtesy of Interamics.)

chip carriers, pin grid arrays), as well as some advanced multichip packages (72–74). Flexible sheets of dielectric (typically 0.010–0.020 in. [0.025–0.050 cm] thick) are cast from a slurry of ceramic (usually Al_2O_3) and glass powders, solvents, and organic binders. After being dried, the flexible sheets can be cut, and via holes are punched. The vias are filled, and conductor patterns are defined on each sheet by screen printing thick-film pastes of a refractory metal (usually tungsten or molybdenum). The required number of sheets are stacked and aligned, laminated under high pressure, and then fired at high temperature (>1500 °C) in a reducing atmosphere to volatilize the organic binders and densify the metal and ceramic into a solid multilayer structure. Cavities for chip mounting can be fabricated in the ceramic. Features such as pins, perimeter leads, or seal rings can be brazed onto metallized pads, and the exposed metal areas can be electroplated with a finish metal, which is usually gold over a nickel base.

The line width and pitch of cofired ceramic is limited by the resolution of the screen-printing processes and the large shrinkage of the laminated structure during firing (typically 15–20% shrinkage, with a tolerance of ±0.8%). The minimum conductor line pitch is equivalent to that of thick-film technology; however, many more layers can be fabricated by the lamination process than by the additive thick-film process. The dielectric constant of Al_2O_3 (typically 9–10) is higher than that of the glass–ceramic mixture used in thick-film technology, but thicker dielectric layers can be fabricated in cofired ceramic to achieve lower interconnection capacitance. The conductor resistance is relatively large because of the higher resistivity of the refractory metals (about three times the sheet resistance of thick-film copper). Cofired ceramic is a relatively expensive technology that involves high capital investment and high tooling costs. Its advantages over thick-film technology are the thicker dielectric layers, better control of dielectric thickness, and a much larger number of layers that can be laminated. The drawbacks of cofired ceramic are its high cost, low-conductivity conductor, and high-dielectric-constant dielectric.

The most advanced implementation of cofired-ceramic-packaging technology is the thermal conduction module (TCM) used in large-scale computers (IBM) (*4, 72, 74*). This package can accommodate over 100 flip-chip-bonded ICs on a 90 by 90 mm cofired ceramic substrate. The multilayer ceramic substrate contains 33 metal layers for chip pad redistribution, signal interconnection, and power distribution (Figure 14). Each chip contains 120 bonding pads, and 1800 pins are brazed to the bottom of the substrate for connection to a PWB.

A few companies have recently developed low-temperature co-firable materials that combine the advantages of thick-film and cofired-ceramic technologies and exclude some of their disadvantages (*75–77*). The material system consists of a low-firing-temperature dielectric tape with a high glass content and compatible conductor and resistor pastes. The materials are processed in the same way as cofired ceramic (Figure 13), but the low firing temperature of the dielectric (~900 °C) permits the use of less-expensive

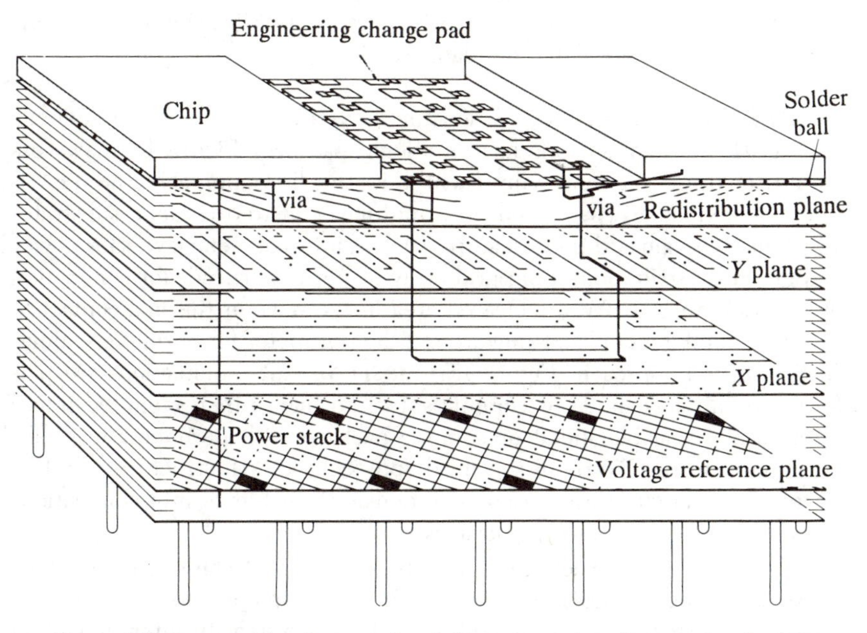

Figure 14. Cross-sectional view of multilayer ceramic substrate for the IBM TCM. (Reproduced with permission from reference 4. Copyright 1983 Scientific American, Inc.)

thick-film furnaces and allows higher conductivity metal pastes, such as palladium–silver and gold, to be fired instead of refractory metals. The dielectric constant of the dielectric, which ranges from 6 to 9, is also low. The process takes advantage of the large number of layers, higher print resolution, and good control of dielectric thickness offered by cofired-ceramic technology. The potential disadvantages of this technology are the low thermal conductivity (0.1 that of Al_2O_3) and low tensile strength of the high-glass-content dielectric and the difficulty of brazing pins or seal rings to the package because of limited brazing temperatures.

Thin-Film Multilayer Packaging. The most significant development in high-density interconnection technology for multichip packaging involves the use of IC fabrication processes (e.g., photolithography, vacuum deposition, and wet and dry etching) to pattern multiple layers of a thin-film conductor and a low-ϵ_r (usually polymeric) dielectric (*21, 36, 78–100*). Copper, gold, and aluminum have been used as conductor materials; copper is especially desirable for its high conductivity. Thin-film resistor materials such as NiCr, TaN, or CrSi may also be incorporated into the package. The most widely used interlayer dielectric materials are polyimides, a class of polymers with exceptional thermal, mechanical, and chemical stability and a low ϵ_r. The thin-film multilayer (TFML) interconnection structures may be patterned on a variety of rigid substrates, including tape-cast or cofired multilayer ceramics, metals (e.g., Cu, Al, Mo, and steel), or silicon wafers. The material options and process technologies for TFML interconnections are described in more detail in a later section.

Packaging Approaches. Various methods are available for implementing TFML interconnections in a multichip package. Figure 15 shows examples of three packaging approaches. In the first approach, the TFML interconnections are patterned on a multilayer cofired ceramic substrate, which may contain additional features such as internal metal layers for power and ground distribution, a pin grid array or perimeter leads for connection of the package to a PWB, a cleared area in the pins for thermal contact to the PWB, and a metallized ring around the perimeter for hermetic sealing. In the second approach (Figure 15), the TFML interconnections are patterned on a blank metal or ceramic substrate that is then mounted into a hermetically sealable metal or ceramic package with perimeter leads for attachment to the PWB. In the third approach, TFML structures are fabricated on a larger substrate for high-density board-level interconnections between single or multichip packages.

In the multichip package, a variety of pretested chips (e.g., bipolar, MOS, and GaAs) and discrete components (decoupling capacitors and termination resistors) may be mounted on the high-density interconnection substrate. This approach is sometimes termed "hybrid-wafer-scale integra-

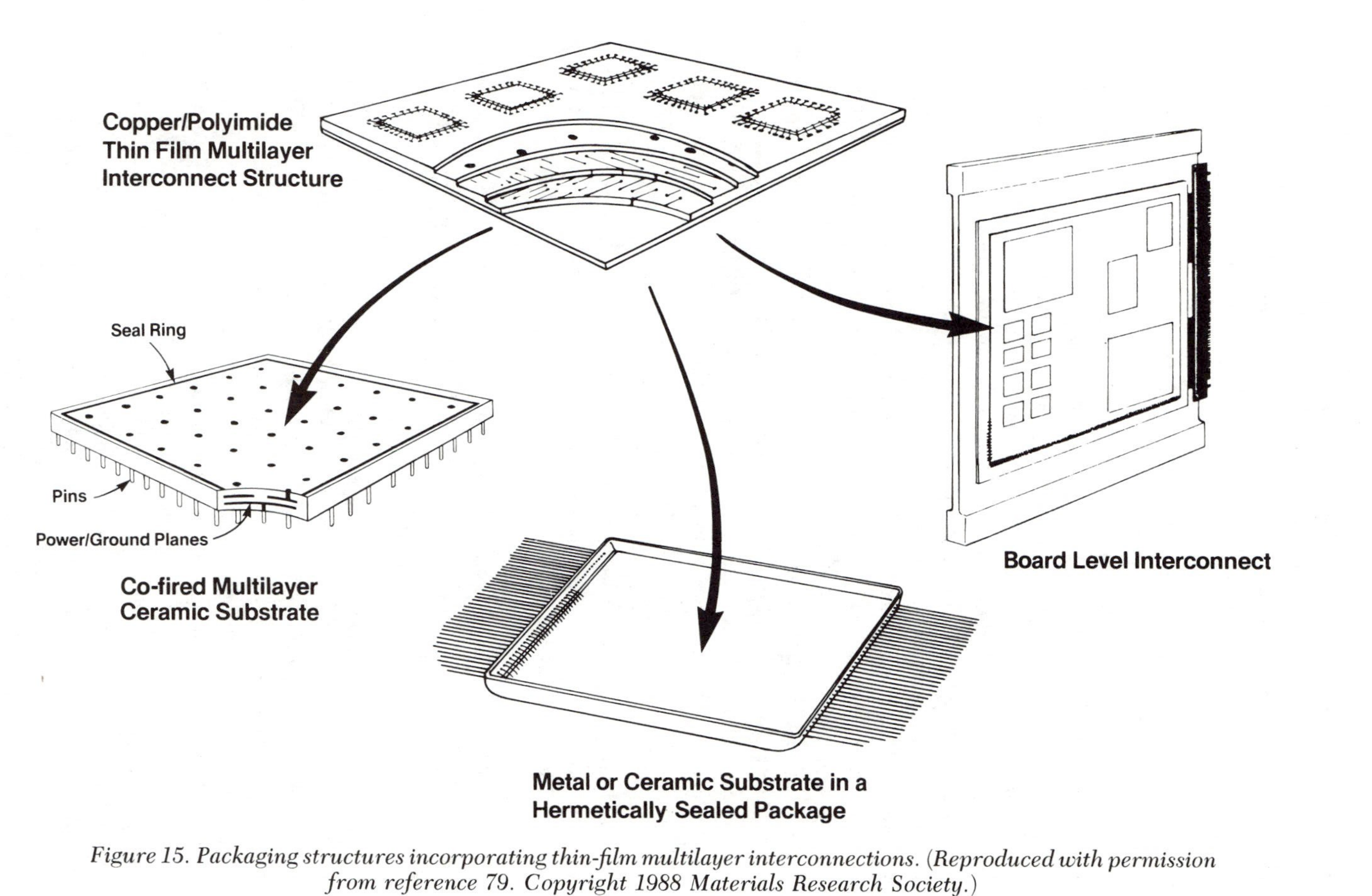

Figure 15. Packaging structures incorporating thin-film multilayer interconnections. (Reproduced with permission from reference 79. Copyright 1988 Materials Research Society.)

tion" (HWSI), in contrast to wafer-scale integration, in which the system is fabricated monolithically on a single wafer. In the MCP or HWSI approach, chips may be bonded by a variety of techniques, including wire bonding, tape-automated bonding (TAB), or flip-chip solder bonding (*89, 97, 98*). The flip-chip approach permits the highest packing density of chips.

For chips mounted face up, heat is transferred to the substrate by conduction through the interconnection layers, and because the polymer dielectric has poor thermal conductivity, heat conduction is often promoted by an array of metallized vias through the interconnection layers (Figure 16) (*100*). For face-down-mounted chips, the heat may be removed from the back side by using pistons (as in the TCM) or conductive fluids, or heat may be conducted through the solder bonds to the interconnection substrate (*98*).

In all of these packaging approaches, the thin-film metal layers are used for high-density interconnections, power and ground distribution, and chip attachment. Figure 16 shows a typical cross section consisting of two layers of signal lines sandwiched between ground or voltage planes and a top metal layer for chip attachment and bonding. This arrangement places the signal lines in an offset stripline configuration with controlled characteristic impedance. Most interconnection geometries are sized for 50-Ω characteristic impedance, which requires a ratio of signal line width to dielectric thickness (between signal line and the nearest plane) of approximately 1:1 in a stripline with an ϵ_r of 3.5.

The greatest difference between TFML technologies as developed by different companies is in the conductor line cross section and the resulting resistive losses or maximum line lengths. For lower performance systems (e.g., <100-MHz clocks), line widths of 10–20 μm and conductor thicknesses of 2–3 μm may be used for lines less than 10 cm long (*97*), whereas for higher speed systems (200–3000 MHz) or longer lines, a conductor line width of >25 μm and a thickness of >5 μm are needed to avoid unacceptable signal degradation or attenuation (*99*).

Advantages of TFML. The TFML interconnection technology offers a number of inherent advantages over the other MCP technologies (Table I). First, compared with the screen-printing and hole-punching processes used for thick films and cofired ceramics, thin-film-patterning processes, such as photolithography and dry etching, can define features with higher resolution and a higher aspect ratio in the conductor and dielectric layers. The high-aspect-ratio features produce a combination of high interconnection density, low interconnection resistance (because of the large conductor cross section), and low interconnection capacitance (because of the thick dielectric layers). Second, low-temperature metal deposition processes such as sputtering permit the use of high-conductivity metals (e.g., Cu and Al) and can achieve nearly bulk resistivity in thin films, as opposed to the low-conductivity refractory metal pastes (W and Mo) used in cofired ceramic or the Cu and Au

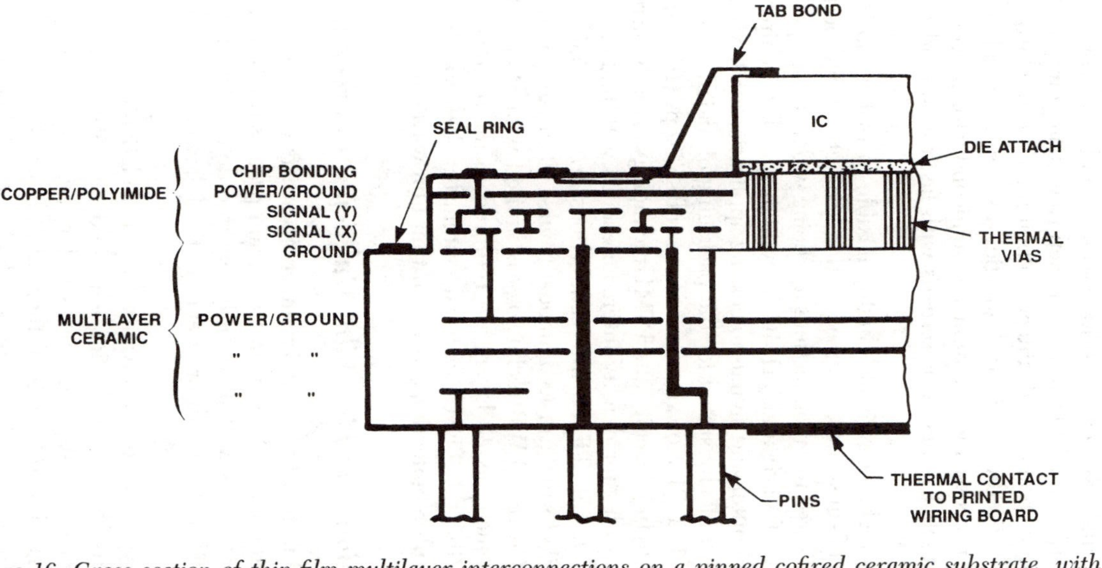

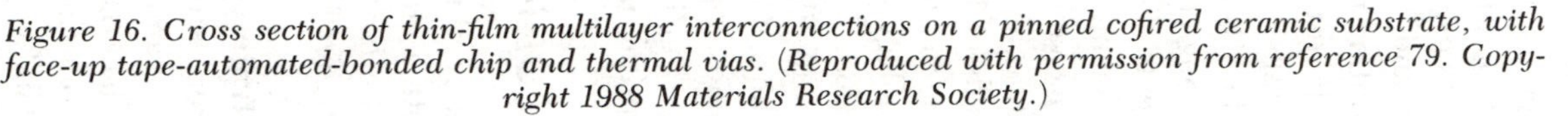

Figure 16. Cross section of thin-film multilayer interconnections on a pinned cofired ceramic substrate, with face-up tape-automated-bonded chip and thermal vias. (Reproduced with permission from reference 79. Copyright 1988 Materials Research Society.)

pastes used for thick films. Third, the polymer dielectrics used in TFML technology have dielectric constants that are significantly lower than those of either cofired ceramic or thick-film glass–ceramic dielectrics. The low dielectric constant results in low propagation delay, low interconnection capacitance (or high characteristic impedance), and low crosstalk, as discussed previously. Fourth, from a manufacturing standpoint, TFML technology offers production cost advantages (albeit higher capital expense) over thick-film and cofired-ceramic technologies by replacing labor-intensive screen-printing processes with automated semiconductor processes and by replacing hard-tooled punches or screens with fast-turnaround photolithographic masks.

The TFML technology is also highly flexible in that it can be applied to a wide range of substrates, interconnection geometries, and performance requirements. TFML interconnections are being widely developed for multichip packaging of VLSI and VHSIC (very-high-speed IC) chips for high-performance computer and signal-processing applications (*80*, *81*, *100*). The technology is also ideally suited for the packaging of digital GaAs ICs with clock speeds of 1–4 GHz or signal rise times of $<$150 ps (*101*). In military and commercial hybrid circuits, the TFML technology extends current thick-film technology to higher density interconnections with higher speed performance (*33*, *80*, *85*). TFML interconnections are potentially useful for higher density PWBs that can accommodate high-I/O surface-mounted SCPs or bare chips. Current PWB technology is limited with respect to wiring density, and as many as 40 layers are required to interconnect high-I/O chips (*32*). Finally, the TFML material system can be extended to high-density optical interconnections based on thin-film waveguides.

Material Options for Thin-Film Multilayer Interconnections

High-density, thin-film-based interconnections are a rapidly emerging technology with a number of different material and process options. These materials and processes, as well as recently reported demonstrations of TFML technology, are discussed in more detail in the following sections.

Substrates. One of the advantages of the relatively low-temperature TFML process is that it can be implemented on a variety of substrates, including ceramics, metals, and silicon wafers (*79*, *91*, *97*, *99*). The ideal substrate material should have high thermal conductivity for good heat dissipation from the ICs. It should have a high modulus of elasticity and a coefficient of thermal expansion (CTE) close to that of the polymer dielectric (typically 20–50 ppm/°C for polyimides) to minimize warpage due to thermal-expansion mismatch (*99*). However, a CTE close to that of silicon (2.3 ppm/°C) is also desirable for flip-chip bonding or attaching large die. The substrate should be chemically inert to the wet etchants and plasma processes used in fabrication. The substrate surface should be smooth, flat, and defect free. Lapping or polishing processes are often needed to achieve the required surface properties. Finally, the ideal substrate should be lightweight, inexpensive, and available in or machinable to a variety of shapes and sizes.

Ceramics (particularly alumina) have been widely used as interconnection substrates, because they are mechanically tough and stiff, chemically inert, and widely available for thick- and thin-film hybrid substrates. Multilayer ceramic has the additional advantages of internal metal layers for power and ground distribution and pin grid arrays or peripheral leads for off-package connections. The biggest drawback to cofired ceramic is the difficulty in achieving the flat, defect-free surface required for thin-film processing. Alternative ceramics such as BeO, AlN, and SiC have thermal conductivities that are significantly higher than that of Al_2O_3 and are currently being developed actively. SiC and AlN also have low CTEs, which are closer to that of silicon. BeO and AlN can be fabricated into multilayer structures, although this process is still being developed for AlN.

A wide variety of metal substrates can be used for TFML interconnections. Metal substrates (and their advantages) that have been investigated include aluminum (light weight, low cost, machinability, and high thermal conductivity); copper (high thermal conductivity and low cost); molybdenum, tungsten, or copper–tungsten (high modulus and low CTE); copper-clad molybdenum or Invar (adjustable CTE); and steel (low cost and machinability).

Silicon is a widely investigated substrate because it has a relatively high thermal conductivity and a perfect CTE match to silicon die. In addition, silicon wafers have a high-quality surface finish and are widely available in standard sizes that are adaptable to IC process equipment. However, silicon has a low modulus and a poor CTE match to polyimides, and thus, very thick substrates will be required to prevent excessive warpage. A final advantage of silicon substrates is that active devices such as driver circuits may be fabricated in the substrate.

Conductors. Copper, gold, and aluminum have been the primary conductor materials used in TFML interconnections. Gold has the advantages of chemical inertness and high electrical conductivity, but gold is expensive and has poor adhesion to polymer dielectrics. Aluminum has good adhesion to polyimides and an inert native oxide and is well understood because of its predominance in IC processing. However, the electrical conductivity of aluminum is significantly lower than that of copper.

Copper has been the most widely used material because of its high conductivity, solderability, low cost, and ability to be electrolytically or chemically plated. However, copper has a weak interaction with polyimides (*102–104*) and, consequently, poor adhesion. Furthermore, recent studies have shown that the polyimide precursor, polyamic acid, oxidizes copper and that the oxides migrate several micrometers into the polyimide (*105*). For these reasons, thin layers (20–100 nm) of interface metals such as Cr, Ti, TiW, Mo, or Ni, which have a strong interaction with polyimides (*102–104, 106, 107*), are usually used as an adhesion layer between copper and a polyimide. Cr, the most extensively studied and widely used interface metal, has excellent adhesion to polyimides but is difficult to pattern. Ti or TiW is easier to pattern but has slightly poorer adhesion compared with Cr.

The top metal layer in TFML structures must be compatible with bonding and assembly processes. Gold, with an underlying barrier metal such as TiW or Ni, is usually patterned over the top metal layer for oxidation protection and wire-bonding compatibility. For solder-bonding processes, a very thin layer (<100 nm) of gold is deposited over a solderable metal such as nickel.

Dielectric Materials. The greatest variety of material options exists for the polymer dielectric layers. Polyimides have been the predominant dielectric material and have been extensively studied and described in the literature (*80, 108–114*). Polyimides encompass a broad class of polymers characterized by a phthalimide ring structure (nitrogen bonded to two carbonyls that are bonded to the *ortho* positions of a benzene ring).

The reaction sequence for the synthesis of polyimides is shown in Scheme I. The imide ring is formed via a thermally initiated condensation reaction from a polyamic acid, a soluble precursor polymer synthesized from

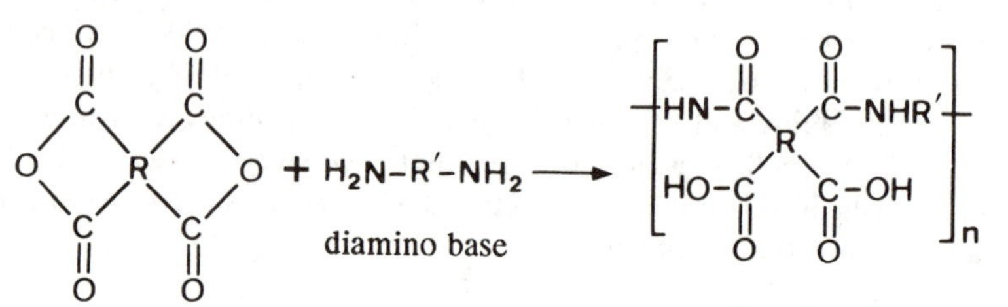

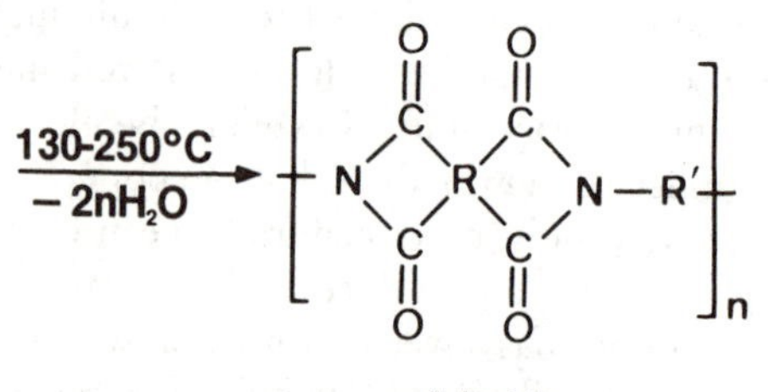

Scheme I. Reaction sequence for the synthesis of polyimides.

an acid dianhydride and a diamine, each of which contains various aromatic substituent groups. The highly aromatic structure gives polyimides excellent thermal stability and radiation resistance. One of the most common polyimides for TFML applications is formed from benzophenone tetracarboxylic acid dianhydride (BTDA) and oxydianiline (ODA) plus *m*-phenylenediamine (MPD) (*see* structure) (*113*).

Polyimides such as BTDA–ODA–MPD possess a combination of physical properties and processing characteristics that make them uniquely suited as a dielectric material for TFML interconnections. The precursor solutions, a polyamic acid or a soluble polyimide, can be deposited by a variety of techniques to produce a wide range of film thicknesses. The ability of the solutions to flow before imidization promotes the planarization of the underlying substrate or conductor topography. After complete imidization at 300–420 °C, polyimides are thermally stable (decomposition temperature of >450 °C) and chemically inert and, thus, resistant to degradation during subsequent processing or assembly steps such as metal deposition, etching, photolithography, or soldering.

Polyimides have a relatively high tensile strength (100–200 MPa) and a large ultimate strain (20–30%). These properties make the films resistant to cracking despite the large stresses created by the large mismatch between the CTE of polyimides and those of most substrates. Finally, polyimides have desirable dielectric properties for high-speed signal propagation (as discussed previously), specifically, a low dielectric constant (typically 3.5 at 50% relative humidity) and a low dissipation factor (typically 0.002–0.007).

Although polyimides such as BTDA–OPA–MPD have many excellent properties, a number of improvements to these materials are desirable. The most significant drawbacks of a polyimide are its high absorption of moisture (typically 3 wt %) and the large effect of absorbed water on its dielectric constant (ϵ_r varies from 3.1 to 4.1 over 0–100% relative humidity [*78, 115*]). The other major disadvantage is the large mismatch between the CTE of polyimides (20–50 ppm/°C) and those of substrates. This mismatch causes high tensile stress in the polyimide and warping of substrates. Recent development efforts for polyimides have been directed toward reducing moisture absorption, CTE, and dielectric constant; increasing mechanical strength and toughness; improving the planarizing capabilities, self-adhesion, and adhesion to substrates; and increasing the shelf life.

A wide variety of new polyimides with improved characteristics have been introduced recently as products. Significant examples are the rigid-

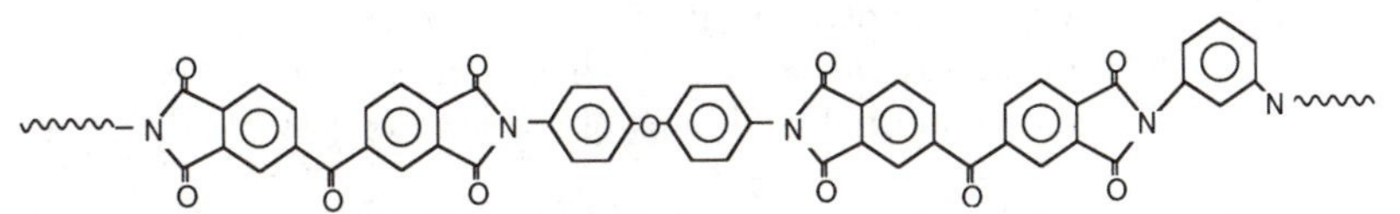

A typical polyimide (BDTA–ODA–MPD) for thin-film multilayer interconnections

rod polyimides with very low CTEs (~2–3 ppm/°C) (*113, 114*). Compared with BTDA–ODA–MPD, these polyimides also have lower moisture absorption, higher tensile strength, a higher glass transition temperature (T_g), and a lower dielectric constant. However, a significant problem with self-adhesion exists for these high-T_g polyimides. Preimidized polyimides that contain fluorocarbon or siloxane groups in the backbone to make them soluble in highly polar solvents have also been developed and have the advantage of longer shelf life and lower curing temperature, because the curing involves solvent evaporation only.

Alternative polymers that have certain advantages over polyimides have also been introduced; they include poly(phenylquinoxaline), poly(phenylquinoline), and poly(benzocyclobutenes) (PBCBs) (*93, 116*). The PBCBs have a low curing temperature (250 °C), low dielectric constant (2.6), low dissipation factor (0.0045), and low moisture absorption (0.3%) The development of specialty polymers for packaging and high-density interconnections will continue to be an active area of research as polymer manufacturers focus on the needs of the microelectronic industry.

Processing of Thin-Film Multilayers

Unique Process Requirements. The fabrication of TFML interconnections involves a repetitive sequence of thin-film processes to deposit and pattern conductor and dielectric layers. Many processes used in IC fabrication, such as vacuum deposition of metals, photolithography, wet and dry etching, and newly emerging processes (such as laser etching and deposition), may be used in the fabrication of TFML interconnections. However, the geometries and substrates required for packaging impose a number of unique requirements on conventional thin-film processes.

Feature Sizes. Although minimum feature sizes in TFML interconnections are large relative to IC feature sizes (i.e., 25-μm versus 1-μm line widths), the conductor and dielectric layers are substantially thicker in TFML structures, a fact that results in high aspect ratios. Conductor layers must be several micrometers thick to keep resistive losses low, and dielectric layers must be 10 to 30 μm thick to maintain low interconnection capacitance. Thus a thickness:width aspect ratio as large as 1:1 is frequently required. This aspect ratio demands anisotropic-etching processes.

High-aspect-ratio features create large topographies that must be planarized when several layers are stacked. The thicker films also require long processing times for dry deposition and etching processes. Finally, the strain energy due to thermal-expansion mismatch increases with increasing film thickness and creates potential problems of cracking or loss of adhesion.

Substrates. The substrates used for thin-film interconnections also present some unique problems. A variety of substrate materials, sizes, and shapes may be used for a wide range of package applications. These substrates require process equipment with more flexibility than most IC equipment, which is designed for automatic handling of a single-size wafer. Many thin-film substrates, particularly ceramics, have greater surface roughness and more camber than do silicon wafers. These characteristics limit photolithographic resolution and minimum feature size. The large shrinkage tolerance in cofired ceramic substrates also creates a problem with pattern registration between the thin-film layers and the cofired ceramic. Finally, substrates may contain structures such as seal rings or pins that must be protected during processing.

Yield. Perhaps the greatest challenge in the processing of high-density interconnections arises from the large substrate area and its effect on yield. The simplest model for describing the effect of randomly distributed defects on process yield assumes a Poisson distribution for defects on the surface, that is, defects are not clustered. The yield (Y), which is the probability of having no fault-producing defects, is given by (*117, 118*)

$$Y = \exp(-A_c D) \tag{23}$$

in which A_c is the critical area in which a defect will cause a fault and D is the area density of defects larger than a given minimum size. In many clean-room environments, D is inversely proportional to the square of particle size, and thus for faults caused by particulates (*21, 36*),

$$Y = \exp[-k(L/x_c)^2] \tag{24}$$

in which L is the side length of a square substrate, x_c is the minimum critical defect size, and k is a proportionality constant. For conductor line faults (e.g., open and shorted lines), x_c is related to the minimum feature size of the conductor pattern, whereas for pinholes creating electrical shorts between layers, x_c can be much smaller. Because the yield depends exponentially on $(L/x)^2$, the yield for patterning 10-μm lines on a 10 by 10 cm substrate is equivalent to patterning 1-μm lines on a 1-cm chip. This patterning approaches the limits of current IC production technology.

Another way of viewing the yield problem is to consider that a silicon wafer may be diced into a number of chips, and only a fraction of these chips needs to be functional, whereas a TFML substrate that is comparable in size to a wafer must be fault free. Because of this severe yield constraint, defect detection and repair techniques may be necessary to obtain acceptable yields for TFML interconnections. Defects may be detected by using probe cards, high-speed capacitance or resistance probes (*119*), automated visual inspection (*120*), or voltage-contrast scanning electron microscopy (*21*). Methods such as laser etching and deposition (*21, 121–129*), ink-jet printing (*130*), or wire bonding may be used to selectively repair conductor faults.

Thin-Film Multilayer Processes. ***Subtractive Approach.*** Conductor patterns in TFML structures may be defined by either subtractive (*78, 80, 87*) or additive (*85, 88, 96, 131*) processes. In the subtractive approach (Figure 17), a blanket coating of metal is deposited on the substrate, and a positive image of the desired pattern is defined in a photoresist layer on the metal by using photolithographic techniques similar to those used in IC processing. Unwanted metal is then removed by wet or dry etching. The conductor pattern is coated with a polyimide to the required thickness. This step planarizes the underlying conductor topography. The via holes that connect the layers are etched in the polyimide film and then filled by depositing a conformal metal film. The via metallization and subsequent conductor layer are then subtractively patterned by using a single photolithographic and etching step.

Additive Approach. In the additive approach (Figure 18), a negative image of the conductor pattern is defined in a photoresist layer that must be thicker than the conductor layer. Metal is then selectively deposited in the open areas of the resist by electroplating or by a lift-off technique. For

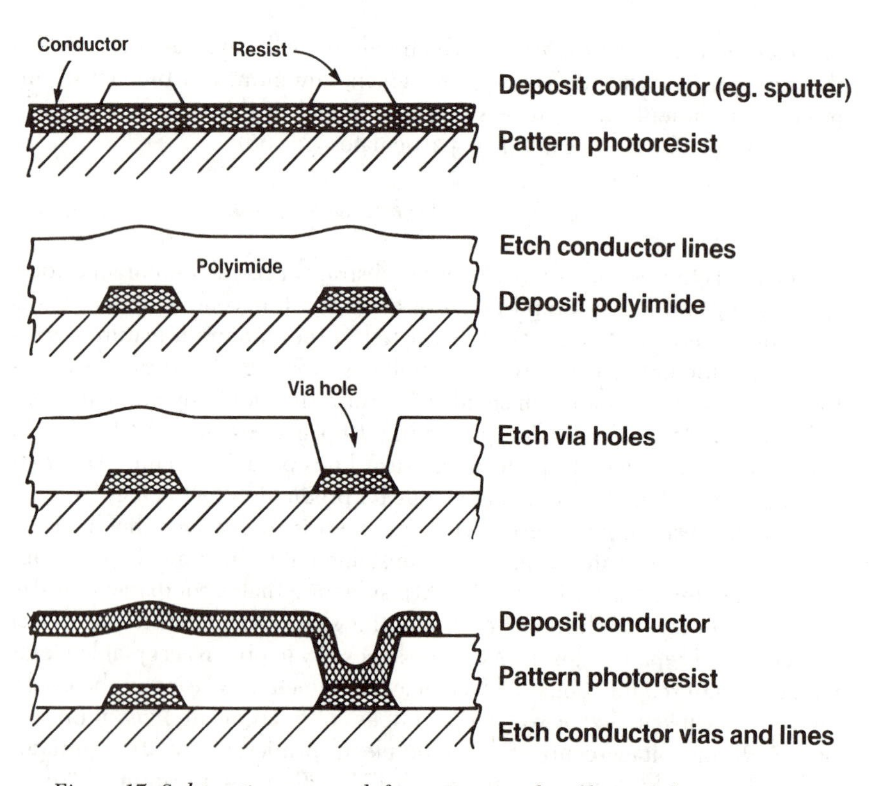

Figure 17. Subtractive approach for processing thin-film multilayer structures.

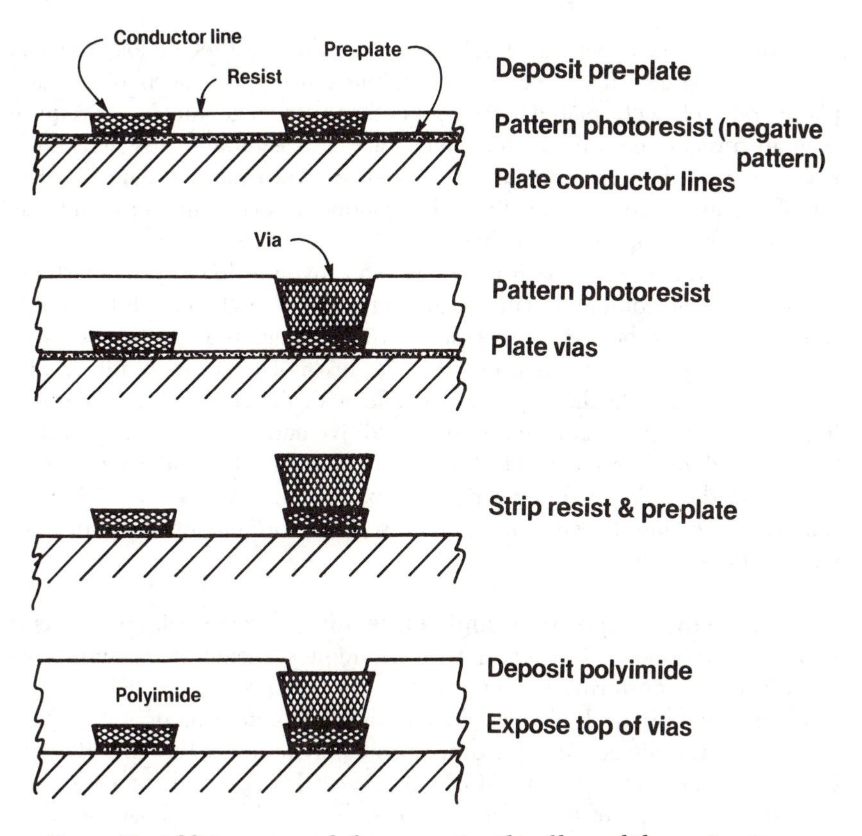

Figure 18. Additive approach for processing thin-film multilayer structures.

electroplating, a thin plating base must be deposited beneath the photoresist to provide electrical contact to all of the conductor features. Both the photoresist and plating base must be stripped after plating.

In the lift-off process, a blanket metal coating is deposited, usually by evaporation, over the photoresist, which is then dissolved to lift off the unwanted metal and leave the desired pattern. The lift-off process may be assisted by depositing and patterning a dielectric layer, a release layer, or both beneath the photoresist (*131, 132*). In both additive approaches, via posts are patterned in a step separate from that used to pattern the conductor lines. The polyimide is then coated over the lines and via posts, and shallow etching or mechanical polishing is done to expose the top of the via posts. The process sequence is then repeated to pattern additional layers.

Comparison of Approaches. The additive processes can achieve conductor features with a large aspect ratio, although selective plating can create a negative side wall angle in the conductor that reduces the spacing between conductor lines and that is is difficult to coat with polymer dielectrics. Electroplating has the advantage of being a mature, low-cost technology that can be used on large substrates. However, the uniformity, morphology, and plating rates of electroplated metals depend on pattern geometry and require accurate process control. Plated metals are also generally less dense and have lower conductivity compared with sputtered or evaporated metals. The lift-off processes are generally limited to conductor layers thinner than those required for package interconnections.

The subtractive approach is simpler and involves fewer process steps, because vias and conductor patterns are defined in a single photolithographic step. However, subtractive processing requires staggered vias through several metallization layers, because the vias are not completely filled. The additive approach has the important advantage of allowing the vertical stacking of vias through several layers. Both additive and subtractive approaches have been demonstrated in TFML interconnections. The following section focuses on the individual deposition and patterning processes used in subtractive processing, because these processes are more universal and present more options.

Conductor Deposition and Patterning Processes. Conductor films may be deposited by a variety of techniques, including vacuum processes (such as sputtering and evaporation) or wet processes (such as electroplating or chemical plating). When the conductor material has poor adhesion to the dielectric (e.g., copper on a polyimide), a thin (20–100 nm) layer of a metal such as Cr, Ti, Mo, or Ni is usually deposited as an adhesion layer on either side of the conductor. The most widely used vacuum deposition process is rf (radio frequency) or dc sputtering, which can be used to deposit any metal (*133*). Sputtering produces conformal films with good adhesion to substrates and provides excellent control of film thickness, stress, and morphology.

Alternative vacuum deposition processes that have recently been developed offer potential advantages such as higher deposition rates, improved adhesion, or better control of film stress or morphology. These processes include ion-beam sputtering (*134*), ion-cluster evaporation (*135*, *136*), and cathodic-arc deposition (*137*).

Possibly the most important development in metal deposition processes is the direct writing of conductor lines by laser-activated processes, including (1) laser decomposition of liquid or gaseous organometallic compounds (*122*, *123*), (2) laser-activated deposition of catalytic metals such as Pd, followed by selective chemical plating (*124*), and (3) laser-jet electroplating (*125*). With these processes, selective repair of conductor lines or maskless, programmable writing of conductor patterns is possible, with significantly enhanced yields compared with subtractive or additive processes based on photolithographic techniques.

A variety of etching processes are available for conductor patterning.

Wet etching is an inexpensive, mature technology that can achieve high etch rates with good selectivity for specific metals. However, because the chemical reactions are isotropic, wet etching undercuts the photoresist and produces a sloped side wall and is therefore limited to conductor features with an aspect ratio of less than ~0.5. Higher aspect ratios can be achieved by using vacuum processes such as ion-beam etching, reactive-ion-beam etching (RIBE), or reactive-ion etching (RIE). These processes etch anisotropically through the action of charged species that are accelerated perpendicular to the substrate. .rp Ion-beam etching uses an inert gas (e.g., Ar) and can etch any material, although etching is usually at a slow rate and with poor selectivity over other materials. Thus, a thick photoresist mask and an effective etch stop are required. The etch rate and selectivity are improved by using a reactive gas (e.g., Cl_2) in RIBE and RIE; however, these processes are more complex, the range of etchable materials is more limited, and high electrode temperatures may be required to volatilize the reaction products, as in the RIE of copper (*138*).

A final alternative for metal patterning is the use of lasers to selectively and accurately etch small metal areas. Lasers are routinely used to trim thin-film resistors (*126*); they may also be used to repair defects or etch conductor lines by direct ablation in an inert atmosphere (*127, 129*) or by laser-initiated etching in a reactive gas or liquid (*128*).

Polyimide Deposition and Patterning Processes. ***Deposition Processes.*** A variety of processes may be used to deposit films of polyimide or other soluble polymers. A polyimide is normally obtained as a solution of the polyamic acid or the fully imidized polyimide in a strong solvent such as *N*-methylpyrrolidone (NMP). These solutions can be deposited by using techniques such as spin or spray coating, screening, dipping, or roll coating to produce a wide range of film thicknesses (1–100 μm). Some planarization of conductor topography is achieved with each coating, and multiple coats are often deposited to improve planarization and to achieve the thick dielectric layers required for high-impedance interconnections.

After deposition, the films are cured by heating at a controlled rate in a convection oven, hot plate, or tube furnace to evaporate solvents and reaction products (primarily H_2O) and to convert the polyamic acid to the polyimide. The final properties and adhesion of polyimide films depend on the curing rate and final curing temperature, which can range from 300 to 450 °C.

The most accurate processes for depositing polyimide solutions are spin coating and spray coating. Spin coating is a well-characterized process that is used frequently to deposit photoresists in IC processing. The thickness of spin-coated films depends on the solution viscosity and solution concentration and can be varied over a wide range by varying the dispense volume and the time and angular speed of spinning (*78, 139*). Spin coating generally produces a coating of uniform thickness; however, it is limited to square or round substrates that are a few inches in diameter and with no large surface topography.

Spray coating is not subject to these limitations. Polyimides can be spray coated on large, odd-shaped substrates in commercial conveyorized systems that provide high throughput and precise control of spray parameters such as the solution flow rate, atomization pressure, nozzle diameter and distance to the substrate, and conveyor speed (*80*). The diluting solvent, solution viscosity, and solution concentration are critical to the uniformity of the sprayed films. Because of the large number of process variables, statistically designed experiments are essential in developing an optimized spray-coating process.

A recently reported alternative to spin or spray coating is screen printing of polyimide solutions (*82, 85, 90*). Screen printing is a low-cost, high-throughput process capable of directly patterning the polyimide films as they are deposited. Another alternative is the vapor deposition of polyimides, which was reported by researchers who co-evaporated the diamine and dianhydride monomers at stoichiometric rates (*140*). The evaporated films had better adhesion, a lower dielectric constant, and a lower dissipation factor compared with spin-coated polyimides. With this process, uniform, defect-free, conformal films can be cured in situ during deposition.

Patterning Processes. Polyimide films may be patterned by a variety of processes, including wet or dry etching through a photolithographically defined mask, direct photopatterning of photosensitive polyimides, or laser ablation. The process steps involved in wet etching, dry etching, and direct photopatterning are shown in Figure 19.

Wet Processes. Wet etching (*141, 142*) is a simple, inexpensive process, but like most wet-patterning processes, it is limited to low-resolution, low-aspect-ratio features. Most wet etchants can dissolve only partially cured polyimide films, and the difficulty in controlling the degree of partial curing limits the reproducibility of etch rates and pattern geometries. Furthermore, the additional shrinkage of the films during the final cure after wet etching may cause further loss of resolution or localized cracking because of the stress that is concentrated at small-radius patterned features.

Dry Processes. Many of the limitations of wet etching are overcome by using dry processes, such as plasma etching or RIE, to etch the polyimides. With plasma etching or RIE, the substrates are placed on the lower electrode in a parallel-plate system sustaining an rf plasma. The polyimide is usually etched in a mixture of oxygen and a fluorine-containing gas such as CF_4 or SF_6 (*78, 143, 144*). Because a photoresist etches at nearly the same rate as a polyimide, the RIE of thick films requires that a masking

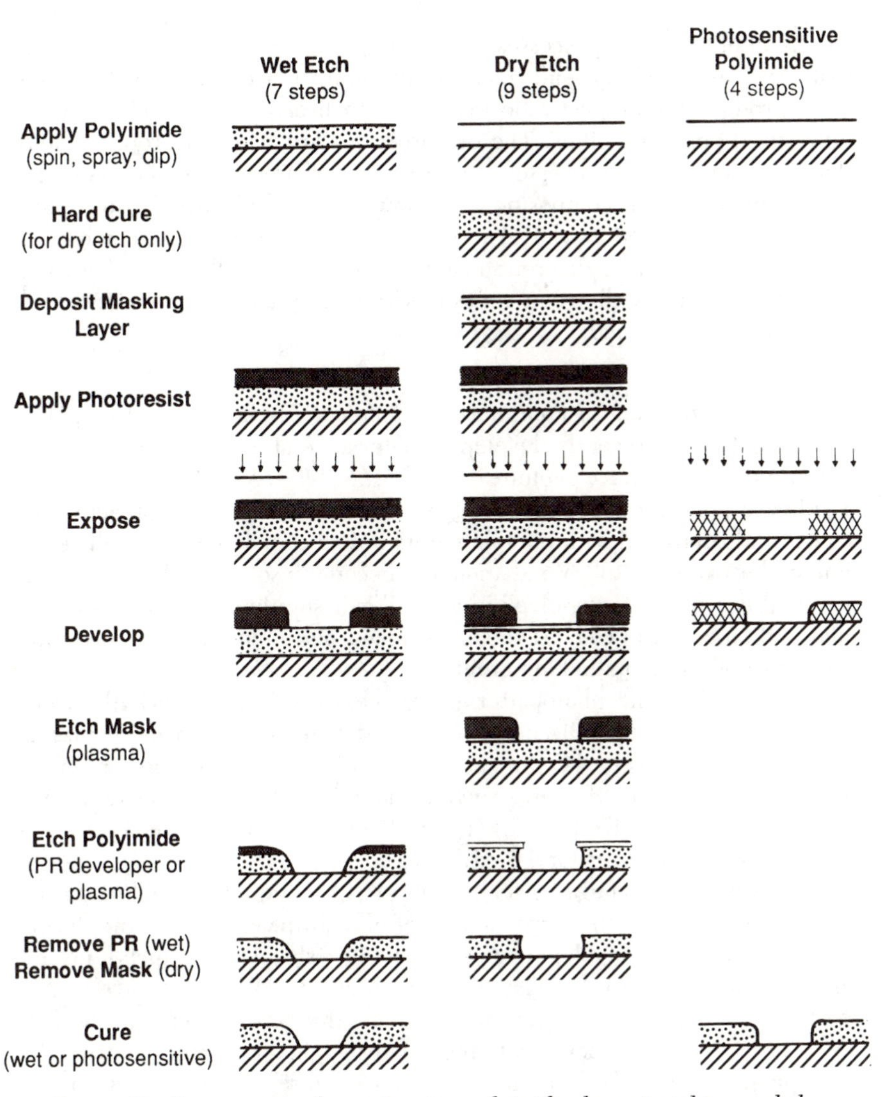

Figure 19. Process steps for patterning polyimides by wet etching and dry etching and for direct photopatterning of a photosensitive polyimide. The abbreviation PR stands for photoresist.

layer of a slow-etching material, such as a metal, silicon nitride, or silicon oxide, be deposited and patterned on top of the polyimide, a procedure that is similar to a trilevel resist process. Alternative dry etch processes include reactive-ion-beam etching (RIBE) or ion-beam-assisted etching, in which a beam of reactive ions (e.g., O^+) or inert ions (e.g., Ar^+) is accelerated toward the substrates, which may be flooded with a reactive gas such as O_2 (*145*).

In all of these dry processes, the etching reactions are initiated by ions that are accelerated perpendicularly to the film surface, and thus dry processes produce high-aspect-ratio features with nearly vertical side walls in fully-cured polyimide films. The dry processes are also more reproducible than the wet-etching processes, and a variety of process parameters (e.g., reactor pressure, gas composition and flow rate, and rf power or ion-beam energy) can be varied to control accurately such characteristics as etch rate, selectivity for etching different materials, and side wall angle of etched features. The side walls of via holes are frequently tapered to ensure good metal step coverage (*80*).

Direct Patterning of Photosensitive Polyimides. Photosensitive polyimides (PSPIs) are recently developed materials that can be directly photopatterned like a negative photoresist (*80, 85, 88, 146–148*). The most common PSPIs are polyamic acids that have been esterified with photoreactive alcohols and combined with photoinitiators to form a polymer that will cross-link under exposure to UV radiation and become insoluble. The unexposed material is selectively dissolved in a developer solution, and the patterned film is then cured to convert the cross-linked polyamic acid to a polyimide and drive off the cross-linking groups.

With PSPIs, the photopatterning process involves significantly fewer process steps (Figure 19) and eliminates the need for expensive plasma equipment. However, the process is currently limited to aspect ratios of about 0.5 for negative features such as via holes. Higher aspect ratios can be obtained for positive features such as lines; unfortunately, holes are the more-prevalent features for TFML structures.

The resolution of negative features in a PSPI is limited by swelling during development and by the large amount of film shrinkage that occurs during the final cure after development (typically 50% shrinkage). Thick films are also difficult to pattern because of the high absorption of UV wavelengths by PSPIs. High UV absorption limits the depth of cross-linking in the film (*88, 148*). During development, the uncross-linked polymer beneath the cross-linked surface layer is dissolved and results in an undercut and an overhanging side wall profile. This problem may be overcome by repeated coating and development of thin films (*80*) or by filtering the 365-nm light (*145*). Improved PSPIs are continually being developed by several manufacturers and are a promising alternative to wet or dry etching of polyimides.

Direct Patterning by Laser Ablation. A promising future technique for patterning polyimide films is direct patterning by laser ablation. These processes are in the initial stages of development at a number of laboratories (*149–152*). A polyimide can be thermally or photochemically decomposed by a variety of lasers and wavelengths, including CO_2, Nd–YAG, Ar^+, or excimer lasers. Pulsed excimer lasers operating at wavelengths of 193–351 nm have produced the best results. These lasers cleanly ablate polyimides without producing by-products of thermal decomposition.

Most of the initial work on excimer laser etching has been concerned with reaction mechanisms, particularly the relative roles of thermal and photochemical dissociations, rather than the practical problems of masking, optics, alignment, and control of feature geometries. In a few studies, fine features have been patterned in polyimide films by exposing a large-area laser beam through a contact mask or by using a focused beam. Laser ablation offers exciting possibilities for maskless, computer-controlled direct writing of via holes or other features in polyimide films, with significant improvements in prototype turnaround time and yield enhancement through elimination of the masks, process steps, and defects associated with photolithographic processes.

Demonstrations of Thin-Film Multilayer Interconnections

Test Vehicles and Prototype Packages. High-density thin-film interconnection technologies are being actively developed at a number of companies and research laboratories for a variety of applications. Several groups have described their process approaches in some detail (*78*, *80*, *85*, *88*, *93*, *96*). In addition, several demonstrations of TFML technology, including test vehicles and functional multichip packages, have been described in the literature.

Test vehicles have been used to determine the process capabilities, electrical and thermal performance, and reliability of TFML interconnections (*78*, *79*, *85*, *88*, *91*, *95*, *98*, *100*, *153*). Examples of test vehicles include (1) simple stripline and microstrip structures to characterize the transmission line properties of TFML interconnections, (2) multichip ring oscillator circuits to determine the propagation delay and dynamic behavior of TFML interconnections (*78*, *95*, *153*), (3) thermal test vehicles to determine the thermal impedance of heat conduction through TFML layers (*98*, *100*) and through the substrate and package (*91*), and (4) a multichip test vehicle to test output driver circuits in submicrometer bipolar ICs (*79*, *100*). The latter test vehicle, which was patterned on a 3 by 3 in. (7.6 by 7.6 cm) pinned cofired ceramic substrate with five metal layers of TFML interconnections, is shown in Figure 20.

Functional prototype packages that demonstrate the feasibility and performance advantages of TFML interconnections have been described. These prototypes illustrate the broad range of applications of high-density TFML interconnections.

The top of Figure 21 shows a microprocessor module developed by Honeywell for computer applications. The module contains nine bipolar gate array ICs, with 174 I/Os per chip that are bonded by TAB to an 80 by 80 mm substrate with 420 off-package connections around the edge of the substrate (*80*). One of the five metal layers of this package, with 37-μm-wide

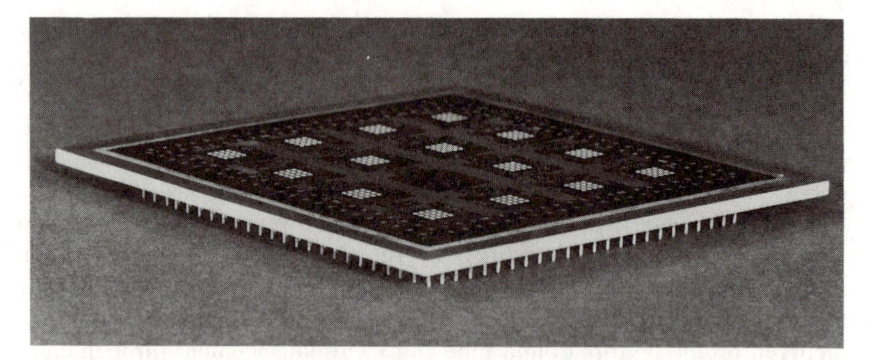

Figure 20. Multichip test vehicle for submicrometer bipolar ICs. The vehicle contains five metal layers of copper–polyimide interconnections on a pinned cofired ceramic substrate (3 by 3 in. [7.6 by 7.6 cm]) (Reproduced with permission from reference 79. Copyright 1988 Materials Research Society.)

lines on a 250-μm pitch (which is half the typical routing density) is shown in the bottom of Figure 21.

Another reported prototype reported by Honeywell is an 18-chip hybrid module for image-processing applications (Figure 22) (*33, 80*). This package contains two 8000-gate CMOS gate arrays and 12 static RAM (random-access-memory) chips on a 2.25 by 2.25 in. (5.1 by 5.1 cm) TFML substrate mounted in a hermetically sealable flatpack. Honeywell has also demonstrated TFML packaging for process control electronics (*99*), as well as a multichip memory module for supercomputer applications and multichip memory and logic modules for space-borne computers. A single-chip package for digital GaAs ICs with 200 I/Os and designed to operate at data rates of up to 3 gigabits per second is currently being developed (R. Jensen, unpublished). This package contains thin-film TaN resistors patterned very close to the chip to permit the termination of high-speed signal lines and thus prevent signal reflections and ringing.

Other functional TFML packages that have been described in the literature include (1) TFML interconnections that are being used by Mitsubishi in hybrid multichip circuits to increase the interconnection density and reduce costs compared with current thick-film technology (*85*), (2) a two-layer Cu–polyimide interconnection approach for digital cipher electronics in a mobile radio communication system (*83*), and (3) a 4-kilobyte ECL (emitter-coupled logic) RAM module operating at 100 MHz (developed by Hewlett–Packard), with four layers of interconnection on a Si substrate (*92*).

Novel Approaches. A number of groups have focused on silicon substrates for thin-film interconnections (*89, 91–93, 97*). The most advanced technology of this type is AT&T's advanced VLSI package (AVP), which uses flip-chip solder bonding of ICs onto a silicon substrate containing high-

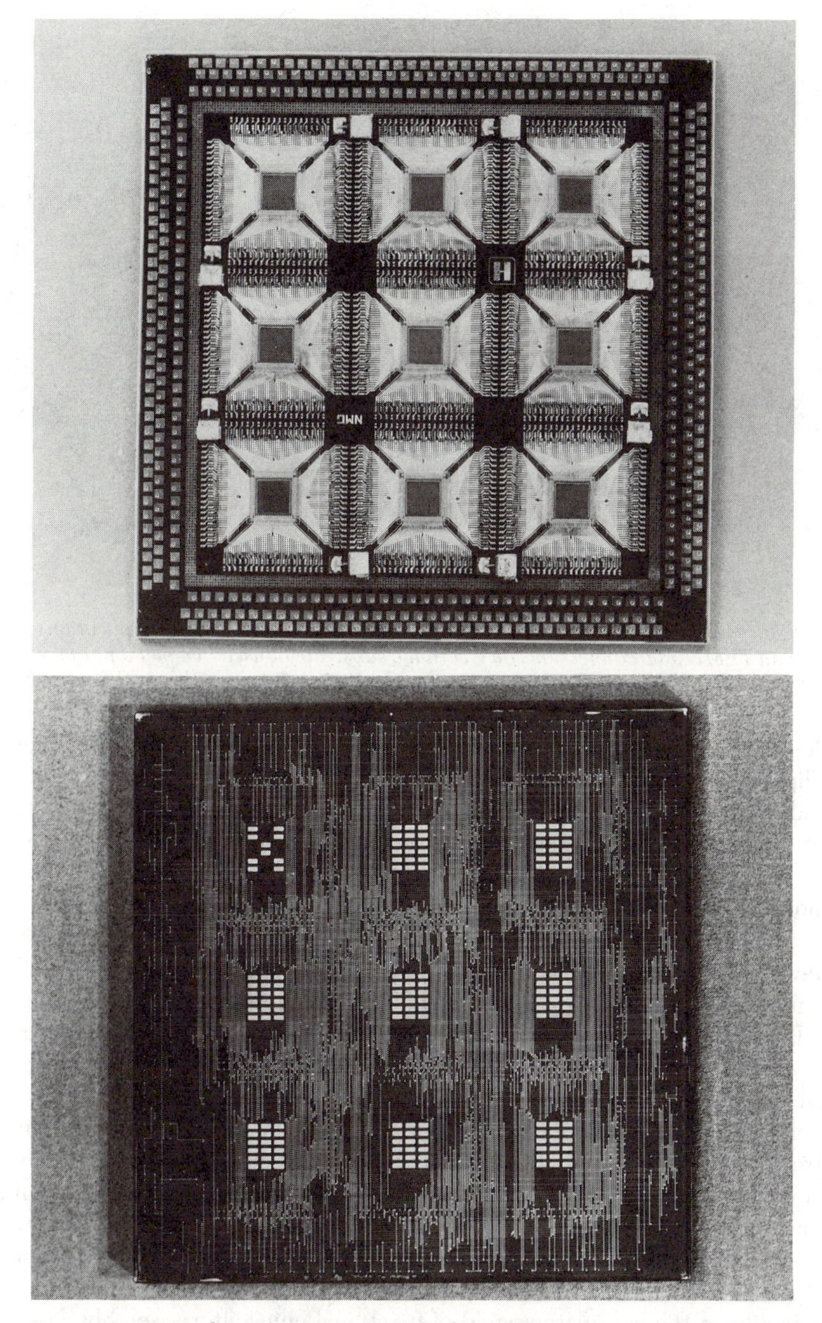

Figure 21. Nine-chip microprocessor module with TFML copper–polyimide interconnections on an 80 by 80 mm ceramic substrate. Top, completed package populated with tape-automated-bonded ICs; bottom, internal signal interconnection layers. (Reproduced from reference 80. Copyright 1987 American Chemical Society.)

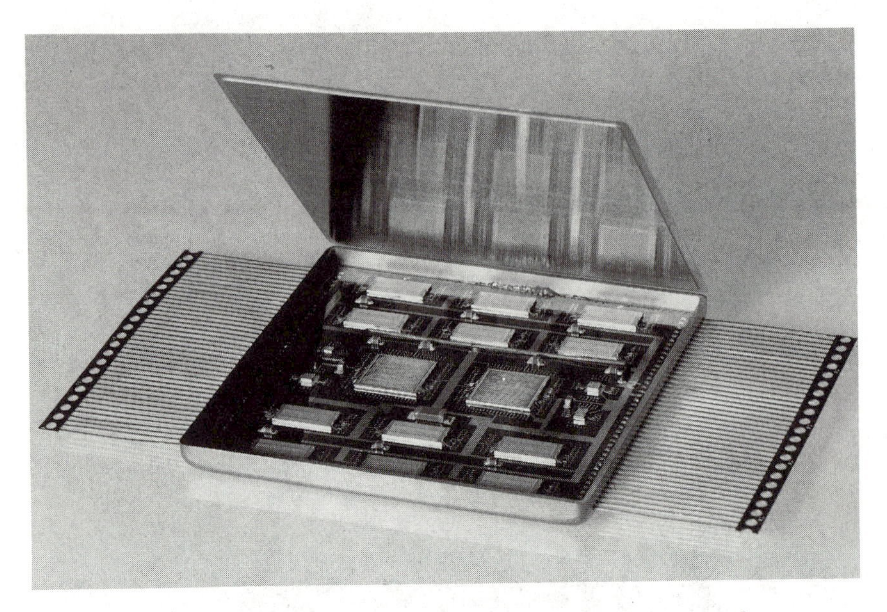

Figure 22. Multichip package for electrooptical-signal-processing applications, with TFML interconnections on a ceramic substrate housed in a metal flatpack. (Reproduced from reference 80. Copyright 1987 American Chemical Society.)

density (20–μm signal line pitch) Cu–polyimide interconnections (*97, 98*). Flip-chip bonding permits very close spacing of the chips, and large power and ground solder bumps are used for heat transfer to the substrate. The AVP has been demonstrated in a three-chip processor circuit operating at 14 MHz; the chips were interconnected on a substrate (1.3 by 3.0 cm) mounted in a 160-I/O PGA package.

In one of the most novel approaches to high-density interconnections developed by General Electric (*94, 95*), chips are bonded to a rigid, thermally conductive substrate, and an overlay of Kapton, a flexible polyimide sheet, is laminated onto the top surface of the chips. Contact to the chips is made by etching vias by laser through the Kapton, and the TFML structure is then patterned on the Kapton overlay, again by using maskless laser-patterning techniques. This approach permits a very high chip packing density, excellent thermal dissipation, and rapid turnaround of prototype circuits through the use of maskless, laser-patterning processes. However, some concerns have been expressed regarding the reliability and temperature limitation of the laminating adhesive, the accessibility to inspection and repair of the interface to the chips, and the functional package yield, which depends on the cumulative yield of functional chips after assembly.

The most advanced application of TFML technology has been reported by NEC for their SX-1 and SX-2 supercomputers that are currently in production (*81*). The logic module for the computer contains 36 ceramic chip carriers on a TFML-personalized pinned cofired ceramic substrate. An unpopulated substrate and a cross section of the chip carrier and substrate are shown in Figure 23. A 1000-gate CML logic chip or four RAM chips are bonded by TAB inside each "flip-TAB" chip carrier, and the carriers are then soldered face down to the multichip substrate through an area array of metallic bumps. The 100 by 100 mm multichip substrate contains five metal layers of TFML interconnections, with 25-μm-wide signal lines on a 75-μm pitch sandwiched between meshed ground planes in an offset stripline con-

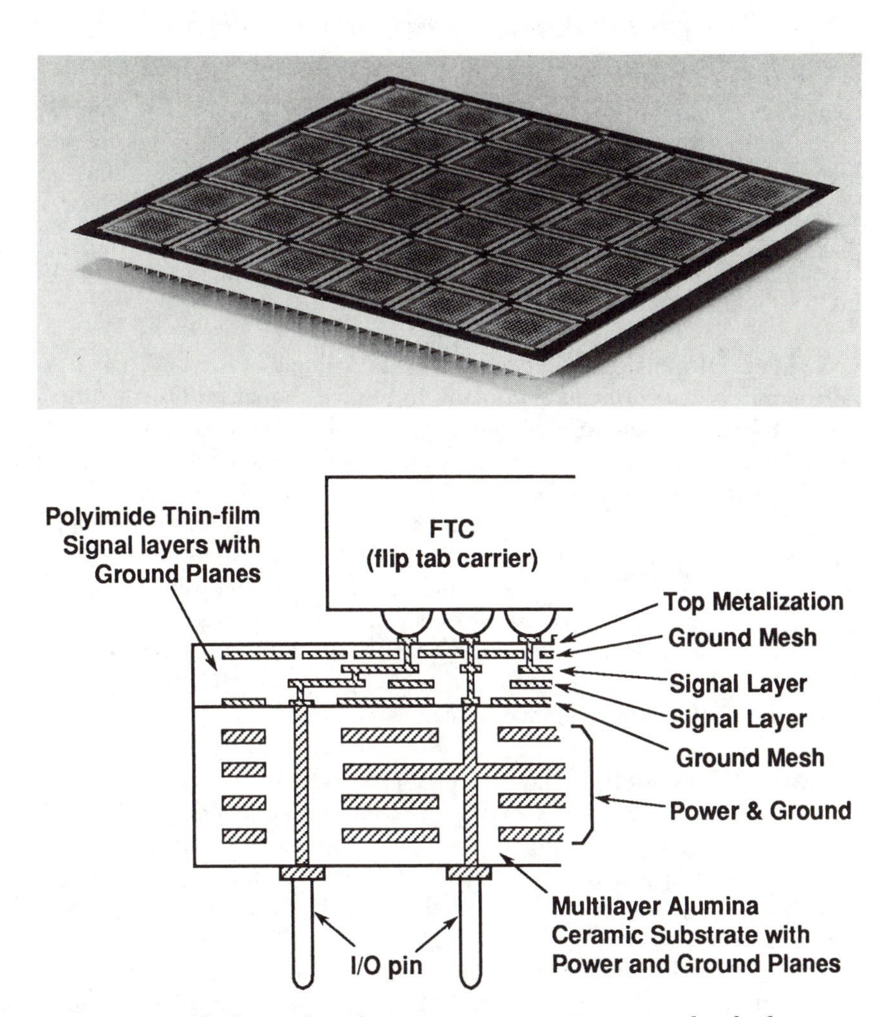

Figure 23. Multichip package for a supercomputer. Top, unpopulated substrate with TFML copper–polyimide interconnections on a 100 by 100 mm multilayer ceramic substrate; bottom, cross section of interconnection structure and flip-TAB carrier. (Reproduced with permission from reference 81. Copyright 1985 Institute of Electrical and Electronics Engineers.)

figuration. The cofired ceramic substrate contains internal metal planes for power and ground distribution, and a grid of 2177 I/O pins on 0.100-in. (0.254-cm) centers for connection to a PWB. The module is liquid cooled, with heat conducted through pistons contacting the back side of the chip carriers, similar to the thermal conduction module. The multichip package permits the system to achieve a performance of 1300 megaflops (million floating point operations per second) with a machine cycle time of 6 ns.

Future Packaging and Interconnection Technologies

Novel packaging and interconnection technologies will be essential for continued advancements in the performance of IC-based systems. For example, many advanced processors use highly parallel architectures requiring very high interconnection density and fan-out. Ultimately, these systems approach the limit of neural networks, which are highly connected networks of simple neuronlike processors in which the interconnections play an integral role in the processing function (*154*). A few examples of newly emerging packaging and interconnection concepts are discussed in this section.

Three-Dimensional Packaging. The volume of most advanced supercomputers is continually shrinking to reduce the speed-of-light propagation delays. To increase the volumetric density of devices, new concepts are needed to increase interconnection density in the third dimension, perpendicular to the plane of chip and board interconnections (*155–159*). An example is the "button board", which uses "buttons" of compressed copper wire for connections between stacked PWBs (*155*). Optical interconnections may also be used for signal propagation in the third dimension (*159*). The volumetric density of chips has also been increased by stacking and laminating a number of chips to form a cube with connections patterned on one edge; this approach is especially attractive for memory ICs that have low power dissipation and redundant circuitry.

Heat Dissipation. The very high power densities resulting from increased circuit density and speed require innovative approaches for dissipating heat, such as heat pipes (*160, 161*), immersion cooling (*64-66*), or structures such as microchannels (*162, 163*) or microcapillaries (*163–165*) etched into the substrate or fabricated internally within a cofired ceramic substrate (*166*).

A novel method for achieving low thermal impedance between the chip and substrate has been developed at Stanford University (*163–165*). Reentrant microcapillaries (3–5 μm wide) are etched into the substrate (either a silicon wafer or a ceramic coated with a polyimide) and partially filled with a fluid such as silicone oil. This structure increases the heat-transfer area, reduces the interfacial thermal resistance, and provides a stress-free, reversible die attachment through capillary forces. Thermal resistances of 0.05–0.07 °C/W have been measured for these microcapillary interfaces.

The group at Stanford has also developed a method for increasing the convective heat transfer from the back side of a silicon substrate by sawing high-aspect-ratio grooves (50 μm wide and 300 μm deep) into the substrate (*162, 163*). Cooling fluid is forced through the grooves and the convective heat transfer increases because of the increased area and larger heat-transfer coefficient. Thermal resistances of 0.1 °C/W have been demonstrated with the microchannels.

High-Density Bonding. A technique for achieving very high density of bonding to an IC has also been developed at Stanford (*167*). Cantilever beams of gold lines on silicon dioxide are formed by anisotropically etching trenches in the underlying silicon substrate. The ICs are bonded face down to the cantilever beams, which bow upward because of internal stress in the oxide film. The stressed beams produce a contact force of ~0.4 mN for each bond. This structure permits a nonpermanent, very high density (40-μm pitch) bond to the ICs. Figure 24 shows Stanford's overall "active substrate" concept combining microchannel cooling, microcapillary die attachment, and cantilever beam contacts on a silicon substrate that may contain active driver circuits.

Wafer-Scale Integration. One of the most well-publicized and widely debated approaches to increase system performance is wafer-scale integration (WSI), in which an entire system or a large functional block of circuits is integrated on a large chip approaching the size of a silicon wafer (*168–171*). WSI can greatly increase circuit density, improve reliability because of fewer interconnections, and increase speed and reduce power be-

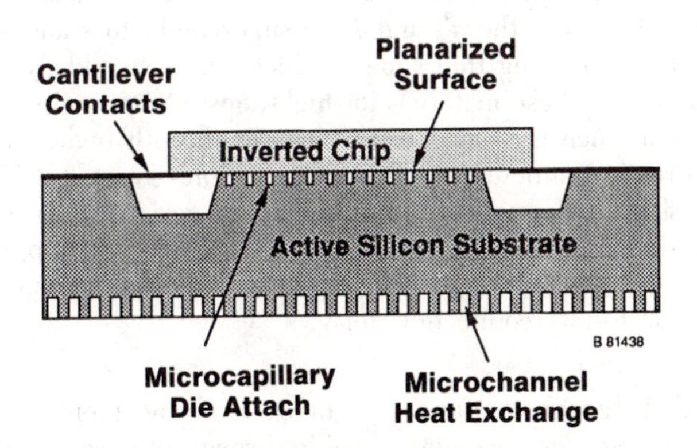

Figure 24. Integrated active-substrate system proposed by Stanford University. The system includes cantilever bonding, microcapillary die attachment, and microchannel heat exchange on a silicon substrate with active circuits. (Figure was based on reference 167.)

cause of fewer off-chip signals. However, the severe yield penalty incurred by the large chip area requires highly redundant circuitry or reconfigurable designs based on self-test and self-repair of circuit elements. Efficient heat removal and power distribution methods are also required for WSI chips.

An alternative approach is to mount individual pretested chips on a silicon or alternative substrate containing high-density interconnections (*91–93*, *97*, *172–173*). This approach, known as hybrid-wafer-scale integration (HWSI), circumvents the yield limitations of WSI but retains many of its advantages. The silicon substrate has several inherent advantages, including a thermal expansion coefficient matched to those of the chips, high thermal conductivity, a good surface finish, and standardized sizes for semiconductor processing. The fabrication of active devices such as chip-to-chip drivers and receivers in the silicon substrate is possible, and the size and power consumption of the ICs can thus be reduced. Finally, with HWSI, the combination of different circuit technologies, such as bipolar, MOS, and GaAs, in a hybrid module is possible. Excellent comparisons of WSI and HWSI approaches have been made (*131*, *171*).

Superconductors. One of the most important limitations to wiring density in HWSI or WSI interconnections is the dispersion and loss caused by the high resistance of long interconnections. As interconnection density increases, the conductor cross section must decrease and resistive losses become limiting. High-critical-temperature (T_c) superconductors have been considered as a solution to this problem, because the dc resistance will be zero below the T_c, and the high-frequency attenuation is negligible at frequencies of up to 10 GHz (*174–177*). However, the critical current density (I_c) in present-day superconducting materials (10^5–10^6 A/cm^2) limits the minimum interconnection cross section. The rapid advances being made now toward increasing the T_c and I_c of superconductors and developing techniques for processing thin superconductor films are likely to improve the applicability of these materials for high-density interconnections. Standard conductors such as Al and Cu also have significantly higher conductivity at liquid-nitrogen temperatures (five to eight times greater at 77 K for Al) and, thus, offer another solution for reducing the resistance or cross section of high-density interconnections. However, none of these approaches for reducing interconnection resistance can alleviate the fundamental speed-of-light limitation to interconnection delay.

Optical Interconnections. Significant advancements in interconnection technology are becoming possible through the use of optical signal transmission. Semiconductor diode lasers or light-emitting diodes are used to convert electrical signals into photonic signals that are transmitted through optical interconnections and reconverted into electrical signals by semiconductor detectors. In general, optical interconnections have a much higher bandwidth than do electrical interconnections because of the higher frequencies of the optical signals. Therefore, multiple data streams can be multiplexed and propagated through a single optical interconnection rather than through many electrical interconnections.

Fiber optics have been used for a number of years in the telecommunication industry for long-distance signal transmission (*178*). The advantages of optical interconnections in these applications include their high bandwidth, low loss, light weight, low volume, security, and freedom from electromagnetic interference. More recently, optical interconnections have been proposed for board-to-board, chip-to-chip, and even intrachip interconnections in computer systems and local-area networks (*179–183*). In these applications, optical interconnections offer several potential advantages over electrical interconnections, including higher density because of the smaller cross section of optical waveguides, higher fan-out because of the absence of capacitive loading, higher bandwidth with less dispersion, and freedom from electromagnetic interference, ground loops, and crosstalk. Optical signals can also cross without interacting and thus permit crossovers within a single layer in planar waveguides.

Fibers may be used for board-to-board communication, but for higher density interconnections within MCPs, fibers are too bulky, and therefore, optical interconnections are fabricated as waveguides in thin films (*183*). Several investigators have reported the use of polyimides as an optical waveguide medium (*184–186*). Because polyimides are also a widely used dielectric for TFML electrical interconnections, a combination of electrical and optical interconnections, which benefits from the advantages of both technologies, is possible within the same material system. Electrical conductor layers could be used for power distribution and low-speed signals, whereas optical interconnections would be used for high-speed, high-fan-out signals such as clock distribution and data bus lines. Optical interconnections may be fabricated also through wafers for three-dimensional interconnection of stacked ICs (*159*).

A number of technological advancements are needed to realize the full performance potential of optical interconnections in computer systems, including the development of lower power lasers and receivers, the monolithic integration of optoelectronic devices with electronic circuits, the integration of optical and electrical interconnections within a package, the improvement of methods for coupling optical signals into waveguides or fibers, the reduction of scattering losses and absorption in waveguides, and the development of new, highly parallel architectures and multiplexing schemes that can exploit the high fan-out, bandwidth, and density of optical interconnections.

References

1. Reid, T. R. *The Chip;* Simon and Schuster: New York, 1984; Chapter 1.
2. Rust, R. D.; Doane, D. A. *Solid State Technol.* **1986,** *29*(*6*), 125–128.
3. Neugebauer, C. A. In *Electronic Packaging Materials Science III;* Jaccodine, R.; Jackson, D. A.; Sundahl, R. C., Eds.; MRS Symposia Proceedings; Materials Research Society: Pittsburgh, PA, 1988; Vol. 108; pp 13–25.
4. Blodgett, A. J. *Sci. Am.* **1983,** *249*(*1*), 86–96.
5. Seraphin, D. P.; Lee, L. C.; Appelt, B. K.; Marsh, L. L. In *Electronic Packaging Materials Science;* Geiss, E. A.; Tu, K. N.; Uhlmann, D. R., Eds.; MRS Symposia Proceedings; Materials Research Society: Pittsburgh, PA, 1985; Vol. 40; pp 21–48.
6. Bupp, J. R.; Chellis, L. N.; Ruane, R. E.; Wiley, J. P. *IBM J. Res. Dev.* **1982,** *26*, 306–317.
7. Wang, D. W. In *Electronic Packaging Materials Science III;* Jaccodine, R.; Jackson, D. A.; Sundahl, R. C., Eds.; MRS Symposia Proceedings; Materials Research Society: Pittsburgh, PA, 1988; Vol. 108; pp 125–139.
8. Manka, H. H. *Soldering Handbook for Printed Circuits and Surface Mounting;* Van Nostrand Reinhold: New York, 1986.
9. Balde, J. W. *Solid State Technol.* **1986,** *29*(*6*), 99–103.
10. *Surface Mount Technology;* International Society for Hybrid Microelectronics: Silver Spring, MD, 1984.
11. *Surface Mount Technology* (*SMT*), A Compendum of Technical Articles Presented at the First, Second, and Third Annual Conferences; International Electronics Packaging Society: Glen Ellyn, IL, 1984.
12. Shukla, R. K.; Mencinger, N. P. *Solid State Technol.* **1985,** *28*(*28*), 67–74.
13. Dehaine, G.; Kurzweil, K.; Lewandowski, P. *Proc. Int. Symp. Microelectron.;* International Society for Hybrid Microelectronics: Reston, VA, 1984; pp 353–357.
14. Berg, H. M.; Mitchell, C. *Proc. 35th Electron. Components Conf.*, Washington DC; IEEE: New York, 1985; pp 98–106.
15. Marshall, J. F., *Solid State Technol.* **1984,** *27*(*27*), 175–179.
16. Hoffman, P. *Solid State Technol.* **1988,** *31*(*31*), 85–88.
17. Scharr, T. A. *Int. J. Hybrid Microelectron.* **1983,** *6*, 561–565.
18. Koopman, N. G.; Totta, P. A. In *Proc. First Electron. Mater. Process. Conf.*, Chicago, IL; ASM International: Metals Park, OH; pp 137–151.
19. Koopman, N. G.; Reiley, T. C.; Totta, P. A. *Proc. Int. Symp. Microelectron.;* International Society for Hybrid Microelectronics: Reston, VA, 1988; pp 295–300.
20. Goldman, L. S.; Totta, P. A. *Solid State Technol.* **1983,** *26*(*26*), 91–97.
21. Ho, C. W. In *VLSI Electronics: Microstructure Science;* Einspruch, N. G., Ed.; Academic: New York, 1982; Vol. 5, Chapter 3.
22. Steidel, C. A. In *VLSI Technology;* Sze, S. M., Ed.; McGraw–Hill: New York, 1983; Chapter 13.
23. Striny, K. M. In *VLSI Technology, Second Edition;* Sze, S. M., Ed.; McGraw–Hill: New York, 1988; Chapter 13.
24. Schroen, W. H.; Rankratz, J. M. In *VLSI Electronics: Microstructure Science;* Einspruch, N. G., Ed.; Academic: New York, 1985; Vol. 9, Chapter 5.
25. Tummala, R. R.; Rymaszewski, E. J., Eds. *Microelectronics Packaging Handbook;* Van Nostrand Reinhold: New York, 1989.
26. Skidmore, K. *Semicond. Int.* **1988,** *11*(*11*), 61–65.
27. Bartlett, C. J. *Solid State Technol.* **1986,** *29*(*29*), 119–123.
28. Bowlby, R. *IEEE Spectrum* **1985,** *22*(*22*), 37–42.

29. Cummings, J. P.; Chase, D. H. *Digest of Papers, 1986 Government Microcircuit Applications Conf.* (*GOMAC*); Defense Technical Information Center: Washington, DC, 1986; pp 281–284.
30. Smeby, J. M. *IEEE Trans. Compon. Hybrids Manuf. Technol.* **1985,** *CHMT–8,* 391–396.
31. Hall, P. M. *Proc. 34th Electron. Compon. Conf.* New Orleans, LA; Institute of Electrical and Electronics Engineers: New York, **1984, pp 107–116.**
32. Blackshaw, M.; Dance, F. *Solid State Technol.* **1986,** *29*(29), 141–145.
33. Kompelien, D.; Moravec, T. J.; DeFlumere, M. *Proc. Int. Symp. Microelectron.*, Atlanta, GA; International Society for Hybrid Microelectronics: Reston, VA, 1986; pp 749–757.
34. Cummings, J. P.; Jensen, R. J.; Kompelien, D. J.; Moravec, T. J. *Proc. 32nd Electron. Compon. Conf.;* Institute of Electrical and Electronics Engineers: New York, 1982; pp 465–475.
35. Davidson, E. E. *IEEE Trans. Compon. Hybrids Manuf. Technol.* **1982,** *CHMT–6,* 272–282.
36. Ho, C. W.; Chance, D. A.; Bajorek, C. H.; Acosta, R. E. *IBM J. Res. Develop.* **1982,** *26,* 286–296.
37. Clatterbaugh, G.; Charles, H. K., Jr. *Proc. Int. Symp. Microelectron.*, Atlanta, GA; International Society for Hybrid Microelectronics: Reston, VA, 1986; pp 766–774.
38. Belcourt, F. J.; Lane, T. A. *Proc. Int. Symp. Microelectron.*, Atlanta, GA; International Society for Hybrid Microelectronics: Reston, VA, 1986; pp 802–808.
39. Lewis, E. T. *Proc. Int. Symp. Microelectron.*, Atlanta, GA; International Society for Hybrid Microelectronics: Reston, VA, 1986; pp 722–729.
40. Gilbert, B. K. In *VLSI Electronics: Microstructure Science;* Einspruch, N. G.; Wisseman, W. R., Ed.; Academic: New York, 1985; Vol. 11, Chapter 8.
41. Katopis, G. A. *Proc. IEEE* **1985,** *73,* 1405–1415.
42. Sainati, R. *Proc. 35th Electron. Compon. Conf.*, Washington, DC; Institute of Electrical and Electronics Engineers: New York, 1985; pp 365–371.
43. Lewis, E. T. *IEEE Trans. Compon. Hybrids Manuf. Technol.* **1979,** *CHMT–2,* 441–450.
44. Scheinfein, M.; Liao, J. C.; Palusinski, O.; Prince, J. *Proc. Int. Electron. Manuf. Technol. Symp.*, San Francisco, CA; Institute of Electrical and Electronics Engineers: New York, 1986; pp 23–29.
45. Palmquist, S. L.; Moravec, T. J. *Proc. Int. Electron. Packag. Conf.*, Dallas, TX; International Electronics Packaging Society: Wheaton, IL, 1988; pp 670–683.
46. VanValkenburg, M. E. *Network Analysis;* Prentice–Hall: Englewood Cliffs, NJ, 1964; p 451.
47. Galbard, O. G. *Electrotechnology* **1969,** *April,* 59.
48. Howe, H., Jr. *Stripline Circuit Design;* Artech House: Dedham MA, 1974.
49. Gupta, K. C.; Garg, R.; Bahl, I. J. *Microstrip Lines and Slotlines;* Artech House: Dedham MA, 1979.
50. Dworsky, L. N. *Modern Transmission Line Theory and Applications;* Wiley: New York, 1979.
51. Kaupp, H. R. *IEEE Trans. Comput.* **1967,** *EC–16,* 185–193.
52. Sainati, R. *Proc. 5th VLSI Packag. Workshop*, Paris, France, Nov. 1986, pp 5–9.
53. Haefner, S. J. *Proc. IRE* **1937,** *25*(25), 434–447.
54. Sinha, A. K., Cooper, J. A., Jr.; Levinstein, H. J. *IEEE Electron. Device Lett.* **1982,** *EDL–3,* 90–92.

55. Carter, D. L., Guise, D. F. *VLSI Design* **1984,** *January,* 63–68.
56. Catt, I. *IEEE Trans. Electron. Comput.* **1967,** *EC–16,* 743–764.
57. Seki, S.; Hasegawa, H. *IEEE Trans. Electron. Devices* **1984,** *EDL–31,* 1948–1953.
58. Simons, R. E. *Solid State Technol.* **1983,** *26(10),* 131–137.
59. Keyes, R. W. In *VLSI Electronics: Microstructure Science;* Einspruch, N. G., Ed.; Academic: New York, 1981; Vol. 1, Chapter 5.
60. Bird, R. B.; Stewart, W. E.; Lightfoot, E. N. *Transport Phenomena;* Wiley: New York, 1960; p 245.
61. Strang, G.; Fix, G. J. *An Analysis of the Finite Element Method;* Prentice–Hall: Englewood Cliffs, NJ, 1973.
62. Mahalingam, M.; Berg, H. M. *Int. J. Hybrid Microelectron.* **1984,** *7(7),* 1–9.
63. Martuza, M. *Electronics* **1982,** *Feb. 10,* 145–148.
64. Simons, R. E., Chu, R. C. *Proc. Int. Symp. Microelectron.;* International Society for Hybrid Microelectronics: Reston, VA, **1985, pp 314–321.**
65. Danielson, R. D., Krajewski, N., Brost, J. *Electron. Packag. Prod.* **1986,** *July,* 44–45.
66. Bar-Cohen, A.; Schweitzer, H. *Proc. 4th Int. Electron. Packag. Conf.*, Baltimore MD, Oct. 1984; pp 596–615.
67. *Handbook of Thick Film Hybrid Microelectronics;* Harper, C. A., Ed.; McGraw–Hill: New York, 1974.
68. Hamer, D. W.; Biggers, J. V. *Thick Film Hybrid Microcircuit Technology;* Wiley–Interscience: New York, 1972.
69. Pitkanen, D. E.; Cummings, J. P.; Speerschneider, C. J. *Solid State Technol.* **1980,** *23(10),* 141–146.
70. *Introduction to Hybrid Thick Film Technology;* Center for Professional Advancement: East Brunswick, NJ, 1982.
71. McIver, C. H. *Electronics* **1982,** *November 3,* 96.
72. Schwartz, B. *J. Phys. Chem. Solids* **1984,** *45,* 1051–1068.
73. Hassler, B. A. *Proc. Int. Symp. Microelectron.*, Atlanta, GA; International Society for Hybrid Microelectronics: Reston, VA, **1986, 741–748.**
74. Blodgett, A. J., Jr. *IEEE Trans. Compon. Hybrids Manuf. Technol.* **1980,** *CHMT–3,* 634–637.
75. Shimada, Y.; Utsumi, K.; Suzuki, M.; Takamizawa, H.; Mitta, M.; Yano, S. *Proc. 33rd Electron. Compon. Conf.;* Institute of Electrical and Electronics Engineers: New York, 1983; pp 314–319.
76. Eustice, A. L.; Horowitz, S. J.; Stewart, J. J.; Travis, A. R.; Sawhill, H. T. *Proc. 36th Electron. Compon. Conf.;* Institute of Electrical and Electronics Engineers: New York, 1986; pp 37–47.
77. Nishigaki, S., Yano, S.; Kawabe, H.; Fukuta, J.; Nonomura, T.; Hebishima, S. *Proc. Int. Symp. Microelectron.*, Minneapolis, MN; International Society for Hybrid Microelectronics: Reston, VA, 1987; pp 400–407.
78. Jensen, R. J.; Cummings, J. P.; Vora, H. *IEEE Trans. Compon. Hybrids Manuf. Technol.* **1984,** *CHMT–7,* 384–393.
79. Jensen, R. J. In *Electronic Packaging Materials Science III;* Jaccodine, R.; Jackson, D. A.; Sundahl, R. C., Eds.; MRS Symposia Proceedings, Vol. 108; Materials Research Society: Pittsburgh, PA, 1988; pp 73–79.
80. Jensen, R. J. In *Polymers for High Technology: Electronics and Photonics;* Bowden, M. J.; Turner, S. R., Eds.; ACS Symposium Series 346; American Chemical Society: Washington, DC, 1987; Chapter 40.
81. Watari, T.; Murano, H. *IEEE Trans. Compon. Hybrids Manuf. Technol.* **1985,** *CHMT–8,* 462–467.
82. Nguyen, P. H.; Russo, F. R. *Proc. Int. Symp. Microelectron.*, Atlanta, GA;

International Society for Hybrid Microelectronics: Reston, VA, 1986; 702–706.
83. Ackerman, K.-P.; Hug, R.; Berner, G. *Proc. Int. Symp. Microelectron.*, Atlanta, GA; International Society for Hybrid Microelectronics: Reston, VA, 1986; pp 519–524.
84. Shiflett, C. C.; Buckholz, D. B.; Fandskar, C. C.; Small, R. D.; Markham, J. L. *Proc. Int. Symp. Microelectron.*, Atlanta, GA; International Society for Hybrid Microelectronics: Reston, VA, 1986; pp 481–486.
85. Takasago, H.; Takada, M.; Adachi, K.; Endo, A.; Yamada, K.; Makita, T.; Gofuku, E.; Onishi, Y. *Proc. 36th Electron. Compon. Conf.;* Institute of Electrical and Electronics Engineers: New York, 1986; pp 481–487.
86. Ecker, M. E.; Olson, L. T. *Proc. Int. Symp. Microelectron.;* International Society for Hybrid Microelectronics: Reston, VA, 1981; pp 251–257.
87. Tsunetsugu, H.; Takagi, A.; Moriya, K. *Int. J. Hybrid Microelectron.* **1985,** *8,* 21–26.
88. Moriya, K.; Ohsaki, K.; Katsura, K. *Proc. 34th Electron. Compon. Conf.*, New Orleans, LA; Institute of Electrical and Electronics Engineers: New York, 1984; pp 82–87.
89. Kimijima, S.; Miyagi, T.; Sudo, T.; Shimada, O. *Proc. Int. Symp. Microelectron.*, Seattle, WA; International Society for Hybrid Microelectronics: Reston, VA, 1988; pp 314–319.
90. Nguyen, P. H.; Falletta, C. E., Russo, F. R. *Proc. Int. Symp. Microelectron.;* International Society for Hybrid Microelectronics: Reston, VA, 1985; pp 258–261.
91. Hagge, J. K. *Proc. 38th Electron. Compon. Conf.*, Los Angeles, CA; Institute of Electrical and Electronics Engineers: New York, 1988; pp 282–292.
92. Chao, C. C.; Scholz, K. D.; Leibovitz, J.; Cobarruviaz, M. L.; Chang, C. C. *Proc. 38th Electron. Compon. Conf.*, Los Angeles, CA; Institute of Electrical and Electronics Engineers: New York, 1988; pp 276–281.
93. Johnson, R. W.; Phillips, T. L.; Jaeger, R. C.; Hahn, S. F.; Burdeaux, D. C. *Proc. 38th Electron. Compon. Conf.*, Los Angeles, CA; Institute of Electrical and Electronics Engineers: New York, 1988; pp 267–273.
94. Neugebauer, C. A.; Carlson, R. O.; Fillion, R. A.; Haller, T. R. *Solid State Technol.* **1988,** *31(31)*, 93–98.
95. Levinson, L. M.; Eichelberger, C. W.; Wojnarowski, R. J.; Carlson, R. O. *Proc. Int. Symp. Microelectron.*, Seattle, WA; International Society for Hybrid Microelectronics: Reston, VA, 1988; pp 301–305.
96. Pan, T. J.; Poon, S.; Nelson, B. *Proc. Int. Electron. Packag. Conf.*, Dallas, TX; International Electronics Packaging Society: Wheaton, IL, 1988; pp 174–189.
97. Bartlett, C. J.; Segelken, J. M.; Teneketges, N. A. *IEEE Trans. Compon. Hybrids Manuf. Technol.* **1987,** *CHMT-12,* 647–653.
98. Lee, Y. C.; Ghaffari, H. T.; Segelhen, J. M. *Proc. 38th Electron. Compon. Conf.*, Los Angeles, CA; Institute of Electrical and Electronics Engineers: New York, 1988; pp 293–310.
99. Jayaraj, K.; Moravec, T. J.; Jensen, R.; Belcourt, F. J.; Sainati, R. In *Proc. First Electron. Mater. Process. Conf.*, Chicago, IL, 1988; ASM International: Metals Park, OH; pp 111–118.
100. Speerschneider, C. J.; Belcourt, F. J.; Jensen, R. J.; Smeby, J. M. *Proc. VHSIC Packag. Conf.*, Houston, TX; Palisades Institute: New York, 1987; pp 131–143.
101. Vidano, R. P.; Cummings, J. P.; Jensen, R. J.; Walters, W. W.; Helix, M. J.

Proc. 33rd Electron. Compon. Conf.; Institute of Electrical and Electronics Engineers: New York, 1983; pp 334–343.
102. Chou, N. J.; Tang, C. H. *J. Vac. Sci. Technol.* **1984,** *A2*, 751–755.
103. Hahn, P. O.; Rubloff, G. W.; Bartha, J. W.; Legoues, F.; Tromp, R.; Ho, P. S. in *Electronic Packaging Materials Science;* Geiss, E. A.; Tu, K. N.; Uhlmann, D. R.; Eds.; MRS Symposia Proceedings, Vol. 40; Materials Research Society: Pittsburgh, PA, 1985; pp 251–263.
104. White, R. C.; Haight, R.; Silverman, B. D.; Ho, P. S. *Appl. Phys. Lett.* **1987,** *51*, 482–483.
105. Kim, Y. H.; Walker, G. F.; Kim, J.; Park, J. *J. Adhesion Sci. Technol.* **1987,** *1*, 331–339.
106. Chou, N. J.; Dong, D. W.; Kim, J.; Liu, A. C. *J. Electrochem. Soc.* **1984,** *131*, 2335–2339.
107. Ohuchi, F. S.; Freilich, S. C. *J. Vac. Sci. Technol.* **1986,** *A4*, 1039–1045.
108. *Polyimides: Synthesis, Characterization and Applications;* Mittal, K. L., Ed.; Plenum: New York, 1984; Vols. 1–2.
109. Jensen, R. J.; Lai, J. H. In *Polymers for Electronic Applications;* Lai, J. H., Ed.; CRC Press: Boca Raton, FL, in press.
110. Bessonov, M. I.; Koton, M. M.; Kudryavtsev, V. V.; Laius, L. A.; *Polyimides: Thermally Stable Polymers;* Consultants Bureau: New York, 1987.
111. *Proc. Second Int. Conf. Polyimides,* Ellenville, NY, Nov. 1985; University Microfilms: Ann Arbor, MI.
112. Senturia, S. D. In *Polymers for High Technology: Electronics and Photonics;* Bowden, M. J.; Turner, S. R., Eds.; ACS Symposium Series 346; American Chemical Society: Washington, DC, 1987; pp 428–436.
113. Schuckert, C. C.; Fox, G. B.; Merriman, B. T. In *Proc. Symp. Packag. Electron. Devices;* Bindra, P.; Susko, R. A., Eds.; Electrochemical Society: Pennington, NJ, **1989, 116–128.**
114. Numata, S.; Miwa, T.; Makino, D.; Imaizumi, J.; Kinjo, N. In *Electronic Packaging Materials Science III;* Jaccodine, R.; Jackson, K. A.; Sundahl, R. C., Eds.; MRS Symposia Proceedings, Vol. 108; Materials Research Society: Pittsburgh, PA, 1988; pp 113–124.
115. Jensen, R. J.; Douglas, R. B.; Smeby, J. M.; Moravec, T. J. *Proc. VHSIC Packag. Conf.*, Houston, TX; Palisades Institute: New York, 1987; pp 193–205.
116. Kirchoff, R. A.; Baker, C. E.; Gilpin, J. A.; Hahn, S. F.; Schrock, A. K. In *Proc. 18th Int. SAMPE Conf.*, October 7–9, 1986.
117. Stapper, C. H. *Proc. IEEE* **1983,** *71*, 453–470.
118. Stapper, C. H. *IBM J. Res. Develop.* **1983,** *27*, 549–557.
119. Spence, C. F. *Proc. Int. Symp. Microelectron.* **1985, 174–182.**
120. Mandeville, J. R. *IBM J. Res. Dev.* **1985,** *29*, 73–86.
121. Ehrlich, D. J.; Tsao, J. Y. In *VLSI Electronics: Microstructure Science;* Einspruch, N. G., Ed.; Academic: New York, 1983; Vol. 7, Chapter 3.
122. Baum, T. H. *J. Electrochem. Soc.* **1987,** *134*, 2616–2619.
123. Gupta, A.; Jagannathan, R. *Appl. Phys. Lett.* **1987,** *51*, 2254–2256.
124. Cole, H. S.; Liu, Y. S.; Rose, J. W.; Guida, R. *Appl. Phys. Lett.* **1988,** *53*, 2111–2113.
125. von Gutfeld, R. J.; Vigliotti, D. R. *Appl. Phys. Lett.* **1985,** *46*, 1003–1005.
126. von Gutfield, R. J. In *Laser Applications;* Ready, J. F.; Erf, R. K., Eds.; Academic: New York, 1984; Vol. 5., pp 1–67.
127. Perry, P. B.; Ray, S. K.; Hodgson, R. *Thin Solid Films* **1981,** *85*, 111–117.
128. Baller, T. S.; VanVeen, G. N. A.; Dieleman, J. *J. Vac. Sci. Technol.* **1988,** *A6*, 1409–1413.
129. Viswanathan, R.; Hussla, I. *J. Opt. Soc. Am.* **1986,** *B3*.
130. Teng, K. F.; Azadpour, M. A.; Yang, H. Y. *Proc. 38th Electron. Compon.*

Conf., Los Angeles, CA; Institute of Electrical and Electronics Engineers: New York, 1988; pp 326–329.
131. McDonald, J. F.; Steckl, A. J.; Neugebauer, C. A.; Carlson, R. O.; Bergendahl, A. S. *J. Vac. Sci. Technol.* **1986,** *A4,* 3127–3138.
132. Rothman, L. B. *J. Electrochem. Soc.* **1983,** *130,* 1131–1136.
133. Vossen, J. L.; Cuomo, J. J. In *Thin Film Processes;* Vossen, J. L.; Kern, W., Eds.; Academic: New York, 1978, Chapter II–1.
134. Takagi, T. *Thin Solid Films* **1982,** *92,* 1–17.
135. Takagi, T.; Yamada, I.; Matsubara, K. *Thin Solid Films* **1979,** *58,* 9–19.
136. Younger, P. R. *Solid State Technol.* **1984,** *27(11),* 143–147.
137. Sanders, D. M.; Pyle, E. A. *J. Vac. Sci. Technol.* **1987,** *A5,* 2728–2731.
138. Schwartz, G. C.; Schaible, P. M. *J. Electrochem. Soc.* **1983,** *130,* 1777–1779.
139. Givens, F. L.; Daughton, W. J. *J. Electrochem. Soc.* **1979,** *126,* 269–272.
140. Salem, J. R.; Sequeda, F. O.; Duran, J.; Lee, W. Y.; Yang, R. M. *J. Vac. Sci. Technol.* **1986,** *A4,* 369–374.
141. Harada, Y.; Matsumoto, F.; Nakakado, T. *J. Electrochem. Soc.* **1983,** *130,* 129–134.
142. Lee, Y. K.; Craig, J. D. In *Polymer Materials for Electronic Applications,* Feit, E. D.; Wilkins, C. W., Jr., Eds.; ACS Symposium Series 184; American Chemical Society: Washington, DC, 1982.
143. Turban, G.; Rapeaux, M. *J. Electrochem. Soc.* **1983,** *130,* 2231–2236.
144. Egitto, F. D.; Emmi, F.; Horwath, R. S.; Vukanovic, V. *J. Vac. Sci. Technol.* **1985,** *B3,* 893–904.
145. Vanderlinde, W. E.; Ruoff, A. L. *J. Vac. Sci. Technol.* **1988,** *B6,* 1621–1625.
146. Rubner, R.; Ahne, H.; Kuhn, E.; Kolodziej, G. *Photogr. Sci. Eng.* **1979,** *23,* 303–309.
147. Rohde, O.; Riediker, M.; Schaffner, A.; Bateman, J. *Advances in Resist Technology and Processing II;* SPIE Vol. 539, 1985, pp 175–180.
148. Clatterbaugh, G. V.; Charles, H. K. *Proc. Int. Symp. Microelectron.,* Seattle, WA; International Society for Hybrid Microelectronics: Reston, VA, 1988; pp 320–332.
149. Brannon, J. H.; Lankard, J. R.; Baise, A. I.; Burns, F.; Kaufman, J. *J. Appl. Phys.* **1985,** *58,* 2036–2043.
150. Yeh, J. T. C. *J. Vac. Sci. Technol.* **1986,** *A4,* 653–658.
151. Srinivasan, V.; Smrtic, M. A.; Babu, S. V. *J. Appl. Phys.* **1986,** *59,* 3861–3867.
152. Treyz, G. V.; Weinman, L.; Scarmozzino, R.; Osgood, R. M., Jr., presented at the 35th National Symposium of the American Vacuum Society, Atlanta, GA, October 1988, paper TC2–ThA7.
153. Lane, T. A.; Belcourt, F. J.; Jensen, R. J. *IEEE Trans. Compon. Hybrids Manuf. Technol.* **1987,** *CHMT–12,* 577–585.
154. Jackel, L. D.; Howard, R. E.; Graf, H. P.; Straughn, B.; Denker, J. S. *J. Vac. Sci. Technol.* **1986,** *B4,* 61–63.
155. Smolley, R. *Proc. Int. Symp. Microelectron.,* Atlanta, GA; International Society for Hybrid Microelectronics: Reston, VA, 1986; pp 326–333.
156. Little, M. J.; Grinberg, J. *Byte* **1988,** *13(11),* 311–319.
157. Grinberg, J.; Nudd, G. R.; Etchells, R. D. *IEEE Trans. Comput.* **1984,** *C–33,* 69–81.
158. Scarlett, J. *Electron. Eng.* **1984,** *June,* 43–47.
159. Hornak, L. A.; Tewksbury, S. K. *IEEE Trans. Electron Devices* **1987,** *ED–34,* 1557–1563
160. Basiulius, A.; Minning, C. P. *Proc. 37th Natl. Aerospace Electron. Conf.,* Dayton, OH, May 1985.
161. Sekhon, K. S. *Proc. 4th Int. Electron. Packag. Conf.,* Baltimore, MD; International Electronics Packaging Society; Wheaton, IL, 1984; 575–585.

162. Tuckerman, D. B.; Pease, R. F. W. *IEEE Electron Device Lett.* **1981,** *EDL–2,* 126–129.
163. Tuckerman, D. B., PhD Thesis, Stanford University, 1983.
164. Paal, A.; Pease, R. F. W. *Proc. Int. Electron. Manuf. Technol. Symp.;* San Francisco, CA; September 1986; Institute of Electrical and Electronics Engineers: New York, pp 169–172.
165. Tuckerman, D. B.; Pease, R. F. W. *1983 Symp. VLSI Technol.*, Maui, HI, September 1983; pp 60–61.
166. Kishimoto, T.; Ohsaki, T. *IEEE Trans. Compon. Hybrids Manuf. Technol.* **1986,** *CHMT–9,* 328–335.
167. Hong, S.; Brauman, J. C.; Weihs, T. P.; Kwon, O. K. In *Electronic Packaging Materials Science III;* Jaccodine, R.; Jackson, K. A.; Sundahl, R. C., Eds.; MRS Symposium Proceedings, Vol. 108; Materials Research Society: Pittsburgh, PA, 1988; pp 309–317.
168. Peltzer, D. L. *VLSI Design* **1983,** *4,* 43–47.
169. Pountain, D. *Byte* **1986,** *11*(*11*), 351–355.
170. McDonald, J. F.; Rogers, E. H.; Rose, K.; Steckl, A. J. *IEEE Spectrum* **1984,** *21*(*10*), 32–39.
171. Carlson, R. O.; Neugebauer, C. A. *Proc. IEEE* **1986,** *74,* 1741–1752.
172. Spielberger, R. K.; Huang, C. D.; Nunne, W. H.; Mones, A. H.; Fett, D. L.; Hampton, F. L. *IEEE Trans. Compon. Hybrids Manuf. Technol.* **1984,** *CHMT–7,* 193–196.
173. Johnson, R. W.; Cornelius, M.; Davidson, J. F.; Jaeger, R. C. *Proc. Int. Symp. Microelectron.*, Atlanta, GA; International Society for Microelectronics: Reston, VA, 1986; pp 758–765.
174. Frye R. C. *Proc. Int. Electron. Packag. Conf.*, Dallas, TX; International Electronics Packaging Society: Wheaton, IL, 1988; pp 759–770.
175. Kroger, H.; Hilbert, C.; Ghoshal, U.; Gibson, D.; Smith, L. *Proc. Int. Electron. Packag. Conf.*, Dallas, TX; International Electronics Packaging Society: Wheaton, IL, 1988; pp 174–189.
176. Frye, R. C. In *Electronic Packaging Materials Science III;* Jaccodine, R.; Jackson, K. A.; Sundahl, R. C., Eds.; MRS Symposium Proceedings, Vol. 108; Materials Research Society: Pittsburgh, PA, 1988; pp 27–38.
177. Kwon, O. K.; Langley, B. W.; Pease, R. F. W.; Beasley, M. R. *IEEE Electron Devices Lett.* **1987,** *EDL–8,* 582.
178. Midwinter, J. E. *J. Lightwave Technol.* **1985,** *LT–3,* 927.
179. Haugen, P. R.; Rychnovsky, S.; Husain, A.; Hutcheson, L. D. *Opt. Eng.* **1986,** *25,* 1076–1084.
180. Haugen, P. R.; Husain, A.; Hutcheson, L. D. In *Optical Computing;* Neff, J. A., Ed.; *Proc. SPIE* **1986,** *625,* 110–116.
181. Hutcheson, L. D.; Haugen, P.; Husain, A. *IEEE Spectrum* **1987,** *24*(*24*), 30–35.
182. Goodman, J. W. *Optical Computing; Proc. SPIE* **1984,** *456,* 72–85.
183. Sullivan, C. T.; Roth, M. C.; Budzynski, T. *Proc. SPIE O–E/Fibers,* San Diego, CA, Aug. 1987; paper 837–10.
184. Sullivan, C. T. *Proc. SPIE OE/Fiber Lase,* Boston, MA, Sept. 1988; paper 994–14.
185. Selvaraj, R.; Lin, H. T.; McDonald, J. F. *IEEE J. Lightwave Technol.* **1988,** *6,* 1034.
186. Franke, H.; Crow, J. D. *Integrated Optical Circuit Enginering; Proc. SPIE* **1986,** *651,* 102.

RECEIVED for review December 30, 1987. ACCEPTED revised manuscript January 23, 1989.

10

Semiconductor Processing Problems Solved by Wet (Solution) Chemistry

Marjorie K. Balazs

Balazs Analytical Laboratory, Sunnyvale, CA, 94086

This chapter gives explicit examples of how the techniques of wet (solution) chemistry can be applied to the production of integrated circuits. The quality control for processed thin films, chemicals, and pure water, along with microcontamination analysis, to resolve production problems are discussed. These examples indicate that wet chemical techniques are the only ones available for absolute standardization and measurement of trace metals and their effect on the devices produced by current very-large-scale-integration (VLSI) technology.

THE SEMICONDUCTOR INDUSTRY USES CHEMICALS and chemistry to produce its products but does not employ large numbers of chemists or chemical engineers. Clearly, the omission of these professionals has hindered the advancement and reduced the competitive edge of the semiconductor industry. The lack of chemical expertise has led to a neglect of advanced chemical processes that can establish a fundamental understanding of integrated-circuit (IC) production. Thus some of the most accurate and sensitive methods for the measurement and identification of organic and inorganic materials for process quality control, analysis of microcontamination, incoming-material evaluation, or production problem evaluation have been ignored.

Although electrical engineers and physicists have resorted to creative, interesting, and often very precise electrical measurements to investigate problems or establish control parameters for IC production, these methods do not yield direct chemical information concerning materials or processes.

0065–2393/89/0221–0505$06.00/0

Electron beam techniques have aided electrical measurements greatly, but these methods often lack sensitivity (X-ray and Auger spectroscopy and ESCA [electron spectroscopy for chemical analysis]) and accuracy (SIMS [secondary-ion mass spectrometry], etc.), two attributes that are of prime importance in IC process technology. Fortunately, materials can be analyzed with both accuracy and sensitivity by wet chemical analysis.

Currently, American semiconductor manufacturers are losing the battle with Japan and Korea in the production and sales of reliable, low-priced IC devices or chips. The failures and low yields that have hindered U.S. efforts are partially caused by the inadequate understanding of the chemical processes in semiconductor production and the lack of good quality controls for chemicals and processes.

For the United States to win this world competition, we must have a better understanding of the materials used in making ICs, that is, their exact composition and trace contamination analysis below part-per-billion levels, and we must be more critical about the accuracy of the measurements we make.

Wet Chemical Analysis in Semiconductor Manufacturing

Wet chemical analysis is especially useful to semiconductor IC manufacturers in the following five areas:

1. measurement of dopant concentrations in dielectric thin films,
2. evaluation of aluminum metallization and other thin films such as silicides and titanium–tungsten (TiW),
3. determination of metals in ultrapure water,
4. quality control of chemicals and their evaluation during and after use, and
5. use of wet chemistry in conjunction with other instrumental methods for the resolution of microcontamination problems.

Analysis of PSG and PBSG

Wet Chemical Methods. Wet chemical methods may be used routinely to determine the phosphorus content (total, P_2O_3, P_2O_5, and PH_3) of phosphosilicate glass (PSG) and phosphoborosilicate glass (PBSG) dielectric thin films, the total boron content of PBSG, and the silicon and copper contents of aluminum films. Methods for the determination of other elements critical to semiconductor manufacturing are still being developed.

Wet chemical colorimetric methods are the choice for the accurate determination of the composition of doped thin films. Since early 1960, colorimetry has been used to measure the percentage of phosphorus in PSG (*1*,

2) and the percentages of phosphorus and boron in PBSG. Colorimetry is the only method that gives accurate and absolute results; it is the primary method of determining the quantity of dopant in thin films, and consequently, it has been used to standardize phosphorus-doping processes. Because wet chemical methods have no matrix effects, the accuracy with these methods far exceeds that by X-ray, ESCA, Auger, or Fourier transform–infrared (FTIR) spectroscopy, which are methods that show matrix effects.

Wet chemical methods involve sophisticated sample preparation and standardization with National Bureau of Standards reference materials but are not difficult for the analytical chemist nor necessarily time consuming (Figure 1). The time from sample preparation to final results for various analytical methods, such as GFAA (graphite furnace atomic absorption), ICP (inductively coupled plasma spectroscopy), ICP–MS (ICP–mass spectrometry), and colorimetry, ranges from 0.5 to 5.0 h, depending on the technique used. Colorimetry is the method of choice because of its extreme accuracy. Typical results of the colorimetric analysis of doped oxides are shown in Tables I and II, which show the accuracy and precision of the measurements.

Because wet chemistry gives absolute results and is more sensitive compared with electron beam methods, it is useful in measuring the uniformity of a dopant across a wafer. ESCA, X-ray spectroscopy, and Auger spectroscopy can measure uniformity more quickly, but these methods give relative, not absolute, measurements. Furthermore, calculations for wet chemical analyses are done in three significant figures (Figure 2). However, because the accuracy of dopant measurements in thin films is generally ±3% of the actual value, data are reported more often in two significant figures. Nevertheless, the reliability of the results from wet analysis given in two significant figures is much greater than that of results from electron beam equipment.

Instrumental Methods. Engineers in the IC industry prefer to use X-ray or FTIR spectroscopy to determine the quantities of phosphorus in thin films because of the speed of these methods. These spectroscopic methods are satisfactory for a relative indication of the dopant level in thin films or additives to metallization layers, but they do have serious drawbacks. X-ray spectroscopy is seriously affected by matrix effects and can easily be off by ±15–20% of the actual concentration of dopant in thin films if the equipment is not properly calibrated against a material that has been analyzed by wet techniques. X-ray spectroscopy is further affected by the film thickness and the dopant profile throughout the film.

FTIR spectroscopy is a more accurate method for a quick analysis of the phosphorus in thin films. However, the P_2O_5 peak in FTIR is seriously affected if the thin film becomes hydrated. Some loss of sensitivity is caused by the movement of part of the peak from the P_2O_5 region to the $P_2O_5 \cdot H_2O$ region upon hydration of the thin film (Figure 3). This shift affects the measured phosphorus content of a thin film. Also, the lack of a significant

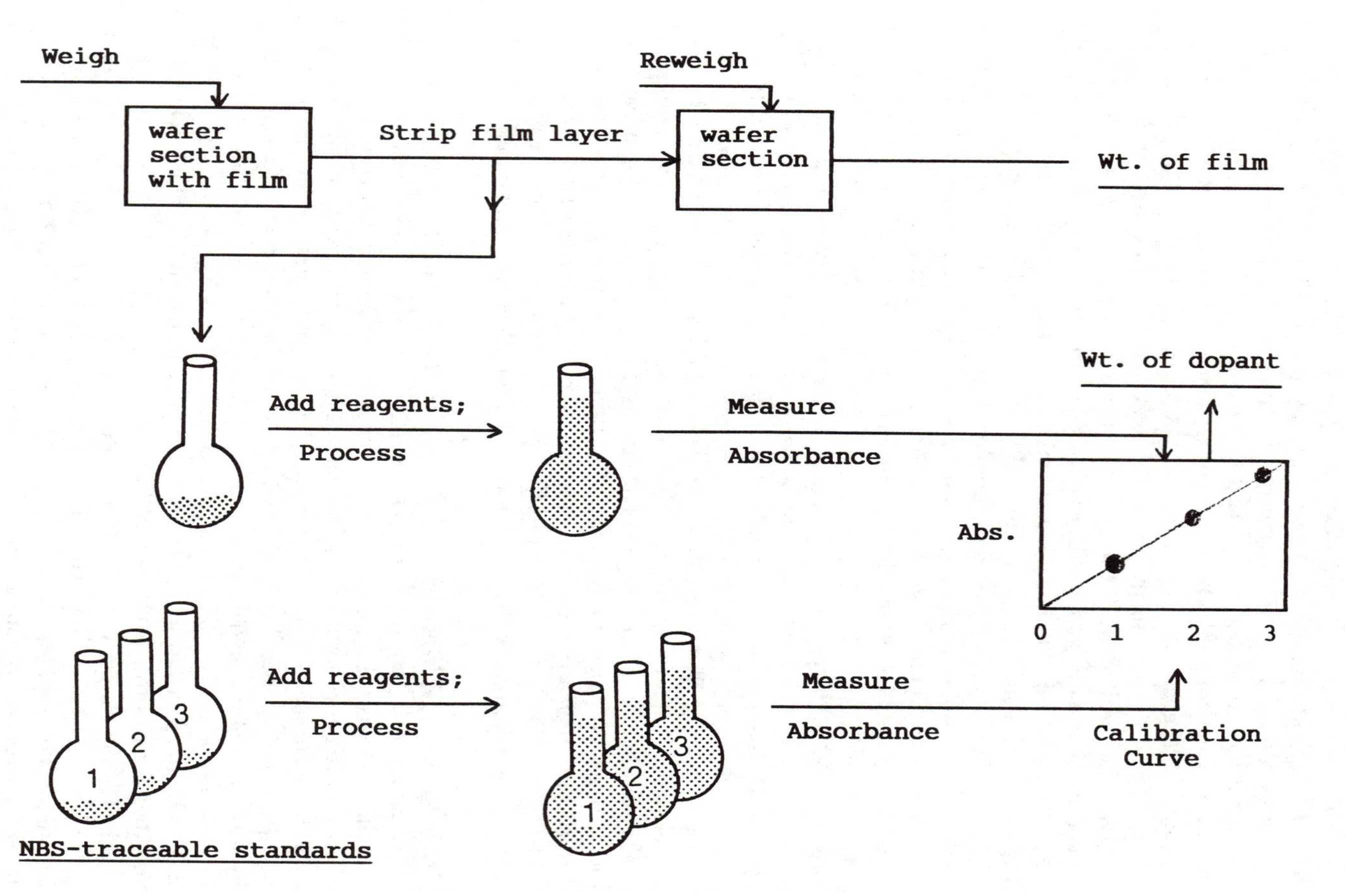

Figure 1. Scheme for wet chemical analysis.

Table I. Phosphorus Analysis of Sections of PBSG Films

Wafer Oxide Thickness (Å)	*Wt. % P*			*Maximum Variation*
	1	*2*	*3*	
6100	1.78	1.79		0.01
5500	1.89	1.98		0.09
6300	1.91	1.96		0.05
4900	1.98	2.04	1.96	0.08
3100	2.48	2.43		0.05
5400	3.63	3.60		0.03
7000	3.69	3.74		0.05
5300	4.39	4.34	4.36	0.05
4700	4.41	4.45	4.60	0.19
5900	5.00	5.02		0.02
4800	5.20	5.16		0.04
3800	5.94	5.74		0.20
3400	5.98	6.02		0.04

Table II. Boron Analysis of Sections of PBSG Films

Wafer Oxide Thickness (Å)	*Wt. % B*		*Maximum Variation*
	1	*2*	
6000	1.35	1.64	0.29
3600	1.51	1.46	0.05
3300	1.86	1.83	0.03
5000	2.56	2.64	0.08
7900	3.54	3.61	0.16
5100	3.68	3.84	0.20
7300	3.77	3.61	0.16
4400	4.99	5.11	0.12
5400	5.27	5.29	0.02
4000	5.36	5.19	0.17
5400	5.80	5.78	0.02

boron peak makes FTIR spectroscopy less reliable for the measurement of boron in PBSG films.

The determination of specific phosphorus compounds in thin films is important. Only through wet chemical analysis was it possible to first discover the presence and then to accurately measure the quantities of P_2O_5, P_2O_3, and phosphine found in plasma, plasma-enhanced, LPO–LTO (low-pressure oxide–low-temperature oxide), and CVD (chemical vapor deposition) processes (3). Methods such as X-ray or FTIR spectroscopy would have seen all phosphorus atoms and would have characterized them as totally useful phosphorus. In plasma and plasma-enhanced CVD films, phosphine is totally useless in doping processes.

A similar problem exists with boron-doped oxides or PBSGs. In this case, boron is frequently found as very small submicrometer crystals sitting

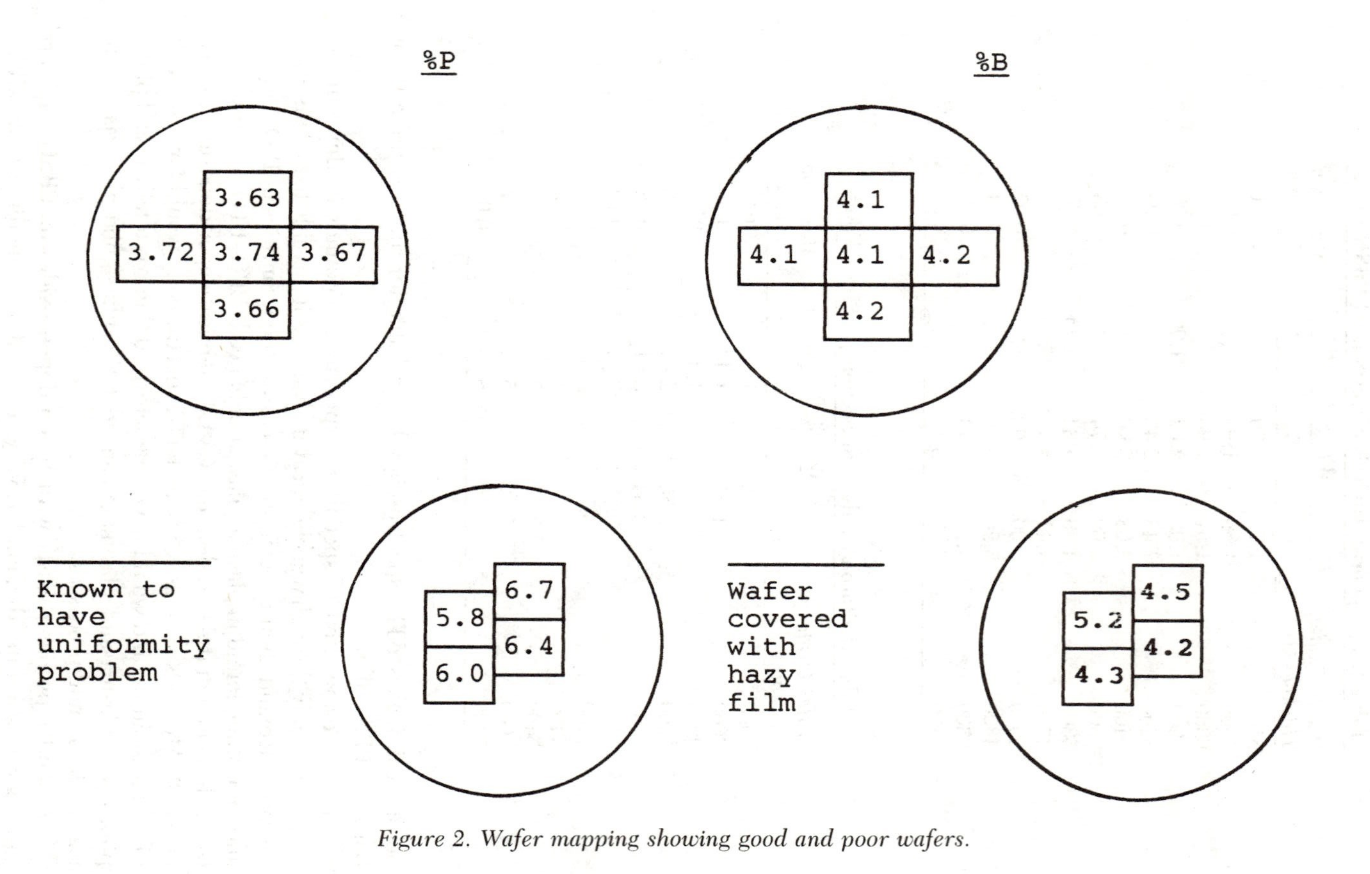

Figure 2. Wafer mapping showing good and poor wafers.

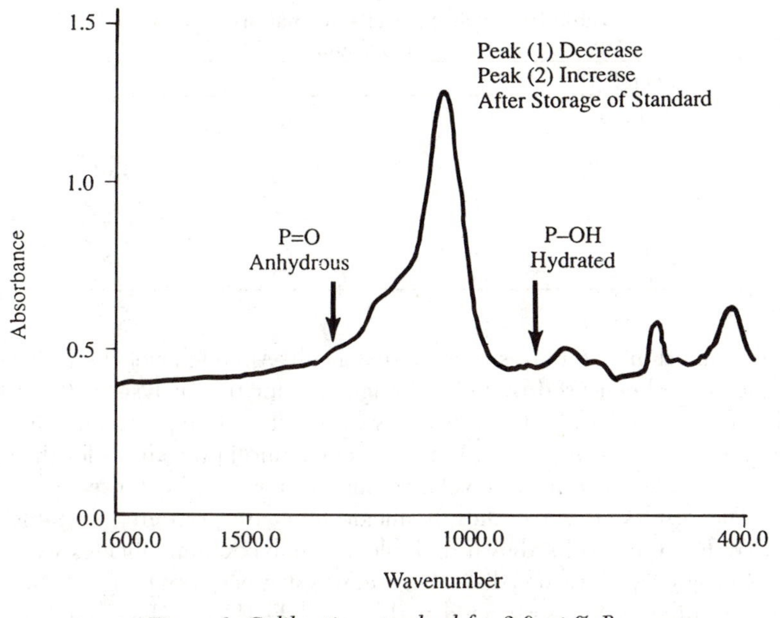

Figure 3. Calibration standard for 3.0 wt % P.

on top of the PBSG. These crystals make the film look somewhat hazy. By using wet chemical techniques, this material can be removed from the top of the film and quantified. A second piece of a wafer can be analyzed for total boron. By subtracting surface boron from total boron, one can determine the useful amount of boron in the thin film that will assist in lowering the melting point of the oxide during reflow. Oftentimes the quantity of boron left on the surface sufficiently reduces the quantity of boron in the thin film to affect the temperature at which reflow will take place.

Analysis of Other Thin Films

Silicon in Aluminum Films. Table III illustrates the typical results of measurements of silicon in an aluminum film compared with that in various targets used to make those thin films. Calculations can be carried out easily in three significant figures. No other method of measuring silicon in aluminum has the sensitivity or the ability to come within less than 20% of the actual value.

Silicides and Titanium–Tungsten. The determination of the ratio of metals to silicon in metal silicides or of titanium to tungsten in tita-

Table III. Typical Results of Analysis of % Si in Aluminum

Sample	*% Si*
Al Film	
1	0.98
2	1.01
Target	
1	0.97
2	1.01
3	1.01

nium–tungsten materials is even more difficult. Although wet chemical techniques have been used by process engineers for the analysis of these thin films for over a decade, these analyses were not routine procedures in semiconductor-processing plants. Routine wet chemical procedures for the analysis of silicides have been developed and are now used extensively.

The results of two studies of silicide films are presented in Tables IV and V. For the results shown in Table IV, molybdenum silicides were deposited on vitreous carbon disks. One set was stripped with base (2 N NaOH), and the other set was stripped with acid ($HF–HNO_3–H_2O$, 1:1:10). The study of chromium silicides (Table V) gave similar results.

To verify the accuracy of an analytical chemical procedure, a material balance or verification by other techniques must be obtained. For the analysis of molybdenum and chromium silicides, a material balance and verification of results by three techniques were the goals. Both acidic and basic stripping procedures were used in sample preparation. Two different colorimetric procedures and atomic absorption techniques were used for quantitative determinations. For both molybdenum and chromium silicides, the different analytical procedures gave comparable results but no material balance (accountability for 100% of the materials in a sample). These results indicated that our data were accurate but that other elements were present that were not being measured. Hydrogen, oxygen, or nitrogen was suspected to be the missing element. Auger analysis revealed the presence of approximately 15% oxygen in samples of molybdenum silicide. Because Auger analysis cannot give accurate data nor measure hydrogen, an absolute material balance was not obtained. Measurements of these elements would require a quantitative gas chromatographic–mass spectrometric study and the design of special testing chambers.

Analysis of Water

Water is one of the most widely used commodities in the production of integrated circuits, and the quality of water is extremely important in this

Table IV. Analysis of Molybdenum Silicides

Sample	*Etchant*	*% Mo*	*% Si*	*Total % Mo + % Si*	*Empirical Formula*
A1	base	51.8	35.9	87.7	
A2	acid	50.3	35.2	85.5	$Mo_{1.0}Si_{2.4}O_{1.7}$
A3	acid	51.5	36.3	87.8	$Mo_{1.0}Si_{2.4}O_{1.4}$
B1	acid	60.5	28.8	89.3	$Mo_{1.0}Si_{1.5}O_{1.2}$
B2	acid	58.8	25.7	84.5	$Mo_{1.0}Si_{1.5}O_{1.6}$

Table V. Analysis of Chromium Silicides

Sample	*% Cr*	*% Si*	*Av % Si*	*Total % Cr + % Si*
A1	31.7	50.2	51.6	83.3
A2		52.9		
B1	24.3	53.7	53.8	78.1
B2		53.9		
C1	25.4	55.2	55.6	81.0
C2		56.0		
D1	19.1	45.7	46.0	65.1
D2		46.3		
Target	50.3	43.9		94.2

industry. Numerous papers have been written about this subject. The quality of water during the past decade has improved significantly. Table VI shows the latest specifications for pure water. Clearly, the attainable levels of purity have improved tremendously. Until 1985, the levels of metallic constituents were not measured even in high-quality water below the part-per-billion range. With the advent of more-sensitive tools for wet chemical analysis, such as ICP–MS (inductively coupled plasma spectroscopy–mass spectroscopy), sub-part-per-trillion levels of metallic constituents in water could be analyzed.

In 1985, the results of a study done at Balazs Analytical Laboratory (Bal-Lab) (*4*) and in Plessey Research (*5*) led to the specification of metallic materials in water. At that time, the metals that were analyzed were those thought most likely to be present in and deleterious to ICs. Because both ion chromatography and GFAA are very time consuming, the evaluation of only nine or ten metals and six to eight nonmetallic species at the part-per-trillion level is all that one would do normally to get results within a reasonable time. Without sample concentration, analysis at this level will be at the limits of these methods.

To better detect metallic residuals in high-purity water, a new method was applied that allows the determination of trace elements in pure water in the part-per-trillion level. Table VII gives the results of a study recently completed at Bal-Lab. Pure water samples from seven sites, Bal-Lab (site A) and six IC producers (sites B–G), were obtained on the same day and analyzed semiquantitatively by ICP–MS to determine the elements (from lithium to uranium) present. Standards were prepared, and the samples were then analyzed quantitatively to determine the exact amount of each metal in the seven samples. More than 30 trace elements were detected by ICP–MS. Of these trace elements, 11 were prominent and 20 were quite measurable. The entire analysis, including the preparation of standards, took less than 4 h.

Table VI. Specifications and Guidelines for Semiconductor-Pure Water

	SPECIFICATIONS			**GUIDELINES**			
		256K DRAM		*1M DRAM*		*4M DRAM*	
Item	Detection Limit**	1985 Specs Attainable	1985 Specs Acceptable	1988 Specs Attainable	1988 Specs Acceptable	<1μ VLSI	ULSI Target
Resistivity @ 25°C	18.2 max.	18.2	17.9	18.2	18.0	18.2	18.2
TOC (ppb)	5	<20	<50	<10	<30	<10	5
THM (ppb)	<1	–	–	–		<3	–
Particle / L by SEM 0.1–0.2 μm		–	–	–	–	<1500	<1000
0.2–0.3 μm		–	–		<2000	<800	<500
0.3–0.5 μm				<200	<200	<50	<10
>0.5 μm				<1	<1	<1	<1
Particle / L by on-line laser 0.3–0.5 μm	<1	–	–	–	–	<50	<10
>0.5 μm	<1	–	–	–	<100	<1	<1
Bacteria / 100mL							
by culture	<1	0	<6	0	<6	0	0
by SEM		–	–	<1	<10	<5	0
by EPI		–	–	<5	<50	<10	<1
Silica-dissolved (ppb)	0.25	<3	<5	<0.4	4	3	1
Boron (ppb)	0.05	–	–	<0.05	2.0	0.005	*
Ions (ppb)							
Na^+	0.05	0.05	0.2	<0.05	0.1	0.025	
K^+	0.1	0.1	0.3	<0.1	0.1	0.05	
Cl^-	0.05	0.05	0.2	<0.05	0.1	0.025	
Br^-	0.1	<0.1	0.1	<0.1	0.1	0.05	
NO_3^-	0.1	<0.1	0.1	<0.1	0.1	0.05	
$SO_4^=$	0.1	0.1	0.3	0.05	0.2	<0.05	
Ions total	0.5	<0.5	1.2	<0.5	<0.7	<0.2	
Residue (ppm)	<0.1	<0.1	<0.3	<0.1	0.1	<.05	*

Table VI. Specifications and Guidelines for Semiconductor-Pure Water—Continued

	SPECIFICATIONS			GUIDELINES			
		256K DRAM		1M DRAM		4M DRAM	
Item	Detection Limit**	1985 Specs Attainable	1985 Specs Acceptable	1988 Specs Attainable	1988 Specs Acceptable	<1μ VLSI	ULSI Target
Metals (ppb)‡							
Li	0.03	—	—	<0.03	0.05	0.003	
Na	0.05	0.05	2.0	<0.05	0.1	0.005	
K	.05	0.1	0.3	<0.05	0.1	0.005	
Mg	0.02	—	—	<0.02	0.05	0.002	
Ca	2	—	—	<2	<2.0	0.002	
Sr	0.01	—	—	<0.01	0.05	0.001	
Ba	0.01	—	—	<0.01	0.05	0.001	
B	0.05	—	—	<0.05	2.0	0.005	
Al	0.05	0.2	2.0	<0.05	0.05	0.005	
Cr	0.02	0.02	0.1	<0.02	0.05	0.002	
Mn	0.02	0.05	0.5	<0.02	0.05	0.002	
Fe	0.1	0.02	0.1	<0.02	0.1	0.002	
Ni	0.02	—	—	<0.02	0.05	0.002	
Cu	0.02	0.02	0.1	<0.02	0.05	0.002	
Zn	0.02	0.02	0.1	<0.02	0.05	0.002	
Pb	0.05	—	—	<0.05	0.05	0.005	

— Not available at this time

* Unknown

** With reasonable conc. where applicable

‡ Using ICP-MS, GFAAS, IC where required for lowest level of detection. These elements represent the metals that are usually found in ultrapure water.

NOTE: These specifications were developed at Balazs Analytical Laboratory. Abbreviations are defined as follows: DRAM, dynamic random access memory; VLSI, very-large-scale integration; ULSI, ultralarge-scale integration; TOC, total oxidizable carbon; THM, trihalomethane; SEM, scanning electron microscopy; and EPI, epifluorescence

The significance of the ICP–MS study is twofold. First, and foremost, the study revealed that the metals that were being analyzed previously by ICP and GFAA were probably not the appropriate metals to be measured in terms of abundance. Second, boron is revealed as a significant contaminant in pure water, a fact that leads immediately to the question of whether this element is being picked up by the wafer as a dopant. Oftentimes, contaminating dopants are detected electrically in an IC. Tracking the source of the contaminating material has been difficult. However, the above investigation indicates that wet chemical analysis using ICP–MS should be able to determine the metals that are affecting the production of integrated circuits.

Table VII. Metal Content of Ultrapure Water

Metal	Detection Limit (ppt)	Quantity in Sample (ppt)						
		A	B	C	D	E	F	G
Li	30	–[a]	50	40	40	–	40	40
B	20	900	–	600	1200	–	2100	6100
Na	90	–	140	140	200	–	230	130
Al	10	–	–	13	–	–	–	–
Mg	4	–	20	9	8	20	10	10
Cu	9	–	–	–	–	–	20	–
Zn	35	–	–	40	–	–	47	57
Sr	2	–	2	2	–	3	–	4
Ba	3	–	7	3	10	10	6	4

NOTE: Detection limits were not determined for Sc, Co, Ge, Ga, and Ag, which should be <0.05 ppb. The following metals were not found in the samples in quantities above the indicated detection limits in parts per trillion: Ti, 6; Mn, 16; Ni, 320; Cd, 5; Sn, 2.

[a]The symbol – indicates that the metal was not present at quantity greater than the detection limit.

Analysis of Chemicals

Most companies producing ICs today have been quite lax in the quality control of the liquid chemicals used in their processes. These chemicals include acids, bases, buffered etches, photoresists, and organic solvents. Essentially, the producers of ICs have left the quality control of these materials to the chemical manufacturers. Specifications for chemicals are set by SEMI (Semiconductor Equipment Manufacturers International) and are generally met by all chemical manufacturers. The level of metallic contaminations that can exist in U.S. chemicals ranges from 0.1 to 1 ppm. If the levels of metallic impurities were actually this high, U.S. manufacturers would not be able to produce integrated circuits.

Analysis by Wet Methods. Contamination problems or yield losses caused by poor chemical quality have often led to a chemical being analyzed in our facility. These chemicals have passed SEMI specifications but cannot produce quality products. The quantity of contaminants that causes yield problems ranges from 10 to 100 ppb, depending on the element.

The case of isopropyl alcohol used in processing ICs is an example of chemical quality specifications. A manufacturer sent the laboratory two bottles of two different lots of isopropyl alcohol. With one lot, the manufacturer produced ICs that passed all electrical tests. With the other lot, the ICs failed. After these two materials were analyzed side by side with a variety of wet chemical techniques, it was found that the two lots differed only in potassium content. Both samples contained potassium in quantities well within SEMI specifications, but one contained almost 10 times as much potassium as the other (10 ppb versus 100 ppb).

The example just given reveals two important points: (1) SEMI specifications are inadequate for the production of ICs and (2) analytical quality control of incoming liquid chemicals for the production of integrated circuits must be done at a level of <100 ppb.

Metal concentrations in liquid chemicals at part-per-billion levels are difficult to determine. Until 1985, large volumes of the chemicals had to be concentrated to raise metal concentrations to levels measurable by the available methods. Measurement at part-per-billion levels was time consuming, exhausting, and frequently questionable because of the possibility of contamination during sample preparation.

At present, with ICP–MS, many chemicals can be diluted with high-purity water and analyzed directly without any further sample preparation. This method, with its enormous sensitivity for many metals at levels of part per billion to part per trillion, is valuable for good quality control of incoming chemicals, particularly acids, buffered etches, and various resist strippers.

Analysis of Sulfuric Acid by ICP–MS. Results of an ICP–MS study of sulfuric acid from nine manufacturers are shown in Table VIII. Twenty-two elements were quantitatively measured. Na, Fe, Ca, Cu, Zn, Al, Cr, K, Mg, Ni, and Co were present in significant quantities and are listed in

Table VIII. Elemental Content of H_2SO_4 from Nine Manufacturers

Element	*1*	*2*	*3*	*4*	*5*	*6*	*7*	*8*	*9*
Al	40	16	17	10	14	5	6	3	8
Ba	1	0.5	0.3	0.5	0.5	0.4	0.3	0.3	0.4
B	2	10	<1	9	<1	5	2	2	<1
Cd	0.8	0.1	<0.1	<0.1	<0.1	<0.1	<0.1	<0.1	<0.1
Ca	37	120	30	63	24	4	37	13	13
Cr	27	4	30	7	10	15	8	5	0.3
Co	3	29	2	0.6	0.3	0.6	0.4	1	0.3
Cu	28	29	180	12	86	18	14	13	5
Ga	<0.1	0.1	<0.1	<0.1	0.5	<0.1	<0.1	<0.1	0.4
Ge	<0.1	0.1	<0.1	<0.1	<0.1	<0.1	<0.1	<0.1	<0.1
Au	0.2	<0.2	0.6	<0.2	5	0.2	10	14	2
Fe	140	103	56	67	100	19	24	17	9
Pb	3	53	2	0.1	4	1	3	1	1
Li	1	0.5	<0.1	<0.1	0.3	<0.1	<0.1	<0.1	<0.1
Mg	42	16	7	9	4	13	6	4	2
Mn	3	0.7	1	0.6	0.5	0.4	0.4	0.7	<0.1
Ni	12	19	13	2	3	3	4	2	0.5
K	20	38	16	14	<10	11	<10	23	<10
Na	140	28	68	110	130	48	43	18	12
Ag	270	240	280	41	–[a]	75	135	56	32
Sr	0.2	0.2	0.2	0.2	0.1	0.2	0.4	<0.1	<0.1
Sn	0.7	0.7	0.4	0.2	0.4	0.1	0.3	0.5	<0.1
Zn	51	11	32	7	36	12	11	4	4
Total Conc.	540	413	451	301.6	510.3	148.6	153.4	103	82.1

NOTE: Values are in parts per billion. Total concentration is the sum of the concentrations of Al, Ca, Cr, Co, Cu, Fe, Mg, Ni, K, Na, and Zn.

[a]The symbol – means the element was not detected.

the approximate order of decreasing concentration. The concentrations of these trace elements in sulfuric acid varied from one manufacturer to another. The two most predominant elements in the sulfuric acid samples were Na and Fe. A high concentration of Ca is observed for manufacturer 2 and high concentrations of Cu are observed for manufacturers 3 and 5.

The concentrations of 11 elements (Na, Fe, Ca, Cu, Zn, Al, Cr, K, Mg, Ni, and Co) in each sulfuric acid sample were summed, and the resulting values are tabulated in Table VIII as the total concentration. The total concentrations of trace elements in the sulfuric acids from manufacturers 1–5 were four to five times higher than those of manufacturers 6–9.

Sulfuric acid from manufacturers 1 and 6 were sent to our laboratory for analysis by a client who had yield problems when using sulfuric acid from manufacturer 1 but no problems with sulfuric acid from manufacturer 6. The total trace element concentration in the sulfuric acid from manufacturer 1 was about four times higher than that from manufacturer 6. Low trace element concentrations were also found in sulfuric acids from U.S. (manufacturer 7) and Japanese (manufacturer 8) manufacturers. Sulfuric acid from manufacturer 9 had the lowest trace metal content. This sulfuric acid was processed through an acid reclaim system and was purer than the unused sulfuric acid put into the reclamation unit.

To verify the accuracy of the measurements from ICP–MS, recovery data for sulfuric acid were obtained. A sample of sulfuric acid (20 g) was spiked with 50 ppb of various elements and prepared under the same experimental conditions as the unspiked samples. The recovery of most elements was 90% or more (Table VIII). The recovery of Cr, Cu, Na, and Sn was >70%. The recovery of B and W was very poor; these elements were almost completely lost. The recovery of these trace elements can be improved by not evaporating the sample to dryness.

Quality Control for Chemicals. Although more work needs to be done to correlate sample failure with chemical quality, the case of H_2SO_4 reveals that users have choices in purchasing quality chemicals, but extra quality control must be exercised to identify the highest quality chemicals. The Japanese have repeatedly improved yield by 10–30% by using high-quality chemicals. Japanese specifications are far more stringent than those of SEMI. It is not that the Japanese have higher quality chemicals; it is that they differentiate between chemicals of the highest quality that should be used for the manufacture of ICs and those of slightly lower quality that should not be used in semiconductor-manufacturing facilities.

Microcontamination Analysis

Wet chemical analytical techniques are useful in resolving problems of microcontamination. Ion chromatography frequently reveals daily changes in the concentrations of metallic and nonmetallic ions in water that cause problems in the manufacture of ICs.

An example showing the usefulness of ion chromatography in microcontamination analysis is the case of a manufacturing plant that was closed because of the drop in resistivity of high-purity water as it passed a UV bactericidal lamp. The resistivity dropped from 18.2 to 17.6 MΩ-cm, which was below the acceptable limit of 17.8 MΩ-cm specified by this particular manufacturer, and thus the plant was closed. Everything from a photosensitive organic compound to the possible breakdown of a material that had worked its way into the UV, was suspected. Samples were taken before and after the UV lamp and analyzed by ion chromatography. The problem was caused by the presence of chloramine in the water at low part-per-billion levels. The photooxidation of chloramine to H^+, NO_3^-, and Cl^- resulted in the tremendous increase of particles that were left on the wafers. When the UV light was turned off, the chloramine did not oxidize, the resistivity was 18.2 MΩ-cm, and the wafers came out clean. When the lamps were turned on, resistivity dropped, and the wafers came out of the water rinse seriously contaminated with particles. A correlation between part-per-billion quantities of ions in solution and particles on a wafer had never been witnessed before. Clearly, this study raises the question of resistivity specifications.

Other Applications

When aluminum–silicon (Al–Si) metallization was deposited on wafers by evaporation, wet chemical analysis was used to study the changes in concentrations of Al to Si in the single evaporating cup. Wet chemical analysis was also used to determine the uniformity of silicon in aluminum and the evaporation patterns on the wafers in a rotating evaporator. The results revealed an accumulation of silicon in the cup and clearly showed the total lack of uniformity within a wafer and from wafer to wafer. The study showed why metallization problems were occurring and how to stop them.

Streaks left on wafers from organic solvents can be identified by using wet chemical sample preparation techniques followed by gas chromatography–mass spectrometry (GC–MS), another sophisticated analytical method used by wet chemists.

Organic materials that have entered DI (deionized) water systems through the upper openings of the degasifier or through its filter have been isolated and identified by wet chemical methods. Materials that plug filters or foul DI resins have also been identified by wet analytical techniques. In addition, processing questions, such as how many rinses are needed to remove sulfuric acid from a wafer, have been answered. To date, our laboratory has resolved over 500 such problems by using wet chemical techniques alone or in conjunction with electron beam equipment.

Conclusion

The present use of wet analytical techniques such as ion chromatography, GFAA, and ICP by many of the larger semiconductor-manufacturing companies has contributed greatly to improved yields, reduced costs, and increased competitiveness in the market place. However, wet chemistry is still underutilized, overlooked, ignored, or simply not understood well enough to be considered a useful technique by many IC manufacturing managers. Improved understanding, better products, and higher yields of ICs will be realized if analytical wet chemists are part of the quality-control team.

References

1. DuRant, P. In *Proceedings of the 5th International Conference on Chemical Vapor Deposition*; Blocker, J. M., Jr.; Hintermann, H. E.; Hall, L. H., Eds.; Electrochemical Society: Princeton, 1975; p 422.
2. Beck, C. *Solid State Technol.* **1977, *20*(3), 71–74.**
3. Houskova, J.; Ho, K.; Balazs, M. K. In *Microelectronics Processing: Inorganic Materials Characterization*; Casper, L. A., Ed.; ACS Symposium Series 295; American Chemical Society: Washington, DC, 1986; pp 320–332.
4. Stewart, D. A.; Brinklow, A. J.; Hewitt, J. In *Proceedings of the 4th Annual Semiconductor Pure Water Conference*; Balazs, M. K., Ed.; Palo Alto, CA, 1985; pp 145–164.
5. Houskova, J.; Chu, T. In *Proceedings of the 4th Annual Semiconductor Pure Water Conference*; Balazs, M. K., Ed.; Palo Alto, CA, 1985; pp 180–196.

RECEIVED for review December 30, 1987. ACCEPTED revised manuscript August 29, 1988.

INDEXES

AUTHOR INDEX

AFFILIATION INDEX

SUBJECT INDEX

A

E

F

G

H

I

M

N

O

Copy editing, indexing, and production: A. Maureen R. Rouhi

Acquisitions Editor: Robin M. Giroux

Managing Editor: Janet S. Dodd

Typeset by Techna Type, Inc., York, PA
Printed and bound by Maple Press, York, PA

Other ACS Books

Polymeric Materials: Chemistry for the Future
By Joseph Alper and Gordon L. Nelson
110 pp; clothbound, ISBN 0–8412–1622–3

The Language of Biotechnology: A Dictionary of Terms
By John M. Walker and Michael Cox
ACS Professional Reference Book
256 pp; clothbound, ISBN 0–8412–1489–1

Cancer: The Outlaw Cell, Second Edition
Edited by Richard E. LaFond
274 pp; clothbound, ISBN 0–8412–1419–0

Chemical Structure Software for Personal Computers
Edited by Daniel E. Meyer, Wendy A. Warr, and Richard A. Love
ACS Professional Reference Book
107 pp; clothbound, ISBN 0–8412–1538–3

Practical Statistics for the Physical Sciences
By Larry L. Havlicek
ACS Professional Reference Book
198 pp; clothbound; ISBN 0–8412–1453–0

The ACS Style Guide: A Manual for Authors and Editors
Edited by Janet S. Dodd
264 pp; clothbound, ISBN 0–8412–0917–0

Personal Computers for Scientists: A Byte at a Time
By Glenn I. Ouchi
276 pp; clothbound, ISBN 0–8412–1000–4